APPLIED ANALYSIS
OF VARIANCE IN
BEHAVIORAL SCIENCE

STATISTICS: Textbooks and Monographs

A Series Edited by

D. B. Owen, Founding Editor, 1972–1991

W. R. Schucany, Coordinating Editor
Department of Statistics
Southern Methodist University
Dallas, Texas

Additional Volumes in Preparation

APPLIED ANALYSIS OF VARIANCE IN BEHAVIORAL SCIENCE

edited by

LYNNE K. EDWARDS

University of Minnesota
Minneapolis, Minnesota

Marcel Dekker, Inc.   New York • Basel • Hong Kong

Library of Congress Cataloging-in-Publication Data

Applied analysis of variance in behavioral science / edited by Lynne K.
Edwards.
  p. cm. -- (Statistics: textbooks and monographs; v. 137)
  Includes bibliographical references and index.
  ISBN 0-8247-8896-6 (acid-free paper)
  1. Psychology--Statistical methods. 2. Analysis of variance. I. Edwards,
Lynne K. II. Series: Statistics, textbooks and monographs; v. 137.
BF39.2.A52A66 1993
150'.1'5195--dc20   93-7738
                       CIP

The publisher offers discounts on this book when ordered in bulk quantities.
For more information, write to Special Sales/Professional Marketing at the
address below.

This book is printed on acid-free paper.

MARCEL DEKKER, INC.
270 Madison Avenue, New York, New York 10016

Current printing (last digit):
10 9 8 7 6 5 4 3 2 1

PRINTED IN THE UNITED STATES OF AMERICA

Preface

This book is devoted to the discussion of analysis of variance (ANOVA) techniques. It presents ANOVA as a research design, a collection of statistical models, an analysis model, and an arithmetic summary of data. Discussions focus primarily on univariate data, but multivariate generalizations are touched on in several chapters with one chapter specifically dedicated to multivariate analysis of variance (MANOVA).

To use this book most effectively requires knowledge of elementary probability, the power of tests, use of the z-, t-, and F-distributions, and the computational aspects of sums of squares and mean squares, all of which are available in most applied inferential statistics courses and first graduate or upper undergraduate courses in ANOVA.

ANOVA has been one of the most pervasive methods in fields of psychology, education, business, agronomy, allied health sciences, and medicine, to name a few. There are many good textbooks available in the area of applied ANOVA. To supplement and complement these textbooks, we have attempted to cover selective yet diverse subareas with their theoretical background, the logic of the respective analysis, practical guidance, and up-to-date research results and implications. We have tried to present concepts, although applied ANOVA is all too often seen as a collection of formulas computed by hand or by statistical packages. We have provided examples and research contexts relevant to social and behavioral sciences, although classic ANOVA tends to be agronomy-based, with technical terms and examples alien to our audience. Furthermore, we have provided extensive detail and numerical examples to help walk the reader step by step through analyses where statistical handbooks and encyclopedias as reference books often cannot.

This book will serve as a quick reference as well as a comprehensive summary of recent research. As the editor, to reflect the interdisciplinary nature of the applied field, I have chosen some of the most prominent and productive researchers in the areas of statistics, biostatistics, psychology, and education. We have written this book with researchers, practitioners, and graduate students in social and behavioral sciences in mind, but statisticians interested in applied fields may also find it useful.

This book is not intended to be an exhaustive and comprehensive collection of ANOVA techniques nor is it intended to be an integrated textbook on the topic. The contributors and I have tried to achieve a balance in three areas: (1) to be comprehensive by dealing with representative topics and issues, (2) to present underrated topics with relevant applications, and (3) to present new topical developments. In addition we have tried to strike a balance in the theory–application continuum, including theoretical issues as well as applied problems. We have tried to provide the reader with an opportunity to select topics and theoretical depth as needed, à la carte, so to speak. Chapter 1, "Analysis of Variance Overview," presents an expanded discussion of the selected topics as well as a schematic diagram outlining these topics.

I would like to acknowledge the late Donald B. Owen for his help with this book. He was instrumental in building the Marcel Dekker Statistics: Texbooks and Monographs Series, serving as its main series editor. Before his untimely death, he invited me to join the project by adding this book to the series. His integrity and commitment, as well as his care for this book, mulling over its title until it was "just right," gave me added impetus during its editing.

I thank Addison-Wesley for permission to publish the adapted F table (Appendix Table 1) from the F distribution tables prepared by the late Donald B. Owen (1962) in his *Handbook of Statistical Tables*.

I want to thank the American Educational Research Association (AERA) for its permission to reproduce the table of critical values for Welsch Step-Down and Shaffer-Welsch Tests (Table 2.2 in Chapter 2), which originally appeared in P. H. Ramsey, and P. P. Ramsey (1990), *J. Educ. Statist.*, *15*, 341–352. The AERA is also acknowledged for their permission to reproduce a portion of L. K. Edwards (1991), *J. Educ. Statist.*, *16*, 53–76, which appears in Chapter 12.

Brooks/Cole is acknowledged for their permission to reproduce a portion of R. E. Kirk (1982), *Experimental Design*, 2nd ed., which appears in Chapter 6.

I acknowledge Dr. Hartley for his kind permission to reproduce a portion of his studentized range tables (Appendix Table 2), which originally appeared in H. L. Harter (1969), *Order Statistics and Their Use in Testing and Estimation, vol. 1: Tests Based on Range and Studentized Range of Samples from a Normal Population*, Aerospace Research Laboratories, U.S. Air Force.

I would like to thank the following for their help with this project: The contributors and, in particular, Anant Kshirsagar for doing double duty as a contributor and one of the series editors; those in my department who provided me with assistance in handling correspondence with contributors and preparing tables in the appendix; Carol Church for her copyediting skills in making chapters readable with simplicity and clarity; Joseph Stubenrauch and Maria Allegra, both of Marcel Dekker, Inc., and their anonymous reviewer for providing me with constructive suggestions; the Minnesota Supercomputer Institute for funding portions of the projects reported in my chapters as well as providing me with a conducive work environment in which I completed the final portion of this book; and finally, my late father for his continuous encouragement.

Lynne K. Edwards

Contents

<cite/>

Contributors

Patricia C. Bland Department of Educational Psychology and Computer Information Services, University of Minnesota, Minneapolis, Minnesota

John W. Cotton Departments of Education and Psychology, University of California, Santa Barbara, California

Harold D. Delaney Department of Psychology, University of New Mexico, Albuquerque, New Mexico

Lynne K. Edwards Department of Educational Psychology and MN Supercomputer Institute, University of Minnesota, Minneapolis, Minnesota

Richard J. Harris Department of Psychology, University of New Mexico, Albuquerque, New Mexico

Ronald R. Hocking Department of Statistics, Texas A&M University, College Station, Texas

H. J. Keselman Department of Psychology, University of Manitoba, Winnipeg, Manitoba, Canada

Joanne C. Keselman Department of Educational Psychology, University of Manitoba, Winnipeg, Manitoba, Canada

Roger E. Kirk   Department of Psychology and Institute of Graduate Statistics, Baylor University, Waco, Texas

Anant M. Kshirsagar   Department of Biostatistics, University of Michigan, Ann Arbor, Michigan

Kinley Larntz   Department of Applied Statistics, University of Minnesota, St. Paul, Minnesota

Scott E. Maxwell   Department of Psychology, University of Notre Dame, Notre Dame, Indiana

Keith E. Muller   Department of Biostatistics, University of North Carolina, Chapel Hill, North Carolina

Ralph G. O'Brien   Department of Statistics (Biostatistics), University of Florida, Gainesville, Florida

Mary F. O'Callaghan   Department of Psychology, University of Notre Dame, Notre Dame, Indiana

Philip H. Ramsey   Department of Psychology, Queens College of the City University of New York, Flushing, New York

Stephen W. Raudenbush   Department of Counseling, Education Psychology, and Special Education, Michigan State University, East Lansing, Michigan

Shayle R. Searle   Biometrics Unit, College of Agriculture and Life Sciences, Cornell University, Ithaca, New York

Rand Wilcox   Department of Psychology, University of Southern California, Los Angeles, California

1
Analysis of Variance Overview

LYNNE K. EDWARDS University of Minnesota, Minneapolis, Minnesota

1.1 INTRODUCTION

The early development of the analysis of variance (ANOVA) was due mainly to Sir Ronald A. Fisher. Originally conceived as a convenient arithmetic arrangement of data, the analysis of variance has come to be associated with research designs, mathematical models, and analytical tools. This book presents ANOVA from all three perspectives. The goal is not so much to accommodate all perspectives in every chapter, but to reflect them in this book as a whole. Accordingly, some chapters are focused more heavily on one aspect than on others.

In the first half of this chapter, we present an overview of the book including the objectives and topics. The analysis of variance techniques appear in almost every field, from agronomy and engineering to psychology and education. Even limiting focus to applied behavioral sciences, the interests in ANOVA range widely. Some topic areas have seen tremendous changes and developments over the years. In some others, classic techniques and designs have been neglected despite their potential contributions to the field. Sometimes there is no consensus on which method is the best; and sometimes one method is preferred (but not uniformly) over the others under certain conditions. It is confusing to understand the status of the field and to evaluate each new method or technique. For this reason we have tried to prepare a book on analysis of variance, including the new techniques and controversies, in a simple and concise manner.

Because this book is not intended as a representative and exhaustive collection of the topics and issues surrounding the analysis of variance, some explanation is required for topic selections and the focus. I have tried to show the integration and logical connection among the chapters in Section 1.2. I have also provided an organization of chapters in Figure 1.1. Obviously, each reader may prefer a different connection or path than the one suggested in Figure 1.1. Each chapter can be read independently of others because we have tried to maintain unity in the book as a whole, while keeping each chapter as autonomous and complete as possible.

Although this book is more of a collection of topics than a unified textbook on analysis of variance, there are some fundamental workings of ANOVA with which we expect our readers to be familiar. Instead of reviewing all the fundamentals, we quickly review in the second half of this chapter four selected issues: an introduction to ANOVA as the general linear model, the expectation of mean squares, $E(MS)$, the computation of sums of squares (SS), and the corresponding degrees of freedom (d.f.). For an extensive review of computational aspects as well as a more detailed discussion on specific designs and analyses, numerous texts are available. The texts by Edwards (1985), Keppel (1982), Kirk (1982), Maxwell and Delaney (1990), Winer (1971), and Winer, Brown, and Michels (1991) present ANOVA from a perspective of applied behavioral sciences. Myers and Milton (1991) introduce ANOVA from a perspective of matrix algebra; and Hocking (1985), Searle (1971), and Graybill (1976) discuss ANOVA in the context of the general linear models. Scheffé's (1959) text still remains a classic on analysis of variance.

1.2 OVERVIEW OF THE BOOK

1.2.1 The Objectives of the Book

It was impossible to include in this book all analysis of variance topics or issues which may be deemed representative of the field. This book, however, is intended as a supplement to various texts available in ANOVA. It is hoped that, based on the general foundations developed through a first course in analysis of variance, graduate students, researchers, and practitioners in applied behavioral sciences will gain a clearer understanding of the current status of the ANOVA procedures and update their knowledge of recent developments. Statisticians interested in gaining knowledge of ANOVA in applied contexts may also benefit from this book. This book presents a starting point in gaining a comprehensive, if not exhaustive, view of the field. It also highlights future topics and unresolved issues for research. In addition, many numerical examples illustrate applications of many techniques and procedures based on actual or realistic data sets. I believe examples rooted in one's own field enhance the understanding of the model or technique in question. This is why there is no example from engineering or agronomy in this book.

Some skeptics may argue that analysis of variance represents a set of designs and techniques which have been overemphasized and outdated, on which there is no need to expand. Some may go yet one step further and claim regression and correlational techniques have replaced the need to consider ANOVA or experimental design as a separate branch of statistical training for our students. We argue that these views may come, in part, from a simplistic view that ANOVA is nothing but a collection of arithmetic rules, that there have been no noticeable developments made in the field since the days of Fisher, that there is no need to understand statistical models behind the designs or techniques because the statistical packages would automatically and correctly incorporate these in the analysis, and that certain designs and techniques are rightfully underused because these have no relevance to the behavioral sciences. We would like to show why the aforementioned views are simplistic and outmoded by discussing some of the misunderstandings in frequently used designs and procedures, some of the exciting new developments in ANOVA techniques, and some realistic applications of underused designs.

1.2.2   The Topics of the Book

The book consists of 16 chapters. These are not meant as an exhaustive collection of analysis of variance techniques. The topics have been selected to present a balanced perspective among the following goals: (1) to provide a quick review of ANOVA as a computational technique as well as a field; (2) to represent familiar topics and procedures in applied behavioral sciences; (3) to illustrate helpful uses of underutilized designs; (4) to encourage proper study planning; (5) to recommend models and techniques to use when data are irregular; (6) to highlight some specialty topics; (7) to provide an ANOVA analog in categorical variables; (8) to present estimation problems; and (9) to give practical help in computer applications area. A schematic chapter organization (Figure 1.1) shows one possible sequence in which these chapters could be read.

Obviously, some issues and topics are left out of this book, such as theoretical issues regarding the assumptions of ANOVA, e.g., the tests of normality or of variances. Some of these topics were included in Krishnaiah's (1980) book on analysis of variance but I felt these topics demanded more theoretical preparation from our readers than what we were trying to accomplish in this book. Furthermore, these issues are far more general than the confines of analysis of variance. Although some of the newer approaches to ANOVA (e.g., exploratory or graphic approaches) are rel-

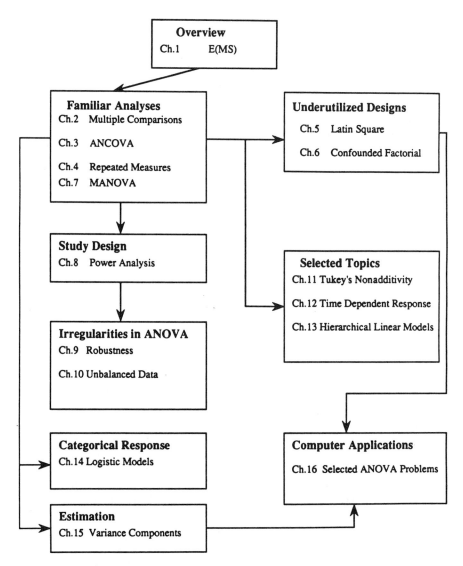

Figure 1.1 Organization of chapters.

evant, I did not select these for this book. These topics, I feel, deserve a book by themselves, such as the book on exploratory analysis of variance by Hoaglin, Mosteller, and Tukey (1991).

1.2.2.1 Familiar Analyses

Chapters 2, 3, 4, and 7 form a segment of designs and procedures frequently used in the applied fields. In Chapter 2, P. Ramsey summarizes the ever-developing area of multiple comparisons of means, concentrating on pairwise comparison methods in the completely randomized group design. Ramsey introduces various multiple comparisons techniques, both established and new. A numerical example is used to compare the outcomes from several multiple comparison methods.

In Chapter 3, S. Maxwell, H. Delaney, and M. Scherzinger present an integrated summary of analysis of covariance (ANCOVA) by emphasizing this design in the context of randomized experiments. Although ANCOVA is one of the most frequently used statistical techniques in behavioral sciences, it is also one most misused and often criticized. Maxwell et al. also summarize recent developments in handling heterogeneous regression slopes and nonparametric alternatives.

H. Keselman and J. Keselman in Chapter 4 present the analysis of repeated measurements with emphasis on multiple comparisons of correlated means. Included are procedures appropriate for testing specific contrasts when either or both the assumption of homogeneous variance for between-subjects factors and the assumption of circularity for within-subjects factors are not met. This is an issue of practical and theoretical interest because the overall test on treatments is rarely the goal in itself and rarely are the aforementioned assumptions satisfied. Included in their numerical examples are SAS commands (SAS, 1990) in transforming data necessary to conduct a series of follow-up tests after a significant omnibus multivariate test is obtained.

A few chapters later (after dealing with two underutilized univariate designs in Chapters 5 and 6), we switch our gears to multivariate applications in Chapter 7. R. Harris discusses multivariate analysis of variance (MANOVA) from a practical and applied perspective. Harris opens his chapter with research situations in which both multiple ANOVAs and a MANOVA are applicable and discusses relative advantages and disadvantages of the two approaches. Using numerical examples in applied research, a practical guide to interpreting discriminant functions and canonical variates is provided.

1.2.2.2 Underutilized Designs

Chapters 5 and 6 deal with designs often underutilized in the behavioral sciences. Latin square designs are presented by J. Cotton and fractional factorial designs by R. Kirk. These designs have been represented in the traditional agronomy-oriented textbooks, but there have been few applications to the behavioral sciences. In Chapter 5, after introducing a classical Latin square design, Cotton delves into a more complex research design where carryover effects are added to the traditional row, column, and treatment effects. The carryover effects may be an important research focus, for example, in learning experiments. He describes the two types of parameter constraints, a sum-to-zero and a parameter-set-to-zero. He illustrates the parameter estimation procedures by SAS and SYSTAT (Wilkinson, 1989) using numerical examples.

In Chapter 6, R. Kirk covers confounded factorial designs, among the least frequently used designs, though useful in research situations in which a block size cannot be freely expanded to allow a completely factorial structure for the treatments (e.g., twin studies in which the block size is restricted to two). The use of modular arithmetic in constructing these confounded factorial designs is explained. In addition, Kirk shows the computational procedures in unbalanced situations from the cell means model perspective using matrix algebra.

1.2.2.3 Study Design

In Chapter 8, R. O'Brien and K. Muller discuss the issue of power estimation and sample size determination as an important study planning process. To plan a sample size with a desired level of power is an extremely important step, although one not widely used. O'Brien and Muller introduce a unique power analysis procedure for both univariate and multivariate hypotheses. They provide four realistic research scenarios to illustrate their power analysis procedure using the modules developed by the authors themselves.

1.2.2.4 Irregularities in ANOVA

Chapters 9 and 10 illustrate models and techniques appropriate under "irregular" conditions. In Chapter 9, R. Wilcox summarizes the impacts of unequal variances and/or nonnormality on the traditional ANOVA tests and introduces recent robust techniques. Among those included are the M-estimators and trimmed t-tests. Wilcox provides an introduction to robustness issues as well as a comprehensive reference list for use in tracking recent developments, and he poses some unresolved issues.

In Chapter 10, S. Searle summarizes cell means models in the context of unequal sample sizes (i.e., unbalanced data). Although this chapter is included in a section dealing with irregularities, having unequal sample sizes is not at all irregular. On the contrary, unbalanced data are almost always the rule rather than the exception. Cell means models, an intuitive approach to dealing with unequal sample sizes, are presented in both all-cells-filled and some-cells-empty situations. Cell means models have been gaining increased popularity in applied behavioral sciences, but they are not yet pervasive in practice. This chapter may serve as an impetus in applying cell means models to actual research situations.

1.2.2.5 Selected Topics

Selected specialized topics are covered in Chapters 11 through 13. Tukey's nonadditivity test is examined in Chapter 11 by A. Kshirsagar. Beginning with a traditional application of Tukey's test as a method of detecting an interaction in a two-way classification with $n = 1$ in each cell, Kshirsagar discusses various generalizations of Tukey's test, including its applications to models with more than one interaction term and to MANOVA and MANCOVA situations.

In Chapter 12, L. Edwards deals with selected analytical models for analyzing time-dependent observations as an extension of classical repeated measures designs. The choice of analytical model is not straightforward, either when the time points are not as extensive as they are in a typical time series analysis or when the observations cannot be randomly sequenced as they are in a true experiment. Noting that different procedures place essentially different assumptions on the variance-covariance matrix, Edwards summarizes a comparison of selected models by computer simulation.

In Chapter 13, the hierarchical linear model (HLM) is introduced by S. Raudenbush as an alternative to several classical experimental design situations. The two-level hierarchical linear model is discussed in full with actual research applications. Using examples from a two-factor crossed model and a randomized block design, both classical and HLM approaches are compared for their model structures and for their estimation results.

1.2.2.6 Categorical Response

An ANOVA analog to categorical dependent variables is discussed in Chapter 14 by K. Larntz. A simple logistic regression model and the logistic general linear model are introduced. Both binary and multinominal responses are examined in this chapter. Numerical examples are taken from actual studies and their fully worked-out results are presented. In particular, Larntz illustrates the use of Akaike Information Criterion in selecting the predictor variables.

1.2.2.7 Estimation

Traditionally in applied behavioral sciences, many thought (somewhat erroneously) that the effects of interest were fixed in most cases. But random effects are more pervasive than we perhaps realize. Estimation of variance components is one of the research foci in studies involving random effects. In Chapter 15 Hocking describes several variance component estimation procedures and their biases. In particular, the problem of negative variance component estimation and a diagnostic procedure developed by Hocking are illustrated in artificial numerical examples with specified data characteristics.

1.2.2.8 Computer Applications

One of the reasons why some of the designs or procedures are less frequently used may be that there is no easy access to computer packages for these. A relatively simple design may require a command structure which is somewhat complex; and a relatively complex design may not even have a corresponding example in the computer manuals. In Chapter 16, L. Edwards and P. Bland illustrate some commands and the printouts from the three major statistical packages for selected ANOVA problems. Some of the differences among these packages in handling the selected examples and their expended CPUs are discussed.

Finally, this is a reminder that the sequence of chapters suggested in Figure 1.1 is just one possible order in which these chapters can be read. The authors have tried to keep each chapter autonomous, so you may read the chapters in any order, based on your interests. It is our hope that this book will acquaint its readers with unfamiliar topics in ANOVA and re-kindle already existing interest in familiar topics as well. Now let us turn our attention to a review of the selected fundamentals of ANOVA.

1.3 ANOVA AS A GENERAL LINEAR MODEL

We will introduce ANOVA in the framework of the general linear model. Before we discuss the relationship between the two, a quick distinction among analysis models, designs, and statistical models may be helpful. Although this is oversimplified, it makes a point that specifying a certain analysis model (e.g., a one-way ANOVA) does not automatically embody a design (e.g., a completely randomized design) with a certain statistical model (e.g., an overparameterized linear model with a sum-to-zero constraint on treatment effects and errors distributed with mean zero and variance σ^2). An analysis model specifies the structure of data (as in a one-way classification), but it does not specify how the data were collected. A design, on the other hand, specifies a plan of the study, e.g., whether subjects were randomly assigned to treatments (as in a completely randomized group design) or were repeatedly observed under all treatment conditions (as in a repeated measures design). A statistical model specifies parameters and their relationship along with their distributional properties and constraints.

The general linear model is a statistical model which embodies ANOVA and helps us identify its similarities with the regression model. The general linear model, as applies to this chapter, includes one dependent variable Y and one or more independent or explanatory variables X. We shall assume that, in general, Y is random whereas X's are known or constant. The general linear model specifies the relationship between Y and Xs as consisting of two components: the deterministic component representing the portion of Y that can be explained as a function of X's, and the random error component representing the portion of Y unexplained by a set of X's.

1.3.1 Regression

The general linear regression model can be expressed as

$$Y_j = \beta_0 + \beta_1 X_{1j} + \beta_2 X_{2j} + \cdots + \beta_p X_{pj} + \varepsilon_j \tag{1.1}$$

$$= \beta_0 + \sum_i^p \beta_i X_{ij} + \varepsilon_j \quad (i = 1, \ldots, p; j = 1, \ldots, N)$$

where Y_j is the jth individual's observation, β's are unknown parameters, X_{ij} is the value of the ith explanatory variable for the jth individual, and ε_j is the error for the jth individual. The model (1.1) is linear in the sense that the coefficients β's for the explanatory X variables are *linear*.

Using matrix algebra, Eq. (1.1) can be expressed as $\mathbf{y} = \mathbf{X}\boldsymbol{\beta} + \boldsymbol{\varepsilon}$ where $\mathbf{y}$ is the vector of observations, $\mathbf{X}$ is a matrix consisting of explanatory variables, $\boldsymbol{\beta}$ is the vector of $(p + 1)$ parameters, and $\boldsymbol{\varepsilon}$ is the vector of errors.

$$
\begin{array}{ccccc}
\mathbf{y} & = & \mathbf{X} & \boldsymbol{\beta} & + & \boldsymbol{\varepsilon}
\end{array}
$$

$$
\begin{bmatrix} Y_1 \\ Y_2 \\ \vdots \\ Y_N \end{bmatrix}
=
\begin{bmatrix}
1 & X_{11} & X_{12} & \cdot & X_{1p} \\
1 & X_{21} & X_{22} & \cdot & X_{2p} \\
\vdots & \vdots & \vdots & \vdots & \vdots \\
1 & X_{N1} & X_{N2} & \cdot & X_{Np}
\end{bmatrix}
\begin{bmatrix} \beta_0 \\ \beta_1 \\ \vdots \\ \beta_p \end{bmatrix}
+
\begin{bmatrix} \varepsilon_1 \\ \varepsilon_2 \\ \vdots \\ \varepsilon_N \end{bmatrix}
$$

The sample estimates $\hat{\boldsymbol{\beta}}$ are obtained by solving $\mathbf{y} = \mathbf{X}\hat{\boldsymbol{\beta}} + \hat{\boldsymbol{\varepsilon}}$, in such a way that these estimates minimize the residual sum of squares, $\hat{\boldsymbol{\varepsilon}}'\hat{\boldsymbol{\varepsilon}} = (\mathbf{y} - \hat{\mathbf{y}})'(\mathbf{y} - \hat{\mathbf{y}})$, where $\hat{\mathbf{y}} = \mathbf{X}\hat{\boldsymbol{\beta}}$ and $\mathbf{Z}'$ is a transpose of $\mathbf{Z}$.

By differentiating the residual sum of squares with respect to each parameter, $\hat{\beta}_0, \hat{\beta}_1, \ldots, \hat{\beta}_p$, we obtain $(p + 1)$ *normal* equations. In matrix notation, they are

$$\mathbf{X}'\mathbf{X}\hat{\boldsymbol{\beta}} = \mathbf{X}'\mathbf{y}.$$

Assuming that the inverse $(\mathbf{X}'\mathbf{X})^{-1}$ exists, the general solution of $\hat{\boldsymbol{\beta}}$ is

$$(\mathbf{X}'\mathbf{X})^{-1}(\mathbf{X}'\mathbf{X})\hat{\boldsymbol{\beta}} = (\mathbf{X}'\mathbf{X})^{-1}\mathbf{X}'\mathbf{y}.$$

Because $(\mathbf{X'X})^{-1}(\mathbf{X'X}) = \mathbf{I}$, the identity matrix, and because $\mathbf{I\hat{\beta}} = \hat{\beta}$, the solution is

$$\hat{\beta} = (\mathbf{X'X})^{-1}\mathbf{X'y}.$$

Aside from obtaining estimates of the parameters, it is of interest to test whether the model (with all independent variables as a set) sufficiently explains the variation found in Y. This is accomplished by decomposing the (adjusted) total sum of squares, $SSTOT$, into the SSR (the SS due to the model), and the SSE (the SS due to error).

$$\sum_{j}^{N} (Y_j - \overline{Y}.)^2 = \sum_{j}^{N} (Y_j - \hat{Y}_j)^2 + \sum_{j}^{N} (\hat{Y}_j - \overline{Y}.)^2$$

or

$$SSTOT \qquad = SSE \qquad + SSR.$$

Each SS is then divided by its corresponding degrees of freedom to obtain MS. We can think of degrees of freedom as the number of observations that are free to vary without any constraint or the number of observations minus the number of parameters to be estimated. The total sum of squares can be thought of having one linear constraint, that $\Sigma_j^N (Y_j - \overline{Y}.) = 0$, or one parameter, μ is estimated by $\overline{Y}.$. Using similar logic, SSE and SSR will each have $(N - p - 1)$ and p degrees of freedom, respectively. Each SS, when divided by its corresponding degrees of freedom, becomes MS, and then the general null hypothesis of H_0: $\beta_1 = \cdots = \beta_p$ can be tested by comparing

$$F = \frac{SSR/p}{SSE/N - p - 1} = \frac{MSR}{MSE}$$

against the central F distribution at a nominal level α, $F[\alpha; p, N - p - 1]$. We reject the general null hypothesis if the obtained F ratio is equal to or larger than the corresponding critical F value.

1.3.2 ANOVA

Let us now express a one-way classification model in the general linear model context. Many parameterizations are possible in ANOVA. One of these, a means model, is

$$Y_{ij} = \mu_i + \varepsilon_{ij} \qquad (i = 1, \ldots, a; j = 1, \ldots, n_i) \qquad (1.2)$$

where Y_{ij} is the observation of the jth individual in the ith group, μ_i is the mean for the ith group, and ε_{ij} is the random error, assumed to be normally distributed with mean zero and variance σ^2. Each individual's response is expressed as the mean of the group plus the error. Assuming a balanced design where equal sample size n is allocated to each group, $n_i = n$ and the total sample size is $an = N$.

An alternative to (1.2) and familiar overparameterized model for a balanced design is

$$Y_{ij} = \mu + \alpha_i + \varepsilon_{ij} \qquad (i = 1, \ldots, a; j = 1, \ldots, n) \qquad (1.3)$$

where μ is the grand mean, α_i is the ith group effect, $\alpha_i = \mu_i - \mu$; and if the group effect is assumed to be fixed or constant, we typically place a sum-to-zero constraint, $\Sigma_i \, \alpha_i = 0$.

Using model (1.3) and translating it to the following expression for the general linear model,

$$
\begin{bmatrix} Y_{11} \\ Y_{12} \\ \vdots \\ Y_{an} \end{bmatrix}
=
\begin{bmatrix}
1 & X_{11} & X_{12} & \cdot & X_{1,a-1} \\
1 & X_{21} & X_{22} & \cdot & X_{2,a-1} \\
\vdots & \vdots & \vdots & & \vdots \\
1 & X_{N1} & X_{N2} & \cdot & X_{N,a-1}
\end{bmatrix}
\begin{bmatrix} \mu \\ \alpha_1 \\ \vdots \\ \alpha_{a-1} \end{bmatrix}
+
\begin{bmatrix} \varepsilon_{11} \\ \varepsilon_{12} \\ \vdots \\ \varepsilon_{an} \end{bmatrix}
$$

Unlike a regression model where the elements of $\mathbf{X}$ are measured, thus uniquely given, there is more than one way to specify $\mathbf{X}$ (sometimes referred to as a design matrix) in ANOVA, with the exception that they are values of some indicator variables. To give an example specification, we will adopt *effect coding* for $\mathbf{X}$ with the model (1.3) where three groups each with $n = 2$ are allocated. Using effect coding, three parameters can be estimated: $\alpha_1 = \mu_1 - \mu$, $\alpha_2 = \mu_2 - \mu$, and the grand mean μ itself. Although α_3 does not appear in $\hat{\boldsymbol{\beta}}$, it can easily be derived from the sum-to-zero constraint as $\alpha_3 = -(\alpha_1 + \alpha_2)$.

$$
\begin{bmatrix} Y_{11} \\ Y_{12} \\ Y_{21} \\ Y_{22} \\ Y_{31} \\ Y_{32} \end{bmatrix}
=
\begin{bmatrix}
1 & 1 & 0 \\
1 & 1 & 0 \\
1 & 0 & 1 \\
1 & 0 & 1 \\
1 & -1 & -1 \\
1 & -1 & -1
\end{bmatrix}
\begin{bmatrix} \mu \\ \alpha_1 \\ \alpha_2 \end{bmatrix}
+
\begin{bmatrix} \varepsilon_{11} \\ \varepsilon_{12} \\ \varepsilon_{21} \\ \varepsilon_{22} \\ \varepsilon_{31} \\ \varepsilon_{32} \end{bmatrix}
$$

For testing the general null hypothesis of no group mean difference, H_0: $\mu_1 = \mu_2 = \cdots = \mu_a$, is accomplished by first decomposing the (adjusted) total sum of squares ($SSTOT$) into the SSA (the SS due to factor A), and the SSE (the SS due to error).

$$\sum_i^a \sum_j^n (Y_{ij} - \overline{Y}..)^2 = \sum_i^a \sum_j^n (Y_{ij} - \overline{Y}_{i.})^2 + n \sum_i^a (\overline{Y}_{i.} - \overline{Y}..)^2$$

$$SSTOT \qquad\qquad = SSE \qquad\qquad\qquad + SSA$$

with corresponding degrees of freedom

$$an - 1 \qquad = a(n - 1) \qquad + \qquad (a - 1).$$

Analogous to the test of zero regression parameters, the general null hypothesis on means can be tested by comparing

$$F = \frac{SSA/(a - 1)}{SSE/a(n - 1)} = \frac{MSA}{MSE}$$

against the central F distribution, $F[\alpha; (a - 1), a(n - 1)]$. We then reject the general null hypothesis if the obtained F ratio is equal to or larger than the corresponding critical F value.

1.4 OVERVIEW OF THE EXPECTED MEAN SQUARES, $E(MS)$

We will discuss some simple rules for obtaining expected mean squares and degrees of freedom (d.f.) and give a quick review of the sum of squares (SS) computation. If one is able to derive expected mean squares $E(MS)$ for the sources of variation, one can more easily understand why a certain pair of mean squares form a valid F test for the hypothesis of interest. The notion that analysis of variance is only a convenient arithmetic representation may be true for simple models while we compute the necessary sums of squares (SS) and mean squares (MS). But the choice of correct denominator MS to form a valid F ratio for the hypothesis of interest is not always obvious. This point becomes clearer when we understand three types of models: fixed effects, mixed effects, and random effects. A model is called a *fixed effects* model when all factors in the model are considered fixed. That is, the levels of each factor are not randomly chosen from a pool of infinite levels and the inference is limited to the exact levels used in the study. We exclude the grand mean (which is always considered fixed) from this consideration. Some statistical computer package such as SPSS (1990), which does not allow us to specify whether each factor is fixed or random, automatically assumes that all factors are fixed and forms F ratios based on this assumption if we do not tell it which term to use as the error. But this assumes that we know which MS to choose for the F denominator.

A model is considered *random effects* when all factors (except the grand mean) are random. That is, the levels of each factor are randomly selected from an infinite pool of levels and the conclusion is generalized to the levels beyond the ones actually used in the study. A *mixed effects* model represents a model with factors (excluding the grand mean), some of which are fixed and some random.

The $E(MS)$ for any source of variation can be expressed as a sum of various terms representing various effects. These are called *variance components* if they belong to random factors and are called *quadratic forms* if they belong to fixed factors. A valid F ratio for the hypothesis of interest is that of a numerator MS containing all the terms identical to those of the denominator MS plus one additional component or quadratic form, whose absence is tested by the hypothesis. Under a regular assumption of normally distributed random variables, the F ratio distributes as a central F distribution with the degrees of freedom for the numerator and the denominator when the component or quadratic form in question is zero, i.e., when the null hypothesis is true.

In addition to forming correct F ratios, identifying the correct components involved in the $E(MS)$ is a crucial process in variance component estimation problems (see Chapter 15). We review only two-way balanced classifications for obtaining $E(MS)$ components. One-way classification situations can easily be reduced from a case of two-way classification; and three-way and higher-order classifications can be generalized from two-way classification examples. The balanced situations (in which all treatment combinations had the same sample sizes) are presented here because the coefficients for the terms involved in the $E(MS)$ are easier to calculate than in unbalanced situations (in which unequal sample sizes are used in treatment combinations), but the general rules still apply. For incomplete or fractional designs in which certain assumptions or structural constraints are introduced, far more complex and idiosyncratic rules are applied. Because of this, such designs as Latin square and confounded factorial designs are not illustrated in the following discussion.

There are several procedures for deriving $E(MS)$. We especially refer the reader to Blackwell, Brown, and Mosteller (1991), Cornfield and Tukey (1956), Edwards (1964), and Henderson (1959, 1969). After a general introduction, we will present the rules based on the concept of sampling fractions introduced by Cornfield and Tukey. We will then illustrate a more intuitive procedure explained in Schultz (1955) and introduced in Edwards (1964) for limited designs, using selected balanced designs in two-way classifications. Readers are referred to a summary presentation by Searle (1971) for various estimation procedures.

1.4.1 Introduction: Two-Way Crossed Classification

In a crossed two-way classification with interaction, let Y_{ijk} be the response of the kth individual in the combination of the ith level of A treatment and the jth level of B treatment. The statistical model can be expressed as

$$Y_{ijk} = \mu + \alpha_i + \beta_j + \alpha\beta_{ij} + \varepsilon_{ijk},$$

$$i = 1, \ldots, a; j = 1, \ldots, b; k = 1, \ldots, n_{ij}, \qquad (1.4)$$

where n_{ij} is the sample size and when it is balanced, $n_{ij} = n$, and the errors are distributed as normal, $\varepsilon_{ijk} \sim N(0, \sigma^2)$. The first term, μ, is the grand mean, α_i is the ith A treatment effect, β_j the jth B treatment effect, and $\alpha\beta_{ij}$ is the ijth interaction of A and B factors. We then need to examine whether each factor is fixed or random. A fixed effects model is one in which both A and B are considered to be fixed (and AB is thus also fixed). Traditionally, we place a series of sum-to-zero constraints on the parameters in (1.4): $\Sigma_i \alpha_i = 0$, $\Sigma_j \beta_j = 0$, and $\Sigma_i \alpha\beta_{ij} = 0$ and $\Sigma_j \alpha\beta_{ij} = 0$. Under the random effects model, the α-, β-, and $\alpha\beta$- effects are assumed to have zero means and variances, σ_α^2, σ_β^2, and $\sigma_{\alpha\beta}^2$, respectively. In a mixed effects model, the sum-to-zero constraints are placed on the fixed factor and only on one direction of the interaction term; i.e., when A is random and B is fixed, the interaction $\alpha\beta$ is assumed to be fixed in the direction of summing across the fixed factor, B. That is, only the sum-to-zero restriction of $\Sigma_j \alpha\beta_{ij} = 0$ is applicable to the interaction term.

While much of the computational aspect of ANOVA is deferred to various texts on analysis of variance, the sum of squares computations and the corresponding d.f. for two-way classifications are summarized in Table 1.1. Briefly, the notations used in Table 1.1 are:

$\overline{Y}_{i..}$ is the ith A factor sample mean,
$\overline{Y}_{.j.}$ is the jth B factor sample mean,
$\overline{Y}_{ij.}$ is the ijth AB treatment sample cell mean, and
$\overline{Y}...$ is the sample grand mean.

In addition, small letters corresponding to factor names are used as the number of levels for these factors. That is, a represents the number of levels associated with factor A and b the number of levels associated with factor B. A more elaborate and rule-based SS computation is presented in Millman and Glass (1967) and summarized in Searle (1971).

Table 1.1   Analysis of Variance Summary Table

Sources	d.f.	Sum of squares
A	$(a-1)$	$SSA = bn \sum_i (\overline{Y}_{i..} - \overline{Y}...)^2$
B	$(b-1)$	$SSB = an \sum_j (\overline{Y}_{.j.} - \overline{Y}...)^2$
AB	$(a-1)(b-1)$	$SSAB = n \sum_i \sum_j (\overline{Y}_{ij.} - \overline{Y}_{i..} - \overline{Y}_{.j.} + \overline{Y}...)^2$
Error	$ab(n-1)$	$SSE = \sum_i \sum_j \sum_k (Y_{ijk} - \overline{Y}...)^2$

A simple rule in determining the d.f. is to use the level of the factor minus 1 for a crossed factor which does not have any nesting structure indicated by parentheses. For example, factor A will have $(a - 1)$ d.f. and factor B, $(b - 1)$. The interaction AB will have the d.f. which is a multiplication of the corresponding factors' d.f. For $S(AB)$, where parentheses indicate that Subjects are nested in AB combinations, the levels of the factors embedded in the parentheses, ab, will be multiplied to the level of the outside factor S minus 1, that is, $(n - 1)$.

1.4.2 Sampling Fractions

One of the pervasive methods for deriving the expected MS components is based on the concept of sampling fractions. A general model using sampling fractions was introduced by Cornfield and Tukey. Recently, Blackwell et al. illustrated the way to derive variance components in a crossed two-way classification. As an example, assume Row factor has R possible levels in the population and Column factor has C possible levels in the population; then assume that each cell of a treatment combination can be sampled from the population of N. If Row factor is fixed, that is, if the population levels are finite and are all sampled, the sampling fraction $r/R = 1$, the coefficient placed in front of a variance component or a quadratic form is $(1 - r/R) = 0$, forcing the term to vanish.

On the other hand, if Row factor is random and r levels are randomly selected from an infinite or semi-infinite pool of levels R, the fraction r/R is very close to zero. Thus $(1 - r/R)$ can be treated as 1 for all practical purposes, causing the term for which this coefficient is associated to stay in the $E(MS)$. Since subjects are almost always treated as random and by assuming n subjects are sampled from an infinite pool of N, the fraction $n/N = 0$ for all practical purposes. The sampling fraction concept works well in building the computer algorithms within the confines of balanced and nonfractional designs. A general case of two-way crossed classification, modifying the notation slightly from Blackwell et al. (1991, p. 258) and from Cornfield and Tukey (1956, p. 926), is presented in Table 1.2. Unfortunately, this procedure uses variance component expressions for all factors. That is, some of the components above are not "variance components" belonging to random factors but are quadratic forms. Another step is required to convert variance components to those of quadratic forms for fixed factors when the exact expressions are desired. But this rule still applies in finding which terms vanish or remain in the $E(MS)$.

Assuming that subjects are random, the error variance component will remain in every row for $(1 - n/N) = 1$. If Row and Column are both fixed, then r/R and c/C are 1, which, in turn, produces no interaction term

Table 1.2 The Average Value of Mean Squares in a Two-Way Classification

Sources	General case (average value of mean squares)
Row	$(1 - n/N)\, \sigma_e^2 + (1 - c/C)\, n\sigma_{rc}^2 + nc\sigma_r^2$
Column	$(1 - n/N)\, \sigma_e^2 + (1 - r/R)\, n\sigma_{rc}^2 + nr\sigma_c^2$
Row by Column	$(1 - n/N)\, \sigma_e^2 + n\sigma_{rc}^2$
Error	σ_e^2

Note: $(1 - n/N)$ is a sampling fraction almost always equal to 1 because subjects are usually considered as random; hence, n/N is 0 for all practical purposes.

in the Row factor and Column factor of $E(MS)$. Similarly, if Row is fixed and Column is random, the interaction component vanishes from source Column but will remain in source Row. Subsequently, the overall hypothesis of no Row effects will be tested by the F ratio of the mean square of Row over the mean square of Column, or $MSR/MSRC$, but the corresponding hypothesis for Column is tested by the mean square of Column over the error mean square, or MSC/MSE.

1.4.3 Modified Schultz's Rules for $E(MS)$

1.4.3.1 Two-Way Crossed Classification

Various efforts have been made in deriving $E(MS)$ components. One can start from the beginning by taking the expectations of random variables (Edwards, 1964). But we would like a shortcut. Henderson (1959, 1969) is often given credit for one of the earliest such efforts. Schultz also gives a comprehensive explanation and rules. Following intuitive Schultz's rules and using some refinements, deriving the $E(MS)$ can be simplified to the point of being done by inspection. Instead of listing all the rules, we now focus on the explanations and direct readers to a rule-by-rule specification presented in Henderson and to other references listed in Searle.

We first assume that we are dealing with a random effects model where all components, except the grand mean, are random. From (1.4), we have sources due to α, β, $\alpha\beta$, and ε. Table 1.3 lists the $E(MS)$ for a two-way crossed classification under the assumption of a random model.

Steps involved in applying Schultz's rules are:

(1) Identify all sources of variation in the $E(MS)$ table by assigning symbols to them. Thus, sources A, B, AB, and $S(AB)$ (the subjects nested in AB combinations or the within cells) are identified. The nesting is indicated by the use of parentheses. Use capitals for the purpose of source identification and use corresponding small letters as subscripts for identi-

Table 1.3 $E(MS)$ Table for a Random Effects Two-Way Model

Sources	d.f.	$E(MS)$	
A	$(a - 1)$	$\sigma_e^2 + n\sigma_{ab}^2$	$+ nb\sigma_a^2$
B	$(b - 1)$	$\sigma_e^2 + n\sigma_{ab}^2 + na\sigma_b^2$	
AB	$(a - 1)(b - 1)$	$\sigma_e^2 + n\sigma_{ab}^2$	
$S(AB)$	$ab(n - 1)$	σ_e^2	

fying the corresponding variances. For example, Source A will include in its $E(MSA)$, σ_a^2.

(2)   Each factor has the number of levels corresponding to the small letter of the source. Hence, A factor has a levels. Each component, excluding the sampling error component, will have as its coefficient all other constants indicating the levels of other factors. That is, all the subscripts in small letters not used in the component are assigned as its coefficient. Therefore, σ_a^2 is given nb, the number of subjects, n, and the level of B factor, b, as its coefficient.

(3)   The within-cells component will be expressed simply as σ_e^2, thus $\sigma_e^2 = \sigma_{s(ab)}^2$, and it will appear in every source. [Following the convention by Kirk (1982, p. 255), σ_e^2 is used as the error component in both randomized group designs and randomized block designs. In randomized group designs, the error term represents the random sampling variance, $\sigma_e^2 = \sigma_Y^2$; and in randomized block designs with homogeneous entities assigned within the same block, $\sigma_e^2 = \sigma_Y^2(1 - \rho)$ where ρ is the common correlation between the observations within the same block.]

(4)   List in the variance components from the bottom of the $E(MS)$ table and then move upward. As discussed, fill in the error variance component in every source by starting at the row for the within cells. For the $S(AB)$ source, the error is the source itself. This then exhausts all the necessary components for that row. However, the next step up, the source AB, will have an additional component (beyond the sampling error) including the source component itself. The source component will be $n\sigma_{ab}^2$. This component will then be copied up to every source listed above if the source in question is a subset of AB. That is, because source A and source B are each a subset of AB, $n\sigma_{ab}^2$ is copied onto both source A and source B. Since $n\sigma_{ab}^2$ is the last component necessary to completely identify the source, filling in $n\sigma_{ab}^2$ completes the variance component estimation process for source AB.

We then move on to the next row up, source B. Source B now has the error variance and the AB component from previous steps. It also needs the component representing the source itself, $na\sigma_b^2$. This component, how-ever, will not be copied onto source A because A is not a subset of B. Now that all components are represented for B, we move up to source A. For source A, the only component still needed is $nb\sigma_a^2$, and this completes the process for A.

(5) The above steps complete the component identification process. However, if one or more factors are fixed (and, therefore, a fixed effects model or a mixed effects model is the model in question), another step is necessary to determine which components should be deleted. Let us assume for illustration that A is fixed and B is random. (In Table 1.4, the com-ponents to be deleted are in brackets.) In a mixed effects model, the interaction component, AB, vanishes from the random source B but not from the fixed source A. An easy rule that accompanies this decision is that for each component (except for the error component and the com-ponent representing the source itself), one must examine all the subscripts aside from the symbol representing the source itself; and if any one of the subscripts represents fixed effects, that component in question vanishes. This is because the interaction between a random and a fixed component results in a component which is, to quote Schultz (1955, p. 125), ". . . 'random in one direction only,' i.e., such a component does *exist* as a part of the expectation of the mean square of the fixed effect (since measured over the random variate) but *does not exist* as a part of the expectation of the random variate (since measured over the fixed effect)." In other words, the AB interaction with A fixed and B random will have the sum-to-zero restriction only in the direction of A, that is, $\Sigma_i \alpha\beta_{ij} = 0$. When the expected sum of squares $E(SS)$ is calculated, we sum across in the direction of the source itself in the very end (Millman and Glass, 1967); therefore, the SS carrying the AB interaction appearing in SSA is obtained in the order of $\Sigma_i \Sigma_j$, where the inner loop is summed before the outer loop; and the same quantity will be summed over in the order of $\Sigma_j \Sigma_i$ for SSB. Hence, the sum-to-zero restriction will take effect only for source B where the inner loop sums across the fixed factor, but not for source A where the inner loop sums across the random factor. This establishes that if there is *any* fixed effects factor to be summed over before the source of interest itself, the component in question vanishes. [This point is made explicit by Schultz and is consistent with the sampling fraction approach. Henderson (1969) approaches this differently from Henderson (1959) and from Schultz. A discussion of this point in relation to the choice of constraints can be found in Searle (1971).]

(6) This deletion rule applies only to "essential" subscripts, the sub-scripts that are not in parentheses. That is, in a nested example in the next section, $A(B)$ has only one essential factor, A. Subsequently, the essential

Table 1.4 $E(MS)$ Table with Subjects Random, A Fixed, and B Random

Sources	d.f.	$E(MS)$ components	
A	$(a - 1)$	$\sigma_e^2 + n\sigma_{ab}^2$	$+ nb\theta_a^2$
B	$(b - 1)$	$\sigma_e^2 + [n\sigma_{ab}^2] + na\sigma_b^2$	
AB	$(a - 1)(b - 1)$	$\sigma_e^2 + n\sigma_{ab}^2$	
$S(AB)$	$ab(n - 1)$	σ_e^2	

Note: $\theta_a^2 = \Sigma_i \alpha_i^2/(a - 1)$ if a sum-to-zero constraint is applied. The bracketed component is to be deleted.

subscript a [but not (b)] will be examined in determining whether the component should vanish. This rule becomes important in nested designs. Finally, for fixed or mixed effects models, variance component expressions can be replaced with quadratic expressions for fixed factors; i.e., the "variance" expression, σ^2, can be replaced by θ^2.

Let us apply the above rules in a mixed effects model with A random and B fixed (Table 1.4). Observe that the AB interaction component would disappear from source A (but not from source B) because the sum-to-zero restriction takes effect during its inner summation process. [If the design were completely fixed (i.e., both A and B were fixed), the AB interaction component would disappear from both source A and source B.]

As discussed earlier, the $E(MS)$ table provides a handy guide for preparing a valid F ratio because the appropriate error term for a test of a given effect is the MS whose expectation contains all the components except the component representing the source itself. Therefore, to test A effects in Table 1.4, the ratio $E(MSA)/E(MSAB)$ is the valid one because $(\sigma_e^2 + n\sigma_{ab}^2 + nb\theta_a^2)/(\sigma_e^2 + n\sigma_{ab}^2)$ has the correct form. Hence, $F = MSA/MSAB$ is a valid F test for the hypothesis H_0: $\theta_a^2 = 0$.

1.4.3.2 Two-Way Classification with One Nested Factor

In a balanced nested model, let B factor be nested in A factor. Therefore, there are A, $B(A)$, and the error but no AB interaction in the model. The statistical model can be written as

$$Y_{ijk} = \mu + \alpha_i + \beta_{j(i)} + \varepsilon_{ijk},$$

$$i = 1, \ldots, a; j = 1, \ldots, b; k = 1, \ldots, n, \qquad (1.5)$$

where n is the sample size for a balanced design, and the errors are distributed as normal, $\varepsilon_{ijk} \sim N(0, \sigma^2)$. The terms, μ and α_i are the grand mean and the ith A treatment effect, respectively. $\beta_{j(i)}$ is the $j(i)$th level of

Table 1.5 $E(MS)$ for a Nested Design with Subjects Random, A Fixed, and B Random

Sources	d.f.	$E(MS)$ components
A	$(a - 1)$	$\sigma_e^2 + n\sigma_{b(a)}^2 + nb^*\theta_a^2$
$B(A)$	$a(b^* - 1)$	$\sigma_e^2 + n\sigma_{b(a)}^2$
$S(AB)$	$ab^*(n - 1)$	σ_e^2

Note: b^* is the levels of B factor *within* each level of A. Strictly speaking, $S(B(A))$ is the correct designation of the error term, but it is represented as $S(AB)$.

$B(A)$ treatment effect. (Although the subscript j ranges from 1 to b, this disregards the nesting. Therefore, if B were crossed with A, b^* would be the maximum levels of B; in other words, b^* is the levels of B *within* each level of A.)

Under a fixed effects model where A and $B(A)$ are fixed, we can constrain the above parameters with $\Sigma_i \, \alpha_i = 0$ and $\Sigma_j^b \, \beta_j = 0$. Under the random effects model, the α- and β- effects are assumed to have zero means and variances, σ_α^2 and σ_β^2, respectively. In a mixed effects model, as before, the sum-to-zero constraint is placed on the fixed factor.

Using the rules described in Section 2.1.3, especially carefully incorporating rule (6) about the essential subscripts, an $E(MS)$ table for a mixed model example with A fixed and B random is presented in Table 1.5.

Because there is no AB interaction, the AB interaction component does not appear anywhere in the table. Factor A can appear by itself but factor B cannot exist in any other form than $B(A)$. For source $B(A)$, the component $B(A)$ represents the source itself so it does not vanish; and for source A, the component $B(A)$ does not vanish because the essential subscript excluding the source itself, b, is random. The effects of A are then tested by $MSA/MSB(A)$ and the effects of $B(A)$ by $MSB(A)/MSS(AB)$.

1.4.3.3 Two-Way Classification: A Randomized Block Design

Let B be a Block, A be a treatment, and BA be the interaction between Block and Treatment. Also let $n = 1$ in each block undergoing i levels of repeated factor A. Without loss of generality, this example can be extended to a case where a block size of a is used for b number of blocks, i.e., there are a number of matched subjects assigned to the levels of A. If we had multiple replications within each block-by-treatment combination, the MS within cells could be estimated.

Table 1.6 $E(MS)$ for a Randomized Block Design with A Fixed and B Random

Sources	d.f.	$E(MS)$ components
B	$(a - 1)$	$\sigma_e^2 + [\sigma_{ba}^2] + a\sigma_b^2$
A	$(b - 1)$	$\sigma_e^2 + \sigma_{ba}^2 + n\theta_a^2$
BA	$(a - 1)(b - 1)$	$\sigma_e^2 + \sigma_{ba}^2$

Note: The two components in $E(MSBA)$, $\sigma_e^2 + \sigma_{ba}^2$, are not separately estimable with $n = 1$ in each block. The bracketed component is to be deleted.

Under the assumption where the block-by-treatment interaction is included, the statistical model is

$$Y_{ik} = \mu + \alpha_i + \pi_k + \alpha\pi_{ik} + \varepsilon_{ik}, \quad i = 1, \ldots, a; \ k = 1, \ldots, n, \qquad (1.4)$$

where n is the number of blocks and the errors are distributed as normal, $\varepsilon_{ik} \sim N(0, \sigma^2)$. The first term, μ, is the grand mean, α_i is the ith A treatment effect, π_k the kth block or subject effect, and $\alpha\pi_{ik}$ the ikth interaction of A repeated factor and blocks. Subjects are assumed to be random, hence π_k is random. If factor A is fixed, this model is a mixed effects model with a constraint placed on A, $\Sigma_i \, \alpha_i = 0$, and the interaction is fixed in one direction, $\Sigma_i \, \alpha\pi_{ik} = 0$. We realize that with $n = 1$ in block, the error variance and the interaction are not simultaneously estimable. When it is reasonable to drop the block-by-treatment interaction, $\alpha\pi_{ik}$ is dropped from Eq. (1.4). We will not drop the interaction here because (1) it is instructive to illustrate the $E(MS)$ rules by designating interaction components because the interaction source then will have its namesake in its $E(MS)$, and if multiple replications are introduced in randomized block designs, the interaction component becomes estimable; and (2) even though the within-cell error variance and the interaction components are not separately estimable with $n = 1$ in each block, it does not alter the appropriate F ratio required for the hypothesis of interest in a mixed effects model.

The $E(MS)$ table starts at the bottom, source BA. As in the randomized group designs, an interaction source should include not only the random sampling error component but also the interaction component itself. Then, source BA consists of $\sigma_e^2 + \sigma_{ba}^2$. With B random and A fixed, the interaction component disappears from source B; therefore, no exact F test exists for the effects of Block. However, often this is not a serious shortcoming because blocking is chosen mainly to remove individual differences from the error term rather than to represent one of the main factors of interest. Furthermore, the interaction may be assumed to be zero when appropriate, which in turn will allow the effects of B to be tested by $MSB/MSBA$.

As for the coefficient, note that the BA interaction term, if it is included in the model, will have 1 as a coefficient rather than the usual n because it is $n = 1$ in this case. Similarly, the block component will have a as the coefficient, and A component will have b as a coefficient.

We have briefly reviewed the process of deriving $E(MS)$ for selected balanced two-way classification situations. Although the designs considered in this chapter are not exhaustive, they illustrate basic principles for deriving $E(MS)$ components and establishing the valid F ratios. Along with the computational aspects of SS and the rules for obtaining d.f., we are ready to pursue more advanced topics in analysis of variance presented in other chapters of this book.

REFERENCES

Blackwell, T., Brown, C., and Mosteller, F. (1991). Which Denominator? *Fundamentals of Exploratory Analysis of Variance*. Hoaglin, D. C., Mosteller, F., and Tukey, J. W. (eds.). Wiley, New York, pp. 252–294.
Cornfield, J. and Tukey, J. W. (1956). Average Values of Mean Squares in Factorials. *Ann. Math. Statist.*, 27: 907–949.
Edwards, A. L. (1964). *Expected Values of Discrete Random Variables and Elementary Statistics*. Wiley, New York.
Edwards, A. L. (1985). *Experimental Design in Psychological Research* (5th ed.). Harper and Row, New York.
Graybill, F. A. (1976). *Theory and Application of the Linear Model*. Duxbury Press, Boston.
Henderson, C. R. (1959). Design and Analysis of Animal Husbandry Experiments. *Techniques and Procedures in Animal Production Research*. American Society of Animal Production, Beltsville, Maryland, pp. 2–56.
Henderson, C. R. (1969). Design and Analysis of Animal Science Experiments. *Techniques and Procedures in Animal Science Research*. American Society of Animal Science, Albany, New York, pp. 1–35.
Hoaglin, D. C., Mosteller, F., and Tukey, J. W. (eds.). (1991). *Fundamentals of Exploratory Analysis of Variance*. Wiley, New York.
Hocking, R. R. (1985). *The Analysis of Linear Models*. Brooks/Cole, Belmont, California.
Keppel, G. (1982). *Design and Analysis: A Researcher's Handbook* (2nd ed.). Prentice-Hall, Englewood Cliffs, New Jersey.
Kirk, R. E. (1982). *Experimental Design* (2nd ed.). Brooks/Cole, Belmont, California.
Krishnaiah, P. R. (ed.). (1980). *Handbook of Statistics* (Vol. 1): *Analysis of Variance*. North-Holland, New York.
Maxwell, S. E. and Delaney, H. (1990). *Designing Experiments and Analyzing Data: A Model Comparison Perspective*. Wadsworth, Belmont, California.

Millman, J. and Glass, G. V. (1967). Rules of Thumb for Writing the Anova Table. *J Educ. Meas.*, *4*: 41–51.

Myers, R. H. and Milton, J. S. (1991). *A First Course in the Theory of Linear Statistical Models*. PWS-Kent, Boston.

SAS User's Guide, Ver. 6 (1990). SAS Institute, Inc., Cary, North Carolina.

Scheffé, H. (1959). *The Analysis of Variance*. Wiley, New York.

Schultz, E. F. (1955). Rules of Thumb for Determining Expectations of Mean Squares in Analysis of Variance. *Biometrics*, *11*: 123–135.

Searle, S. R. (1971). *Linear Models*. Wiley, New York.

SPSS Reference Guide. (1990). SPSS, Inc., Chicago.

Wilkinson, L. (1989). *SYSTAT: The System for Statistics*. SYSTAT Inc., Evanston, Illinois.

Winer, B. J. (1971). *Statistical Principles in Experimental Design* (2nd ed.). McGraw-Hill, New York.

Winer, B. J., Brown, D. R., and Michels, K. M. (1991). *Statistical Principles in Experimental Design* (3rd ed.). McGraw-Hill, New York.

2
Multiple Comparisons of Independent Means

PHILIP H. RAMSEY Queens College of the City University of New York, Flushing, New York

2.1 INTRODUCTION

2.1.1 General Introduction

In any statistical test one must define the probability of falsely rejecting a null hypothesis, α, known as a Type I error rate (also known as the level of significance). If a single statistical test is conducted at level α, the expected Type I error rate for that test is α. However, a large number of statistical tests (or comparisons among statistics), each run at level α, will increase the probability of a false rejection beyond the designated level α. The general problem of multiple comparisons can be stated as the problem of determining an appropriate method for selecting a level α' for each comparison or test.

Perhaps the simplest solution to the multiple comparison problem is the *error rate per comparison* in which $\alpha' = \alpha$. That is, no matter how many tests are applied, each test is run at level α as if no other tests were involved. There is a considerable literature suggesting that this approach is not adequate due to the likelihood of some false rejection (Hochberg, 1988; Hochberg and Tamhane, 1987; Miller, 1981). However, there remains some controversy. It can be argued that running a large number of tests in a single experiment is no different from running a large number of separate experiments, each with one test run at level α (Saville, 1990, 1991). Such arguments have not been accepted as convincing (Holland, 1991; Lea, 1991).

Saville argues that excessive error rates due to multiple testing are no problem and can be ignored. His proposal essentially abolishes the entire field of multiple comparisons. In some areas of research a large number of experiments could be run with one statistical test each. In evaluating such research one needs to consider whether the number of significant results in a large collection of experiments exceeds the number expected by chance. In other words, multiple comparison procedures must be applied in summarizing the large number of statistical tests run in separate experiments. One example in which the error rate per comparison is routinely used is planned orthogonal comparisons. This topic will be considered in detail later in this chapter.

One alternative to the error rate per comparison is to collect a group of tests into a family and limit the Type I error rate for the family. For a family of C statistical tests, the *error rate per family* would require that $\alpha' = \alpha/C$. If the family includes all the tests within an experiment, this approach could be called the *error rate per experiment*. This approach ensures that the likelihood of a Type I error is not excessive. However, it is likely to provide α' levels which are unnecessarily stringent, resulting in a loss of power (Hochberg and Tamhane, 1987; Toothaker, 1991). However, as noted below, it provides a useful lower limit for α' when other methods of Type I error control are used.

Perhaps the most widely accepted error rate for a family of tests is the *familywise error rate* in which α' is determined so that the probability of any Type I errors is limited to α; that is, the probability that the number of Type I errors is one or more is bounded by α, $P(\#TI \geq 1) \leq \alpha$. If the family includes all the tests in an experiment, this error rate is known as the *experimentwise error rate*.

2.1.2 Analysis of Variance

For I independent groups with N_i observations in group i, we have X_{ni} as observation n in group i. The equality of $I \geq 3$ independent means $(\overline{X}_i, i = 1, \ldots, I)$ is routinely tested with the independent groups, analysis of variance (ANOVA), F test. The null hypothesis tested by the ANOVA F test is the full null hypothesis that all I population means are equal (i.e., $H_0: \mu_1 = \mu_2 = \cdots = \mu_I$).

To apply the independent groups ANOVA we calculate for each group $SS_i = \Sigma_{n=1}^{N_i}(X_{ni} - \overline{X}_i)^2$, the sum of squared deviations of the scores in that group about the group mean. We determine SS_E, the sum of squares within groups, by $SS_E = \Sigma_{i=1}^{I} SS_i$. With N_i scores in group i there will be $df_i = N_i - 1$. Combining all I groups, we have $df_E = \Sigma_{i=1}^{I} df_i$. The variances in I groups can be averaged or "pooled" by calculating MS_E, the mean square within groups, by $MS_E = SS_E/df_E$. If the populations from which these groups are drawn have a common variance, σ^2, then MS_E is an unbiased estimator of σ^2.

Taking $N_T = \Sigma_{i=1}^I N_i$ and $\overline{X}_T = \Sigma_{i=1}^I \Sigma_{n=1}^{N_i} X_{ni}/N_T$, we have the between-groups or treatment sum of squares, $SS_{\text{TREAT}} = \Sigma_{i=1}^I N_i(\overline{X}_i - \overline{X}_T)^2$. Expressed as a variance, we have $MS_{\text{TREAT}} = SS_{\text{TREAT}}/df_{\text{TREAT}}$, where $df_{\text{TREAT}} = I - 1$, and the resulting F statistic, $F = MS_{\text{TREAT}}/MS_E$. Running the test at level α we reject H_0 if $F \geq$ the critical value, $CV = F_{1-\alpha}(df_{\text{TREAT}}, df_E)$.

Three assumptions are required for the ANOVA F test:

1. Each score is independent of all the other scores.
2. The population distribution for each group is normal.
3. The population variances are equal.

If these assumptions are satisfied perfectly, the overall F test is an exact test with α being its Type I error rate. If they are approximately correct, the F test is no longer exact with α being only its approximate Type I error rate.

Accepting or rejecting the full null hypothesis is not usually adequate to provide a complete interpretation of an experiment. Tests of specific or focused hypotheses may be more useful, and contrasts among means can be used to examine such specific hypotheses.

2.2   CONTRASTS

2.2.1   Planned Comparisons

Suppose we want to know if the first of three groups differs from the mean of the last two groups. That is, we want to test the hypothesis

$$H_0: \mu_1 = \frac{\mu_2 + \mu_3}{2}.$$

The alternative hypothesis is simply

$$H_1: \mu_1 \neq \frac{\mu_2 + \mu_3}{2}.$$

It is instructive to consider various forms of the null hypothesis. In particular, we could write this same H_0 as

$$H_0: \mu_1 - (1/2)\mu_2 - (1/2)\mu_3 = 0$$

or

$$H_0: 2\mu_1 - \mu_2 - \mu_3 = 0.$$

These three forms of the null hypothesis are identical. If one of them is true, then the other two *must* also be true.

The last form of H_0 could be expressed simply by identifying the coefficients $(2, -1, -1)$. The order is important because 2 is the coefficient of μ_1, the first -1 applies to μ_2, and the second -1 applies to μ_3. A *contrast* or *comparison*, Ψ, among population means is defined by a set of ordered coefficients applied to the respective means. We can identify the coefficients as $(c_1, c_2, \ldots, c_I)$, respectively. The coefficients must satisfy two requirements:

1. At least two coefficients must be nonzero (i.e., $c_i \neq 0$ for at least two values of i).
2. The sum of the coefficients must equal zero, i.e., $\Sigma\, c_i = 0$.

The contrast discussed above is $\Psi_1 = 2\mu_1 - \mu_2 - \mu_3$ and is identified by the coefficients $(2, -1, -1)$. We now have yet another way of expressing the null hypothesis as

H_0: $\Psi_1 = 0$.

Of course, the alternative hypothesis can be expressed as

H_1: $\Psi_1 \neq 0$.

To test this hypothesis, we consider the sample estimate, $\hat{\Psi}_1 = 2\overline{X}_1 - \overline{X}_2 - \overline{X}_3$. The expected value of $\hat{\Psi}_1$ will be $E(\hat{\Psi}_1) = \Psi_1$. That is, $\hat{\Psi}_1$ is an unbiased estimator of Ψ_1. If $\hat{\Psi}_1$ is significantly different from zero, then we can reject H_0: $\Psi_1 = 0$.

With unequal sample sizes, the sum of squares of $\hat{\Psi}$ is given by

$$SS_{\hat{\Psi}} = \frac{\hat{\Psi}^2}{\Sigma(c_i^2/N_i)}. \tag{2.1}$$

With equal sample sizes, N, the sum of squares of $\hat{\Psi}$ is given by

$$SS_{\hat{\Psi}} = \frac{N\hat{\Psi}^2}{\Sigma c_i^2}. \tag{2.2}$$

The general rule for degrees of freedom is that *df* is the number of objects being compared minus the number of linear constraints placed on the objects (or the number of parameters being estimated). Although *I* means are involved, linear contrasts always compare one group of means with a second group of means with a typical constraint that these sum to zero. Thus, there are always two things being compared (i.e., two groups of means). All such contrasts have $df_{\hat{\Psi}} = 1$.

An *F* test for the hypothesis, $\Psi = 0$, is given by

$$F_{\hat{\Psi}} = \frac{MS_{\hat{\Psi}}}{MS_E}, \tag{2.3}$$

where $MS_{\hat{\psi}}$ has $df_{\hat{\psi}}$, which is 1, MS_E has df_E, and the critical value is $CV = F_{1-\alpha}(1, df_E)$.

A particular comparison, if selected independently of the data, is known as a *planned* or an *a priori* comparison. If an experiment was originally planned because previous literature showed that there should be a difference between the first group and the second and third groups, then the contrast, $(2, -1, -1)$, would be a good choice for a planned comparison. Identifying the contrast before the experiment is run would guarantee that the choice is independent of the data and therefore a planned comparison.

Any comparison among means which is identified by observing characteristics of the data is known as *post hoc* or *a posteriori*. For example, suppose we examined the data and noticed that the first mean appeared to differ markedly from the second and third means. The contrast $(2, -1, -1)$ would not be a planned comparison because it would be determined from examination of the data. We will consider post hoc comparisons in more detail in a later section.

A single planned comparison can be tested for significance using the F test of Eq. 2.3. If that test is run at level α and the usual ANOVA assumptions are satisfied, then the probability of a Type I error is limited to α. However, there may be more than one planned comparison to be tested. When this occurs, the risk of a Type I error is increased. The *familywise Type I error rate* limits the probability that the number of Type I errors is at least one, $P(\#\text{TI} \geq 1)$. Thus, to control the Type I error rate for a collection of comparisons to a level α, we require $P(\#\text{TI} \geq 1) \leq \alpha$.

2.2.2 Orthogonal Contrasts

In addition to the contrast, $\Psi_1 = (2, -1, -1)$, discussed above, suppose we were interested in a second contrast testing the equality of μ_2 and μ_3. Such a contrast could be designated as Ψ_2 and identified by the coefficients $(0, 1, -1)$. The first contrast asks if the first group differs from the last two groups, while the second contrast asks if the last two groups differ from each other. When two questions are asked of the same data, it becomes important to know whether these questions are independent. To evaluate this question of independence directly, we consider the orthogonality of the contrasts.

To define orthogonality of contrasts, we first identify two contrasts. The first contrast, Ψ_1, is identified by the coefficients $(c_{11}, c_{12}, \ldots, c_{1I})$. The second contrast, Ψ_2, is identified by the coefficients $(c_{21}, c_{22}, \ldots, c_{2I})$. We define the pair of contrasts as *orthogonal* if

$$\sum_{i=1}^{I} \frac{c_{1i}c_{2i}}{N_i} = 0. \tag{2.4}$$

With the same sample size, N, in all I groups, we have two orthogonal contrasts if

$$\sum_{i=1}^{I} c_{1i}c_{2i} = 0. \tag{2.5}$$

This means that, with equal sample sizes, two contrasts are orthogonal if the sum of products of corresponding coefficients is zero. Applying Eq. (2.5) to the two contrasts discussed above, we have

$$\sum c_{1i}c_{2i} = (2)(0) + (-1)(1) + (-1)(-1)$$
$$= 0 - 1 + 1$$
$$= 0.$$

Since the result is zero, we conclude that the two contrasts are orthogonal. This means that the questions being asked by the two contrasts are independent questions. It might be helpful to remember that in geometry orthogonal lines are perpendicular lines. Orthogonal contrasts ask independent questions just as orthogonal lines run in perpendicular directions.

Another aspect of independent questions is that knowing the answer to one question tells you nothing about the answer to the other question. Suppose that we paused after finding the first contrast, $\hat{\psi}_1$, significant before considering the calculations for the second contrast. Would we be in a position to predict a significant result for the second contrast? The orthogonality or independence of the two contrasts indicates that, knowing only the results of the first contrast, we would *not* be able to predict the significance or nonsignificance of the second contrast.

It is quite easy to find contrasts which are not orthogonal. Consider the contrast $\Psi_3 = \mu_1 - \mu_2$. This contrast has coefficients $(1, -1, 0)$. It addresses the question of the equality of the first two groups. Applying Eq. (2.5) to Ψ_1 and Ψ_3, we have

$$\sum c_{1i}c_{3i} = 2(1) + (-1)(-1) + (-1)(0)$$
$$= 2 + 1 + 0$$
$$= 3 \neq 0.$$

Thus, we see that these two contrasts are *not* orthogonal and therefore ask somewhat related questions. It is left as an exercise to show that Ψ_2 and Ψ_3 are also nonorthogonal.

Among I means there can never be more than $I - 1$ mutually orthogonal contrasts. In the above discussion we had three means and therefore no more than two mutually orthogonal contrasts. The two orthogonal contrasts, Ψ_1 and Ψ_2, represent a complete set of mutually orthogonal contrasts. This means that there is no additional contrast that is orthogonal to both these contrasts. It is possible to generate additional pairs of contrasts among the three means which are orthogonal to each other but no additional contrast orthogonal to these two. For example, we could define the contrast, $\Psi_4 = \mu_1 + \mu_2 - 2\mu_3$, which can be shown to be orthogonal to Ψ_3. However, Ψ_4 is not orthogonal to Ψ_1 or Ψ_2.

Among four means there would be sets of, at most, three mutually orthogonal contrasts. It might be helpful to remember that $df_{TREAT} = I - 1$. There are $I - 1$ degrees of freedom in comparing I means and $I - 1$ mutually orthogonal contrasts among I means. If we restrict our attention to independent questions (or orthogonal contrasts), only as many questions are possible as there are independent pieces of information in the set of means.

With unequal sample sizes, the task of finding orthogonal contrasts becomes more difficult. Suppose we want to test the hypothesis $\mu_1 = (\mu_2 + \mu_3)/2$ in an experiment in which the sample sizes are $(N_1, N_2, N_3) = (9, 10, 7)$. With equal N the contrast could be $(1, -1/2, -1/2)$, where the second and third groups can be combined to form a single coefficient, -1. With unequal N this would be equivalent to assigning 1 to the first group, $-10/(10+7)$ to the second group, and $-7/(10+7)$ to the third group. These coefficients $(1, -10/17, -7/17)$ satisfy the second requirement for a contrast,

$$\sum c_i = 1 - 10/17 - 7/17 = 0.$$

Similarly, to test the hypothesis $\mu_2 = \mu_3$, we simply take the coefficients $(0, 1, -1)$, which also sum to zero. To demonstrate the orthogonality of these two contrasts using Eq. (2.4) we obtain

$$\sum_{i=1}^{I} \frac{c_{1i}c_{2i}}{N_i} = \frac{(1)(0)}{9} + \frac{(-10/17)(1)}{10} + \frac{(-7/17)(-1)}{7}$$

$$= 0 - \frac{1}{17} + \frac{1}{17} = 0.$$

In conducting an experiment, there is no requirement that contrasts be orthogonal. In fact, in an experiment which replicates some parts of earlier experiments, one may need to examine nonorthogonal contrasts of several previous experiments. However, there are several advantages to using orthogonal contrasts whenever possible:

1. Each contrast can be interpreted independently of all other contrasts orthogonal to it.
2. Orthogonal contrasts extract the maximum amount of information using a minimum number of contrasts.

The second advantage of orthogonal contrasts over nonorthogonal contrasts relates to efficiency. If there are $I - 1$ independent pieces of information, then $I - 1$ orthogonal contrasts use all that information. Nonorthogonal contrasts require more than $I - 1$ such contrasts to use all the information.

2.2.3 Type I Error Rates for Comparisons

If two comparisons are each run at level α', the probability that the number of Type I errors is at least one, $P(\#TI \geq 1)$, can be almost as high as $2\alpha'$. For C orthogonal contrasts we have

$$P(\#TI \geq 1) = 1 - (1 - \alpha')^c. \tag{2.6}$$

That is, two independent tests run at level $\alpha' = .05$ will have $P(\#TI \geq 1) = 1 - .95^2 = 1 - .9025 = .0975$. For C not too large, $C\alpha'$ is close to the true error rate. Clearly, $.0975 < 2(.05) = .10$. That is, C independent tests run at α' each will have

$$P(\#TI \geq 1) < C\alpha'. \tag{2.7}$$

For nonorthogonal comparisons, the value of $P(\#TI \geq 1)$ is somewhat more difficult to determine. In the most extreme case, the same comparison could be used twice. That would be the case of perfect dependence. Since the use of the same contrast twice cannot lead to any more rejections than a single application of that contrast, we have $P(\#TI \geq 1) = \alpha'$. Of course, nonorthogonal contrasts will always be between the two extremes of perfect dependence and perfect independence. Therefore, we can generally conclude that for C nonorthogonal contrasts, each run at level α', we have

$$\alpha' \leq P(\#TI \geq 1) \leq 1 - (1 - \alpha')^c. \tag{2.8}$$

In most cases the degree of dependence between two contrasts cannot easily be used to determine a precise upper limit for $P(\#TI \geq 1)$. Therefore, we must usually use Eq. (2.6) or (2.7) even with nonorthogonal comparisons.

2.2.4 Bonferroni Limits for $P(\#TI \geq 1)$

When testing $C \geq 2$ contrasts it has become customary to place some limit on the combined error rate. There are a number of ways of placing limits on $P(\#TI \geq 1)$. For three orthogonal contrasts tested with $\alpha' = .05$, $P(\#TI \geq 1)$ is .1426. With six orthogonal contrasts each tested at $\alpha' = .05$, $P(\#TI \geq 1)$ is .2649.

A simple method of limiting $P(\#TI \geq 1)$ is to run each test at a stringent α level. For example, with three orthogonal contrasts, each could be run at $\alpha' = .05/3 = .01667$. Similarly, with six orthogonal contrasts, each would be run at $\alpha' = .05/6 = .008333$. This method of adjusting the α level by dividing by the number of tests is known as the *Bonferroni procedure* or Dunn's (1961) procedure. It is not necessary to run all tests at the same level, α'. The only requirement is that the separate levels add to α. However, with no a priori reason to choose unequal levels for individual tests, it is simplest to use equal allocations.

The Bonferroni method is based on the inequality 2.7. To limit $P(\#TI \geq 1)$ to α for C comparisons, we run each F test at level α' where

$$\alpha' = \alpha/C. \tag{2.9}$$

From Eq. (2.7) we know that

$$P(\#TI \geq 1) < C\alpha' = C(\alpha/C) = \alpha.$$

This method works with both orthogonal and nonorthogonal comparisons.

A slightly more powerful version of the Bonferroni procedure is obtained using Eq. (2.6). In this case, to limit $P(\#TI \geq 1)$ to level α for C comparisons, we take

$$\alpha' = 1 - (1 - \alpha)^{1/C}. \tag{2.10}$$

We run our C tests at level α' based on Eq. (2.10). For both orthogonal and nonorthogonal comparisons we have from Eq. (2.6)

$$P(\#TI \geq 1) \leq 1 - (1 - \alpha')^C$$
$$= 1 - (1 - [1 - \{1 - \alpha\}^{1/C}])^C$$
$$= 1 - (\{1 - \alpha\}^{1/C})^C$$
$$= \alpha.$$

$$P(\text{nsTI} \geq 1) \leq \alpha.$$

The equality holds if all C comparisons are independent.

The Bonferroni approach can be used in many different applications. For example, the two contrasts discussed above could each be run at $\alpha' = .025$ following the Bonferroni rule $.05/2 = .025$. This guarantees a limit of .05 for $P(\#TI \geq 1)$.

2.2.5 Orthogonal Polynomials

For equally spaced treatment conditions, one set of contrasts is provided by *orthogonal polynomials*, defined as a set of orthogonal contrasts among I means, which tests for trends in a data set. The contrasts start with linear trends and progress to the highest of $I - 1$ trends. The trends are usually designated linear, quadratic, cubic, quartic, etc.

Although it is possible to apply orthogonal polynomials with unequal N, we will restrict our discussion to cases in which sample sizes are equal. We will also assume that the treatments are equally spaced. Table 2.1 gives the appropriate coefficients for orthogonal polynomials up to $I = 8$ means. With $I = 3$ means, the only possible trends in a data set are linear and quadratic. A linear trend indicates a constant change either constantly increasing or constantly decreasing. A quadratic trend indicates a single change either starting with increasing values and then changing to decreasing or vice versa. The cubic trend indicates two changes. The trends continue up to the upper limit of $I - 1$ (i.e., df_{TREAT}).

Table 2.1 Coefficients of Orthogonal Polynomials

k	Polynomial	Coefficients								Σc^2
3	Linear	-1	0	1						2
	Quadratic	1	-2	1						6
4	Linear	-3	-1	1	3					20
	Quadratic	1	-1	-1	1					4
	Cubic	-1	3	-3	1					20
5	Linear	-2	-1	0	1	2				10
	Quadratic	2	-1	-2	-1	2				14
	Cubic	-1	2	0	-2	1				10
	Quartic	1	-4	6	-4	1				70
6	Linear	-5	-3	-1	1	3	5			70
	Quadratic	5	-1	-4	-4	-1	5			84
	Cubic	-5	7	4	-4	-7	5			180
	Quartic	1	-3	2	2	-3	1			28
	Quintic	-1	5	-10	10	-5	1			252
7	Linear	-3	-2	-1	0	1	2	3		28
	Quadratic	5	0	-3	-4	-3	0	5		84
	Cubic	-1	1	1	0	-1	-1	1		6
	Quartic	3	-7	1	6	1	-7	3		154
	Quintic	-1	4	-5	0	5	-4	1		84
	Sextic	1	-6	15	-20	15	-6	1		924
8	Linear	-7	-5	-3	-1	1	3	5	7	168
	Quadratic	7	1	-3	-5	-5	-3	1	7	168
	Cubic	-7	5	7	3	-3	-7	-5	7	264
	Quartic	7	-13	-3	9	9	-3	-13	7	616
	Quintic	-7	23	-17	-15	15	17	-23	7	2184
	Sextic	1	-5	9	-5	-5	9	-5	1	264

Source: The entries in this table were computed by the author.

Example 2.1

Example 2.1 presents data from an experiment by Nelson, Rosenthal, and Rosnow (1986) which illustrates the use of orthogonal polynomials. In the experiment, authors and editors of journal articles were asked to rate confidence in results of a statistical test. Ratings ranged from 0 (no confidence) to 5 (maximum confidence). The subjects were asked to rate results when the exact probability of a test statistic was .04, .05, .06 or .07. The authors expected to find an unusually high drop in confidence when the *p* value or exact probability increased from .05 to .06. They reasoned that a small drop in confidence would occur when the *p* value increased from .04 to .05, so the large drop in confidence with the change from .05 to .06 would represent one change in the trend. Similarly, the large drop from .05 to .06 would change to a small drop with the change from .06 to .07. Thus, two changes would occur and a significant cubic trend would be found. The small drop in confidence for changes from .04 to .05 and .06 to .07 would be expected due to the lack of change in conclusions from statistical tests run at the .05 level.

p values	.04	.05	.06	.07
Mean ratings	3.158	2.915	2.291	2.132
N	78	78	78	78

The error term is $MS_E = 0.292$. Because the experiment was part of a larger experiment, the error *df* was $df_E = 72$. As we will consider in detail below, a number of testing strategies could be used. The simplest would be to test only the cubic trend because it was determined by the investigators on an a priori basis. However, some researchers might cautiously require other tests as well. For all four means, we have $SS_{TREAT} = 56.3776$ and $F = 64.36 > 2.76 = F_{.95}(3,60) \approx F_{.95}(3,72)$. We reject the full null, $H_0: \mu_1 = \mu_2 = \mu_3 = \mu_4$ at $\alpha = .05$.

Linear Trend. From Table 2.1 we have the linear coefficients $(-3, -1, 1, 3)$. Therefore, we have

$$\hat{\Psi}_{LIN} = (-3)3.158 + (-1)2.915 + (1)2.291 + (3)2.132 = -3.702.$$

$$SS_{LIN} = N(\hat{\Psi}_{LIN})^2/\Sigma \, c_i^2 = 78(-3.702)^2/20 = 53.4487.$$

$$F_{LIN} = 53.4487/.292 = 183.04 > 4.00 = F_{.95}(1,60).$$

It should be no surprise that there is a strong linear trend. Most of the treatment *SS* is due to the linear component. The authors and editors in the study show progressively lower ratings as the *p* values go up from .04 to .07.

Departure from Linearity. Since linear trend accounts for the trend in many experiments, it is sometimes desirable to combine the higher trends and test for departure from linearity.

$SS_{\text{DEP from LIN}} = SS_{\text{TREAT}} - SS_{\text{LIN}} = 56.3776 - 53.4487 = 2.9289.$

$df_{\text{DEP from LIN}} = df_{\text{TREAT}} - df_{\text{LIN}} = 3 - 1 = 2.$

$F_{\text{DEP from LIN}} = MS_{\text{DEP from LIN}}/MS_E = (2.9289/2)/.292.$

$F_{\text{DEP from LIN}} = 1.4645/.292 = 5.0153.$

We reject H_0 because $5.02 > 3.15 = F_{.95}(2,60).$
There is a significant trend beyond the linear trend.
Quadratic Trend. Again from Table 2.1 we have the coefficients $(1, -1, -1, 1)$

$\hat{\Psi}_{\text{QUAD}} = (1)3.158 + (-1)2.915 + (-1)2.291 + (1)2.132 = 0.084.$

$SS_{\text{QUAD}} = N(\hat{\Psi}_{\text{QUAD}})^2/\Sigma\, c_i^2 = 78(.084)^2/4 = 0.1376.$

$F_{\text{QUAD}} = 0.1376/.292 = 0.47 < 4.00 = F_{.95}(1,60).$

There is no significant quadratic trend at $\alpha = .05.$
Cubic Trend. From Table 2.1 we have $(-1, 3, -3, 1)$

$\hat{\Psi}_{\text{CUBIC}} = (-1)3.158 + (3)2.915 + (-3)2.291 + (1)2.132 = 0.846.$

$SS_{\text{CUBIC}} = N(\hat{\Psi}_{\text{CUBIC}})^2/\Sigma\, c_i^2 = 78(0.846)^2/20 = 2.7913.$

$F_{\text{CUBIC}} = 2.7913/.292 = 9.56 > 4.00 = F_{.95}(1,60).$

There is a significant cubic trend as predicted by Nelson et al. There is a significantly greater decline in ratings from p values changing from .05 to .06.

There are several strategies for testing orthogonal contrasts. The most common is the one used in Example 2.1, where tests start with the overall F test followed by such tests as those for linear, quadratic, and cubic trends. Traditionally, all these tests are run at a common α level, such as 0.05.

Another strategy is to use the Bonferroni procedure where each of the $I - 1$ contrasts is tested at level $\alpha/(I - 1)$. In Example 2.1 we would use $.05/3 = .01667$. In most cases this second approach is probably less powerful than the first because it requires a more stringent test of the cubic trend.

A third approach is to test only those contrasts which are of interest. In Example 2.1 Nelson et al. were interested only in the cubic trend. They could have tested only that trend and ignored all other tests. This is probably the most powerful approach of all. Because it also requires only a single statistical test, there is less likelihood of a Type I error due to multiple testing.

2.2.6 Post Hoc Comparisons

In our discussion of contrasts (both orthogonal and nonorthogonal), we have assumed that their selection is done on an a priori basis. That is, we assumed the researcher knew which contrasts were of interest before the data were collected or at least before results were examined. With post hoc contrasts, we select certain specific contrasts out of an infinite number of contrasts after we observe certain relationships in the data.

To illustrate the problem, consider Ψ_1 introduced in Section 2.2.1. That contrast compares the first mean to the average of means two and three, $2\mu_1 - \mu_2 - \mu_3$. If there is no effect, then in 100 such experiments, each run at $\alpha = .025$, we should get two or three significant results due to chance alone. This is our risk of a Type I error.

Now suppose we selected Ψ_1 by observing in our data that means two and three look much larger than mean one. In 100 replications we might sometimes define Ψ_1 as explained above; and, in other experiments, where the pattern of means looks different, we might use a different definition for Ψ_1. That is, in those experiments where means one and two were close and mean three was different, we might define Ψ_1 as $\mu_1 + \mu_2 - 2\mu_3$. In the 100 experiments we would be likely to get two or three significant results when Ψ_1 was defined as $2\mu_1 - \mu_2 - \mu_3$. We would also be likely to get two or three different results when the definition of Ψ_1 is changed to $\mu_1 + \mu_2 - 2\mu_3$. We would be likely to get additional significant results when other definitions are used. The total number of significant results in the 100 experiments would far exceed the 2.5% to be expected using the chosen α level. This process of selecting the particular contrast to be tested by examining the data is known as post hoc (or a posteriori) selection. Clearly, the actual Type I error rate can be much higher than the nominal level.

There is an appropriate method for testing post hoc comparisons. A method developed by Scheffé (1959) can be used to test any contrast regardless of the way the contrast is selected. All that is needed is a simple modification of Eq. (2.3). In particular, we have the Scheffé F test $F_{\hat{\Psi}}$ as

$$F_{\hat{\Psi}\text{Scheffé}} = \frac{MS_{\hat{\Psi}}}{MS_E}, \tag{2.11}$$

where $MS_{\hat{\Psi}} = SS_{\hat{\Psi}}/df_{\text{TREAT}} = SS_{\hat{\Psi}}/(I - 1)$; i.e., the numerator df is always $I - 1$ and the denominator df is always df_E; thus the critical value is $CV = F_{1-\alpha}(I - 1, df_E)$.

Scheffé's procedure is very flexible because it can be used to test any contrast and as many contrasts as desired while limiting $P(\#\text{TI} \geq 1) \leq \alpha$. However, as might be expected, there is a cost to this great flexibility.

Scheffé's procedure is quite conservative. It is often nonsignificant when many other procedures are significant. In other words, Scheffé's procedure is not very powerful. However, Scheffé proved that if the overall F test is significant, then at least one contrast among the means must be significant by the Scheffé procedure. Unfortunately, the only significant contrast may be very difficult to interpret.

Another advantage of Scheffé's procedure is the availability of confidence intervals. For any contrast, $\hat{\Psi}$, we have

$$\hat{\Psi} \pm CV \sqrt{MS_E \sum_{i=1}^{I} \frac{c_i^2}{N_i}},$$
(2.12)

where $CV = F_{1-\alpha}(I - 1, df_E)$.

2.3 PAIRWISE TESTING

Testing a group of I means two at a time is known as *pairwise testing*. A group of only two means requires a single test. A t test run at level α will limit the probability of a Type I error to α. However, among I means there are many pairs of means to be compared. In particular, the number of pairs of means will be $I(I - 1)/2$. In a group of $I = 3$ means, there are $3(2)/2 = 3$ pairs. If a limited number of the possible pairs are selected on an a priori basis, the Type I error rate needs to be limited only to those tests. However, if the pairs to be tested are selected on a post hoc basis, the Type I error rate is determined by all possible pairs.

2.3.1 Multiple *t* Tests

One procedure for testing differences among I means is known as multiple t tests or the unprotected t procedure. This procedure would require a separate t test to determine the significance of the difference between each pair of means. A case of $I = 3$ means would require three t tests. If each test were run at $\alpha = .05$ level, $P(\#\text{TI} \geq 1)$ would be almost as large as $0.15 (= 3 \times 0.05)$ if we apply Eq. (2.7). Allowing for the nonindependence of pairwise testing, Games (1971) showed that the value of $P(\#\text{TI} \geq 1)$ is closer to 0.12. Most researchers consider the experimentwise error rate of 0.12 to be much too large for pairwise testing when the desired nominal level is 0.05. Thus, the multiple t approach is generally considered to be unacceptable.

The original multiple t procedure uses a single, pooled error term based upon the within-groups sum of squares from all I groups. An alternative approach is to use separate error terms for each pair based on the sum of squares for the two groups involved. This approach would be considered when the assumption of equal population variances is not satisfied. We will consider several such procedures that deal with unequal variances in a later section.

2.3.2 Tukey's HSD Procedure

A popular procedure for pairwise testing was proposed by Tukey (1953), who designated it as the honestly significant difference (HSD) procedure. In testing I means from groups with N observations each, the critical difference for each pair of means is obtained from the studentized range distribution. A q statistic can be calculated from the difference between two means. Critical values for that statistic are available in the Appendix. Alternatively, a critical difference, CD, which must be equaled or exceeded by the difference between two means for that difference to be declared significant, can be determined. In particular, we have

$$CD = q_{1-\alpha}(I, df_E) \sqrt{\frac{MS_E}{N}}, \qquad (2.13)$$

where N is the number of observations contributing to *each* mean and $q_{1-\alpha}(I, df_E)$ is the $(1 - \alpha)$th quantile point of the studentized range distribution with parameters I and df_E.

Each pair of means is significantly different if their difference is equal to or exceeds CD from Eq. (2.13). The HSD procedure does *not* require a significant overall F test. However, if the usual ANOVA assumptions are satisfied, the HSD will always ensure that $P(\#TI \geq 1) \leq \alpha$.

Although the HSD provides adequate control of Type I errors, other procedures provide the same Type I error control and are more powerful (Einot and Gabriel 1975; Ramsey, 1978, 1981; Welsch, 1977). There are various ways to define power for detecting differences between pairs of $I \geq 3$ means. The clearest distinction between testing procedures is provided by all-pairs power. We define *all-pairs power* as the probability of correctly rejecting all false nulls between pairs of means in a set of means. A number of procedures limit $P(\#TI \geq 1) \leq \alpha$ but have substantially greater all-pairs power than HSD.

Tukey (1953) proposed one modification to the HSD which he designated the wholly significant difference (WSD) procedure. That method has also been called the Tukey (b) procedure with the HSD designated the Tukey (a) procedure (Winer, 1971). Although the WSD has been popular in the behavioral sciences, there are two problems with that method. First, it has never been shown to ensure $P(\#TI \geq 1) \leq \alpha$. Second, the WSD has been shown to be less powerful than other methods such as Welsch's (1977) procedure, which is just as easy to apply and does satisfy $P(\#TI \geq 1) \leq \alpha$ (Ramsey, 1978).

2.3.3 Fisher's LSD Procedure

One of the first alternatives for multiple t tests was suggested by Fisher (1935). He proposed a method known as the protected t procedure, which he called the least significant difference (LSD) procedure. In Fisher's original LSD procedure, a nonsignificant overall F test indicates the plausibility of the full null hypothesis with no further testing allowed. However, a significant F is followed by t tests applied to each individual pair of means. The F test in the LSD procedure protects against the case wherein the full null is true, making it superior to the multiple t or unprotected t procedure.

An important difference exists between the multiple t procedure and the LSD procedure. When the full null hypothesis is true, the multiple t procedure will have an excessively high Type I error rate. The LSD procedure avoids this problem. However, there is still a problem with the Type I error control of the original LSD procedure.

Hypothesizing a configuration of population means in which some, but not all, means are equal is called a *partial null hypothesis*. For example, with three means we might suspect that means one and two are equal but that mean three is different. Among three means there are exactly three, possible, partial null hypotheses:

$H_{01}: \mu_1 = \mu_2,$

$H_{02}: \mu_1 = \mu_3,$

$H_{03}: \mu_2 = \mu_3.$

Thus among three means there are five possible scenarios. First, the full null can be true. Next, three possibilities are that any one of the three partial nulls can be true. Finally, no null is true and all three population means are different. In Fisher's LSD procedure the full null is tested with the F test and each pair of means is tested with a t test. With exactly $I = 3$ means, Fisher's LSD procedure will always limit $P(\#TI \geq 1) \leq \alpha$.

However, Fisher's LSD procedure itself has trouble controlling $P(\#TI \geq 1)$ with $I > 3$. For example, one partial null hypothesis with four means would be $H_0: \mu_1 = \mu_2 = \mu_3 \neq \mu_4$. Suppose this particular partial null hypothesis is true. The overall F test is likely to be significant because the full null is *not* true. All possible pairwise comparisons will include three pairwise tests among the first three means. Because all three population means are equal, these three tests are all testing true null hypotheses. The probability of one or more Type I errors is approximately 0.12. Thus, the problem for Fisher's LSD procedure with four means is the same as that for multiple t tests with three means.

Fortunately, Hayter (1986) proposed a simple modification of Fisher's LSD procedure which can be applied to any number of means. In the

Hayter-Fisher modified LSD procedure, the full null hypothesis is tested with the overall F test just as in the original LSD procedure. However, the pairwise tests are done using special critical values from the q statistic. In particular, in testing I means from groups with N observations each, a significant F test is followed by testing each pair. The pairs are all tested exactly as in Tukey's HSD procedure except that CD is taken to be

$$CD = q_{1-\alpha}(df_{\text{TREAT}}, df_E) \sqrt{\frac{MS_E}{N}}, \tag{2.14}$$

where N is the number of observations contributing to *each* mean and $q_{1-\alpha}(df_{\text{TREAT}}, df_E)$ is the $(1 - \alpha)$ quantile point of the studentized range distribution with parameters df_{TREAT} and df_E.

With three means, the modified Hayter-Fisher LSD procedure gives identical results to the original LSD procedure. With $I \geq 4$ the two procedures can give different results. However, the modified LSD always limits $P(\#TI \geq 1) \leq \alpha$.

2.3.4 Newman-Keuls Step-Down Procedures

An early attempt at a powerful pairwise procedure was suggested by Newman (1939) and Keuls (1952). They proposed a method which could be considered a modification of the HSD. The Newman-Keuls approach is a step-down procedure in which sets of paired-mean differences are tested sequentially from the largest to the smallest range involved in these differences. For a set of I means, the full set of all I means is tested just as in using the HSD. The difference between the largest and smallest means must exceed CD given by Eq. (2.13). Nonsignificance leads to acceptance of the full null and no additional testing is allowed. Significance of this maximum range test leads to rejection of the full null hypothesis and rejection of the partial null for the difference between the largest and smallest means. In addition, the significant maximum range test allows the testing of smaller sets.

Two sets of $I - 1$ means are considered next. The smallest mean and the next-to-largest mean are the most extreme means in one set. The largest mean and the next-to-smallest mean are the most extreme means in the other set. Each of the sets is tested by comparing the difference between the two most extreme means to CD from Eq. (2.13) where I has been replaced by $I - 1$. Nonsignificance in either subset requires nonsignificance for all pairwise differences between any pair of means within that subset. Significance leads to rejection of the pairwise difference between the two most extreme means in the subset and additional testing is conducted for smaller subsets within the significant subset. This process is continued until subsets, including only two means, are tested and Eq. (2.13) is applied with $I = 2$.

The Newman-Keuls procedure must be applied with great care. For example, suppose we have four ordered means, $(1, 2, 3, 4)$. The two most extreme means, 1 and 4, are tested by comparing their difference to CD from Eq. (2.13) with $I = 4$. If that result is significant, we proceed to testing 1 versus 3 and 2 versus 4 with CD from Eq. (2.13), but I is now replaced with $I - 1 = 3$. Now suppose that 1 versus 3 is significant while 2 versus 4 is nonsignificant. One might first wonder whether it is appropriate to proceed to test 2 versus 3. In fact, one must *not* test 2 versus 3 because 2 versus 4 was nonsignificant. The testing of 2 versus 3 would be allowed only if 1 versus 3 *and* 2 versus 4 were both significant. We are required to consider 2 versus 3 to be nonsignificant if 1 versus 3 is nonsignificant *or* if 2 versus 4 is nonsignificant.

It might appear that the Newman-Keuls approach would ensure that $P(\#\text{TI} \geq 1) \leq \alpha$ when all tests are applied at level α. The procedure does limit the Type I error rate for all possible partial nulls. However, there are other types of nulls. We define a *multiple null hypothesis* as any hypothesis in which two or more distinct subsets have true nulls. For example, with $I = 4$, we might have $\mu_1 = \mu_2 \neq \mu_3 = \mu_4$. It is well known that the possibility of true multiple null hypotheses ensures the possibility that $P(\#\text{TI} \geq 1) > \alpha$ (Hartley, 1955; Einot and Gabriel, 1975).

The Newman-Keuls procedure can also be applied using the F distribution (Einot and Gabriel). In that case the full null hypothesis is tested with the overall F test. Nonsignificance implies no significant pairwise differences. However, a significant overall F test does *not* justify the rejection of the partial null of the equality of the largest and smallest means. Each pair of means can be determined to be significantly different if and only if *every* subset of means including the pair results in a significant F test. However, the F test version of the Newman-Keuls also *fails* to ensure that $P(\#\text{TI} \geq 1) \leq \alpha$.

2.3.5 Welsch Step-Down Procedure

Welsch (1977) proposed a modification of the q test version of the Newman-Keuls procedure which does ensure that $P(\#\text{TI} \geq 1) \leq \alpha$. The Welsch procedure is applied just as the Newman-Keuls procedure, starting with the test of the largest and smallest means using CD from Eq. (2.13). The modification occurs with smaller subsets. Subset sizes range from $p = 2$ to $p = I - 1$. Since the subsets are taken of ordered means, each subset of size p can be called a stretch of size p. For a stretch of size p from I means and level α we use a more stringent α level by taking

$$\alpha_{I-1} = \alpha, \tag{2.15}$$
$$\alpha_p = 1 - (1 - \alpha)^{p/I} \quad \text{for } p = 2, \ldots, I - 2.$$

Using the α levels determined from Eq. (2.15), each stretch of size p is tested for significance using the critical difference

$$CD_p = q_{1-\alpha_p}(p, df_E) \sqrt{\frac{MS_E}{N}}, \tag{2.16}$$

where N is the number of observations contributing to *each* mean and $q_{1-\alpha_p}(p, df_E)$ is the $(1 - \alpha_p)$ quantile point of the studentized range distribution with parameters p and df_E.

Values of $q_{1-\alpha_p}$ required by Eq. (2.16) are given in the Appendix and Table 2.2. Using these table values, the Welsch step-down procedure is as easy to perform as the original Newman-Keuls procedure. It has the advantage that it ensures $P(\#TI \geq 1) \leq \alpha$.

(*Text continues on page 49.*)

Table 2.2 Critical Values for Welsch Step-Down and Shaffer-Welsch Tests

No. of $\overline{X}$ I	Stretch size p	\multicolumn{8}{c}{$\alpha = .05$ Error degrees of freedom}
		1.	2.	3.	4.	5.	6.	7.	8.
3	2	18.0	6.09	4.50	3.93	3.64	3.46	3.34	3.26
4	3	35.5	8.72	5.91	5.04	4.60	4.34	4.16	4.04
4	2	35.5	8.72	5.88	4.92	4.46	4.18	4.01	3.88
5	4	44.5	10.81	7.13	5.89	5.28	4.93	4.69	4.53
5	3	44.5	10.81	7.13	5.89	5.28	4.93	4.69	4.53
5	2	44.3	9.77	6.38	5.27	4.74	4.43	4.22	4.08
6	5	53.3	12.04	7.91	6.50	5.80	5.39	5.12	4.93
6	4	53.3	12.04	7.91	6.50	5.80	5.39	5.12	4.93
6	3	53.3	11.86	7.62	6.22	5.54	5.14	4.88	4.70
6	2	53.1	10.72	6.82	5.57	4.98	4.63	4.41	4.25
7	6	62.1	13.01	8.47	6.94	6.18	5.73	5.44	5.23
7	5	62.1	13.01	8.47	6.94	6.18	5.73	5.44	5.23
7	4	62.1	13.01	8.36	6.79	6.03	5.58	5.29	5.08
7	3	62.1	12.82	8.05	6.51	5.76	5.33	5.05	4.85
7	2	61.9	11.60	7.21	5.83	5.18	4.80	4.56	4.39
8	7	70.9	13.92	8.91	7.29	6.49	6.01	5.69	5.47
8	6	70.9	13.92	8.91	7.29	6.49	6.01	5.69	5.47
8	5	70.9	13.92	8.88	7.21	6.39	5.91	5.59	5.37

Table 2.2 Continued

No. of $\overline{X}$	Stretch size	$\alpha = .05$ Error degrees of freedom							
I	p	1.	2.	3.	4.	5.	6.	7.	8.
8	4	70.9	13.92	8.76	7.06	6.24	5.76	5.44	5.22
8	3	70.9	13.72	8.45	6.76	5.96	5.49	5.19	4.98
8	2	70.7	12.41	7.57	6.07	5.36	4.96	4.70	4.52
9	8	79.6	14.77	9.29	7.58	6.74	6.24	5.91	5.67
9	7	79.6	14.77	9.29	7.58	6.74	6.24	5.91	5.67
9	6	79.6	14.77	9.29	7.54	6.68	6.17	5.83	5.59
9	5	79.6	14.77	9.26	7.46	6.58	6.07	5.73	5.49
9	4	79.6	14.77	9.14	7.30	6.42	5.91	5.57	5.34
9	3	79.6	14.56	8.81	6.99	6.14	5.64	5.32	5.10
9	2	79.4	13.17	7.89	6.28	5.52	5.09	4.82	4.63
10	9	88.4	15.58	9.64	7.83	6.96	6.43	6.09	5.84
10	8	88.4	15.58	9.64	7.83	6.96	6.43	6.09	5.84
10	7	88.4	15.58	9.64	7.82	6.92	6.38	6.03	5.78
10	6	88.4	15.58	9.64	7.77	6.85	6.31	5.96	5.71
10	5	88.4	15.58	9.61	7.69	6.75	6.21	5.85	5.60
10	4	88.4	15.58	9.48	7.52	6.59	6.05	5.69	5.45
10	3	88.4	15.35	9.14	7.21	6.30	5.77	5.44	5.20
10	2	88.2	13.89	8.19	6.47	5.67	5.22	4.93	4.73

		9.	10.	11.	12.	13.	14.	15.	16.	17.
3	2	3.20	3.15	3.11	3.08	3.06	3.03	3.01	3.00	2.98
4	3	3.95	3.88	3.82	3.77	3.73	3.70	3.67	3.65	3.63
4	2	3.79	3.71	3.66	3.61	3.57	3.54	3.51	3.49	3.47
5	4	4.41	4.33	4.26	4.20	4.15	4.11	4.08	4.05	4.02
5	3	4.40	4.31	4.23	4.17	4.12	4.08	4.04	4.01	3.98
5	2	3.98	3.90	3.83	3.78	3.74	3.70	3.67	3.64	3.62
6	5	4.79	4.68	4.59	4.52	4.46	4.41	4.37	4.34	4.30
6	4	4.79	4.68	4.59	4.52	4.46	4.41	4.37	4.34	4.30
6	3	4.57	4.46	4.38	4.31	4.26	4.21	4.17	4.14	4.11
6	2	4.13	4.04	3.97	3.92	3.87	3.83	3.80	3.77	3.74
7	6	5.07	4.95	4.86	4.78	4.72	4.66	4.61	4.57	4.54
7	5	5.07	4.95	4.86	4.78	4.72	4.66	4.61	4.57	4.54
7	4	4.93	4.81	4.72	4.64	4.58	4.53	4.48	4.44	4.41
7	3	4.71	4.59	4.51	4.43	4.38	4.33	4.28	4.25	4.21
7	2	4.27	4.17	4.10	4.03	3.98	3.94	3.91	3.87	3.85
8	7	5.30	5.17	5.07	4.99	4.92	4.86	4.81	4.77	4.73
8	6	5.30	5.17	5.07	4.99	4.92	4.86	4.81	4.77	4.73
8	5	5.20	5.07	4.97	4.89	4.82	4.76	4.71	4.67	4.63

Table 2.2 Continued

No. of $\overline{X}$ I	Stretch size p	$\alpha = .05$ Error degrees of freedom								
		9.	10.	11.	12.	13.	14.	15.	16.	17.
8	4	5.06	4.93	4.83	4.75	4.68	4.63	4.58	4.54	4.50
8	3	4.83	4.71	4.61	4.54	4.48	4.42	4.38	4.34	4.31
8	2	4.38	4.28	4.20	4.14	4.08	4.04	4.00	3.97	3.94
9	8	5.49	5.36	5.25	5.16	5.09	5.02	4.97	4.92	4.88
9	7	5.49	5.36	5.25	5.16	5.09	5.02	4.97	4.92	4.88
9	6	5.42	5.28	5.17	5.08	5.01	4.95	4.90	4.85	4.81
9	5	5.32	5.18	5.07	4.98	4.91	4.85	4.80	4.75	4.71
9	4	5.17	5.04	4.93	4.85	4.77	4.72	4.66	4.62	4.58
9	3	4.94	4.81	4.71	4.63	4.56	4.51	4.46	4.42	4.38
9	2	4.49	4.38	4.29	4.22	4.17	4.12	4.08	4.05	4.02
10	9	5.66	5.51	5.40	5.31	5.23	5.16	5.11	5.06	5.02
10	8	5.66	5.51	5.40	5.31	5.23	5.16	5.11	5.06	5.02
10	7	5.60	5.45	5.34	5.25	5.17	5.10	5.05	5.00	4.96
10	6	5.52	5.38	5.26	5.17	5.09	5.03	4.97	4.93	4.88
10	5	5.42	5.27	5.16	5.07	4.99	4.93	4.88	4.83	4.79
10	4	5.27	5.13	5.02	4.93	4.86	4.79	4.74	4.70	4.66
10	3	5.03	4.90	4.80	4.71	4.64	4.59	4.54	4.49	4.46
10	2	4.58	4.47	4.38	4.30	4.25	4.20	4.15	4.12	4.09
		18.	19.	20.	24.	30.	40.	60.	120.	∞
3	2	2.97	2.96	2.95	2.92	2.89	2.86	2.83	2.80	2.77
4	3	3.61	3.59	3.58	3.53	3.49	3.44	3.40	3.36	3.31
4	2	3.45	3.43	3.42	3.37	3.33	3.29	3.24	3.20	3.16
5	4	4.00	3.98	3.96	3.90	3.85	3.79	3.74	3.68	3.63
5	3	3.96	3.94	3.92	3.86	3.80	3.75	3.69	3.64	3.58
5	2	3.60	3.58	3.57	3.51	3.47	3.42	3.37	3.33	3.28
6	5	4.28	4.25	4.23	4.17	4.10	4.04	3.98	3.92	3.86
6	4	4.28	4.25	4.23	4.16	4.09	4.03	3.96	3.90	3.84
6	3	4.08	4.06	4.04	3.97	3.91	3.85	3.79	3.73	3.68
6	2	3.72	3.70	3.68	3.63	3.58	3.52	3.47	3.42	3.38
7	6	4.51	4.48	4.46	4.38	4.31	4.23	4.16	4.10	4.03
7	5	4.51	4.48	4.46	4.38	4.31	4.23	4.16	4.09	4.02
7	4	4.38	4.35	4.33	4.26	4.18	4.11	4.05	3.98	3.91
7	3	4.19	4.16	4.14	4.07	4.00	3.94	3.87	3.81	3.75
7	2	3.82	3.80	3.78	3.72	3.67	3.61	3.56	3.51	3.46
8	7	4.69	4.66	4.64	4.56	4.47	4.40	4.32	4.24	4.17
8	6	4.69	4.66	4.64	4.56	4.47	4.40	4.32	4.24	4.17
8	5	4.60	4.57	4.54	4.46	4.38	4.31	4.23	4.16	4.09
8	4	4.47	4.44	4.42	4.34	4.26	4.19	4.12	4.05	3.98

Table 2.2 Continued

No. of $\overline{X}$	Stretch size	$\alpha = .05$ Error degrees of freedom								
I	p	18.	19.	20.	24.	30.	40.	60.	120.	∞
8	3	4.28	4.25	4.23	4.15	4.08	4.01	3.95	3.88	3.82
8	2	3.91	3.89	3.87	3.81	3.75	3.69	3.63	3.58	3.52
9	8	4.85	4.82	4.79	4.70	4.61	4.53	4.45	4.37	4.29
9	7	4.85	4.82	4.79	4.70	4.61	4.53	4.45	4.37	4.29
9	6	4.77	4.74	4.72	4.63	4.54	4.46	4.38	4.30	4.22
9	5	4.68	4.65	4.62	4.54	4.45	4.37	4.29	4.22	4.14
9	4	4.55	4.52	4.49	4.41	4.33	4.25	4.18	4.11	4.03
9	3	4.35	4.33	4.30	4.22	4.15	4.08	4.01	3.94	3.87
9	2	3.99	3.96	3.94	3.88	3.82	3.75	3.70	3.64	3.58
10	9	4.98	4.95	4.92	4.83	4.73	4.65	4.56	4.47	4.39
10	8	4.98	4.95	4.92	4.83	4.73	4.65	4.56	4.47	4.39
10	7	4.92	4.89	4.86	4.77	4.68	4.59	4.50	4.42	4.34
10	6	4.85	4.81	4.79	4.69	4.61	4.52	4.44	4.35	4.27
10	5	4.75	4.72	4.69	4.60	4.52	4.43	4.35	4.27	4.19
10	4	4.62	4.59	4.56	4.48	4.39	4.31	4.23	4.16	4.08
10	3	4.42	4.39	4.37	4.29	4.21	4.14	4.06	3.99	3.92
10	2	4.06	4.03	4.01	3.94	3.88	3.81	3.75	3.69	3.63

$\alpha = .01$
Error degrees of freedom

		1.	2.	3.	4.	5.	6.	7.	8.
3	2	90.0	14.04	8.26	6.51	5.70	5.24	4.95	4.75
4	3	179.6	19.90	10.62	8.12	6.98	6.33	5.92	5.64
4	2	179.6	19.90	10.53	7.91	6.75	6.10	5.70	5.42
5	4	224.6	24.58	12.67	9.33	7.86	7.04	6.54	6.20
5	3	224.6	24.58	12.67	9.33	7.86	7.04	6.52	6.17
5	2	224.5	22.26	11.38	8.41	7.11	6.39	5.95	5.64
6	5	269.4	27.32	13.99	10.23	8.56	7.64	7.05	6.65
6	4	269.4	27.32	13.99	10.23	8.56	7.64	7.05	6.65
6	3	269.4	26.93	13.49	9.80	8.19	7.30	6.75	6.37
6	2	269.2	24.39	12.11	8.83	7.42	6.64	6.16	5.83
7	6	314.2	29.51	14.96	10.90	9.09	8.08	7.45	7.01
7	5	314.2	29.51	14.96	10.90	9.09	8.08	7.45	7.01
7	4	314.2	29.51	14.75	10.66	8.87	7.87	7.25	6.83
7	3	314.2	29.09	14.22	10.22	8.48	7.53	6.94	6.54
7	2	313.9	26.35	12.77	9.21	7.68	6.85	6.34	5.99
8	7	359.0	31.56	15.71	11.43	9.51	8.44	7.76	7.30
8	6	359.0	31.56	15.71	11.43	9.51	8.44	7.76	7.30

Table 2.2 Continued

No. of $\overline{X}$	Stretch size	$\alpha = .01$ Error degrees of freedom							
I	p	1.	2.	3.	4.	5.	6.	7.	8.
8	5	359.0	31.56	15.65	11.29	9.36	8.30	7.63	7.17
8	4	359.0	31.56	15.44	11.05	9.14	8.08	7.43	6.98
8	3	359.0	31.11	14.88	10.59	8.74	7.73	7.11	6.68
8	2	358.8	28.18	13.37	9.54	7.92	7.04	6.49	6.13
9	8	403.7	33.47	16.36	11.87	9.86	8.74	8.03	7.54
9	7	403.7	33.47	16.36	11.87	9.86	8.74	8.03	7.54
9	6	403.7	33.47	16.36	11.79	9.76	8.64	7.93	7.44
9	5	403.7	33.47	16.29	11.65	9.61	8.49	7.79	7.31
9	4	403.7	33.47	16.07	11.40	9.38	8.27	7.58	7.11
9	3	403.7	33.00	15.49	10.92	8.97	7.92	7.26	6.82
9	2	403.7	29.89	13.92	9.85	8.13	7.21	6.64	6.25
10	9	448.7	35.29	16.96	12.25	10.16	9.00	8.26	7.75
10	8	448.7	35.29	16.96	12.25	10.16	9.00	8.26	7.75
10	7	448.7	35.29	16.96	12.21	10.09	8.92	8.18	7.67
10	6	448.7	35.29	16.96	12.13	9.99	8.82	8.08	7.57
10	5	448.7	35.29	16.89	11.98	9.84	8.67	7.93	7.43
10	4	448.7	35.29	16.66	11.72	9.60	8.44	7.72	7.24
10	3	448.7	34.78	16.06	11.23	9.19	8.08	7.40	6.93
10	2	448.3	31.51	14.43	10.13	8.33	7.36	6.76	6.36

		9.	10.	11.	12.	13.	14.	15.	16.	17.
3	2	4.60	4.48	4.39	4.32	4.26	4.21	4.17	4.13	4.10
4	3	5.43	5.27	5.15	5.05	4.96	4.89	4.84	4.79	4.74
4	2	5.22	5.06	4.94	4.85	4.77	4.70	4.65	4.60	4.56
5	4	5.96	5.77	5.62	5.50	5.40	5.32	5.25	5.19	5.14
5	3	5.92	5.72	5.57	5.45	5.35	5.27	5.20	5.13	5.08
5	2	5.42	5.25	5.12	5.02	4.93	4.86	4.80	4.75	4.70
6	5	6.36	6.14	5.97	5.84	5.73	5.63	5.56	5.49	5.43
6	4	6.36	6.14	5.97	5.83	5.72	5.62	5.54	5.47	5.41
6	3	6.09	5.89	5.72	5.59	5.49	5.40	5.32	5.26	5.20
6	2	5.59	5.41	5.27	5.16	5.07	4.99	4.92	4.87	4.82
7	6	6.69	6.45	6.27	6.12	5.99	5.89	5.80	5.72	5.66
7	5	6.69	6.45	6.27	6.12	5.99	5.89	5.80	5.72	5.66
7	4	6.52	6.28	6.10	5.96	5.84	5.74	5.65	5.58	5.52
7	3	6.25	6.03	5.85	5.72	5.61	5.51	5.43	5.36	5.31
7	2	5.73	5.54	5.40	5.28	5.18	5.10	5.03	4.97	4.92
8	7	6.96	6.71	6.51	6.34	6.21	6.10	6.01	5.93	5.86
8	6	6.96	6.71	6.51	6.34	6.21	6.10	6.01	5.93	5.86
8	5	6.83	6.58	6.39	6.23	6.10	5.99	5.90	5.82	5.75

Table 2.2 Continued

No. of $\overline{X}$	Stretch size	$\alpha = .01$ Error degrees of freedom								
I	p	9.	10.	11.	12.	13.	14.	15.	16.	17.
8	4	6.65	6.41	6.22	6.07	5.94	5.84	5.75	5.67	5.61
8	3	6.38	6.15	5.97	5.82	5.71	5.61	5.53	5.46	5.39
8	2	5.86	5.66	5.51	5.38	5.28	5.19	5.12	5.06	5.01
9	8	7.19	6.92	6.71	6.54	6.40	6.28	6.18	6.10	6.02
9	7	7.19	6.92	6.71	6.54	6.40	6.28	6.18	6.10	6.02
9	6	7.09	6.82	6.61	6.44	6.31	6.19	6.09	6.01	5.94
9	5	6.96	6.70	6.49	6.33	6.19	6.08	5.98	5.90	5.83
9	4	6.78	6.52	6.32	6.16	6.03	5.92	5.83	5.75	5.69
9	3	6.50	6.26	6.07	5.92	5.80	5.70	5.61	5.54	5.47
9	2	5.97	5.76	5.60	5.47	5.37	5.28	5.20	5.14	5.09
10	9	7.38	7.10	6.88	6.70	6.56	6.43	6.33	6.24	6.16
10	8	7.38	7.10	6.88	6.70	6.56	6.43	6.33	6.24	6.16
10	7	7.30	7.02	6.80	6.63	6.48	6.36	6.26	6.17	6.09
10	6	7.20	6.93	6.71	6.53	6.39	6.27	6.17	6.08	6.01
10	5	7.07	6.80	6.59	6.41	6.27	6.16	6.06	5.97	5.90
10	4	6.89	6.62	6.42	6.25	6.12	6.00	5.91	5.83	5.76
10	3	6.60	6.35	6.16	6.00	5.88	5.77	5.68	5.61	5.54
10	2	6.07	5.86	5.69	5.55	5.45	5.35	5.28	5.21	5.15
		18.	19.	20.	24.	30.	40.	60.	120.	∞
3	2	4.07	4.05	4.02	3.96	3.89	3.82	3.76	3.70	3.64
4	3	4.70	4.67	4.64	4.55	4.45	4.37	4.28	4.20	4.12
4	2	4.52	4.49	4.46	4.37	4.28	4.20	4.12	4.04	3.97
5	4	5.09	5.05	5.02	4.91	4.80	4.70	4.59	4.50	4.40
5	3	5.04	5.00	4.96	4.85	4.74	4.64	4.54	4.44	4.35
5	2	4.66	4.63	4.60	4.50	4.41	4.32	4.23	4.15	4.07
6	5	5.38	5.33	5.29	5.17	5.05	4.93	4.82	4.71	4.60
6	4	5.36	5.31	5.27	5.15	5.02	4.91	4.79	4.68	4.57
6	3	5.15	5.11	5.07	4.95	4.84	4.73	4.62	4.52	4.42
6	2	4.78	4.74	4.71	4.61	4.51	4.41	4.32	4.23	4.15
7	6	5.60	5.55	5.51	5.37	5.24	5.11	4.99	4.87	4.76
7	5	5.60	5.55	5.51	5.37	5.23	5.10	4.98	4.86	4.74
7	4	5.46	5.41	5.37	5.24	5.11	4.98	4.86	4.75	4.64
7	3	5.25	5.21	5.17	5.04	4.92	4.81	4.70	4.59	4.49
7	2	4.88	4.84	4.80	4.70	4.59	4.49	4.40	4.31	4.22
8	7	5.79	5.74	5.69	5.54	5.40	5.26	5.13	5.01	4.88
8	6	5.79	5.74	5.69	5.54	5.40	5.26	5.13	5.00	4.87
8	5	5.69	5.64	5.59	5.45	5.31	5.17	5.04	4.92	4.80

Table 2.2 Continued

No. of $\overline{X}$	Stretch size	$\alpha = .01$ Error degrees of freedom								
I	p	18.	19.	20.	24.	30.	40.	60.	120.	∞
8	4	5.55	5.50	5.45	5.31	5.18	5.05	4.93	4.81	4.69
8	3	5.34	5.29	5.25	5.12	5.00	4.88	4.76	4.65	4.54
8	2	4.96	4.92	4.88	4.77	4.66	4.56	4.46	4.37	4.27
9	8	5.96	5.90	5.85	5.69	5.54	5.39	5.25	5.12	4.99
9	7	5.96	5.90	5.85	5.69	5.54	5.39	5.25	5.11	4.98
9	6	5.87	5.82	5.77	5.61	5.46	5.32	5.18	5.05	4.92
9	5	5.77	5.71	5.66	5.51	5.37	5.23	5.10	4.97	4.84
9	4	5.63	5.57	5.53	5.38	5.24	5.11	4.98	4.86	4.74
9	3	5.42	5.37	5.32	5.19	5.06	4.94	4.82	4.70	4.59
9	2	5.04	4.99	4.96	4.84	4.73	4.62	4.52	4.42	4.32
10	9	6.10	6.04	5.98	5.82	5.66	5.50	5.36	5.21	5.08
10	8	6.10	6.04	5.98	5.82	5.66	5.50	5.36	5.21	5.07
10	7	6.03	5.97	5.92	5.75	5.60	5.45	5.30	5.16	5.02
10	6	5.94	5.89	5.83	5.67	5.52	5.37	5.23	5.09	4.96
10	5	5.84	5.78	5.73	5.58	5.43	5.28	5.15	5.01	4.88
10	4	5.69	5.64	5.59	5.44	5.30	5.16	5.03	4.90	4.78
10	3	5.48	5.43	5.39	5.25	5.12	4.99	4.87	4.75	4.63
10	2	5.10	5.06	5.02	4.90	4.79	4.67	4.57	4.47	4.37

Source: Adapted from Ramsey and Ramsey (1990). Used with permission of the American Educational Research Association.

2.3.6 Ryan Procedures

Ryan (1960) proposed a general method modifying the Newman-Keuls procedure. In his method all subsets are tested with adjusted α levels such as those in Eqs. (2.15). When subsets are tested with the q statistic based on Eqs. (2.15), Ryan's procedure is equivalent to that of Welsch. Both the q test and F test version of Ryan's procedure ensure that $P(\#TI \geq 1) \leq \alpha$. The F test version has been found to result in markedly greater all-pairs power (Ramsey, 1981). However, the large number of subsets generated for even moderate values of I requires a computer program for the application of the Ryan F procedure.

2.3.7 Peritz Procedures

Peritz (1970) proposed a blend of the Ryan and Newman-Keuls approaches, providing even greater power while maintaining $P(\#TI \geq 1) \leq$

α. Peritz's procedure can be applied with either q tests or F tests, both versions testing all possible hypotheses that can be formed from subsets of the I means. Peritz's procedure calls for the acceptance of all hypotheses accepted by both Ryan's procedure [i.e., testing at levels given by Eq. (2.15)] and the Newman-Keuls procedure (i.e., testing at level α), and rejection of all hypotheses rejected by both. In addition, it deals with contentious hypotheses, those rejected by the Newman-Keuls procedure and accepted by Ryan's procedure.

The Peritz approach to contentious hypotheses is to test the contentious sets from largest to smallest. For any subset, P (of size p), whose hypothesis is contentious, the hypothesis can be rejected if

(a) P is not contained in any previously contentious set that has subsequently been accepted *and*

(b)   the test of P is critical at level α_p [given by Eq. (2.15)] *or* for all sets R in the complement of P (i.e., not in P), the test of R is critical at α_p [given by Eq. (2.15)].

To gain a better understanding of Peritz's procedure, it may be helpful to compare it carefully with Ryan's procedure. These two procedures differ only in the way they test composite hypotheses. As an illustration, with $I = 4$ we might sometimes be required to test the composite hypothesis, $(\mu_1 = \mu_2) \cap (\mu_3 = \mu_4)$, since that hypothesis would have to be rejected before we would be able to reject either simple hypothesis, $\mu_1 = \mu_2$ or $\mu_3 = \mu_4$. For Ryan's procedure, each of the simple hypotheses must be critical at the specified level, α_2, thus automatically providing a test of the composite hypothesis.

To be explicit, suppose we are testing at $\alpha = .05$ and the above composite hypothesis is true. That is, $\mu_1 = \mu_2$ *and* $\mu_3 = \mu_4$. Applying Eq. (2.15), we have $\alpha_2 = 1 - (1 - .05)^{2/4} = 1 - (.95)^{0.5} = 1 - 0.9747 = .0253$. By Ryan's procedure, each of the simple hypotheses is tested at the .0253 level. Equation (2.6) gives an upper limit to $P(\#TI \geq 1) = 1 - (1 - .0253)^2 = 1 - .9747^2 = 1 - .9500 = 0.0500$. Thus, Ryan's procedure ensures an upper limit of .05 for $P(\#TI \geq 1)$.

Applying Peritz's procedure to the composite hypothesis above, we reject that composite hypothesis if either simple hypothesis is critical at α_2, and therefore the remaining simple hypothesis can be tested at level α. In particular, suppose the test of $\mu_1 = \mu_2$ is critical at α_2, whereas the test of $\mu_3 = \mu_4$ is critical at α but not at α_2. The test of $\mu_1 = \mu_2$ would be rejected by both Ryan's procedure and the Newman-Keuls procedure (and therefore also by Peritz's procedure), but the test of $\mu_3 = \mu_4$ would be contentious. Since μ_3 and μ_4 form the set whose hypothesis is contentious, the complement of that set is formed by μ_1 and μ_2. But since we know that $\mu_1 = \mu_2$ is critical at α_2, we can also reject $\mu_3 = \mu_4$ according to Peritz's procedure.

Peritz's procedure is clearly more complex than Ryan's and requires a computer program for either the q test or F test version. A program for the q test version has been published (Martin and Toothaker, 1989).

2.3.8 Shaffer Modification of q Test Procedures

Shaffer (1979) proved that any q test procedure can be modified to usually increase its power by testing the full null hypothesis with the ANOVA F test. Her proof showed the modified procedure maintaining $P(\#TI \geq 1) \leq \alpha$. That is, additional power may be obtained without increasing the Type I error rate.

Applying Shaffer's modification to the Welsch step-down procedure requires only two changes in that procedure. First, the full null hypothesis is tested with the F test. Nonsignificance means that all pairs fail to differ significantly. A significant F requires testing of individual pairs to determine significance. Second, the pairs are tested just as in the original Welsch procedure with the exception that the largest and smallest means (i.e., stretch size I) must differ by an amount equal to or greater than the value of CD for $I - 1$ means.

The modified version of the Welsch procedure has been called the Shaffer-Welsch Fq procedure. It can be applied with the F test and the critical values from Table 2.3. The Shaffer-Welsch Fq procedure has been shown to have good all-pairs power (Ramsey, 1981).

We can also apply the Shaffer modification to the Peritz q test procedure. The Shaffer-Peritz Fq procedure would require testing the full null hypothesis with the F test and then using a computer program such as that of Martin and Toothaker for all additional testing. Of course, the largest and smallest means would need to differ by at least the value of CD for $I - 1$. The Peritz q procedure has been shown to have very good power (Ramsey, 1978), and Shaffer's (1979) proof ensures that the Shaffer-Peritz Fq procedure will limit $P(\#TI \geq 1) \leq \alpha$. It should usually have power as high as or higher than that of the original Peritz q procedure.

2.3.9 Holm SRB Procedure

Holm (1979) proposed a sequentially rejective Bonferroni (SRB) procedure which is modification of the Bonferroni approach. The SRB is a general method which can be applied to any contrast including pairwise testing. In testing I means, there are $C = I(I - 1)/2$ pairwise tests. In the original Dunn approach run at level α, each pair of means would be tested with a t test run at level $\alpha' = \alpha/C$. In SRB the C pairs are placed in order with exact probabilities (i.e., p values) of the t test for each pair being ordered from smallest to largest. The smallest p value would be associated with the pair whose means are most likely to be detected as significantly different. In SRB, as in the Dunn approach, this most extreme pair would be considered significantly different if this smallest $p \le \alpha' = \alpha/C$. If the first p is significant, we continue to the next smallest p value. However, the second smallest p would lead to a rejection if and only if that $p \le \alpha' = \alpha/(C - 1)$. Testing is continued for all values of c ($c = 0, \ldots, C - 1$) where $\alpha' = \alpha/(C - c)$. Any nonsignificant p requires the end of testing. All larger p values are nonsignificant. SRB ensures $P(\#\text{TI} \ge 1) \le \alpha$ but gives greater power than does the Dunn procedure.

2.3.10   Shaffer MSRB Procedures

Shaffer (1986) proposed a modification of SRB designated the modified sequentially rejective Bonferroni (MSRB). She suggested that rather than dividing α by C, the number of pairs among I means, we should divide α by C_c, ($c = 0, \ldots C - 1$) the number of possible true null hypotheses after c hypotheses have been rejected. That is, after c significant p values among I means, the number of possible null hypotheses left to be true is presented in Table 2.3 based on results from Shaffer (1986). Taking $\alpha' = \alpha/C_c$ where C_c values are given in Table 2.3, the SRB procedure becomes the MSRB procedure. Values of C_c are provided in Table 2.3 for $I = 3$ to 7.

2.3.11   Dunnett Procedure

Certain specialized procedures have been developed for specific situations. Suppose a researcher wants to compare I groups where $I - 1$ groups are treatment groups and one group is a control group. Dunnett (1955) developed a procedure specifically for such situations. A special table is required for this method. The table is available in a number of sources (Keppel, 1982; Kirk, 1982; Winer, Brown, and Michels, 1991). For the specific type of experiment for which it was designed, Dunnett's procedure appears to be more powerful than any other. A modification is provided by Bechhofer, Dunnett, and Tamhane (1989) and by Tamhane (1987).

2.3.12   Unequal Sample Sizes

As a rule, whenever possible, it is desirable to design experiments with an equal number of observations in each group. The ANOVA F test can be applied with unequal N as easily as with equal N. All the above multiple comparison procedures which are based exclusively on F tests can be ap-

Table 2.3   Number of Possible True Pairwise Hypotheses Among I Means with c Rejections

No. of groups I	c (the number of hypotheses already rejected)																				
	0	1	2	3	4	5	6	7	8	9	10	11	12	13	14	15	16	17	18	19	20
3	3	1	1																		
4	6	3	3	3	2	1															
5	10	6	6	6	6	4	4	3	2	1											
6	15	10	10	10	10	10	7	7	7	6	4	4	3	2	1						
7	21	15	15	15	15	15	15	11	11	11	11	10	9	7	7	6	5	4	3	2	1

Source: The entries in this table were computed by the author.

plied with unequal N as easily as with equal N. Similarly, the SRB and MSRB methods can be applied with unequal N. However, all the procedures using the q test require an assumption of equal N.

With unequal N we identify the I sample sizes as $N_1, N_2, \ldots, N_I$. The harmonic mean of sample sizes is defined as

$$\tilde{N} = \frac{I}{\dfrac{1}{N_1} + \dfrac{1}{N_2} + \cdots + \dfrac{1}{N_I}}. \qquad (2.17)$$

Replacing N in any equation for pairwise testing with $\tilde{N}$ of Eq. (2.17) is the only change needed with unequal N. However, this is only an approximate solution. It can be considered acceptable provided that the largest sample size is no more than twice the smallest sample size. In more extreme cases, a procedure such as the Tukey-Kramer procedure must be applied (Dunnett, 1980a).

2.3.13 Tukey-Kramer Procedure

In cases where sample sizes are too unequal to apply the harmonic mean of Eq. (2.17), we can resort to a modification of Tukey's HSD known as the Tukey-Kramer procedure. In this method, Eq. (2.13) is modified to provide a separate critical difference for each pair of means. In particular, when testing means $\overline{X}_i$ and $\overline{X}_j$ of I means, we have

$$CD_{ij} = q_{1-\alpha}(I, df_E) \sqrt{\frac{MS_E}{2} \left[\frac{1}{N_i} + \frac{1}{N_j} \right]}. \qquad (2.18)$$

Dunnett (1980a) showed that the Tukey-Kramer procedure controls Type I errors even with extreme differences in sample sizes. However, failure of the assumption of equal population variances can easily lead to very high Type I error rates when unequal N are involved. This problem is considered in greater detail below.

2.3.14 Choosing a Pairwise Procedure

No one pairwise testing procedure will always be more powerful than all others. Scheffé's procedure is almost always less powerful than Tukey's HSD procedure when used for pairwise testing (Ury and Wiggins, 1975). Some researcher might prefer the HSD on that basis. Each researcher is free to choose either procedure, but the choice must be made independently of the data values. In particular, it would be *inappropriate* to try one

procedure and then, if some pairs of means are not significant, to switch
to another procedure. Although both the HSD and Scheffé's procedures
limit Type I error rates to α, their rejection regions have different shapes
(Games, 1971). Applying both methods increases the risk of Type I errors
beyond the specified α level.

There are situations in which seemingly different procedures can be
applied to the same data without increasing the limit on Type I errors. The
Hayter-Fisher LSD and the Shaffer-Welsch *Fq* are closely related procedures. In fact, LSD has a rejection region that is a subset of that of the
Shaffer-Welsch. Any result that is significant by LSD *must* also be significant by the Shaffer-Welsch. Any pair of means which fail to differ significantly by the Shaffer-Welsch must also fail to differ by LSD. Thus, a
researcher could begin applying LSD; then if all pairwise differences are
significant, there is no need for the Shaffer-Welsch. However, if some pairs
are *not* significantly different, the Shaffer-Welsch could be applied to the
same data without any increase in Type I errors. Both LSD and the Shaffer-
Welsch are subsets of the Shaffer-Peritz *Fq* procedure. These three procedures could be used as flexible applications of a single method. For some
data sets, the simplicity of the LSD would be adequate to detect all pairwise
differences. If any pairwise differences remain nonsignificant, one could
apply the Shaffer-Welsch. Only if both methods lead to nonsignificance
will it be necessary to resort to the Shaffer-Peritz.

Another set of procedures is HSD, Welsch, and Peritz *q*. These three
can be used sequentially from the simplest HSD to the complex but powerful Peritz *q*. Similarly, the Dunn, SRB and MSRB procedures can be
used sequentially (Holland and Copenhaver, 1988).

Example 2.2
Hypothetical data showing recovery scores of 16 individuals treated with
one of four drugs, with high scores indicating better recovery, follow:

Drug 1	Drug 2	Drug 3	Drug 4
$\overline{X}_1 = 5.6$	$\overline{X}_2 = 10.3$	$\overline{X}_3 = 15.0$	$\overline{X}_4 = 10.4$
$s_1^2 = 6.3$	$s_2^2 = 6.4$	$s_3^2 = 5.2$	$s_4^2 = 8.0$
$N_1 = 4$	$N_2 = 4$	$N_3 = 4$	$N_4 = 4$

The ANOVA $F = 9.099 > 3.49 = F_{.95}(3,12) = CV.$ $MS_E = 6.475$. The
significant F test would justify the rejection of H_0.

The Shaffer-Welsch procedure, Table 2.3, and $p = 3$ give

$$CD = q_{.95}(p, df_E)\sqrt{MS_E/N},$$
$$CD = q_{.95}(3,12)\sqrt{6.475/4},$$
$$CD = 3.77(1.2723) = 4.797.$$

Applying the same formula with $p = 2$, we have

$$CD = q_{.95}(2,12)\sqrt{6.475/4},$$
$$CD = 3.61(1.2723) = 4.593.$$

Thus, the table of differences between *ordered* means is:

	$\overline{X}_1$ 5.6	$\overline{X}_2$ 10.3	$\overline{X}_4$ 10.4	$\overline{X}_3$ 15.0	CD
$\overline{X}_1 = 5.6$	—	4.7*	4.8*	9.4*	4.80 ($p = 4$)
$\overline{X}_2 = 10.3$		—	0.1	4.7	4.80 ($p = 3$)
$\overline{X}_4 = 10.4$			—	4.6	4.59 ($p = 2$)
*p < .05					

Using underline notation we obtain:

$\overline{X}_1$ 5.6	$\overline{X}_2$ 10.3	$\overline{X}_4$ 10.4	$\overline{X}_3$ 15.0

Thus, drug 1 is significantly less effective than any of the other three drugs but drugs 2, 3, and 4 are not significantly different from each other. If we had applied the Hayter-Fisher LSD, all pairwise differences would be compared to a CD of 4.8. In that case, the only significant differences would be between the first mean and means three and four. However, to apply Tukey's HSD all pairwise differences would be compared to $CD = q_{.95}(I, df_E)\sqrt{MS_E/N} = q_{.95}(4,12)1.2723 = (4.20)(1.2723) = 5.34$. Thus, only the largest and smallest means would differ significantly by HSD. The Peritz procedure would determine as significant all pairwise differences except for the smallest difference of 0.1.

There is a simple way to keep track of the requirement that significant differences cannot be found within nonsignificant pairs. Any nonsignificant

Table 2.4 Applying the Dunn, SRB, and MSRB Procedures to Example 2.2

t	Means	p	Dunn	SRB	MSRB
9.4	1 vs. 3	.00011	.00833*	.00833*	.00833*
4.8	1 vs. 4	.01025	.00833	.01000	.01667*
4.7	2 vs. 3	.01336	.00833	.01250	.01667*
4.7	1 vs. 2	.01336	.00833	.01250	.01667*
4.6	3 vs. 4	.01258	.00833	.025	.025*
0.1	2 vs. 4	.47829	.00833	.05	.05

*Experimentwise rate set at .05.

difference in the table of ordered means cannot have a significant difference within that table which is directly below it, directly to the left, or both below and to the left of it. Thus, the nonsignificant difference of 4.7 has a difference of 4.6 just below it which cannot be significant even though it exceeds the $p = 2\ CD$ of 4.59. The difference of 0.1 which is just to the left of the nonsignificant 4.7 would also not be significant even if it exceeded 4.59 because it is immediately to the left of a nonsignificant difference.

The Dunn and SRB procedures give the same outcome of a single significant difference just as shown above for Tukey's HSD. However, the MSRB procedure gives significant differences for all but the smallest difference. Only the Peritz procedure, which is much more complex, was able to give as many rejections as the MSRB.

It should be noted that Shaffer proposed several versions of MSRB and the one presented in Table 2.4 is not the most powerful version. Table 2.3 gives the maximum possible number of possible, true, pairwise nulls. In any data set, one can usually gain greater power by taking the maximum for a given configuration. However, that makes the procedure much more difficult to apply.

2.4 PAIRWISE TESTING WITH UNEQUAL VARIANCES

All procedures in Sections 2.1 and 2.2 require an assumption of equal population variances. Failure of that assumption can lead to serious failure of the Type I error control (Dunnett, 1980b). The overall F test of Eq. (2.3) can be modified to allow for separate error estimates from each group. However, that modification is not necessary for pairwise testing.

The problem of unequal variances is especially difficult in the presence of unequal sample sizes. Among I means any pair $\overline{X}_j$ and $\overline{X}_k$ with $j \neq k$, respective sample sizes N_j and N_k, and variances s_j^2 and s_k^2, can be tested for significance by comparing the difference in means to a *CD* designed for that particular pair. Games and Howell (1976) proposed a procedure which has been designated G-H using the statistic

$$CD = q_{1-\alpha}(I, df_{jk}) \sqrt{\frac{s_j^2}{2N_j} + \frac{s_k^2}{2N_k}}, \qquad (2.19)$$

where $q_{1-\alpha}(I, df_{jk})$ is the $(1 - \alpha)$th quantile point of the studentized range distribution with parameters I and df_{jk} and

$$df_{jk} = \frac{(s_j^2/N_j + s_k^2/N_k)^2}{[(s_j^2/N_j)/(N_j - 1) + (s_k^2/N_k)/(N_j - 1)]}. \qquad (2.20)$$

Dunnett (1980b) found that the G-H procedure can have a rather high .084 empirical Type I error rate when run at $\alpha = .05$ level in the case that population variances are actually equal. With population variances actually unequal, the maximum empirical error rate was .065. Some researchers may find the G-H procedure to be adequate, especially since it is likely to be used only when population variances are actually different. Dunnett (1980b) considered some other procedures which give better control of Type I errors. A good summary is provided by Toothaker (1991).

2.5 FACTORIAL DESIGNS

Factorial designs lead to special problems with regard to multiple comparisons. Those problems are relatively minor in the absence of any interaction effect. Overall means for each factor can be compared in exactly the same manner as a one-way design. For example, in a two-factor design in which factor A has I levels and factor B has J levels, we might have N observations in each cell. The I overall means for factor A will each be based upon NJ observations. Similarly, the J overall means of factor B will each be calculated from NI observations. Treating each factor separately, we can ensure $P(\#TI \geq 1) \leq \alpha$ by running any method at level α provided that method maintains the required limit in one-way designs. Thus, each effect, A, B, and AB, can be limited to a specified familywise level. This is probably the most common approach.

An alternative approach would be to place a single experimentwise level to all three effects simultaneously. If we want to limit the Type I error rate for both factors simultaneously in a two-factor design, then each factor should be tested at level $\alpha/2$. To include both factors and the interaction with an overall limit of α, we run each of the three tests at level $\alpha/3$.

When interaction effects are present, the problem of multiple comparisons becomes more complex. Methods for direct testing of interaction effects have been described by Boik (1979). This is the approach which is likely to be favored by those who accept the arguments about interpretation of interactions advocated by Marascuilo and Levin (1970).

Another alternative is to test for simple main effects. For example, the I levels of factor A can be tested at each level factor B. In that case each mean would be based on N observations but there would be J sets of means. Running the multiple comparison procedure at level α/J for each set would ensure that $P(\#TI \geq 1) \leq \alpha$ for the entire experiment.

A number of detailed methods have been developed for testing various effects in factorial designs (Copenhaver and Holland, 1988; Holland and Copenhaver, 1988; Keselman, Keselman, and Games, 1991). In applying these methods, one must decide how errors are to be controlled and what design is to be applied. Testing simple main effects in addition to interaction effects results in the loss of some benefits of the orthogonality of overall main effects and interaction effects. The Boik approach avoids this problem. Additional research will be necessary to ensure wide acceptance of a multiple comparison procedure for interaction effects.

REFERENCES

Bechhofer, R. E., Dunnett, C. W., and Tamhane, A. C. (1989). Two-Stage Procedures for Comparing Treatments with a Control: Elimination at the First Stage and Estimation at the Second Stage. *Biomet. J., 31*: 545–561.

Boik, R. J. (1979). Interactions, Partial Interactions, and Interaction Contrasts in the Analysis of Variance. *Psych. Bull., 86*: 1084–1089.

Copenhaver, M. D. and Holland, B. (1988). Computation of the Distribution of the Maximum Studentized Range Statistic with Application to Multiple Significance Testing of Simple Effects. *J. Statist. Comput. Simul., 30*: 1–15. (Also see erratum, *31*:67.)

Dunn, O. J. (1961). Multiple Comparisons Among Means. *J. Amer. Statist. Assoc., 56*: 52–64.

Dunnett, C. W. (1955). A Multiple Comparison Procedure for Comparing Several Treatments with a Control. *J. Amer. Statist. Assoc., 50*: 1096–1121.

Dunnett, C. W. (1980a). Pairwise Multiple Comparisons in the Homogeneous Variance, Unequal Sample Size Case. *J. Amer. Statist. Assoc., 75*: 796–800.

Dunnett, C. W. (1980b). Pairwise Multiple Comparisons in the Unequal Variances Case. *J. Amer. Statist. Assoc., 75*: 789–795.

Einot, I. and Gabriel, K. R. (1975). A Study of the Powers of Several Methods of Multiple Comparisons. *J. Amer. Statist. Assoc., 70*: 574–583.

Fisher, R. A. (1935). *Design of Experiments*. Oliver and Boyd, London.

Games, P. A. (1971). Multiple Comparisons of Means. *Amer. Ed. Res. J., 8*: 531–565.

Games, P. A. and Howell, J. F. (1976). Pairwise Multiple Comparison Procedures with Unequal n's and/or Variances: A Monte Carlo Study. *J. Ed. Statist., 1*, 113–125.

Hartley, H. O. (1955). Some Recent Developments in Analysis of Variance. *Comm. Pure Appl. Math., 8*: 47–74.

Hayter, A. J. (1986). The Maximum Familywise Error Rate of Fisher's Least Significant Difference Test. *J. Amer. Statist. Assoc., 81*: 1000–1004.

Hochberg, Y. (1988). A Sharper Bonferroni Procedure for Multiple Tests of Significance. *Biometrika, 75*: 800–802.

Hochberg, Y. and Tamhane, A. C. (1987). *Multiple Comparison Procedures.* Wiley, New York.

Holland, B. (1991). Comment on Saville. *Amer. Statist., 45*: 165.

Holland, B. S. and Copenhaver, M. D. (1988). Improved Bonferroni-Type Multiple Testing Procedures. *Psych. Bull., 104*: 145–149.

Holm, S. (1979). A Simple Sequentially Rejective Multiple Test Procedure. *Scand. J. Statist., 6*: 65–70.

Keppel, G. (1982). *Design and Analysis: A Researcher's Handbook* (2nd ed.). Prentice-Hall, Englewood Cliffs, New Jersey.

Keselman, H. J., Keselman, J. C., and Games, P. A. (1991). Maximum Familywise Type I Error Rate: The Least Significant Difference, Newman-Keuls, and Other Multiple Comparison Procedures. *Psych. Bull., 110*: 155–161.

Keuls, M. (1952). The Use of the 'Studentized Range' in Connection with an Analysis of Variance. *Euphytica, 1*: 112–122.

Kirk, R. E. (1982). *Experimental Design: Procedures for the Behavioral Sciences* (2nd ed.). Brooks/Cole, Belmont, California.

Lea, P. (1991). Multiple Comparisons. *Amer. Statist., 45*: 165–166.

Marascuilo, L. A. and Levin, J. R. (1970). Appropriate Post Hoc Comparisons for Interaction and Nested Hypotheses in Analysis of Variance Designs: The Elimination of Type IV Errors. *Amer. Educ. Res. J., 7*: 397–421.

Martin, S. A. and Toothaker, L. E. (1989). PERITZ: A FORTRAN Program for Performing Multiple Comparisons of Means Using the Peritz Q Method. *Behav. Res. Meth. Instr. Comp., 21*: 465–472.

Miller, R. G., Jr. (1981). *Simultaneous Statistical Inference* (2nd ed.). Springer-Verlag, New York.

Nelson, N., Rosenthal, R., and Rosnow, R. L. (1986). Interpretation of Significance Levels and Effect Sizes by Psychological Researchers. *Amer. Psych., 41*: 1299–1301.

Newman, D. (1939). The Distribution of the Range in Samples from a Normal Population, Expressed in Terms of an Independent Estimate of Standard Deviation. *Biometrika, 31*: 20–30.

Peritz, E. (1970). A Note on Multiple Comparisons (unpublished manuscript). Hebrew University, Israel.

Ramsey, P. H. (1978). Power Differences Between Pairwise Multiple Comparisons. *J. Amer. Statist. Assoc., 73*: 479–485.

Ramsey, P. H. (1981). Power of Univariate Pairwise Multiple Comparison Procedures. *Psych. Bull., 90*, 352–366.

Ramsey, P. H. and Ramsey, P. P. (1990). Critical Values for Two Multiple Comparison Procedures Based upon the Studentized Range Distribution. *J. Ed. Statist., 15*: 341–352.

Ryan, T. A. (1960). Significance Tests for Multiple Comparison of Proportions, Variance, and Other Statistics. *Psych. Bull.*, 57: 318–328.

Saville, D. (1990). Multiple Comparison Procedures: The Practical Solution. *Amer. Statist.*, 44: 174–180.

Saville, D. (1991). Reply to Holland and Lea. *Amer. Statist.*, 45: 166–167.

Shaffer, J. P. (1979). Comparison of Means: An *F* Test Followed by a Modified Multiple Range Procedure. *J. Ed. Statist.*, 4: 14–23.

Shaffer, J. P. (1986). Modified Sequentially Rejective Multiple Test Procedures. *J. Amer. Statist. Assoc.*, 73: 826–831.

Scheffé, H. (1959). *The Analysis of Variance*. Wiley, New York.

Tamhane, A. C. (1987). An Optimal Procedure for Partitioning a Set of Normal Populations with Respect to a Control. *Sankhya, A49*: 335–346.

Toothaker, L. E. (1991). *Multiple Comparisons for Researchers*. Sage, Beverly Hills, California.

Tukey, J. W. (1953). The Problem of Multiple Comparisons (unpublished manuscript). Princeton University, Princeton, New Jersey.

Ury, H. K. and Wiggins, A. D. (1975). A Comparison of Three Procedures for Multiple Comparisons. *Brit. J. Math. Statist. Psych.*, 28: 88–102.

Welsch, R. E. (1977). Stepwise Multiple Comparison Procedures. *J. Amer. Statist. Assoc.*, 72: 566–575.

Winer, B. J. (1971). *Statistical Principles in Experimental Design* (2nd ed.). McGraw-Hill, New York.

Winer, B. J., Brown, D. R., and Michels, K. M. (1991). *Statistical Principles in Experimental Design* (3rd ed.). McGraw-Hill, New York.

3
Analysis of Covariance

SCOTT E. MAXWELL and MARY F. O'CALLAGHAN   University of
Notre Dame, Notre Dame, Indiana

HAROLD D. DELANEY   University of New Mexico, Albuquerque,
New Mexico

3.1   INTRODUCTION

The analysis of covariance (ANCOVA) is often considered to be one of
the most misunderstood and misused statistical techniques commonly employed by behavioral researchers. Although Fisher (1932) developed the
principles of ANCOVA more than 50 years ago, its use has been fraught
with controversial arguments over its appropriateness in the behavioral
sciences. In fact, we believe that many researchers today continue to fail
to appreciate the potential benefits of using ANCOVA. It is worth noting
that ANCOVA continues to be a source of active interest for statisticians,
as reflected by special issues of *Biometrics* devoted to ANCOVA in 1957
and 1982. The major purpose of this chapter is to attempt to explain some
of the advantages offered by ANCOVA and to clarify situations where,
in our judgment, ANCOVA is currently underutilized.

One reason ANCOVA has been so widely misunderstood is that the
technique can be used for two very different purposes. Fisher originally
developed ANCOVA as a method for increasing the precision of estimated
treatment effects in randomized experiments. However, over the years
ANCOVA has taken on another very different role, as a method of adjusting for preexisting group differences in nonrandomized studies. In fact,
until the late 1960s, ANCOVA was often advocated as a solution to identifying causal effects in nonrandomized designs. Over the next few years,
however, a number of methodologists (Campbell and Boruch, 1975; Campbell and Erlebacher, 1970; Elashoff, 1969) raised serious concerns about
the adequacy of ANCOVA in nonrandomized studies. These concerns led

Cronbach and Furby to conclude in their widely cited 1970 article that "Application of analysis of covariance to studies where initial assignment was nonrandom, which was widely recommended 10 years ago, is now in bad repute" (p. 78). Along similar lines, Smith (1957), in an excellent discussion of the interpretational difficulties associated with the use of ANCOVA in nonrandomized designs, stated that adjusted means obtained from ANCOVA might frequently more accurately be referred to as "fictitious means." The skeptical view toward ANCOVA in the 1970s was reinforced by explications of the stringent assumptions required in order for ANCOVA to completely remove the bias otherwise inherent in a nonrandomized comparison (Cronbach, Rogosa, Floden, and Price, 1977; Reichardt, 1979). Most researchers realized that these stringent conditions were unlikely to be met in practice, causing them to abandon the use of ANCOVA. While we would argue that these warnings appropriately dampened some of the previously uncritical applications of the analysis of covariance, at the same time they may have had the unfortunate effect of encouraging researchers to shy away from using ANCOVA altogether. It seems that many researchers may have overgeneralized the problems identified with using ANCOVA in nonrandomized studies and thereby inappropriately attributed the same problems to using ANCOVA in randomized designs as well. In particular, researchers may have realized that they were unlikely to meet the stringent set of conditions identified by Cronbach et al. and Reichardt. Although these authors were clearly explicit in their focus on nonrandomized studies, some researchers may nevertheless have avoided ANCOVA in randomized studies because of these stringent conditions. However, the stringent conditions required for ANCOVA to remove bias completely in nonrandomized studies are much more restrictive than the assumptions required for the proper use of ANCOVA in randomized studies; consequently, in many cases the avoidance of ANCOVA may have been misguided.

It is especially ironic that ANCOVA has continued to be underutilized despite a spate of attention given in the last two decades to the importance of statistical power in psychological and educational research (Cohen, 1962, 1988). While there can be no argument that power has deserved increased attention, it is nevertheless unfortunate that issues of power all too often seem to be translated immediately into questions of sample size. Researchers seem not to realize that several factors other than sample size can also have a sizable influence on power. In particular, power can often be substantially increased through the design and analysis of the study, such as by using ANCOVA, just as Fisher originally intended. It may offer some relief to recognize that behavioral researchers are not alone in their failure to consider such avenues for increasing power. In fact, even as far back as the 1941 edition of *Statistical Methods for Research Workers*, Fisher expressed surprise at the lack of utilization of ANCOVA: "In most kinds of experimentation, however, the possibilities of obtaining greatly increased precision from comparatively simple supplementary observations are almost entirely unexplored, and, indeed, in many fields the possibility of making a critically valid use of such observations is scarcely recognized" (Fisher, 1941, p. 278). The same statement appears in the last edition of the book published while Fisher was alive (1958), indicating Fisher's judgment that many researchers were still apparently unaware of the potential advantages of ANCOVA. Fisher is hardly alone among eminent statisticians in his view that ANCOVA has long been underutilized. For example, in discussing the role of sample size and statistical power in biological experimentation, Finney (1982) stated, "I am always surprised by the relative neglect of the technique [ANCOVA] in situations where a covariate potentially useful for increasing precision either is available or could have been for little extra cost" (p. 559). Similar statements have occurred in the psychological literature. Porter and Raudenbush (1987) stated "While keeping the general nature of counseling psychology research in mind and by taking into account the various alternatives to ANCOVA, we conclude that most randomized experiments would profit from having a covariable and use of ANCOVA" (pp. 391–392).

Because we agree that ANCOVA has largely been underutilized in experimental designs, our focus in this chapter will be on the potential benefits of ANCOVA in true experiments, where subjects have been randomly assigned to groups. Readers who are interested in the use of ANCOVA in nonexperimental settings may want to consult such sources as Huitema (1980), Maxwell and Delaney (1990), Porter and Raudenbush (1987), and Reichardt (1979) for more detail.

Why would a researcher ever want to use ANCOVA in a true experiment? After all, if subjects were randomly assigned to groups, shouldn't the groups be "the same"? One common answer to this question is that ANCOVA might be useful in case "random assignment didn't work." However, this answer fails to appreciate that the real benefit of ANCOVA in randomized designs is not its adjustment of between-group differences but instead its ability to explain within-group variance and thereby reduce the size of the error term and increase statistical power. Some researchers may avoid ANCOVA in this situation because they believe that it unnecessarily complicates interpretation of treatment effects. However, as Porter and Raudenbush show, in randomized designs, the interpretation of effects obtained with ANCOVA is identical to the interpretation in the more common ANOVA model. Another reason for avoiding ANCOVA may be the belief that its assumptions are unlikely to be met. For example, some methodologists have argued that a severe limitation to ANCOVA is its stringent assumption of homogeneity of regression. However, we will argue later in the chapter that this may very well be a strength of ANCOVA, since the possibility of detecting heterogeneity of regression may reveal important substantive patterns in the data.

3.1.1 Organization of the Remainder of the Chapter

The next section on the theoretical foundation of ANCOVA is intended to serve primarily as a review, although for some readers it may also introduce ANCOVA from a model comparison perspective. Readers who need additional background may wish to consult an experimental design text such as Keppel (1982), Kirk (1982), Maxwell and Delaney (1990), or Winer (1971).

After providing the theoretical foundation for ANCOVA, we discuss four advanced topics: factorial ANCOVA, ANCOVA with heterogeneous regression slopes, nonparametric ANCOVA, and alternative design procedures. We have chosen these specific topics not only because of their importance but also because some researchers may not be familiar with current research developments. For readers interested in other advanced topics not covered here, we recommend Huitema's (1980) excellent book on ANCOVA, which provides a very readable survey of a number of issues.

3.2 THEORETICAL BACKGROUND

3.2.1 Linear Models for ANCOVA

As with other commonly used parametric tests such as those employed in ANOVA and regression, the hypothesis tests of interest in ANCOVA can be approached from the perspective of comparing linear models. Indeed, introductions to ANCOVA typically note that the ANCOVA models include both the treatment effect parameters used in ANOVA and the slope and intercept parameters used in regression models. The slope parameters represent the contributions by the covariates. We have focused mainly on continuous covariates. Designs involving discrete concomitant variables would typically be treated as special cases of randomized block or factorial between-subject designs. We will now begin with the simplest statistical models that are appropriate for the case of a single covariate and a single between-subjects factor. More complex situations will be considered shortly.

Letting X_{ij} and Y_{ij} denote the scores for the ith individual in the jth group on the covariate and on the dependent variable, respectively, the test of the treatment effects in a one-way between-subjects ANCOVA involves comparing the following two models:

$$\text{Full:} \qquad Y_{ij} = \mu + \alpha_j + \beta X_{ij} + \varepsilon_{ij}, \qquad\qquad (3.1)$$

$$\text{Restricted:} \qquad Y_{ij} = \mu + \beta X_{ij} + \varepsilon_{ij}. \qquad\qquad (3.2)$$

Here α_j is the effect parameter associated with the jth treatment group, where $j = 1, 2, 3, \ldots a$, with a being the number of groups in the design; β is a population regression coefficient indicating for each model the extent of change in Y associated with a one unit change in X; and ε_{ij} is an error of prediction indicating for each model the extent to which it fails to account for the score on the dependent variable for the ith individual in the jth group. Note that μ, instead of being a grand mean as in ANOVA models, indicates the intercept of a regression line, that is, the mean Y score associated with an X score of zero. Although in within-subjects designs it is desirable to express the covariate in deviation score form so that μ in fact may be interpreted as the population grand mean of the dependent variable (Delaney and Maxwell, 1982), in between-subjects designs one is typically interested in the α_j and β parameters. Implicit in the full model is the presumption that the α_j are fixed treatment effects subject to the constraint that $\Sigma_j \, \alpha_j = 0$. Similarly, the X_{ij} are presumed to be known values that would be fixed as well over hypothetical replications of the study.

The null hypothesis to be tested is that there are no treatment effects. This restriction that there be no treatment effect, that is, that $\alpha_j = 0$ for all j, is imposed on the full model to yield the restricted model. The plausibility of this restriction may be investigated, as usual, by estimating the parameters of the two models and evaluating the magnitude of the increase in errors resulting from the adoption of a simpler model.

3.2.1.1 Parameter Estimates

One can estimate the unknown parameters μ, α_j, and β by ordinary least squares. In the restricted model, the resulting estimates describe the regression line relating Y to X when group membership is ignored. In particular, $\hat{\mu} = \overline{Y} - \hat{\beta}\overline{X}$ is the intercept of the regression line and $\hat{\beta} = b_T$ the slope when the total sample is treated as one group. The parameter estimates for the full model describe a parallel regression lines, with the intercepts being $\hat{\mu} + \hat{\alpha}_j = \overline{Y}_j - \hat{\beta}\overline{X}_j = a_j$ and the slope estimate being b_W, the pooled within-group regression coefficient relating Y and X. Note that this within-group estimate of the slope will almost certainly be different from the estimate derived from the total sample. This illustrates the point, familiar to users of multiple regression, that the estimate of a regression coefficient associated with a particular predictor is affected by adding other terms to the model unless they correlate zero with that predictor. However, from the ANOVA perspective, the more natural explanation is in terms of sampling error: b_W and b_T would be the same if the $\overline{X}_j$ were all identical to one another. However, even with random assignment, some variation in the covariate group means is practically inevitable.

3.2.2 Assumptions

In order to carry out the conventional parametric tests, it is necessary to make the same sorts of assumptions about the ε_{ij} term in the linear model that one makes in ANOVA. In particular, in the ANCOVA model,

$$Y_{ij} = \mu + \alpha_j + \beta X_{ij} + \varepsilon_{ij},$$

it is assumed that the residuals are independently and normally distributed with zero means and the same variance; that is, the ε_{ij} are independent $N(0, \sigma^2)$.

Several aspects of these assumptions deserve comment. First, although the normality assumption applies to distributions of residuals at each possible combination of a treatment level and a particular value of a covariate, sufficient replications to determine the form of the distribution in any one sample will generally *not* be available for any given X_{ij}. However, residuals can be pooled for each of the treatment groups and examined for normality within each group and for homogeneity of variance across groups. Fortunately, Monte Carlo studies suggest that ANCOVA is generally robust to moderate violations of the assumption of normality, particularly when sample sizes are equal (Glass, Peckham, and Sanders, 1972; Levy, 1980).

Second, the standard ANCOVA model presumes that there is a linear relationship between the covariate and the dependent variables. Examination of X-Y scatterplots within treatment groups allows one to form an impression of the plausibility of this assumption for a particular set of data. In terms of statistical tests evaluating this assumption, the availability of replications within each group at a given level of X would again affect how one assesses the reasonableness of this assumption. In the most straightforward case, if there were b levels of the covariate represented one or more times at each of the a levels of the treatment factor, then one could carry out a test for the lack of fit of the linear regression line by comparing the following two models:

Full: $Y_{ijk} = \mu_{jk} + \varepsilon_{ijk}$

Restricted: $Y_{ijk} = \mu + \alpha_j + \beta X_{jk} + \varepsilon_{ijk}$

where μ_{jk} is the population mean of Y in the jth treatment group and at the kth level of the covariate, $j = 1, 2, 3 \ldots a$, $k = 1, 2, 3 \ldots b$, and $i = 1, 2, 3 \ldots n_{jk}$. Such a test would have ab-a-1 and N-ab degrees of freedom, with a significant result indicating a systematic nonlinear relation between Y and X within treatment groups. Such a test is the natural generalization of tests of lack of fit discussed in standard regression texts (e.g., Neter, Wasserman, and Kutner, 1985, pp. 123–132).

In the much more common situation, replications at individual values of the covariate will not be available. One can still carry out a statistical test of linearity of sorts by seeing whether particular higher-order trends add significantly to the prediction of the dependent variable (Maxwell and Delaney, 1990, pp. 390–391). An alternative, approximate test of linearity is discussed by Kirk (1968, p. 471). In most behavioral science research, the linear relationship between Y and X will account for the bulk of the variability in Y associated with X.

A third point is that the ANCOVA model restricts the population slope β to having the same value for all groups. This condition is called homogeneity of regression. We shall consider some of the issues in analyzing data with models allowing for heterogeneous regressions of Y on X in Section 3.3.2 of this chapter.

Finally, it is typically assumed that the values of X are fixed, thus restricting one's statistical inferences to the set of hypothetical replications involving the same set of X values, in the same way that one's inferences in a fixed effects design are restricted to the particular treatment levels included in the study. (For further discussion of fixed vs. random covariates, see Huitema, 1980, pp. 86, 121; Rogosa, 1980, p. 308; Scheffé, 1959, p. 196; and Winer, 1971, p. 765.)

3.2.3 Standard Tests and Results

Given that the standard assumptions discussed above are met, tests of hypotheses of interest can be carried out. For example, letting E_F and E_R denote the sum of squared errors associated with the full and restricted models in Eqs. (3.1) and (3.2), respectively, and letting $df_F = N - a - 1$ and $df_R = N - 2$ denote the corresponding degrees of freedom for the full and restricted models, then the test of the treatment effect can be carried out as a standard F test:

$$F = \frac{(E_R - E_F)/(df_R - df_F)}{E_F/df_F}.$$

An adjusted treatment mean $\overline{Y}_j'$ estimates what performance in the jth group would have been if the group mean on the covariate $\overline{X}_j$ had been equal to the grand mean for the covariate $\overline{X}$:

$$\overline{Y}_j' = \overline{Y}_j - b_W (\overline{X}_j - \overline{X}).$$

The standard error of such an adjusted mean is estimated by $s_{\overline{Y}_j'}$:

$$s_{\overline{Y}_j} = \sqrt{\frac{E_F}{df_F}} \sqrt{\frac{1}{n_j} + \frac{(\overline{X}_j - \overline{X})^2}{\Sigma_j \Sigma_i (X_{ij} - \overline{X}_j)^2}}.$$

In cases in which there are more than two groups, contrasts of interest in the adjusted means are estimated through the appropriate linear combination of adjusted means, e.g., $\hat{\Psi} = \Sigma_j c_j \overline{Y}_j'$, and may be tested for significance using as the estimated standard error of $\hat{\Psi}$:

$$S_{\hat{\Psi}} = \sqrt{\frac{E_F}{df_F}} \sqrt{\sum_j \frac{c_j^2}{n_j} + \frac{[\Sigma c_j(\overline{X}_j - \overline{X})]^2}{\Sigma_j \Sigma_i (X_{ij} - \overline{X}_j)^2}}.$$

Details of the derivation of these results are presented by Cochran (1957). With a fixed covariate, standard post hoc procedures are available, including the Tukey test for pairwise comparisons of treatment groups and the Scheffé and Bonferroni tests of complex comparisons (see Maxwell and Delaney, 1990, for details). With a random covariate, Tukey's test for pairwise comparisons is no longer appropriate; for these comparisons, as discussed by Kirk (1982) and Maxwell and Delaney, the Bryant-Paulson modification of Tukey's method should be used.

3.2.4 Numerical Example

At this point it may be helpful to illustrate the benefits of ANCOVA in randomized pretest-posttest designs by presenting a numerical example. For pedagogical purposes, ANCOVA will be compared to two other approaches often used with similar data, ANOVA on posttest scores and analysis of gain scores. The summary tables, rather than the actual calculations, will be presented here. For computational formulas, the reader is referred to Keppel (1982), Kirk (1982), and Maxwell and Delaney (1990).

The hypothetical data for this analysis are presented in Table 3.1. We assume that these data came from a study conducted by a developmental psychologist examining the effect of two early intervention programs on children's IQ scores. We also assume that subjects were randomly assigned to one of two intervention conditions and were given the Stanford-Binet IQ assessment at both pretest and posttest. An ANOVA of the posttest IQ scores, an ANOVA of gain scores, and an ANCOVA using pretest IQ as the covariate were performed. Again, we present the three analyses in conjunction for illustrative purposes only; in actual research the investigator should choose only one of these methods.

The results of the three analyses of these data are presented in Table 3.2. As can be seen, the F value for the ANCOVA is larger than the F value for either of the two other approaches and is the only method showing significant differences between the conditions on IQ scores. What accounts

Table 3.1 Data for Comparison of ANOVA, Gain Scores Analysis, and ANCOVA

Subject	Group	Pretest	Posttest
1	1	80	87
2	1	125	124
3	1	103	105
4	1	101	107
5	1	125	93
6	1	89	93
7	1	111	93
8	1	116	127
9	1	110	118
10	1	95	106
11	2	101	103
12	2	125	121
13	2	105	109
14	2	104	116
15	2	111	123
16	2	110	139
17	2	125	135
18	2	68	101
19	2	101	121
20	2	95	97

Table 3.2 Alternative Analyses of Data in Table 3.1

Source Table for ANOVA					
Source	SS	d.f.	MS	F	p
Within cells	3564.60	18	198.03		
Group	627.20	1	627.20	3.17	.092

Source Table for ANOVA of Gain Scores					
Source	SS	d.f.	MS	F	p
Within cells	3077.60	18	170.98		
Group	744.20	1	744.20	4.35	.051

Source Table for ANCOVA					
Source	SS	d.f.	MS	F	p
Within cells	2222.19	17	130.72		
Regression	1342.41	1	1342.41	10.27	.005
Group	690.23	1	690.23	5.28	.035

for the advantage of ANCOVA here? We know that there are two general consequences of ANCOVA. First, the estimated magnitude of the treatment effect may be adjusted. The amount of adjustment, however, is primarily determined by the differences of the groups on the covariate. An ANOVA of the pretest for these data revealed no significant differences in this regard ($F_{(1,18)}$ = .02, p = .887), as would generally be true of randomized studies. Consistent with this nonsignificant difference, examination of the source tables reveals relatively small differences between the treatment sums of squares for the three approaches.

The second consequence of ANCOVA, a reduction in within-group variability, is primarily responsible for the results found here. Indeed, in Table 3.2 we see that the within-group sum of squares in ANCOVA is substantially smaller than those of either the ANOVA or the analysis of gain scores. The reason for the decrease in within-group sums of squares, as we move from the ANOVA to the ANCOVA, is fairly easily ascertained. First, conceptually, the two approaches differ in that the ANCOVA takes pretest performance into account when testing for group differences on posttest IQ scores, while ANOVA does not. The statistical advantage of using this additional information in ANCOVA is easily demonstrated. The models compared in an ANOVA are

$$\text{Full:} \quad Y_{ij} = \mu + \alpha_j + \varepsilon_{ij},$$

$$\text{Restricted:} \quad Y_{ij} = \mu + \varepsilon_{ij}.$$

Earlier, we presented the models for ANCOVA as

$$\text{Full:} \quad Y_{ij} = \mu + \alpha_j + \beta X_{ij} + \varepsilon_{ij},$$

$$\text{Restricted:} \quad Y_{ij} = \mu + \beta X_{ij} + \varepsilon_{ij}.$$

As we move from the full model in ANOVA to the full model in ANCOVA, the variability that is due to the relationship between the pretest and posttest is removed from the error term. Referring back to Table 3.2, we see that the within-cells sum of squares for ANOVA is indeed equal to the sum of the within-group sum of squares and the sum of squares due to regression in ANCOVA. The reduction in within-group variability in ANCOVA results in greater power, allowing a group difference to be detected.

It may not be as readily apparent why ANCOVA is superior to the ANOVA of gain scores. The approaches are conceptually similar, in that both methods include the pretest in the full model. The difference lies in the way that initial status on the covariate is taken into account. Again, the models for ANCOVA can be written as

$$\text{Full:} \quad Y_{ij} = \mu + \alpha_j + \beta X_{ij} + \varepsilon_{ij},$$

$$\text{Restricted:} \quad Y_{ij} = \mu + \beta X_{ij} + \varepsilon_{ij},$$

while the models for ANOVA of gain scores can be written as

Full: $Y_{ij} = \mu + \alpha_j + X_{ij} + \varepsilon_{ij},$

Restricted: $Y_{ij} = \mu + X_{ij} + \varepsilon_{ij}.$

The sets of models are identical except that the gain score models assume that β is equal to one. To the extent that β differs from one, the within group variability will be smaller in the ANCOVA full model than in the gain score full model because the ANCOVA model allows β to be estimated so as to minimize the error sum of squares. Examination of Table 3.2 demonstrates this point. The ANCOVA within-group sum of squares is considerably smaller than the error in the gain scores analysis, which again explains why a significant difference was found only when ANCOVA was used.

The general point to be made here is that the primary reason for the advantage of ANCOVA in randomized studies is not an inflation of between-group differences, but instead a reduction in within-group variability which results in greater power. It also must be stressed that although this example demonstrated that ANCOVA often has power to detect group differences where other methods do not, using ANCOVA never guarantees significance. Rather, using ANCOVA in randomized designs increases the chances of finding treatment effects where they exist.

3.2.5 Generalizations

The basic ANCOVA model can readily be generalized to incorporate additional covariates or nonlinear relationships between the dependent variable and one or more covariates.

3.2.5.1 Multiple Covariates

If more than one covariate is available for prediction of the dependent variable, adding the additional predictors to the model changes the basic procedures of the analysis little and may result in a greater reduction in within-cell error. We consider below design strategies to use in making decisions about which potential covariates should actually be employed. Naturally, the linear regression component of the model takes on the features of a multiple regression problem. Since computers will likely be used for any covariance problem, the additional computational burden matters little. However, the interpretation of the regression coefficient associated with any particular predictor must be made with the standard caveat of multiple regression, namely that the weight given any one covariate depends not only on its relationship with the dependent variable but also on its relationship with the other covariates.

If a second covariate denoted Z is to be utilized as well as X, the ANCOVA full model would become

$$Y_{ij} = \mu + \alpha_j + \beta_X X_{ij} + \beta_Z Z_{ij} + \varepsilon_{ij}.$$

The adjusted means in such a situation naturally take into account the departure of the covariate group mean from the covariate grand mean on both X and Z, that is,

$$\overline{Y}'_j = \overline{Y}_j - b_{w_x} (\overline{X}_j - \overline{X}) - b_{w_z} (\overline{Z}_j - \overline{Z})$$

where b_{w_x} and b_{w_z} are the within-group regression coefficients for X and Z, respectively, that is, the estimates of the β_X and β_Z parameters in the full ANCOVA model above. The standard error of the adjusted mean in the multiple covariate case involves a considerably more complex expression than the single covariate case since it depends not only on the residual variance in Y but also on the within-group variability in X and Z and on their covariance. Kirk (1982, p. 737) and Cochran (1957, p. 278) provide formulas for hand calculation of tests of contrasts in the two-covariate situation.

3.2.5.2 Nonlinear Relationships

Cochran (1957, p. 278) suggested that, with regard to possible violations of the assumptions of ANCOVA, "Perhaps the most common error to be anticipated is that linear regressions will be used when the true regression is curvilinear." Although in our view it is debatable whether some other complication such as heterogeneity of regression might not be more common, it is certainly easy to incorporate potential higher-order terms into one's model. For example, if there is reason to suspect a quadratic relationship, then one can use X^2 as well as X in the model, e.g.,

$$Y_{ij} = \mu + \alpha_j + \beta_L X_{ij} + \beta_Q X_{ij}^2 + \varepsilon_{ij}.$$

(Depending on the precision of the computer program being used to analyze the data, it may be advisable to express the covariate in deviation score form to lessen problems of collinearity among the predictors.) Higher-order trends may be investigated in a similar fashion.

3.3 RECENT DEVELOPMENTS

As indicated in the introduction, ANCOVA is a topic of considerable contemporary research. In the remainder of the chapter, we describe four issues related to ANCOVA that have received recent attention from statisticians: factorial ANCOVA, ANCOVA with heterogeneous slopes, nonparametric ANCOVA, and designing ANCOVA studies and choosing covariates. We chose to discuss these four specific topics for three reasons: (1) we believe they are important to behavioral researchers; (2) they are typically at most only briefly mentioned in many standard experimental design texts; and (3) they involve developments more recent than the publication of Huitema's (1980) excellent detailed coverage of ANCOVA.

3.3.1 Factorial ANCOVA

Although ANCOVA is often regarded as one of the most widely misused statistical techniques, it might seem that researchers who understand ANCOVA in one-way designs should be able to generalize their knowledge easily to factorial designs. For example, Kirk (1982) states that "The analysis of covariance for a factorial experiment is a straightforward generalization of the procedures discussed in connection with a completely randomized analysis of covariance design" (p. 743). Similarly, Cliff (1987) states that "It may be obvious that ANCOVA generalizes to factorial and other designs. . . . When the group sizes are equal, one could straightforwardly perform analyses of variance of these intercepts" (p. 285).

Unfortunately, although the use of ANCOVA in factorial designs is based on the same basic logic as its use in single-factor designs, there is an additional complication that is unappreciated by most researchers. This complication caused Pollane and Schnittjer (1977) to doubt the accuracy of the SPSS algorithm for performing ANCOVAs because its results differed from those of other programs. However, as Llabre and Ware (1980) pointed out, the difference was not due to lack of accuracy but instead reflected a difference in the way in which nonorthogonal effects were tested.

The source of the difficulty is that *even in an equal n design*, the effects in a factorial ANCOVA design are no longer orthogonal to one another, as they would have been in the comparable ANOVA design. Instead, as Bingham and Fienberg (1982) point out, the effects are nonorthogonal, just as they would have been if the ANOVA design had an unequal number of subjects per cell. Thus, even in the most straightforward factorial design, researchers must be aware of the nonorthogonal nature of the effects to be tested.

The ANOVA literature is fraught with controversial arguments over how to analyze data from nonorthogonal designs. However, the ANCOVA literature is bereft of such arguments, largely because, in our opinion, so little attention has been paid to the basic fact that nonorthogonality occurs in factorial ANCOVA designs. Although Llabre and Ware showed empirically that nonorthogonality occurs, they made no recommendations for how to resolve it. Bingham and Fienberg, on the other hand, argued that researchers should be consistent. They believe that researchers who advocate a hierarchical approach in unequal n designs (typically referred to as Type II sum of squares) should also use a hierarchical approach in factorial ANCOVA. Similarly, they state that researchers who advocate a nonhierarchical approach in unequal n designs (typically referred to as Type III sum of squares) should use a nonhierarchical approach in factorial ANCOVA.

Although Llabre and Ware (1980) and Bingham and Fienberg (1982) showed that differences can occur between the various approaches for analyzing data from factorial ANCOVA designs, the authors did not explain in terms of models why a covariate creates such nonorthogonality. In other words, why should effects that are orthogonal without a covariate become nonorthogonal in the presence of a covariate?

To understand why the covariate produces nonorthogonality even with equal n, we will consider a 2×2 equal n factorial design. A model for such a design can be written as

$$Y_i = \beta_0 + \beta_1 X_{1i} + \beta_2 X_{2i} + \beta_3 X_{3i} + \beta_4 X_{4i} + \varepsilon_i,$$

where X_1 is a dummy variable representing the level of the first factor, X_2 is a similar dummy variable for the second factor, X_3 is the product of X_1 and X_2 and thus represents the interaction, and X_4 is the covariate. For simplicity, we will assume without loss of generality that effect coding has been used, in which case X_1, X_2, and X_3 are all uncorrelated in a 2×2 design. As a result, when the covariate is not in the model, the two main effects and the interaction are all mutually orthogonal. However, when the covariate is included, it may be correlated with some or all of the other X variables. Of course, if groups are randomly constituted, any such correlation simply reflects sampling error. In either case, however, the sample correlations between X_4 and other predictor variables will inevitably be somewhat different from zero.

Although X_1, X_2, and X_3 will be uncorrelated with equal n, their partial correlations controlling for X_4 will be nonzero to the extent that the first three variables correlate with X_4. As Theil (1971, pp. 548–549) states, the various effects represented by the X variables will be orthogonal to the extent that the partial correlations among the variables are zero. Thus, even though X_1, X_2, and X_3 are uncorrelated, their effects will generally not be orthogonal because the partial correlations between any two of these variables controlling for X_4 will typically not be equal to zero. As a result, the test of X_1, for example, will depend on whether X_2 and/or X_3 is included in the model along with X_4. When both X_2 and X_3 are included, Type III sum of squares will be obtained. However, if X_2 is included but X_3 is not, Type II sum of squares will be obtained. In general, these two types of sums of squares will not be identical in the ANCOVA model even with equal n. Similarly, tests of X_2 and of X_3 will also depend on which other effects are included in the model. For this reason, effects are partially confounded with one another, although the design is an equal n design.

In contrast, when the covariate is not included in the model, not only are the zero order correlations among X_1, X_2, and X_3 all equal to zero, but so are all the partial correlations between any pair of variables controlling for the remaining variable. Only when X_4 is also included in the equation do these partial correlations become nonzero, and this, in turn, creates nonorthogonality among the effects.

Neither Llabre and Ware nor Bingham and Fienberg considered the difference in the effects being tested by the different approaches. It is well known that the difference in approaches to testing main effects in the unequal n ANOVA design is due to a difference in the definitions of marginal means; specifically, the approaches apply different weights to the cell means in order to define each marginal mean. However, Maxwell and Delaney (1988) showed that the reason for the difference between approaches in equal n factorial ANCOVA design arises from another source. Specifically, the difference between Type II and Type III sums of squares in factorial ANCOVA occurs because the regression weight for the covariate is estimated differently in the two cases. For Type III sum of squares, the regression weight is estimated from the within-groups regression coefficient. However, for Type II sum of squares, the regression coefficient is a weighted average of the within-groups coefficient and a second component, which reflects the between-groups interaction term regression coefficient. Maxwell and Delaney showed that when there is no interaction effect on the covariate (considered for this purpose as a dependent variable), both Type II and Type III regression weights are appropriate. However, when such an interaction is nonzero, only the within-groups weight of the Type III approach is appropriate.

Most statisticians agree that Type II sums of squares are misleading in unequal n factorial ANOVA designs. Similarly, Type II sums of squares and their associated significance tests may also be misleading in equal n factorial ANCOVA designs. We should acknowledge that the problem in ANCOVA is less serious when subjects have been randomly assigned to groups than when the group composition is nonrandom. When groups are randomly formed, the Type II and Type III regression weights differ from one another only because of sampling error, so the two approaches test the same hypothesis in the population. However, different results are still entirely possible with these two approaches. The difference is likely to be most dramatic when the population cell means on the covariate are different from one another, as they may well be in the absence of random assignment. In this case, the hypotheses tested by the two approaches are different from one another, and only the Type III sum of squares is likely to test an appropriate hypothesis of scientific interest.

3.3.2 ANCOVA with Heterogeneous Regressions

Homogeneity of regression slopes across groups is often considered to be a crucial assumption that must be met in order to use the covariate in assessing treatment effects. According to this logic, one starts off analyses of designs with a concomitant variable by testing for heterogeneity of regression, with "the often recommended decision rule [being] to use ANCOVA when a significant difference in slopes is not detected and to shun ANCOVA whenever a significant difference is detected" (Rogosa, 1980, p. 314). In other words, this strategy implies that the investigator should drop the covariate and instead perform an ANOVA when heterogeneity of regression is detected. The supposed superiority of ANOVA in this situation is frequently used as an argument against the general utility of ANCOVA. The rationale here seems to be that homogeneity of regression is an ANCOVA assumption that may often be violated, so researchers should typically forego ANCOVA in favor of ANOVA in order to avoid this restrictive assumption. This perspective regards homogeneity of regression as yet another statistical assumption, such as normality or independence of errors. Indeed, simulation studies of the consequences of violating the homogeneity assumption have been conducted similar to those for violations of normality (Hamilton, 1977; Levy, 1980).

This is not the best way to think about heterogeneity of regression. A revealing analogy is provided by two-way ANOVA designs. There, one regards additivity of the factors (i.e., the absence of an interaction) as a hypothesis of interest, not as a fundamental statistical assumption on which the validity of the analyses rests. In fact, if the logic of preferring ANOVA to ANCOVA because of a possible concern regarding heterogeneity of regression were applied consistently, researchers should also prefer one-way to two-way designs so as to avoid the possibility of an interaction. In reality, of course, the major benefit of two-way designs is that they provide a method for testing interaction hypotheses, making it all the more ironic that the possibility of a similar interaction is frequently regarded as a major liability of including a covariate in the analysis.

One would hope that, by regarding the covariate in a one-way ANCOVA as if it were a second factor, one would have a correspondingly richer set of hypotheses to investigate. And, in fact, with the appropriate models and analysis procedures, heterogeneity of regression may be viewed as a blessing rather than a curse. As Henderson (1982, p. 637) has suggested, "the situation where this homogeneity pattern does not hold should often be regarded as an aid to interpretation, rather than—as is so often the case—as a problem."

If interactions exist between treatments and variables reflecting individual differences, it is potentially important, both theoretically and practically, to be able to characterize those interactions. Interest in such attribute-by-treatment interaction (ATI) research has waxed and waned over the years. Cronbach's (1957) American Psychological Association presidential address was one early call to action, with ATI research being held up as an example of the effective integration of the correlational and experimental approaches to psychology. Twenty years later, Cronbach and Snow (1977) gave a heavy dose of realism to the area by their extensive review of difficulties encountered in attempts to identify ATIs. A principal difficulty was the low power of studies attempting to detect ATIs. When both the treatment and attribute factors are discrete, a partial remedy to the low power malady is to covary another predictor of outcome (Delaney and Maxwell, 1980a, 1980b).

Our concern at present is with the test of an interaction between a discrete treatment factor and a continuous covariate. The test of heterogeneity of regression pits the typical full ANCOVA model with a single slope against a model with a slope parameter for each group:

Full: $\quad Y_{ij} = \mu + \alpha_j + \beta_j X_{ij} + \varepsilon_{ij},$
Restricted: $\quad Y_{ij} = \mu + \alpha_j + \beta X_{ij} + \varepsilon_{ij}.$

Computational procedures are described in any number of sources (Kirk, 1982, p. 732ff; Maxwell and Delaney, 1990, p. 404ff; Winer, 1971, p. 773). To increase the likelihood of detecting whether the analysis should allow for heterogeneity of regression, one may use an α value greater than .05, for example, $\alpha = .20$, for the test of heterogeneity (Hendrix, Carter, and Scott, 1982, p. 643; Maxwell and Delaney, 1990, p. 419). If such a test is significant, instead of deciding to ignore the covariate, one should almost always investigate the interaction more fully. Glass et al. (1972, p. 276) have suggested that one should follow up a significant heterogeneity of regression test by blocking on the covariate so that follow-up tests of the interaction can be done in the context of a factorial ANOVA. Although a valid approach, we would not recommend using post hoc blocking, which amounts to discretizing a continuous covariate. As we have discussed elsewhere (Maxwell, Delaney, and Dill, 1984; see also Section 3.3.4), one is throwing away information by blocking in the analysis and, as a result, has less power for detecting effects.

We recommend instead that the covariate be maintained in continuous form. By so doing, even if one were to ignore the interaction and proceed with a standard ANCOVA, one would have greater power for detecting treatment effects (assuming the simple effects of the covariate do not exactly cancel each other by being of the same size but opposite in sign) than in an unadjusted ANOVA of Y. Admittedly, though, the procedure would be more conservative than a procedure that allows for heterogeneity of regression (Glass et al., 1972, Table 17; Rogosa, 1980, p. 316).

Introduced over 50 years ago, the Johnson-Neyman technique (J-N) (Johnson and Fay, 1950; Johnson and Neyman, 1936) provides a methodology for characterizing treatment by covariate interactions. Once it has been decided that the within-group regressions are, in fact, heterogeneous, J-N can be used to determine regions of significance, i.e., ranges of values of X within which the predictions made by the various within-groups regression lines are significantly different from each other. The question of whether predictions of various within-group regression lines are significantly different is equivalent to whether the conditional Y means for a given X are significantly different and, thus, is the same question addressed by what is termed a simple main effect in factorial ANOVA. Simple effects are the most common way of following up a significant interaction in factorial ANOVA (Kirk, 1982, p. 365ff; Maxwell and Delaney, 1990, p. 262ff), and the logic of the J-N procedure is the same. Basically, the two questions asked are "Does the treatment effect depend on X?" and "If so, for what values of X is the treatment effect significant?" Although Levin and Marasciulo (1972) recommend that only interaction contrasts be used as followup tests of significant interactions in factorial ANOVA, the two-stage approach of testing for heterogeneity of regression followed by J-N addresses the questions that are typically of greatest scientific interest. Furthermore, since the test for heterogeneity of regression in the two-group case is a single degree-of-freedom test (of the A by X_{linear} interaction), it would not be possible to partition the interaction into subeffects in this situation.

What we see as the more legitimate concern about J-N as a follow-up to an ATI is that of the possible inflation of α. Given that conditional means can be estimated for an arbitrarily large number of X values, it is a valid concern that one be protected against falsely rejecting the null hypothesis because of the multiplicity of test. [This is the "other" kind of Type IV error that Levin and Marascuilo (1972, p. 370) regarded as less serious than using simple effect tests after interactions, though they presumably would want the simple effect tests of treatments to be performed only after finding a significant main effect of treatment.] The J-N procedure was originally developed for nonsimultaneous inference, that is, where statements about the difference in predictions or conditional means could be made only at a single X value within the region of significance and not simultaneously at all X's within the region. Fortunately, Potthoff (1964) presented the appropriate revision of

J-N to allow one to be able to make inferences at an infinite number of X values without inflating α. Rogosa provides an excellent overview of the theory of determining regions of significance as well as developing the rationale for single "pick-a-point" tests that provide for an overall test of the treatment effect, for example, by choosing a point near the center of the X distribution at which to perform the test.

Rogosa's explication of J-N and its variants seems to have increased interest in such ATI methodology. Some researchers have begun reporting results of using the methodology (Reid and Borkowski, 1987; Sawyers, 1991) while others have restated the methods in different contexts (Griffey, 1982) or developed computer programs to determine the boundaries of the region of significance (Lautenschlager, 1987). Hendrix et al. have discussed the interesting case of modeling heterogeneity of regression in multifactor designs (see also Searle, 1979). Henderson considers the case of heterogeneity of regression in designs involving random factors. Finally, a paper by Schafer (1991) combined J-N with models allowing for quadratic relationships between the covariate and criterion variables.

One compelling aspect of Rogosa's methods for nonparallel regression lines is his argument about the appropriateness of overall tests of treatment effects. The standard ANCOVA test of treatments does not follow an F distribution whenever the population within-group regressions are at all heterogeneous, but it can be replaced by a safer test. Rogosa shows that the standard ANCOVA estimate of the treatment effect corresponds exactly to the estimate of the treatment effect at a particular X value in a model with heterogeneous slopes. The particular X value is denoted C_a, for center of accuracy, and is the point at which the estimate of the vertical distance between the regression lines has the greatest precision (i.e., smallest standard error). When the vertical distance at C_a is divided by its standard error and squared, one gets a "safer ANCOVA," that is, one that will be exactly distributed as a central F when other assumptions are satisfied regardless of the extent of heterogeneity of regression.

A recently published Monte Carlo study produced results apparently at odds with these analytical results. Harwell and Serlin (1988) examined the conventional ANCOVA and Rogosa's safer ANCOVA in conjunction with three nonparametric procedures, looking at both Type I error rates and power for different degrees of heterogeneity of regression, sample sizes, distributional forms and α levels. In their Table 8, which reports Type I error rates for a normally distributed dependent variable but with unequal slopes, the standard ANCOVA whose homogeneous regression assumption has been violated yields empirical α levels that are much closer to nominal levels than Rogosa's safer ANCOVA procedure whose assumptions supposedly have not been violated.

When simulation results are at odds with analytical results, it seems reasonable to ask where things went wrong in the simulation. A plausible explanation seems to be a violation of the assumption that the X values are fixed. Rogosa stresses that his methods are derived, as is typical in regression analysis, by assuming that inferences are to be made to sub-populations having the same X values (1980, p. 308).

If X did not interact with treatment, there would be no problem with its being random (see Harwell and Serlin's Table 4, where Type I errors in both standard and safer ANCOVA are unaffected by random X when the β_j are all equal). With heterogeneity of regression, however, the situation is different; it is analogous to a mixed ANOVA design. With one fixed factor and one random factor, an interaction intrudes upon or inflates the expected mean square for the fixed factor (Maxwell and Delaney, 1990, Ch. 10). In models with heterogeneous regressions, a similar phenomenon occurs whereby variability in $\overline{X}_j$ over samplings translates into increased variability in adjusted means, and thus heterogeneity of regression can masquerade as a treatment main effect.

This phenomenon is schematized in the simple illustration of Figure 3.1. The figure illustrates three hypothetical replications of an experiment where the same values of $\overline{Y}_j$ and $\hat{\beta}_j$ were obtained. Thus, the two components of the variability of adjusted means in a fixed effects design with heterogeneity of regression, namely sampling variability in $\overline{Y}_j$ and $\hat{\beta}_j$ (Maxwell and Delaney, 1990, p. 411ff) are being held constant so that they cannot contribute to the variability in the adjusted means over these particular replications. Nonetheless, varying estimates of the treatment effect, indicated by the braces ("{") in the figure, are obtained. The reason for variability in treatment effect estimates even when $\overline{Y}_j$ and $\hat{\beta}_j$ are held constant is that with a random covariate, different values of $\overline{X}_j$ cause the adjusted means to vary, even though they do not affect the within-group residual variability used to judge the significance of the difference between the adjusted means in a fixed effect design.

One suspects that the reason the standard ANCOVA led to empirical α levels close to the nominal α with heterogeneity of slopes is the effect of two offsetting factors. The inflation of the estimate of the treatment effect resulting from violating the assumption of a fixed covariate is offset by the fact that residual error variance is increasingly overestimated with increasing heterogeneity of regression. The former tendency to produce liberal test results is thus moderated by the overly conservative results for the standard ANCOVA noted by Glass et al. (1972) in which heterogeneity of regression was accompanied by a fixed covariate.

In summary, the issue involved in dealing with heterogeneity of regression in ANCOVA is conceptually the same as in dealing with an interaction

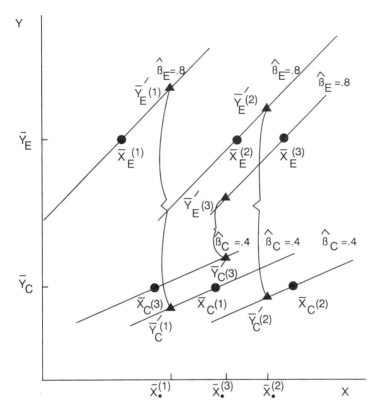

Figure 3.1 Illustrating the effects of a random covariate with heterogeneity of regression. Over three replications (indicated in parentheses) having identical estimates of the slope in the experimental group, the slope in the control group, and the group means on Y, variability in $\overline{X}_j$ produces different estimates of the adjusted treatment effect.

in ANOVA. When heterogeneity is observed, one is well advised to proceed with tests that yield insight into the nature of the varying treatment effects indicated thereby. One also wants to use test statistics that are appropriately distributed. If the investigator is willing to restrict inferences to the X values observed in the study (as is conventional in regression), then the J-N procedure modified for simultaneous inference and Rogosa's pick-a-point test provide an elegant, appropriate methodology for assessing, respectively, simple and overall treatment effects. Some may argue that the covariate should be regarded as a random effect, particularly when it is a pretest form of the same measure used as the dependent variable. If so, this needs to be made explicit and reckoned with in the analysis.

From this perspective, Monte Carlo results showing a liberal bias for Rogosa's procedures when used with a random covariate might be taken as an empirical confirmation of what theoretical considerations would also suggest, namely that interactions between a fixed and random factor result in the fixed effects model test of the fixed factor no longer being appropriate.

3.3.3 Nonparametric ANCOVA

Nonparametric alternatives to analysis of covariance are often recommended when the assumptions underlying parametric analysis of covariance are violated. These assumptions, as described above, include (1) normal distribution of the dependent variable, Y, at each level of the covariate, X, within each group (conditional normality); (2) equality of conditional variances within groups (homoscedasticity); and (3) the ANCOVA model being of the appropriate form, e.g., being correct in utilizing a prediction of Y that is a linear function of X (linearity) and that has a common, within-group regression coefficient (homogeneity of regression). When one or more of these does not hold, the researcher may want to consider nonparametric ANCOVA models which are less restrictive in that they require only an increasing monotonic relationship between the dependent variable and the covariate. Conditional normality, homoscedasticity, homogeneity of regression slopes, and linearity do not have to be assumed.

The choice between a parametric and a nonparametric approach to analysis of covariance is a complex one. It is certainly not the case that all violations of the assumptions underlying parametric ANCOVA require use of a nonparametric analysis. In fact, in a number of situations parametric techniques are superior to nonparametric alternatives even when certain assumptions are no longer valid. When making a decision about the appropriate statistic to use, the researcher must weigh considerations of Type I error rate and power in light of knowledge about such factors as the distributional properties of the variables under investigation, strength and form of the relationship between the dependent variable and covariate, equivalence of within groups regression slopes, and effect size.

Choosing between parametric and nonparametric techniques is complicated by the fact that these techniques often do not test the same hypothesis. When the parametric assumptions hold, both parametric and nonparametric ANCOVA test the same hypothesis of location. The procedures may not test equivalent hypotheses, however, if conditional normality is not met. The parametric analysis continues to test the hypothesis of location, but the nonparametric procedures test the hypothesis that the conditional distribution and/or marginal distributions are identical (Olejnik and Algina, 1985).

The bulk of this section will be devoted to several nonparametric statistical tests based on ranks which are appropriate when the parametric assumptions are not met. A comparison between these techniques and parametric ANCOVA based on considerations of power and robustness to Type I error under a variety of violations of parametric assumptions will be provided. Other topics include the special case of heterogeneity of regression, nonparametric methods for testing for parallel slopes, nonparametric tests based on matching, testing an ordered alternative hypothesis, and methods for performing ANCOVA on categorical data.

3.3.3.1 Nonparametric Analyses Based on Rank Transformation

An overview of selected nonparametric techniques based on ranks is provided. These representative techniques are well researched and have been found to have reasonable power and robustness. The procedures proposed by Quade (1967), Puri and Sen (1969), McSweeney and Porter (1971), and Hettmansperger (1984) will be reviewed. The interested reader is referred to Burnett and Barr (1977) and Shirley (1981) for other alternatives.

Quade's (1967) procedure was one of the earliest nonparametric alternatives to ANCOVA. It requires measurement of variables on at least an ordinal scale, independence of observations, continuity of underlying observations, a monotonic relationship between X and Y, and, in addition, that X be random and have the same distribution in each group (Harwell and Serlin). The computation is relatively straightforward. First, both the X and Y are ranked independently from 1 to N, ignoring group membership. Then, ordinary least squares regression is performed, regressing the ranks of Y on the ranked covariate. The residuals, Z_{ij}, are then calculated by subtracting each subject's actual rank, R, from its predicted rank, $\hat{R}$, on Y, using the expression $Z_{ij} = R_{ij} - \hat{R}_{ij}$. Finally, to test the null hypothesis of no group differences, a parametric analysis of variance is performed on the residuals using the grouping variable, and the observed F is compared to the critical F with $(a - 1, N - a)$ degrees of freedom. Quade's procedure can be used for factorial designs and can accommodate more than one covariate. The procedure assumes homogeneity of regression slopes but does not provide a test of this assumption. Finally, the procedure uses only the total-group regression line in calculation of the test statistic, in contrast to parametric ANCOVA, which uses both the within-group and total-group slopes (Olejnik and Algina, 1985). Maxwell, Delaney, and Manheimer's (1985) comments about such an analysis of variance of residuals using b_T in the parametric case would suggest that this procedure might be overly conservative. Fortunately, as discussed below, Harwell and Serlin introduced a modification of the Puri-Sen procedure that uses both within-group slopes and a total sample slope.

Puri and Sen's method was introduced as a more general case of Quade's procedure. It is appropriate for ranked data as well as other transformations of rank scores, such as normal scores. Perhaps the most straightforward computational version of Puri and Sen's method was that presented by Harwell and Serlin. First, X and Y are ranked separately across groups. The following test statistic is then computed, substituting the ranked data for the original scores:

$$L = \frac{SSY_B - 2r_{s(T)} * SCP_{XY(T)} + r_{s(T)}^2 * SSX_{(B)}}{SSY_{(T)} * (1 - r_{s(T)}^2)},$$

where SSY_B is the between-groups sum of squares for Y, $r_{s(T)}$ is the Spearman rank order correlation coefficient for the entire sample, $SCP_{XY(T)}$ represents the total group sum of cross-products, $SSX_{(B)}$ is the between-groups sum of squares for X, and $SSY_{(T)}$ is the total sum of squares for Y. The observed statistic is compared to a χ^2 value with $a - 1$ degrees of freedom. As can be seen from this formula, Puri and Sen also use only a single, total-sample slope $(r_{s(T)})$. The assumptions made are identical to those made by Quade. Again, this technique can be generalized to the multiple covariate and factorial ANCOVA case. It assumes homogeneity of regression, although no explicit test of this assumption is available.

Harwell and Serlin offered the Puri-Sen-Harwell-Serlin (PSHS) test as a computationally simpler version of Puri and Sen's test. The calculations for this procedure are similar to those for the McSweeney and Porter statistic described below. First, scores on both X and Y are ranked from 1 to N, and a standard parametric ANCOVA is performed on the ranks. The test statistic, $TS = (N - 2)\eta^2$, is then computed, where $\eta^2 = (SS_{Between})/(SS_{Total})$ represents the variance in the covariate-adjusted ranks explained by the grouping variable, and N equals the total sample size. The PSHS statistic, like the Puri-Sen, is compared to a χ^2 distribution.

McSweeney and Porter have proposed probably the simplest nonparametric analysis of covariance procedure. Data on the covariate and dependent variable are ranked independently from 1 to N, and a parametric analysis of covariance is carried out on the ranks. The calculated F value is compared to a critical F with $a - 1$ and $N - a - 1$ degrees of freedom. The assumptions made are again the same as those made in Quade's procedure. The McSweeney and Porter statistic, like the parametric analysis of covariance, uses both the within- and total-group slopes in computation of the test statistic. This approach to nonparametric analysis of covariance was also advocated by Conover and Iman (1982), who generally support use of rank scores for original data in standard parametric analyses when parametric assumptions are not met. In keeping with Conover and Iman, then, it is also possible to generalize McSweeney and Porter's test to the multiple covariate and factorial ANCOVA case.

Hettmansperger offers a general framework for testing hypotheses in linear regression models using rank scores. This approach takes the form of aligned rank tests in which the raw data are aligned, or adjusted, by removing the effects of nuisance parameters before estimating the relationship between the dependent variable and the grouping factor. Harwell and Serlin provide a clear summary of the computational steps necessary for generalizing Hettmansperger's approach to the analysis of covariance case. First, Y is regressed on X, using the total-group slope, and the residuals, $(Y_{ij} - b_T X_{ij})$, are calculated. The residuals are then ranked from 1 to N, and then the weighted ranks, R'_{ij}, are calculated, using the formula $R'_{ij} = 12^{1/2}(R_{ij}/(N - 1) - 1/2)$, where R_{ij} is the unweighted rank. Finally, a standard parametric analysis of variance is carried out on the weighted ranks, using the grouping variable as the independent factor. In the case of a single factor and covariate, this statistic is asymptotically distributed as a χ^2 with $a - 1$ degrees of freedom. Hettmansperger, however, points out that the nominal significance level is better maintained when the χ^2 critical value is replaced by $(a - 1) F_\alpha (a - 1, N - a)$.

3.3.3.2 Comparisons of Rank Transform Approaches

The methods described above have in common the transformation of the raw data into rank scores which are used in the calculation of the test statistic. There are, however, important differences among the nonparametric techniques which have implications for choosing the most appropriate analysis. Perhaps the most important distinction to be made between the procedures is whether they take into account the total-group regression only or both the within-group and total-group regressions when adjusting the dependent variable for the covariate. Only McSweeney and Porter and the Harwell-Serlin modification of the Puri-Sen procedure make an adjustment based on both within-group and total-group regressions, which is consistent with parametric analysis of covariance. The nonparametric methods which use residuals, the Quade, Puri and Sen, and Hettmansperger tests, adjust the dependent variable only for the relationship between the dependent measure and covariate for the entire sample. For both the parametric and nonparametric cases, analysis of the residuals from the regression of Y on X, using only the within-group or total group-slopes, is not equivalent to standard analysis of covariance. Performing an analysis of variance on the residuals will not provide a correct test because the test of hypothesis in analysis of covariance takes into account both the within-

and total-group slopes, while analysis of residuals does not (Maxwell et al., 1985). In the case in which the total-group slope alone is used (which is the case with the rank transform tests using residuals), the test will generally be too conservative. While substituting an analysis of residuals for an analysis of covariance may be generally problematic, it is especially troublesome for the nonexperimental situation in which intact groups differ on the covariate.

The approaches also differ in the reference distributions used to test the null hypothesis. Quade (1982) and McSweeney and Porter (1971) use the F distribution, while Hettmansperger (1984) and Puri and Sen (1969) use the χ^2. Because of these differences and those mentioned above, it is reasonable to expect that the techniques may differ in efficiency and robustness. Several studies have compared the nonparametric techniques with respect to their power and Type I error rate. For example, McSweeney and Porter found that, under a variety of conditions, their rank transformation approach was slightly more powerful than Quade's statistic. Harwell and Serlin also found that the Hettmansperger test produced consistently higher power than the Puri-Sen technique or McSweeney-Porter test across a range of violations, including nonnormality and unequal slopes. However, Olejnik and Algina (1985) concluded that, in general, these procedures demonstrate similar power and robustness and typically provide similar conclusions regarding tests of null hypotheses.

3.3.3.3   Robustness and Power Comparisons of Parametric ANCOVA and Rank Transformation Approaches

The first comparisons of parametric to nonparametric techniques centered on the robustness of the methods. Olejnik and Algina (1987) provide a review of the literature on robustness of parametric techniques. Parametric ANCOVA was found to maintain empirical Type I error rates close to the nominal significance levels under violations of the assumption of equal within-group regression slopes, under moderate departures from conditional normality and when the distributions were exponential or uniform. ANCOVA was found to be conservative, however, when the dependent measures were lognormal or Cauchy with a very heavy tail. However, the form of the covariate distribution does not appear to influence the Type I error rate of the parametric test. ANCOVA was also found to be robust to within-group heteroscedasticity under equal and unequal sample sizes. The empirical Type I error rates under the simultaneous violations of conditional normality and homoscedasticity closely matched those with heteroscedasticity alone. Studies which compared the Type I error rates for parametric analysis of covariance to nonparametric alternatives demonstrated that the alternatives do not provide a substantial advantage over the parametric counterpart. That is, the nonparametric tests are robust under the same conditions as the parametric approach and provide inappropriate Type I error rates in situations where the parametric method is also likely to do so.

Based on considerations of robustness alone, it would seem that violations of assumptions do not necessitate use of nonparametric alternatives. More recently, however, there has been a shift of emphasis from Type I error rates to power (Olejnik and Algina, 1987). Blair (1981) argued that, in general, although parametric techniques are often robust, the choice between parametric and nonparametric tests should also consider statistical power. When power is used as a criterion for selecting an appropriate procedure, there are a number of situations, discussed below, in which a nonparametric ANCOVA test might be preferred.

Olejnik and Algina (1987) compared empirically generated power estimates for the parametric ANCOVA to the McSweeney and Porter rank transformation approach under conditions of homoscedasticity and heteroscedasticity in the randomized two-group pretest-posttest situation. Their results revealed that moderate violations of the assumptions of homoscedasticity and conditional normality have little effect on the power of parametric ANCOVA. The rank transformation approach, however, was affected by these violations, rendering it more or less powerful than the parametric approach depending on the shape of the conditional distribution, the degree of heteroscedasticity, and strength of correlation between the covariate and dependent variable. When homoscedasticity was satisfied and there was a moderate pretest-posttest correlation ($\rho = .7$), the rank transformation approach demonstrated a power advantage relative to the parametric test only under extreme departures from conditional normality. With a weaker pretest-posttest relationship ($\rho = .3$), this advantage was evidenced under milder violations of normality. When the assumption of equal conditional variances was violated, the power advantage of the rank transform grew increasingly larger as the data became more heteroscedastic for normal distributions as well as for distributions with some degree of skew and/or kurtosis. The power difference between parametric and nonparametric techniques was again influenced by the strength of the pretest-posttest relationship, with greater differences favoring the rank transform method when the correlation was weak.

Olejnik and Algina (1985) have also summarized the effects of other violations, including nonlinearity of relationship between X and Y and heterogeneity of regression slopes, on power for parametric and nonparametric tests. Briefly, the nonparametric approaches demonstrated power advantages over the parametric method under nonlinearity, specifically when the true relationship between X and Y was exponential while a linear X-Y relation was assumed (McSweeney and Porter, 1971). The parametric test, however, was most powerful in the presence of heterogeneity of regression, except when the conditional distribution was double exponential or when sample sizes were unequal and the distribution was either conditional or translated exponential. In these latter two cases, the nonparametric tests had a slight power advantage. Seaman, Algina, and Olejnik (1985) also investigated the power of parametric and rank transformation tests when the conditional distributions differed in skew and/or scale and found that, in general, the parametric approach provided greater power than the rank-based approaches. These results and those of others support Olejnik and Algina's (1987) general conclusion that parametric analysis of covariance is preferable to nonparametric under moderate violations of conditional normality and homoscedasticity assumptions when a reasonably good covariate is available in experimental research. When these conditions do not hold, or the relationship between the dependent variable and covariate is nonlinear, a nonparametric alternative may be preferable.

3.3.3.4   Heterogeneity of Regression Slopes

An issue which merits further attention is the special case of the violation of the homogeneity of regression assumption. Up until this point the discussion of nonparametric techniques has implicitly considered only one approach to dealing with violations of the assumption of homogeneity of within-group regression. This approach is to regard it simply as any other assumption and search for techniques which remain robust and powerful when it has been violated. Others have argued, however, that the assumption of homogeneous regressions is a critical one, and it is theoretically meaningless to perform an ANCOVA with unequal within-group regression slopes. A third and quite valid way of dealing with heterogeneity of regression, Rogosa's (1980) ANCOHET model, was discussed earlier. The choice of one of these three approaches to dealing with heterogeneity of regression may generally be a difficult one, when homogeneity alone is violated, but it becomes even more complex in the presence of other violations of assumptions. Although Rogosa's method is appropriate when homogeneity of regression is the only assumption violated, there is evidence (Harwell and Serlin, 1988) that this procedure is less robust and powerful than other nonparametric techniques when conditional normality and homoscedasticity are not satisfied at least with random covariates, as we discussed in the preceding section. Unfortunately, nonparametric alternatives to the Rogosa procedure have not been developed. Thus, a researcher must take into account theoretical issues as well as power and Type I error rates when deciding how to approach the heterogeneity of regression case.

Whichever approach to the heterogenous regression situation is taken, it is first necessary to perform a test for equality of regression slopes. If the assumption of parallel regression slopes is the only one violated, parametric ANCOVA allows a test of heterogeneity of regressions. Under other violations, the rank transformation procedure proposed by Mc-Sweeney and Porter can accommodate this test (Conover and Iman, 1982). Penfield and Koffler (1986) have also developed a nonparametric procedure based on Puri and Sen's general L statistic which explicitly tests this assumption. The properties of these last two tests relative to the parametric test for homogeneity of regression slopes have received little attention in the literature. Furthermore, the robustness and power of the nonparametric techniques in this situation have not been investigated.

3.3.3.5 Other Nonparametric Alternatives to Analysis of Covariance

In addition to the nonparametric rank transformation techniques formulated for use under violation of parametric assumptions, other nonparametric alternatives to analysis of covariance have been developed. Several of these tests are presented here, including Quade's matching test, tests for ordered alternatives, and tests which are appropriate for categorical data.

Quade (1982) proposed a technique for nonparametric ANCOVA by matching subjects. This method was proposed as a compromise between standard analysis of covariance, which requires several assumptions, and a standard analysis of matched pairs, which requires fewer assumptions but at the cost of loss of information. The analysis offered by Quade requires that subjects be matched on a concomitant variable using caliper matching, which specifies that all subjects be matched whose differences on the covariate do not exceed some tolerance, ε, which is the maximum amount by which two covariate values are allowed to differ. Each subject's score on Y is compared with the average of Y scores for those subjects with similar values (within tolerance) of the covariate. A standard analysis of variance is then carried out on these residuals.

Like the rank transformation approaches, Quade's matching procedure does not require normality or a linear relationship between X and Y. Unlike the rank procedures, however, it does not assume homogeneity of regression. Matching also has greater flexibility than the rank transform approaches in that the covariate can be arbitrary, even categorical, in form, and the relationship between X and Y can take on any form. The matching procedure does require, however, that the marginal distribution of X be the same in all populations. Moreover, Quade acknowledges that matching may be less powerful than other nonparametric approaches. In addition to these pragmatic differences between matching and rank transformation procedures, there are some underlying philosophical differences related to the concept of control. The rank transformation approaches, and, more generally, analysis of covariance control for the covariate through statistical adjustment in the analysis of a study, while matching procedures control for the covariate by altering the design of the study.

Nonparametric alternatives have also been developed to test for a possible ordering of treatment effects while adjusting for covariates. This test is similar to standard analysis of covariance except that instead of testing for covariate-adjusted mean differences, the null hypothesis, H_0: $\mu_1 = \mu_2 = \cdots = \mu_a$, is tested against an ordered alternative, H_1: $\mu_1 \leq \mu_2 \leq \cdots \leq \cdots \mu_a$, where the μ's are adjusted means and there is at least one strict inequality. Again, when the usual parametric assumptions do not hold, an alternative may be considered. Marcus and Genizi (1987) suggest two nonparametric tests of ordered alternatives which may be especially useful when the functional form of the relationship between the response variable and covariate is unknown. These two test statistics are based on an extension of Quade's caliper matching method. One is Quade's matching procedure generalized to the a group case, which is appropriate when the covariate is continuous. The other is the Terpstra-Jonckheere statistic generalized for grouped data, which is applicable for discrete covariates. Monte Carlo studies suggest that these two procedures are fairly robust, maintaining significance levels near nominal values with a variety of distributions, including normal, exponential, and highly skewed. Empirically generated power estimates demonstrated that power was highest for the skewed distributions, it increased rapidly to a maximum as tolerance decreased, and then tended to decrease again. The tolerance value that maximizes the power depends on (1) the distributions of X and Y; (2) the functional form of their relationship; (3) the number of groups; and (4) sample sizes. Boyd and Sen (1986) propose another alternative for testing an ordered hypothesis which merges the rank-permutation principle in Puri and Sen (1976) and the union intersection principle of Roy (1953). This procedure is more generally appropriate in the presence of violations of parametric assumptions.

A nonparametric alternative to analysis of covariance may also be considered when the dependent variable is an ordered categorical response variable. This type of variable is typical of outcome variables in clinical research. For example, the effectiveness of a treatment may be evaluated on a four-point ordered scale where symptoms are rated as either none, mild, moderate, or severe (Francom and Chaung-Stein, 1989). In such a case, none of the nonparametric procedures thus far discussed would be suitable, as they assume that the underlying response variable is continuous. Koch, Amara, Davis, and Gillings (1982) review three methods appropriate for categorical data. These include randomization-model non-parametric procedures, maximum likelihood logistic regression, and weighted least squares analysis of correlated marginal functions. They also discuss a fourth method, unweighted least squares, which is computationally simple but provides only an approximate test of treatment effects. The authors demonstrate that these procedures produce comparable results for similar problems. They stress, however, that choice of a procedure for a particular research problem should rest on several methodological issues, such as the sampling framework, assumptions of linearity and homogeneity of regression, the equivalence of covariate distributions across treatment groups, the interpretations one wishes to make regarding treatment effects and generalizability of results, and sample sizes.

3.3.4 Designing ANCOVA Studies and Choosing Covariates

In this section we will consider several variations of the manner in which one or more covariates may be utilized in randomized studies. The focus here is more on the design of the study than on the analysis. For example, how should a covariate be selected? Must it be identified prior to the study or can it be selected post hoc? Can an experimenter's knowledge of subjects' scores on a covariate be used appropriately to influence the manner in which subjects are assigned to groups? Finally, are there other alternatives to ANCOVA which may be preferable in some situations?

We will first consider how a covariate should be selected. In the simplest case, a researcher has identified a single baseline variable that he or she believes may be relevant to the outcome of the study. When should that variable be used as a covariate? In a more complicated situation, several such covariates have been identified. Which, if any, of these measures should be used as covariates? How many of the measures should be included in a particular analysis?

Many researchers use a covariate in randomized experiments for the sole purpose of ascertaining that the randomized process "worked." With this strategy, the researcher typically tests to see whether the randomly constituted groups differ significantly on each baseline variable. If the groups do not differ, then that variable can safely be ignored, and there is no need to use it as a covariate. If, on the other hand, the groups are different (in which case there is said to be "covariate imbalance"), that baseline measure must be used as a covariate in an ANCOVA to correct for the initial imbalance between the groups. Although there may be considerable intuitive rationale for adopting this strategy, Permutt has shown that, in fact, it leads to undesirable consequences. The procedure is necessarily conservative in that the actual Type I error rate is less than the nominal alpha level. Furthermore, the power of the procedure is almost identical to the power of an ANOVA, simply ignoring any information about covariates, leading Permutt (1990, p. 1456) to conclude that ". . . the preliminary-test procedure confers essentially none of the benefits of the analysis of covariance." In fact, Permutt showed that when a set of covariates is available, choosing the specific covariates to enter into the analysis on the basis of this procedure is worse than simply choosing these covariates randomly. That this preliminary test procedure is a failure should hardly seem surprising because, as Altman (1985) put it, "Performing a significance test to compare baseline variables is to assess the probability of something having occurred by chance when we know that it did occur by chance. Such a procedure is clearly absurd" (p. 126).

One inference that might be drawn from Permutt's findings and Altman's statement is that ANCOVA is simply of little value in randomized studies. However, the problem of the preliminary test procedure is not with ANCOVA but instead with the manner in which a covariate is selected. Thus, a better solution must propose a better mechanism for choosing relevant covariates. Fortunately, a number of such mechanisms have been suggested.

Permutt compared several strategies for selecting covariates and concluded that "Adjusting for the covariate most correlated with the outcome is the clear winner" (p. 1460). This again is hardly surprising because, in a randomized study, the primary reason for including a covariate is to reduce the error variance and thereby increase power. However, error variance is decreased most directly by including in the model variables other than treatment condition that are also predictive of the outcome. Thus, in randomized studies, researchers should choose covariates on the basis of the extent to which they uniquely account for variance in the dependent variable.

Although this conclusion sounds straightforward, there is potentially an additional complication. Schluchter and Forsythe (1985) compared 16 strategies for selecting one or more covariates in randomized studies. Each strategy fell into one of three broad categories: (1) select covariates on the basis of their correlation with the response; (2) select covariates whose means differ across the groups; (3) select covariates satisfying both the first and second conditions. Consistent with Permutt, they found that the second and third strategies "should not be used" (Schluchter and Forsythe, 1985, p. 698) because they resulted in a conservative bias. The first strategy, on the other hand, often produced a more efficient estimate of the treatment effect than an analysis of covariance that simply used all available covariates. However, Schluchter and Forsythe have also found that when the same data are used not only to select the covariate(s) but also to perform the ANCOVA test of treatment differences on the outcome measure, the resultant Type I error rate is inflated. The degree of inflation is generally unpredictable, but it clearly worsens as the number of potential covariates increases relative to the size of the sample. Their conclusion is that the best strategy is probably to adjust for all available covariates as long as the sample size is large enough. Their rough guideline is that the number of subjects per group should be at least as large as $2 + 3p$, where p is the number of potential covariates.

Permutt's as well as Schluchter and Forsythe's findings suggest that, in randomized studies, researchers should choose a relatively small number of covariates which they anticipate will make a sizable unique contribution to the prediction of the dependent measure. Once these variables have been identified, they should generally all be included as covariates in an ANCOVA, regardless of their sample correlations with the dependent variable or the magnitude of their group differences.

Although nothing in this proposed strategy suggests the need to test group differences on the covariate(s), some researchers will nevertheless inevitably feel the need for such a test. The question then arises as to how a statistically significant difference between the groups—referred to as "unhappy randomization" by Kenny (1975)—should be interpreted, and what should be done about it in order to obtain a proper group comparison on the dependent variable of interest. The answer to the first part of the question is simple. Assuming that subjects were truly randomly assigned to conditions, a significant difference between groups can reflect only a sampling error. This might seem to suggest that the answer to the second half of the question is that sampling error can simply be ignored. However, because this issue turns out to involve some interesting dilemmas, it deserves some additional discussion.

To address the issue of how to proceed when randomly constituted groups differ on a covariate, it is helpful to distinguish between the unconditional size and the conditional size of the test. The unconditional size of the test refers to the Type I error rate that would be obtained from multiple replications of performing the experiment repeatedly, starting each time at the very beginning of the entire procedure. The conditional size, on the other hand, refers to the Type I error rate for the particular covariate values as they were assigned to groups for the particular study that has already been conducted. Thus, the distinction is that the unconditional size reflects the Type I error rate that would be obtained over infinitely many randomizations of subjects to conditions, whereas the conditional size reflects the Type I error rate that would be obtained from repeating the experiment for the specific configuration of covariate values already obtained.

From the perspective of this distinction, randomization guarantees that the unconditional size of the test will be controlled appropriately as long as the requisite statistical assumptions have been met. The fundamental insight of the benefits of randomization is that across multiple replications, groups will be probabilistically equated on the average. However, such equating is not guaranteed for any single configuration of values on a baseline measure, which is what is called for in conditional size. If an ANOVA were performed on the dependent variable (simply ignoring the baseline variable), the conditional size would be inflated for some configurations of scores on the baseline variable. Whether this is a problem is a source of some debate. For example, although Permutt (1990) and Senn (1989) agree that covariates should not be chosen on the basis of group differences, they disagree about the importance of conditional size. Permutt (1990) argues that "If a clinician rejects the 'consolation of the marathon experimenter' that he makes a type I error only 5 percent of the time, I do not see why he should be comforted by making it only 5 percent of the time conditional on the values of certain covariates. In my view all type I errors result from unlucky randomization. Sometimes seeming evidence of such bad luck appears as imbalance in covariates that happen to have been measured, and sometimes not" (p. 1461). Senn (1989), on the other hand, states that ". . . it is also common experience that arguments of power do not satisfy clinicians who are unimpressed by justification in terms of averages over all experiments" (p. 467). Cox (1982) apparently takes a position similar to Senn's, as reflected in his statement that "It is no defence of a particular experiment which is seriously unbalanced to show that the realized design is atypical" (p. 199). Senn and Cox seem to believe that an optimal data analysis technique should control the Type I error rate, given all of the information available to the experimenter such as covariate imbalance. In any case, Senn shows that analysis of covariance can be used to control the conditional as well as the unconditional size of the test of group differences. Thus, for different reasons, Senn and Permutt arrive at the same conclusion, namely that ANCOVA should be used for relevant covariates that are identified on theoretical grounds prior to examination of the actual data.

Although it would be ideal to conclude that this strategy completely solves the problem of both unconditional and conditional size, the issue is in fact somewhat more complicated. Senn's demonstration that ANCOVA properly controls the conditional size of the test of group differences was based on an assumption that the proper underlying statistical model was used to control for baseline differences. For example, he implicitly assumed that the relationship between the dependent variable and the covariate was linear. However, if this assumption is in fact false but a linear relationship is nevertheless assumed, the conditional size will not necessarily be controlled.

This observation leads to another interesting question. Suppose a researcher has access to scores on a baseline measure for all subjects prior to the introduction of the planned intervention. Further suppose that subjects have been randomly assigned to groups, but that an examination of the group difference on the baseline variable shows that there is a statistically significant difference between the means of the two groups. Should the researcher simply ignore this difference and proceed as planned, based on the fact that as long as the baseline measure is used as a covariate (and the usual assumptions hold), both the unconditional and the conditional size of the test of group differences will be appropriately controlled? Or should the researcher deem this particular outcome of the randomization process to be unfair, repeating the process until a better balance is achieved?

Savage (1962) reports asking a similar question of Fisher. Regarding the possibility of randomly drawing a type of checkerboard pattern when selecting a Latin square, "Sir Ronald said he thought he would draw again and that, ideally, a theory explicitly excluding regular squares should be developed" (p. 88). The implicit conclusion that some outcomes of random assignment are better than others has caused some statisticians to conclude that randomization is not a crucial requirement in designing experiments. For example, Howson and Urbach (1989) state that ". . . the fact that in practically all randomized trials the experimental groups generated by chance are then carefully inspected to ensure that they are similar in relevant aspects suggests strongly that the randomization was not after all essential, and that the real aim in such trials is not to distribute nuisance variables at random, but to compare properly matched groups" (p. 152).

As a solution to this dilemma, Maxwell et al. (1984) point out that baseline variables can be used for two distinct but related purposes. First, baseline variables can be used as covariates in an ANCOVA in the analysis phase of a study. Second, such variables can also be used as a basis for stratifying subjects and then randomly assigning subjects to groups within strata. A basic premise of our chapter is that baseline measures are typically not incorporated into analyses in the behavioral sciences, unless possibly in the case in which there is demonstrated covariate imbalance. However, in our estimation, the use of baseline variables as a basis for assignment of subjects to groups is even more unusual in the behavioral sciences.

When a researcher has access to all subjects' scores on one or more baseline variables prior to introducing the intervention to any of the subjects, there are potentially two major advantages of incorporating the baseline measures into the process for assigning subjects to treatment conditions.

First, Cox (1957), Feldt (1958), and Maxwell et al. (1984) have shown that the precision of the estimated treatment effect can be enhanced by stratified random assignment of subjects to conditions. Using a baseline measure in this manner thus accomplishes the same basic goal as using a covariate in the analysis, namely to obtain a more efficient estimate and thereby increase the power to detect a treatment effect. Specifically, Bellhouse (1986) showed that balancing on baseline measures (i.e., equating group means on the measures) minimizes the variance of any linear contrast of adjusted means in the analysis of covariance.

Second, stratified random assignment reduces the dependence of proper control of conditional size with ANCOVA on the accuracy of model assumptions. To the extent that the various treatment groups have identical distributions on the baseline measures in question, the conditional size and the unconditional size will be the same, even without introducing covariates into the analysis. Thus, such influences as nonlinearity will be lessened to the extent that the groups are identically distributed on the baseline measures.

A number of methods are available for accomplishing stratified random assignment in the case of a single baseline measure. The randomized block design has historically provided a popular approach for carrying out stratified random assignment in the behavioral sciences. Blocking and ANCOVA are often regarded as competitors, because blocking provides a methodology for data analysis as well as experimental design. However, blocking and ANCOVA can also be viewed as complementary because blocking can be used to design the study while ANCOVA can be used to analyze the resultant data obtained from this design. Cox (1957), Feldt (1958), and Maxwell et al. (1984) have compared several methods of assignment combined with several methods of analysis. In general, differences in efficiency between the various methods are small, especially within the class of methods that use the baseline both for assignment and analysis. To consider a strategy that generally works well, consider a specific situation in which na subjects are to be assigned to a treatment groups. As a first step, the researcher should rank order all na subjects on the baseline measure. Then, the a highest ranking subjects should be formed into one block (or stratum), the a next highest ranking subjects into a second block, and so forth, until all subjects have been placed into a block. Then, within each block, each subject should be randomly assigned to one of the a treatment groups. Once all subjects have been assigned in this manner, the interventions themselves can be introduced. In the analysis phase of the study, the baseline measure can be used as a covariate in exactly in the same manner as if subjects had been assigned to groups irrespective of blocks. Cox (1982), Bellhouse (1986), and Press (1987) describe more elaborate strategies that can be employed with multiple baseline measures.

It is important to point out that alternatives to ANCOVA exist in terms of designing a randomized study to maximize the power to detect treatment effects. For example, suppose that a developmental psychologist plans to evaluate the effectiveness of an intervention for improving parenting skills in parent-infant interactions. Further suppose that the psychologist plans to employ a 5-minute behavioral assessment at posttest to obtain measures that can be used to compare the treatment group with a control group. One method of increasing the power to detect a treatment effect would be to use a similar 5-minute behavioral assessment as a pretest measure, which could then be incorporated into the design and/or the analysis of the experiment.

While the use of a pretest is likely to be a reasonable strategy, there may be a better way to accomplish the goal of increasing power. For example, a single 10-minute posttest assessment without any pretest measure might provide greater statistical power than the pretest-posttest design and at the same time impose less burden on both subjects and the experimenter. Maxwell, Cole, Arvey, and Salas (1991) have identified conditions under which the lengthened posttest design is either more or less powerful than the pretest-posttest design. In essence, the lengthened posttest design is likely to be more powerful than the pretest-posttest design in situations where the reliability of the dependent measure is low (.8 or less) and the correlation between the pretest and the posttest is low (.3 or less). On the other hand, when both the reliability and the pretest-posttest correlation are high, the pretest-posttest design is likely to yield more power than the lengthened posttest design. Of course, the choice of optimal design must involve more considerations than just statistical power, but the general message here is that researchers have a variety of methods at their disposal to increase power by carefully choosing not just how best to analyze their data but also how to design their study in the first place.

3.4 CONCLUDING REMARKS

The diversity of topics considered in this chapter underscores the vitality and wide applicability of the analysis of covariance, even when the range of topics is largely restricted to randomized experiments. At the very least, we hope we have convinced readers that covariates should not be regarded as nuisance variables whose only role in randomized experiments is to serve as a check on the randomization. As we have shown from numerous perspectives, covariates can serve an important role in randomized studies in many situations beyond those in which statistically significant group differences on the covariate happen to be found.

REFERENCES

Altman, D. G. (1985). Comparability of Randomized Groups. *Statist., 34*: 125–136.

Bellhouse, D. R. (1986). Randomization in the Analysis of Covariance. *Biometrika, 73*: 207–211.

Bingham, C. and Fienberg, S. E. (1982). Textbook Analysis of Covariance—Is It Correct? *Biometrics, 38*: 747–753.

Blair, C. (1981). A Reaction to 'Consequences of Failure to Meet Assumptions Underlying the Fixed Effects Analysis of Variance and Covariance'. *Rev. Educ. Res., 51*: 499–508.

Boyd, M. N. and Sen, P. K. (1986). Union-Intersection Rank Tests for Ordered Alternatives in ANOCOVA. *J. Amer. Statist. Assoc., 81*: 526–532.

Burnett, T. D. and Barr, D. R. (1977). A Nonparametric Analogy of Analysis of Covariance. *Educ. Psych. Meas., 37*: 341–348.

Campbell, D. T. and Boruch, R. F. (1975). Making the Case for Randomized Assignment to Treatments by Considering the Alternatives: Six Ways in Which Quasi-Experimental Evaluations Tend to Underestimate Effects. *Evaluation and Experiment: Some Critical Issues in Assessing Social Programs.* Bennett, C. A. and Lumsdaine, A. A. (eds.). Academic Press, New York, pp. 195–296.

Campbell, D. T. and Erlebacher, A. E. (1970). How Regression Artifacts in Quasi-Experimental Evaluations Can Mistakenly Make Compensatory Education Look Harmful. *Compensatory Education: A National Debate*, Vol. 3, *Disadvantaged Child.* Hellmuth, J. (ed.). Brunner/Mazel, New York, pp. 185–210.

Cliff, N. (1987). *Analyzing Multivariate Data.* Harcourt Brace Jovanovich, San Diego, California.

Cochran, W. G. (1957). Analysis of Covariance: Its Nature and Uses. *Biometrics, 13*: 261–281.

Cohen, J. (1962). The Statistical Power of Abnormal-Social Psychological Research: A Review. *J. Abn. Soc. Psych., 65*: 145–153.

Cohen, J. (1988). *Statistical Power Analysis for the Behavioral Sciences* (2nd ed.). Erlbaum, Hillsdale, New Jersey.

Conover, W. J. and Iman, R. L. (1982). Analysis of Covariance Using the Rank Transformation. *Biometrics, 38*: 715–724.

Cox, D. R. (1957). The Use of a Concomitant Variable in Selecting an Experimental Design. *Biometrika, 44*: 150–158.

Cox, D. R. (1982). Randomization and Concomitant Variables in the Design of Experiments. *Statistics and Probability: Essays in Honor of C. R. Rao.* Kallianpur, G., Krishnaiah, P. R., and Ghosh, J. K. (eds.). North-Holland, New York, pp. 197–202.

Cronbach, L. J. (1957). The Two Disciplines of Scientific Psychology. *Amer. Psych., 12*: 671–684.

Cronbach, L. J. and Furby, L. (1970). How Should We Measure Change—or Should We? *Psych. Bull., 74*: 68–80.

Cronbach, L. J. and Snow, R. E. (1977). *Aptitudes and Instructional Methods: A Handbook for Research on Interactions.* Irvington, New York.

Cronbach, L. J., Rogosa, D. R., Floden, R. E., and Price, G. G. (1977). Analysis of Covariance in Nonrandomized Experiments: Parameters Affecting Bias. Occasional Paper, Stanford University, Stanford Evaluation Consortium, Berkeley, California.

Delaney, H. D. and Maxwell, S. E. (1980a). The Use of Analysis of Covariance in Tests of Attribute-by-Treatment Interactions. *J. Educ. Statist.*, *5*: 191–207.

Delaney, H. D. and Maxwell, S. E. (1980b). Significant Attribute-by-Treatment Interactions: The Importance of Considering Multiple Attributes. *JSAS Cat. Sel. Doc. Psych.*, *10*: 67–68.

Delaney, H. D. and Maxwell, S. E. (1982). On Using Analysis of Covariance in Repeated Measures Designs. *Multivar. Behav. Res.*, *16*: 105–123.

Elashoff, J. D. (1969). Analysis of Covariance: A Delicate Instrument. *Amer. Educ. Res. J.*, *6*: 383–401.

Feldt, L. S. (1958). A Comparison of the Precision of Three Experimental Designs Employing a Concomitant Variable. *Psychometrika, 23*: 335–353.

Finney, D. J. (1982). Discussion on the Paper by Professor Cox and Dr. McCullagh, *Biometrics, 38*: 557–559.

Fisher, R. A. (1932). *Statistical Methods for Research Workers*. Oliver and Boyd, Edinburgh.

Fisher, R. A. (1941). *Statistical Methods for Research Workers* (8th ed.). Y. E. Stechert, New York.

Fisher, R. A. (1958). *Statistical Methods for Research Workers* (13th ed.). Hafner, New York.

Francom, S. F. and Chaung-Stein, C. (1989). A Log-Linear Model for Ordinal Data to Characterize Differential Change Among Treatments. *Statist. Med.*, *8*: 571–582.

Glass, G. V., Peckham, P. D., and Sanders, J. R. (1972). Consequences of Failure to Meet Assumptions Underlying the Analysis of Variance and Covariance. *Rev. Educ. Res.*, *42*: 237–288.

Griffey, D. C. (1982). Alternate Approaches to Analysis of Covariance: Nonsimultaneous and Simultaneous Inference. *Res. Quart. Exer. Sport, 53*: 20–26.

Hamilton, B. L. (1977). An Empirical Investigation of the Effects of Heterogeneous Regression Slopes in Analysis of Covariance. *Educ. Psych. Meas.*, *37*: 701–712.

Harwell, M. R. and Serlin, R. C. (1988). An Empirical Study of a Proposed Test of Nonparametric Analysis of Covariance. *Psych. Bull.*, *104*: 268–281.

Henderson, C. R., Jr. (1982). Analysis of Covariance in the Mixed Model: Higher-Level, Nonhomogeneous and Random Regression. *Biometrics, 38*: 623–640.

Hendrix, L. J., Carter, M. W., and Scott, D. T. (1982). Covariance Analyses with Heterogeneity of Slopes in Fixed Models. *Biometrics, 38*: 641–650.

Hettmansperger, T. P. (1984). *Statistical Inference Based on Ranks*. Wiley, New York.

Howson, C. and Urbach, P. (1989). *Scientific Reasoning: The Bayesian Approach*. Open Court, LaSalle, Illinois.

Huitema, B. E. (1980). *The Analysis of Covariance and Alternatives*. Wiley, New York.

Johnson, P. O. and Fay, L. C. (1950). The Johnson-Neyman Technique: Its Theory and Application. *Psychometrika, 15*: 349–367.

Johnson, P. O. and Neyman, J. (1936). Tests of Certain Linear Hypotheses and Their Application to Some Educational Problems. *Statist. Res. Mem., 1*: 57–93.

Kenny, D. A. (1975). A Quasi-Experimental Approach to Assessing Treatment Effects in the Nonequivalent Control Group Design. *Psych. Bull., 82*: 345–362.

Keppel, G. (1982). *Design and Analysis: A Researcher's Handbook* (2nd ed.). Prentice-Hall, Englewood Cliffs, New Jersey.

Kirk, R. E. (1968). *Experimental Design: Procedures for the Behavioral Sciences.* Brooks/Cole, Monterey, California.

Kirk, R. E. (1982). *Experimental Design: Procedures for the Behavioral Sciences* (2nd ed.). Brooks/Cole, Monterey, California.

Koch, G. G., Amara, I. A., Davis, G. W., and Gillings, D. B. (1982). A Review of Some Statistical Methods for Covariance Analysis of Categorical Data. *Biometrics, 38*: 563–595.

Lautenschlager, G. J. (1987). JOHN-NEY: An Interactive Program for Computing the Johnson-Neyman Confidence Region for Nonsignificant Prediction Differences. *Appl. Psych. Meas., 11*: 174–194.

Levin, J. R. and Marascuilo, L. A. (1972). Type IV Errors and Interactions. *Psych. Bull., 78*: 368–374.

Levy, K. J. (1980). A Monte Carlo Study of Analysis of Covariance Under Violations of the Assumptions of Normality and Equal Regression Slopes. *Educ. Psych. Meas., 40*: 835–840.

Llabre, M. M. and Ware, W. B. (1980). Equal Cell Size and Nonorthogonality in ANCOVA. *Educ. Psych. Meas., 40*: 91–94.

Marcus, R. and Genizi, A. (1987). Nonparametric Analysis of Covariance with Ordered Alternatives. *J. Roy. Statist. Soc., B49*: 102–111.

Maxwell, S. E., Cole, D. A., Arvey, R. D., and Salas, E. (1991). A Comparison of Methods for Increasing Power in Randomized Between-Subjects Designs. *Psych. Bull., 110*: 328–337.

Maxwell, S. E. and Delaney, H. D. (1990). *Designing Experiments and Analyzing Data: A Model Comparison Perspective.* Wadsworth, Belmont, California.

Maxwell, S. E., Delaney, H. D., and Dill, C. A. (1984). Another Look at ANCOVA Versus Blocking. *Psych. Bull., 95*: 136–147.

Maxwell, S. E., Delaney, H. D., and Manheimer, J. M. (1985). ANOVA of Residuals and ANCOVA: Correcting an Illusion by Using Model Comparisons and Graphs. *J. Educ. Statist., 10*: 197–209.

McSweeney, M. and Porter, A. C. (1971). Small Sample Properties of Nonparametric Index of Response and Rank Analysis of Covariance. Occasional Paper No. 16. Office of Research Consultation. Michigan State University, East Lansing, Michigan.

Neter, J., Wasserman, W., and Kutner, M. H. (1985). *Applied Linear Statistical Models: Regression, Analysis of Variance, and Experimental Designs.* Richard D. Irwin, Homewood, Illinois.

Olejnik, S. F. and Algina, J. (1985). A Review of Nonparametric Alternatives to Analysis of Covariance. *Eval. Rev., 9:* 51–83.

Olejnik, S. F. and Algina, J. (1987). An Analysis of Statistical Power for Parametric ANCOVA and Rank Transform ANCOVA. *Comm. Statist.—Theor. Meth., A16*: 1923–1949.

Penfield, D. A. and Koffler, S. L. (1986). A Nonparametric k-Sample Test for Equality of Slopes. *Educ. Psych. Meas., 46*: 537–542.

Permutt, T. (1990). Testing for Imbalance of Covariates in Controlled Experiments. *Statist. Med., 9*: 1455–1462.

Pollane, L. P. and Schnittjer, C. J. (1977). The Relative Performance of Five Computer Program Packages Which Perform Factorial Univariate Analysis of Covariance. *Educ. Psych. Meas., 37*: 227–231.

Porter, A. C. and Raudenbush, S. W. (1987). Analysis of Covariance: Its Model and Use in Psychological Research. *J. Couns. Psych., 34*: 383–392.

Potthoff, R. F. (1964). On the Johnson-Neyman Technique and Some Extensions Thereof. *Psychometrika, 29*: 241–256.

Press, S. J. (1987). The MISER Criterion for Imbalance in the Analysis of Covariance. *J. Statist. Plan. Inf., 17*: 375–388.

Puri, M. L. and Sen, P. K. (1969). Analysis of Covariance Based on General Rank Scores. *Ann. Math. Statist., 40*: 610–618.

Quade, D. (1967). Rank Analysis of Covariance. *J. Amer. Statist. Assoc., 62*: 1187–1200.

Quade, D. (1982). Nonparametric Analysis of Covariance by Matching. *Biometrics, 38*: 597–611.

Reichardt, C. S. (1979). The Statistical Analysis of Data from Nonequivalent Group Designs. *Quasi-Experimentation: Design and Analysis Issues for Field Settings.* Cook, T. D. and Campbell, D. T. (eds.). Rand McNally, Chicago, pp. 147–205.

Reid, M. K. and Borkowski, J. G. (1987). Causal Attributions of Hyperactive Children: Implications for Teaching Strategies and Self-Control. *J. Educ. Psych., 79*: 296–307.

Rogosa, D. R. (1980). Comparing Non-Parallel Regression Lines. *Psych. Bull, 88*: 307–321.

Roy, S. N. (1953). On a Heuristic Method of Test Construction and Its Use in Multivariate Analysis. *Ann. Math. Statist., 24*: 220–238.

Savage, L. J. (1962). A Prepared Contribution to the Discussion of Savage. *The Foundations of Statistical Inference: A Discussion.* Barnard, G. A. and Cox, D. R. (eds.). Wiley, New York, pp. 88–89.

Sawyers, P. (1991). Optimal Matching of Clients to Alcoholism Treatments: Post-Hoc Analyses of the Community Reinforcement Approach Replication Study. Unpublished Master's Thesis, University of New Mexico.

Schafer, W. (1991). Graphical Description of Johnson-Neyman Outcomes for Linear and Quadratic Regression Surfaces. Paper Presented at the Annual Meeting of the American Educational Research Association (Chicago, April).

Scheffé, H. (1959). *The Analysis of Variance.* Wiley, New York.

Schluchter, M. D. and Forsythe, A. B. (1985). Post-Hoc Selection of Covariates in Randomized Experiments. *Comm. Statist.—Theor. Meth., A14*: 679–699.

Seaman, S. L., Algina, J., and Olejnik, S. F. (1985). Type I Error Probabilities and Power of the Rank and Parametric ANCOVA Procedures. *J. Educ. Statist., 10*: 345–367.

Searle, S. R. (1979). Alternative Covariance Models for the 2-Way Crossed Classification. *Comm. Statist.—Theor. Meth., A8*: 799–818.

Senn, S. J. (1989). Covariate Imbalance and Random Allocation in Clinical Trials. *Statist. Med., 8*: 467–475.

Shirley, E. A. (1981). A Distribution-Free Method for Analysis of Covariance Based on Ranked Data. *J. Appl. Statist., 30*: 158–162.

Smith, F. W. (1957). Interpretation of Adjusted Treatment Means and Regressions in Analysis of Covariance. *Biometrics, 13*: 282–307.

Theil, H. (1971). *Principles of Econometrics*. Wiley, New York.

Winer, B. J. (1971). *Statistical Principles in Experimental Design* (2nd ed.). McGraw-Hill, New York.

4

Analysis of Repeated Measurements

H. J. KESELMAN and JOANNE C. KESELMAN University of
Manitoba, Winnipeg, Manitoba, Canada

4.1 INTRODUCTION

The repeated measures design is a very frequently used paradigm in educational and psychological research (Barcikowski and Robey, 1984; Edgington, 1974; Looney and Stanley, 1989). In a typical repeated measures design, subjects are selected randomly for each combination of the between-subjects factors (when the design contains at least one between-subjects factor) and are exposed to each combination of the within-subjects factors (Winer, 1971). Researchers typically adopt the repeated measures design as a means of reducing error variability and/or as the natural way of measuring certain phenomenon (e.g., developmental changes over time, learning and memory tasks). In such designs, treatment effects for a given subject are measured relative to the average response made by that subject on all treatments. In essence, each subject serves as his or her own control and, accordingly, variability due to differences in the average responsiveness of the subjects is eliminated from the extraneous error variance (Winer, 1971). As a consequence, the power to detect true within-subjects treatment effects is typically larger than would be the case were these effects to be tested in a between-subjects design.

In addition to the relative power advantage of the repeated measures design, because each subject is measured at each level of the within-subjects factor(s), fewer subjects are required in these designs in comparison to between-subjects designs. In this regard, repeated measures designs are advantageous, particularly when it is difficult and/or expensive to acquire subjects.

On the other hand, repeated measures designs also possess a number of potential disadvantages, the most serious being differential carryover, the effects of which can result in biased estimates of the within-subjects treatment effects. As a consequence, certain factors, specifically those which are likely to have effects that persist over time, are generally more accurately assessed in between-subjects designs (Maxwell and Delaney, 1990).

Despite these potential disadvantages, the repeated measures design remains a popular design among researchers because of its sensitivity and efficiency. Accordingly, in recent years the literature has focused on the proper analysis strategy for such designs (Barcikowski and Robey, 1984; Collier, Baker, Mandeville, and Hayes, 1967; Davidson, 1972; Huynh, 1978; Huynh and Feldt, 1976; Keselman, 1982, 1991; Keselman, Keselman, and Shaffer, 1991; Lewis and van Knippenberg, 1984; Looney and Stanley, 1989; Maxwell and Arvey, 1982; McCall and Appelbaum, 1973; Mendoza, Toothaker, and Nicewander, 1974; O'Brien and Kaiser, 1985; Rogan, Keselman, and Mendoza, 1979; Stoloff, 1970; Wallenstein and Fleiss, 1979). Essentially, the data from repeated measures designs can be analyzed by either a univariate or multivariate approach. The valid use of either of these approaches, however, rests on the condition that the data conform to the derivational assumptions of these procedures. Unfortunately, this is seldom the case in educational and psychological research (Davidson, 1972; Wilson, R., 1975; Greenwald, 1976).

The purpose of this chapter, therefore, is to present analysis strategies for repeated measures designs that generally are robust, that is, insensitive to assumption violations. An important feature of our presentation is a discussion of analysis procedures for unbalanced higher-order repeated measures designs, that is, designs in which the between-subjects group sizes are unequal. These designs warrant special consideration as it is well known that the effects of certain assumption violations (e.g., variance heterogeneity) are exacerbated in the presence of unequal group sizes.

Our discussion of analysis strategies for repeated measures designs is restricted to designs in which all factors, except for subjects, are considered to be fixed. Our experience indicates that this configuration typifies repeated measures designs used by educational and psychological researchers. For a discussion of repeated measures designs involving random factors, readers are referred to Kirk (1982, pp. 247, 499) and Maxwell and Delaney (1990, pp. 422–433).

4.2 THE SIMPLE REPEATED MEASURES DESIGN: NO BETWEEN-SUBJECTS AND ONE WITHIN-SUBJECTS FACTOR

Consider the simplest repeated measures design in which a single group of subjects ($i = 1, \ldots, N$) is exposed to and observed under all levels of a single treatment variable ($k = 1, \ldots, B$). In such designs, the repeated measures data are modeled by assuming that the observations $X_{ik}, i = 1, \ldots, N, k = 1, \ldots, B$, are normal, independent, and identically distributed, with common mean vector μ and covariance matrix Σ.

4.2.1 The Univariate Approach

Traditionally, a test of the within-subjects hypothesis that the B population means are equal (H_0: $\mu_1 = \mu_2 = \ldots \mu_B$) has been accomplished by use of the univariate within-subjects F statistic

$$F_B = \frac{MS_B}{MS_{N \times B}} \sim F[\alpha; (B-1), (N-1)(B-1)]. \tag{4.1}$$

For F_B to provide a valid or exact test of the null hypothesis, the data must satisfy three assumptions: (1) normality, (2) independence of errors, and (3) circularity or sphericity. Sphericity is satisfied if and only if $C^T \Sigma C = \lambda I_{(B-1)}$, where C is a normalized matrix of $B - 1$ orthogonal contrasts among the B repeated measurements, C^T is its transpose, Σ is the population covariance matrix, λ is a scalar > 0, and I is an identity matrix of order $(B - 1)$. More specifically, the sphericity condition is met when the product $C^T \Sigma C$ yields a matrix in which all the diagonal and off-diagonal elements are equal to λ and zero, respectively. As the diagonal and off-diagonal elements of $C^T \Sigma C$ equal the variance and covariance of the $B - 1$ orthogonal contrasts, the sphericity assumption is satisfied if and only if the $B - 1$ contrasts are independent and equally variable. For a comprehensive discussion of the sphericity assumption, readers are referred to Huynh and Feldt (1970) and Rouanet and Lépine (1970).

An equivalent way of stating the sphericity assumption is that the treatment-difference variances are homogeneous. Specifically, for any two treatment conditions, k and k', the variance of the difference scores is equal to

$$\sigma_k^2 + \sigma_{k'}^2 - 2 \operatorname{Cov}(X_k, X_{k'}) = \sigma_k^2 + \sigma_{k'}^2 - 2\rho_{kk'}\sigma_k\sigma_{k'}, \tag{4.2}$$

where σ_k^2 and $\sigma_{k'}^2$ are the population variances for treatments k and k', respectively, and $\rho_{kk'}$ is the population correlation coefficient between the scores in treatments k and k'. In this formulation, sphericity requires that the variances of the $\binom{B}{2}$ treatment differences be equal.

The effects of violating the sphericity assumption were first postulated by Kogan (1948), who suggested that positive intercorrelations between measurements would result in a liberal F test of the repeated measures effect. Box (1954) and Imhof (1962) found Kogan's conjecture to be correct; that is, as the degree of nonsphericity increased, the traditional within-subjects F test became increasingly liberal.

4.2.1.1 Adjusted Univariate Procedures

In order to circumvent the problems resulting from violating the sphericity assumption, various adjusted univariate approaches have been suggested. These approaches are based on the work of Box, who demonstrated that, under the null hypothesis, F_B is approximately distributed as an F variable with $(B - 1)\varepsilon$ and $(B - 1)(N - 1)\varepsilon$ degrees of freedom (d.f.). The parameter ε is an index of sphericity and equals

$$\varepsilon = \frac{[\mathrm{tr}(\mathbf{C}^T\mathbf{\Sigma}\mathbf{C})]^2}{(B - 1)\mathrm{tr}[(\mathbf{C}^T\mathbf{\Sigma}\mathbf{C})]^2}, \tag{4.3}$$

where tr refers to the trace operator and $\mathbf{C}$, $\mathbf{\Sigma}$, and B are as previously defined. A traditional definition of ε can be found in Maxwell and Delaney (1990, p. 476). When the sphericity assumption is satisfied, $\varepsilon = 1.0$, and F_B provides an exact test of the repeated measures null hypothesis. When $\mathbf{\Sigma}$ departs from sphericity, the value of ε decreases from one to a lower bound of $(B - 1)^{-1}$ and F_B is approximately distributed with a reduced d.f.

The adjusted univariate approaches described in this section involve (1) computing a sample estimate of the parameter ε and (2) adjusting, that is, multiplying the numerator and denominator d.f. of F_B by this sample estimate.

The $\hat{\varepsilon}$-Adjusted Univariate Procedure

In order to combat the effects of nonsphericity, Greenhouse and Geisser (1959) recommended adjusting the traditional d.f. by the sample estimate $\hat{\varepsilon}$ where

$$\hat{\varepsilon} = \frac{[\mathrm{tr}(\mathbf{C}^T\mathbf{S}\,\mathbf{C})]^2}{(B - 1)\mathrm{tr}[(\mathbf{C}^T\mathbf{S}\,\mathbf{C})]^2}, \tag{4.4}$$

with $\mathbf{S}$ denoting the sample covariance matrix. Thus, according to this procedure,

$$F_B \sim F[\alpha; (B - 1)\hat{\varepsilon}, (N - 1)(B - 1)\hat{\varepsilon}]. \tag{4.5}$$

The ε̄-Adjusted Univariate Procedure

Based on survey results which indicated that ε values greater than 0.75 are most typical of educational and psychological data and coupled with the findings of Collier et al. (1967) that under such conditions ε̂ may be seriously biased, particularly when sample size is small, Huynh and Feldt (1976) proposed that an alternate estimate of ε, ε̄, be used to adjust the d.f. of F_B, where

$$\bar{\varepsilon} = \frac{N(B - 1)\hat{\varepsilon} - 2}{(B - 1)[N - 1 - (B - 1)\hat{\varepsilon}]}. \tag{4.6}$$

Huynh and Feldt (1976) have shown that for any value of N and B, $\bar{\varepsilon} \geq \hat{\varepsilon}$ with the equality holding when $\hat{\varepsilon} = (B - 1)^{-1}$. In addition, unlike $\hat{\varepsilon}$, values of $\bar{\varepsilon}$ can exceed unity. However, because ε cannot assume a value larger than 1.0, whenever $\bar{\varepsilon}$ exceeds one, the estimate is equated to one. According to the results of Huynh and Feldt's (1976) sampling study, $\bar{\varepsilon}$ is less biased and less dependent on sample size than $\hat{\varepsilon}$ when the departure from sphericity is moderate. Applying this procedure,

$$F_B \sim F[\alpha; (B - 1)\bar{\varepsilon}, (N - 1)(B - 1)\bar{\varepsilon}]. \tag{4.7}$$

It should be noted that sample estimates of sphericity are provided as output by all three major statistical packages: BMDP (Dixon, Brown, Engelman, Hill, and Jennrich, 1988, 1990), SAS (SAS Institute, 1990a,b,c), and SPSS (SPSS Inc., 1990).

4.2.2 The Multivariate Approach

As mentioned earlier, the data from repeated measures designs can also be analyzed using multivariate methods. Indeed, Cole and Grizzle (1966) argued that when the observations on successive measurements are correlated, the data, while arising from an apparently univariate design, are essentially multivariate in nature and should be analyzed as such.

Briefly, multivariate methods arrive at tests of the within-subjects effect by transforming the B repeated measurements associated with each subject into a set of $B - 1$ difference scores (e.g., $D_i = X_{ik} - X_{ik'}, k \neq k'$), with the only restriction being that the $B - 1$ comparisons underlying these difference scores are linearly independent. When $B > 2$ (and therefore $B - 1 > 1$), this transformation yields a multivariate problem because there will be more than one difference score for each subject. (When $B = 2$, $B - 1 = 1$ and the univariate and multivariate approaches are equivalent.) A test of the within-subjects effect is accomplished by testing the null hypothesis that the $B - 1$ difference scores have a population mean vector equal to the null vector, that is,

$$H_0: \begin{bmatrix} \mu_1 - \mu_B \\ \vdots \\ \mu_{B-1} - \mu_B \end{bmatrix} = \begin{bmatrix} 0 \\ \vdots \\ 0 \end{bmatrix}. \tag{4.8}$$

This hypothesis, which is equivalent to that tested by the univariate approach, i.e., $H_0: \mu_1 = \mu_2 \cdots = \mu_B$, is assessed using Hotelling's (1931) T^2 statistic, where

$$T^2 = N\overline{\mathbf{D}}^T\mathbf{S}_D^{-1}\overline{\mathbf{D}}, \tag{4.9}$$

$\overline{\mathbf{D}}$ is a vector of the $B - 1$ mean difference scores and $\mathbf{S}_D$ is the sample covariance matrix of these difference scores. Upper $100(1 - \alpha)$ percentage points of Hotelling's (1931) T^2 distribution can be obtained from the relationship

$$F = \frac{N - B + 1}{(N - 1)(B - 1)} T^2 \sim F[\alpha; B - 1, N - B + 1]. \tag{4.10}$$

For detailed discussions of the multivariate approach to the analysis of repeated measures designs, the reader is referred to Maxwell and Delaney (1990, pp. 552–611), Scheffé (1959, pp. 270–274), and Timm (1975, pp. 229–237).

As with the univariate approach, the validity of the multivariate approach rests on the assumption of normality. Unlike the univariate approach, however, an exact multivariate test of the within-subjects effect does not require that the data satisfy the sphericity assumption.

4.2.3   A Comparison of Analysis Strategies

The existing literature indicates that the adjusted univariate procedures are generally robust to violations of the sphericity assumption (Collier et al., 1967; Huynh, 1978; Huynh and Feldt, 1976; Keselman and Keselman, 1990; Mendoza et al., 1974; Noe, 1976; Rogan et al., 1979; Stoloff, 1970; Wilson, K., 1975). For moderate departures from sphericity ($\varepsilon > 0.75$), the $\tilde{\varepsilon}$-adjusted procedure is the better of the adjusted univariate procedures as its empirical Type I error rates are closer to α, the nominal significance level. When $\varepsilon \leq 0.75$, the reverse is the case.

With respect to the assumption of normality, both the adjusted univariate and multivariate approaches are generally robust to departures from this assumption, with the multivariate approach being slightly more sensitive to nonnormality than the adjusted univariate procedures (Rogan et al., 1979).

On the issue of test sensitivity, when the sphericity assumption is satisfied, the traditional univariate approach is uniformly more powerful than the multivariate approach. In the more common situation in which the sphericity assumption is not met, neither an adjusted univariate nor the multivariate approach is uniformly more powerful. Rather, for a given sample size and significance level, the relative power of these approaches is dependent on the nature of the alternative hypothesis, the variance-covariance structure of the data, and the relationship between these two factors (Davidson, 1972).

4.2.4   Multiple Comparison Procedures

4.2.4.1   Simultaneous Multiple Comparison Procedures

Pairwise and/or complex comparisons among the repeated measures means are traditionally carried out using the statistic

$$t = \frac{(\hat{\psi} - \Psi)}{\sqrt{MS_{N \times B} (\Sigma_k c_k^2/N)}}, \tag{4.11}$$

which is distributed as a t variable with degrees of freedom, ν, equal to $(N - 1)(B - 1)$. $\hat{\psi} = \Sigma_k c_k \overline{X}_k$ and $\Psi = \Sigma_k c_k \mu_k$ are the sample and population contrast values, respectively, $\overline{X}_k$ and μ_k are the sample and population means for treatment level k, and c_k is the contrast coefficient, subject to the restriction that $\Sigma_k c_k = 0$.

To test a set of pairwise contrasts, the traditional approach for maintaining the maximum familywise Type I error rate (MFWER) at α is to employ a Tukey (1953) critical value, $q[\alpha; B, (N - 1)(B - 1)]/\sqrt{2}$, where q_α is a $100(1 - \alpha)$ percentile point of the studentized range distribution with parameters B and $\nu = (N - 1)(B - 1)$. For a set of nonpairwise contrasts, the MFWER typically has been protected with a Scheffé critical value, $\{(B - 1) F[\alpha; (B - 1), (N - 1)(B - 1)]\}^{1/2}$, where F_α is the $100(1 - \alpha)$ percentile point of the sampling distribution of F with $\nu_1 = (B - 1)$ and $\nu_2 = (N - 1)(B - 1)$.

Unfortunately, the statistic in (4.11) does not provide a valid test of the multiple comparison hypothesis when the data do not meet the sphericity assumption. Indeed, Boik (1981) has shown that contrast tests are even more sensitive than their omnibus counterparts to violations of this assumption. Thus, the traditional multiple comparison procedures (MCPs) which use the statistic in (4.11) should not be employed with educational and psychological research data as these data rarely conform to the requirements of sphericity.

To understand why the MCPs which use the statistic in (4.11) rest on the sphericity assumption, consider a single contrast j among B repeated measures means that is represented by the coefficients, c_{jk}, in a single column of the previously defined contrast matrix C. Let $\Psi_j = \Sigma_k c_{jk}\mu_k = \mathbf{c}_j^T\boldsymbol{\mu}$ define a population contrast, where $\mathbf{c}_j$ is a vector of the B contrast coefficients such that $\Sigma_k c_{jk} = 0$ and $\Sigma_k c_{jk}^2 = 1$ and $\boldsymbol{\mu}$ is a vector of length B of the μ_ks. Let $\hat{\psi}$ denote the corresponding sample value of the contrast j, where $\hat{\psi}_j = \Sigma_k c_{jk}\overline{X}_k = \mathbf{c}_j^T\overline{\mathbf{X}}$, where $\overline{\mathbf{X}}$ is the vector of the B sample means, $\overline{X}_k$. Finally, let $\sigma^2(\hat{\psi}_j)$ refer to the variance of $\hat{\psi}_j$.

Since $\hat{\psi}_j = \mathbf{c}_j^T\overline{\mathbf{X}}$, using the theory of linear combinations, it is possible to show that

$$\sigma^2(\hat{\psi}_j) = (1/N)\mathbf{c}_j^T\boldsymbol{\Sigma}\,\mathbf{c}_j. \tag{4.12}$$

These values are exactly $(1/N)$ times the diagonal values of the matrix product $\mathbf{C}^T\boldsymbol{\Sigma}\,\mathbf{C}$. For the sphericity assumption to be satisfied, therefore, the variances of each of the $B - 1$ orthonormalized contrasts must equal the constant, λ/N. When this assumption is not met, the $\sigma^2(\hat{\psi}_j)$ values are not equal and will vary from contrast to contrast. Unfortunately, this fact is not reflected in the test statistic in (4.11) which uses a constant estimate of the variance for any contrast j, that is,

$$\sigma^2(\hat{\psi}.) = MS_{N \times B} \left(\frac{\Sigma_k c_{jk}^2}{N} \right). \tag{4.13}$$

When the sphericity assumption is not met, $\sigma^2(\hat{\psi}.)$ will be too large for the contrasts with small $\sigma^2(\hat{\psi}_j)$, and too small for the contrasts with large $\sigma^2(\hat{\psi}_j)$, resulting in conservative and liberal tests, respectively. To circumvent this problem, individual estimates of $\sigma^2(\hat{\psi}_j)$ that vary from contrast to contrast should be used. These estimates are easily obtained by substituting $\mathbf{S}$ for $\boldsymbol{\Sigma}$ in (4.12). Using this approach, a robust multiple comparison test statistic is

$$t = \frac{(\hat{\psi}_j - \Psi)}{\sqrt{\mathbf{c}_j^T\mathbf{S}\mathbf{c}_j/N}}, \tag{4.14}$$

which is distributed as a t variable with $\nu = (N - 1)$. For pairwise comparisons, Eq. (4.14) can be expressed simply as

$$t = \frac{(\overline{X}_k - \overline{X}_{k'})}{\sqrt{(s_k^2 + s_{k'}^2 - 2s_{kk'})/N}}, \tag{4.15}$$

where s_k^2 and $s_{k'}^2$ are the unbiased sample variances for treatment levels k and k', and $s_{kk'}$ is the unbiased sample covariance between treatments k and k'.

In a multivariate approach to contrast testing, one defines a set of transformed D variables, where each D variable represents a contrast of interest (e.g., $D_i = X_{ik} - X_{ik'}$, $k \neq k'$) and computes Hotelling's (1931) T^2 statistic, as shown in (4.9), on each of the D variables. Using this approach, a contrast test is a univariate test of a given D variable and, in fact, is equivalent to the univariate statistic given in (4.14). Accordingly, the multivariate approach uses an individual rather than a common estimate of the variance of a contrast and therefore does not rest on the sphericity assumption.

The literature indicates that for pairwise contrasts among repeated measures means, MFWER protection can be achieved by using the statistic given in (4.14) or (4.15) with either a studentized maximum modulus or a Bonferroni critical value (Alberton and Hochberg, 1984; Maxwell, 1980). For a simple repeated measures design, the studentized maximum modulus critical value is $M[\alpha; c, \nu]$, where M is the upper $100(1 - \alpha)$ percentile point of the studentized maximum modulus distribution with parameters $c = \binom{B}{2}$ and $\nu = (N - 1)$. Tables of M can be found in Hochberg and Tamhane (1987) and Maxwell and Delaney (1990). The Bonferroni critical value is $t[\alpha/(2c); \nu]$, where t is the upper $100(1 - \alpha/(2c))$ percentile point of Student's t distribution with $\nu = (N - 1)$. This critical value is slightly larger than the studentized maximum modulus critical value and therefore will provide a slightly more conservative test. Note that the Bonferroni procedure can also be applied to nonpairwise contrasts using the statistic given in (4.14), where c would equal the number of contrasts being subjected to a test of significance.

4.2.4.2 Stepwise Multiple Comparison Procedures

Another class of MCPs that can be used to test pairwise contrast hypotheses are stepwise MCPs. Unlike simultaneous MCPs (e.g., Dunn, 1961; Scheffé, 1959; Tukey, 1953), which use a constant critical value to assess statistical significance, stepwise MCPs involve a succession of testing stages in which the significance criterion is adjusted throughout the stages. With respect to tests of pairwise contrasts of repeated measures means, Keselman (1991) reported that several of these strategies can be used to limit the MFWER to α for repeated measures data that do not meet the sphericity assumption. For a detailed discussion of stepwise MCPs for repeated measures designs, the reader is referred to Keselman (1991). Proofs of the stepwise strategies discussed below can be found in the original articles.

Hochberg's (1988) Step-up Bonferroni Procedure

In this procedure, the p values corresponding to the c statistics for testing hypotheses $H_1, \ldots, H_c$ are ordered from smallest to largest, i.e., $p_1 \leq p_2 \leq, \ldots, \leq p_c$, where $c = \binom{B}{2}$ for pairwise contrasts. Then, for any

$j = c, c - 1, \ldots, 1$, if $p_j \leq \alpha/(c - j + 1)$ the Hochberg procedure rejects all $H_{j'}$ ($j' \leq j$). According to this procedure, therefore, one begins by examining the largest p value, p_c. If $p_c \leq \alpha$, all hypotheses are rejected. If $p_c > \alpha$, then H_c is retained and one proceeds to compare $p_{(c-1)}$ to $\alpha/2$. If $p_{(c-1)} \leq \alpha/2$, then all H_j ($j = c - 1, \ldots, 1$) are rejected. If not, then $H_{(c-1)}$ is retained and one proceeds to compare $p_{(c-2)}$ with $\alpha/3$, and so on. For tests of pairwise contrasts of repeated measures means in designs in which the data do not meet the sphericity assumption, the test statistics are computed using the formula given in (4.14), in order to limit the MFWER to α.

A Multiple Range Procedure

One of the most popular stepwise strategies for examining pairwise differences between means is that due to Newman (1939) and Keuls (1952). In this procedure, the means are rank ordered from smallest to largest, and the difference between the smallest and largest means is first subjected to an α-level test of significance, typically with a range statistic. If this difference is not significant, testing stops and all pairwise differences are regarded as null. If, on the other hand, this first range test is found to be statistically significant, one "steps down" to examine the two $B - 1$ subsets of ordered means, that is, the smallest mean versus the next-to-largest mean and the largest mean versus the next-to-smallest mean, each tested at an α level of significance. At each stage of statistical testing, only significant subsets of ordered means are subjected to further testing.

Although the Newman-Keuls (NK) procedure is very popular among applied researchers, it is well known that it does not limit the MFWER to α when $B > 3$ (Hochberg and Tamhane, 1987, p. 69). Ryan (1960) and Welsch (1977), however, have shown how to adjust the subset levels of significance in order to limit the MFWER to α. Specifically, in order to maintain strict Type I error control, a set of p ($p = 2, \ldots, B$) ordered means should be tested for significance at a level equal to

$$\alpha_p = 1 - (1 - \alpha)^{p/B} \quad (2 \leq p \leq B - 2), \quad \alpha_{B-1} = \alpha_B = \alpha.$$

In order to circumvent the sphericity assumption, these multiple range tests are computed using the t statistic in (4.14) and are referred to studentized range critical values, $q[\alpha_p; p, (N - 1)]/\sqrt{2}$, to assess statistical significance.

4.3 HIGHER-ORDER REPEATED MEASURES DESIGNS: ONE BETWEEN-SUBJECTS AND ONE WITHIN-SUBJECTS FACTOR

One of the most popular types of repeated measures designs in educational and psychological research is the split-plot design, a combination of between-subjects and within-subjects designs. The simplest of these designs involves a single between-subjects factor and a single within-subjects factor, in which subjects ($i = 1, \ldots, n_j$, $\Sigma_j n_j = N$) are selected randomly for each level of the between-subjects factor ($j = 1, \ldots, A$) and observed and measured under all levels of the within-subjects factor ($k = 1, \ldots, B$). In this design, the repeated measures data are modeled by assuming that the observations X_{ijk}, $i = 1, \ldots, n_j$, $j = 1, \ldots, A$, $k = 1, \ldots, B$ are normal, independent, and identically distributed within each level j, with common mean vector μ_j and covariance matrix Σ_j.

Like the simple repeated measures design, both univariate and multivariate approaches may be used to analyze the data from higher-order repeated measures designs. The valid use of either approach rests on the condition that the data meet the derivational assumptions underlying these procedures.

Before proceeding with a description of each of these approaches, a few specific comments about unbalanced higher-order repeated measures designs, that is, designs in which the between-subjects group sizes are unequal, are in order. In such designs, certain of the effects to be tested are correlated. This raises the question as to the choice of the appropriate nonorthogonal solution. This choice is critical because (1) different solutions test different hypotheses and (2) different solutions are affected differently by assumption violations (Milligan, Wong, and Thompson, 1987). A detailed discussion of the various nonorthogonal solutions is beyond the scope of this chapter. For such a discussion, the reader is referred to Appelbaum and Cramer (1974), Carlson and Timm (1974), and Herr and Gaebelein (1978). In this chapter, we have assumed that any unequal group sizes are the result of random subject loss and, therefore, that a standard parametric solution, which tests unweighted hypotheses, is in order (Maxwell and Delaney, 1990; O'Brien and Kaiser, 1985; Searle, 1987).

4.3.1 The Univariate Approach

4.3.1.1 Tests of the Omnibus Between-Subjects Effect

Traditionally, a test of the omnibus between-subjects null hypothesis (i.e., $H_0: \mu_1 = \mu_2 = \cdots = \mu_A$) is accomplished by use of the statistic

$$F_A = MS_A/MS_{S/A} \sim F[\alpha; (A - 1), (N - A)]. \tag{4.16}$$

The validity of F_A rests on the assumptions of independence of observations, normality, and homogeneity of variances. The use of randomization typically assures us that the scores for different subjects are independent. Further, the empirical literature indicates that violation of the normality assumption has a negligible effect on F_A. Violation of the homogeneity of variance assumption, however, can be quite serious, particularly when group sizes are unequal, that is, when the design is unbalanced. Under these conditions, we recommend an alternative test procedure developed by Welch (1951).

In the Welch (1951) procedure, the statistic $F(W)$ is referred to a $F[\alpha; (A - 1), v_w]$ critical value, where

$$F(W) = \frac{\dfrac{\Sigma_j w_j(\overline{X}_j - \tilde{X})^2}{(A - 1)}}{1 + \dfrac{2(A - 2)}{A^2 - 1} \Sigma_j \left(\dfrac{1}{n_j - 1}\right)\left(1 - \dfrac{w_j}{\Sigma_j w_j}\right)^2}, \tag{4.17}$$

$w_j = n_j/s_j^2$, $\tilde{X} = \Sigma_j w_j \overline{X}_j / \Sigma_j w_j$, and v_w, the degrees of freedom, are equal to

$$v_w = \frac{A^2 - 1}{3 \Sigma_j \left(\dfrac{1}{n_j - 1}\right)\left(1 - \dfrac{w_j}{\Sigma_j w_j}\right)^2}. \tag{4.18}$$

For a review of additional statistics for the heterogeneous variance case, the reader is referred to Clinch and Keselman (1982), Tomarkin and Serlin (1986), and Wilcox (1987).

4.3.1.2 Tests of Omnibus Within-Subjects Effects

Tests of the within-subjects main and interaction effects traditionally have been accomplished by the respective use of the univariate F statistics

$$F_B = MS_B/MS_{B \times S/A} \sim F[\alpha; (B - 1), (N - A)(B - 1)] \tag{4.19}$$

and

$$F_{A \times B} = MS_{A \times B}/MS_{B \times S/A}$$
$$\sim F[\alpha; (A - 1)(B - 1), (N - A)(B - 1)]. \tag{4.20}$$

Like the simple repeated measures design, the validity of F_B and $F_{A \times B}$ rests on the assumptions of normality, independence of errors, and homogeneity of the treatment-difference variances (i.e., sphericity). Further, the presence of a between-subjects grouping factor requires that the data meet an additional assumption, namely that the covariance matrices of these treatment differences are the same for all levels of this grouping factor. Jointly, these latter two assumptions have been referred to as multisample sphericity (Huynh, 1978; Mendoza, 1980).

As one would expect, the literature indicates that the traditional within-subjects F ratios are not robust to departures from the multisample sphericity assumption (Noe, 1976; Rogan et al., 1979). However, for balanced designs, adjusted univariate procedures provide a robust alternative to the traditional test statistics. These procedures are as follows:

1. The $\hat{\varepsilon}$-approximate F test (Greenhouse and Geisser, 1959) where

$$F_B \stackrel{\sim}{} F[\alpha; (B - 1)\hat{\varepsilon}, (N - A)(B - 1)\hat{\varepsilon}] \tag{4.21}$$

and

$$F_{A \times B} \stackrel{\sim}{} F[\alpha; (A - 1)(B - 1)\hat{\varepsilon}, (N - A)(B - 1)\hat{\varepsilon}], \tag{4.22}$$

$\hat{\varepsilon}$ is as given in (4.4), and $\mathbf{S}$ is the pooled sample covariance matrix, that is,

$$\mathbf{S} = \sum_j \frac{n_j - 1}{N - A} \mathbf{S}_j,$$

where $\mathbf{S}_j$ is the sample covariance matrix for group j.

2. The $\bar{\varepsilon}$-adjusted procedure (Huynh and Feldt, 1976) where

$$F_B \stackrel{\sim}{} F[\alpha; (B - 1)\bar{\varepsilon}, (N - A)(B - 1)\bar{\varepsilon}] \tag{4.23}$$

and

$$F_{A \times B} \stackrel{\sim}{} F[\alpha; (A - 1)(B - 1)\bar{\varepsilon}, (N - A)(B - 1)\bar{\varepsilon}] \tag{4.24}$$

and

$$\bar{\varepsilon} = \frac{N(B - 1)\hat{\varepsilon} - 2}{(B - 1)[N - A - (B - 1)\hat{\varepsilon}]}. \tag{4.25}$$

It should be mentioned that Huynh (1978) developed "generalized approximate" (GA) and "improved generalized approximate" (IGA) procedures which are extensions of the $\hat{\varepsilon}$-approximate and $\bar{\varepsilon}$-approximate procedures, respectively, and are designed for conditions of heterogeneous covariance matrices and arbitrary group sizes. These procedures involve referring a within-subjects F ratio to a weighted critical value, where the degrees of freedom associated with this critical value are a function of the group sizes and the sample covariance matrices; the weighting factor is based on these factors as well as the total sample size and the number of groups in the design. The computation of the degrees of freedom and the weighting factor, however, is rather complex. Further, the simpler $\bar{\varepsilon}$-

approximate procedure performs as well as the IGA test in many situations and, therefore, is preferable (Huynh, 1978).

4.3.2 The Multivariate Approach

As with the simple repeated measures design previously discussed, in order to conduct tests of significance, the multivariate analysis of a split-plot design requires the formation of transformed variables which are linear combinations of the original variables. The nature of these transformed variables varies as a function of the hypothesis of interest.

4.3.2.1 Tests of the Omnibus Between-Subjects Effect

In our one between- and one within-subjects split-plot design, the between-subjects main effect involves a comparison of the A marginal means, averaged or collapsed over the B levels of the within-subjects factor. Accordingly, in order to test this effect, one creates a new variable, AVG, which represents the average score for each subject, i.e., AVG = $\Sigma_k X_{ijk}/B$. A test of the between-subjects main effect is achieved by computing a one-way between-subjects ANOVA F-test, where the dependent variable is the AVG score for each subject. Note that this test is equivalent to the univariate test given in (4.16). Indeed, for tests of between-subjects effects in split-plot designs, the multivariate and univariate approaches are equivalent. For a discussion of the assumptions underlying between-subjects effects tests and of robust alternatives, the reader is referred back to Section 4.3.1.1.

4.3.2.2 Tests of Omnibus Within-Subjects Effects

Unlike tests of between-subjects effects, the univariate and multivariate approaches do not yield equivalent tests of within-subjects effects.

The multivariate test of the within-subjects main effect is performed by creating $B - 1$ D variables, following the procedures outlined in Section 4.2.2, and testing the null hypothesis that the $B - 1$ D variables have a population mean vector equal to the null vector. The null hypothesis is testing using Hotelling's (1931) T^2 statistic. Note that for unbalanced designs, in order to arrive at tests of unweighted hypotheses, this statistic must be calculated in such a way that the between-subjects groups are weighted equally (Timm, 1975, p. 451). The upper $100(1 - \alpha)$ percentage points of the T^2 distribution can be obtained from the relationship

$$F = \frac{N - A - B + 2}{(N - A)(B - 1)} T^2 \sim F[\alpha; B - 1, N - A - B + 2]. \quad (4.26)$$

Thus, the multivariate test of the within-subjects main effect is a test of whether the means of all the $B - 1$ D variables, collapsed over the levels of A, equal zero. The multivariate test of the within-subjects interaction effect, on the other hand, is a test of whether the means of all the $B - 1$ D variables differ at different levels of A. Such a test is accomplished by conducting a between-subjects multivariate analysis, where factor A is the between-subjects variable and the $B - 1$ D variables are the dependent variables.

While there are many multivariate test criteria when $A > 2$, the four major contenders are (1) Wilks' (1932) likelihood ratio, (2) the Pillai (1955)–Bartlett (1939) trace statistic, (3) Roy's (1953) largest root criterion, and (4) the Hotelling (1951)–Lawley (1938) trace criterion. When $A = 2$, all criteria are equivalent to Hotelling's T^2 statistic.

Like the univariate approach, the validity of the multivariate approach rests on the assumption of normality and independence of observations across subjects. Unlike the univariate approach, however, valid multivariate tests of the within-subjects hypotheses depend not on the sphericity assumption but only on the equality of the covariance matrices at all levels of the between-subjects factor. For tests of the within-subjects effects, the literature suggests that the multivariate approach generally is robust to violation of the assumption of homogeneity of the between-subjects covariance matrices, provided that the design is balanced.

4.3.3 Multiple Comparison Procedures

4.3.3.1 Contrast Tests on Between-Subjects Marginal Means

As with tests of omnibus between-subjects effects, the univariate and multivariate approaches for testing contrasts on between-subjects marginal means in split-plot designs yield equivalent results. In either approach, the choice of a test statistic for contrasts on the between-subjects marginal means rests on the tenability of the homogeneity of variance assumption. If this assumption is tenable, then a test statistic which uses a pooled estimate of error variance (i.e., $MS_{S/A}$) in estimating the standard error of a contrast is appropriate and will provide the most powerful test. On the other hand, if the homogeneity of variance assumption is untenable, then the more appropriate test statistic is one which allows an individual estimate of the contrast variance (Welch, 1947). Given that one seldom knows whether the variance homogeneity assumption is tenable, the safest course of action is to adopt uniformly a test statistic that is based on the separate variance approach. Indeed, research has indicated that this strategy results in only slight losses in power when the homogeneity of variance assumption is satisfied (Best and Rayner, 1987; Games and Howell, 1976). Test statistics based on the separate variance approach, often referred to as non-pooled statistics, can be obtained from programs in the popular mainframe statistical packages (e.g., the BMDP 3D program, the SAS PROC TTEST, and the SPSS T-TEST and ONEWAY procedures). For pairwise contrasts among means, either simultaneous or stepwise MCPs can be used to limit the MFWER to α (Seaman et al., 1991).

4.3.3.2   Contrast Tests on Within-Subjects Marginal Means

As indicated in our discussion of the simple repeated measures design, MCPs that use a constant, that is, a pooled estimate of error variance in obtaining the standard error of a contrast, do not limit the MFWER to α when the data do not satisfy the sphericity assumption (Alberton and Hochberg, 1984; Maxwell, 1980). With respect to higher-order repeated measures designs, similar results have been reported by Keselman and Keselman (1988). That is, when the assumption of multisample sphericity is not satisfied, the use of various types of pooled estimates of error variance in estimating the standard error of within-subjects contrasts results in biased tests of significance, particularly when the design is unbalanced. Fortunately, a test procedure due to Keselman et al. (1991) is available which provides a robust test of pairwise contrasts of repeated measures means for unbalanced nonspherical data. This procedure involves the use of a nonpooled statistic (hereafter referred to as the KKS statistic) and Satterthwaite's (1941, 1946) solution for degrees of freedom.

To introduce the KKS statistic, consider the following definitions. Let $\mu_k = \Sigma_j a_j \mu_{jk}$, where $\Sigma_j a_j = 1$; that is, μ_k is a weighted mean of the kth components of the vectors $\boldsymbol{\mu}_j$. A contrast among the levels of the repeated measures means is given by $\psi = \Sigma_k c_k \underline{\mu_k} = \Sigma_j a_j (\Sigma_k c_k \mu_{jk})$, where $\Sigma_k c_k = 0$, and is estimated by $\hat{\psi} = \Sigma_j a_j (\Sigma_k c_k \bar{X}_{jk})$. Given that $\boldsymbol{\Sigma}_j$ and $\mathbf{S}_j$ denote the respective population and sample covariance matrices for group j,

$$\sigma^2(\hat{\psi}) = \frac{\Sigma_j a_j^2 (\mathbf{c}^\mathrm{T} \boldsymbol{\Sigma}_j \mathbf{c})}{n_j}, \tag{4.27}$$

which is estimated by

$$\hat{\sigma}^2(\hat{\psi}) = \frac{\Sigma_j a_j^2 (\mathbf{c}^\mathrm{T} \mathbf{S}_j \mathbf{c})}{n_j}. \tag{4.28}$$

For pairwise contrasts and for $a_j = 1/A$, i.e., for tests of hypotheses involving unweighted means,

$$\hat{\psi} = \frac{\Sigma_j(\overline{X}_{jk} - \overline{X}_{jk'})}{A} \tag{4.29}$$

and

$$\hat{\sigma}^2(\hat{\psi}) = \frac{\Sigma_j[(s_{jk}^2 + s_{jk'}^2 - 2s_{jkk'})/n_j]}{A^2}, \tag{4.30}$$

where s_{jk}^2 and $s_{jk'}^2$ are the unbiased sample variances of treatments k and k' for group j and $s_{jkk'}$ is the unbiased sample covariance of treatments k and k' for group j.

To test the hypothesis $\psi = 0$, one uses the statistic

$$\frac{\hat{\psi}}{\hat{\sigma}(\hat{\psi})}, \tag{4.31}$$

which, while not distributed as t, can be approximated by Student's t distribution with estimated Satterthwaite's d.f. given by

$$v_s = \frac{[\hat{\sigma}(\hat{\psi})]^4}{\displaystyle\sum_j \frac{(a_j^2/n_j)^2(\mathbf{c}^\mathrm{T}\mathbf{S}_j\mathbf{c})^2}{n_j - 1}}. \tag{4.32}$$

Like the simple repeated measures design, the multivariate approach to contrast testing in higher-order repeated measures designs involves defining a set of transformed D variables which reflect the contrasts of interest and calculating a series of T^2 statistics on each of the D variables. As previously mentioned, the T^2 statistic allows for individual estimates of contrast variances and therefore is not dependent on the sphericity assumption. In calculating the contrast variances, however, the T^2 statistic involves pooling across the levels of the between-subjects factor and is dependent, therefore, on the homogeneity of covariance assumption. The literature indicates that the T^2 statistic is not uniformly robust to violation of this assumption. Accordingly, this approach cannot be recommended when the assumption of covariance homogeneity is not tenable. In these situations, we recommend that contrast tests on within-subjects marginal means be conducted using the KKS statistic, which does not pool over the levels of either the between- or within-subjects factor in estimating the standard error of the contrast.

Simultaneous Multiple Comparison Procedures

Using their KKS statistic, Keselman et al. (1991) investigated the robustness of various simultaneous MCPs in nonspherical unbalanced repeated measures designs. The results of their simulations indicated that the KKS statistic generally limited the MFWER to α when used in conjunction with a Bonferroni $\{t[\alpha/(2c); v_s]\}$, studentized range $\{q[\alpha; B, v_s]/\sqrt{2}\}$ or a studentized maximum modulus $\{M[\alpha; c, v_s]\}$ critical value. In general, Keselman et al. (1991) found that a Bonferroni critical value provided the best Type I error control, followed by a studentized maximum modulus critical value and, finally, a studentized range critical value.

Stepwise Multiple Comparison Procedures

The stepwise strategies introduced in the discussion of the simple repeated measures design can also be applied to higher-order repeated measures designs. Specifically, Hochberg's step-up Bonferroni procedure will limit the MFWER to α in unbalanced nonspherical designs when used with the KKS statistic and the degrees of freedom given in (4.31) and (4.32), respectively.

Regarding the multiple range procedure, for unbalanced nonspherical data, robust pairwise contrast tests can be achieved using the KKS statistic and the degrees of freedom given in (4.31) and (4.32) with the critical value $q[\alpha_p; p, v_s]/\sqrt{2}$, where p is the size of the set of ordered means ($p = 2, \ldots, B$) and α_p is the Ryan-Welsch significance level.

4.3.4 Assessing the Interaction Effect

Typically, interaction effects are assessed by one of two methods: (1) tests of simple effects and/or (2) interaction contrasts. The choice between these two methods depends on the hypotheses of interest.

4.3.4.1 Simple Between-Subjects Effects

In our $A \times B$ split-plot design, the simple effect of factor A refers to the effect of the between-subjects factor A at a particular or fixed level of the within-subjects factor B. By restricting our attention to a particular level of B, we have essentially eliminated the within-subjects factor from the design and are left with a single-factor between-subjects design. It follows then that the univariate and multivariate approaches to testing simple between-subjects effects are identical.

The simple effects of the between-subjects factor A traditionally have been examined using the statistic

$$F_{A\,\text{at}\,B_k} = \frac{MS_{A\,\text{at}\,B_k}}{MSW} \sim F[\alpha; (A - 1), B(N - A)], \qquad (4.33)$$

where $MSW = [SS_{S/A} + SS_{B \times S/A}]/[B(N - A)]$. This statistic cannot be recommended, however, because its validity rests on the seldom satisfied assumption of variance homogeneity across the levels of both A and B.

A less restrictive approach is to use a test statistic that estimates error variance on the basis of the data at a fixed level of B. Using this method,

$$F_{A \, \text{at} \, B_k} = \frac{MS_{A \, \text{at} \, B_k}}{MS_{S/A \, \text{at} \, B_k}} \sim F[\alpha; (A - 1), (N - A)]. \tag{4.34}$$

Essentially, this approach is equivalent to conducting a simple between-subjects analysis of the grouping factor A at a particular level of B and, therefore, is dependent on the assumptions of independence of observations, normality, and homogeneity of only the A group variances. If this variance homogeneity assumption is untenable, a heterogeneous variance procedure such as the one by Welch or by Brown and Forsythe (1974) should be used. In order to limit the MFWER to α, each simple effect should be assessed at a reduced significance level. For a discussion of MFWER control in simple effect testing, readers are referred to Kirk (1982, pp. 365–371) and Maxwell and Delaney (1990, pp. 265–266).

4.3.4.2 Simple Within-Subjects Effects

In our $A \times B$ split-plot design, the simple effect of factor B refers to the effect of the within-subjects factor B at a particular level of the between-subjects factor A. By focusing our attention on a fixed level of A, the between-subjects factor is effectively eliminated and we are left with a single-factor within-subjects design. Accordingly, the univariate and multivariate approaches to the analysis of simple within-subjects effects are not equivalent.

Adopting a univariate approach, the simple effects of the within-subjects factor B traditionally have been tested using the statistic

$$F_{B \, \text{at} \, A_j} = \frac{MS_{B \, \text{at} \, A_j}}{MS_{B \times S/A}} \sim F[\alpha; (B - 1), (N - A)(B - 1)], \tag{4.35}$$

where $MS_{B \times S/A}$ is an estimate of error variance based on all the cells in the design. The use of this error term, however, rests on the assumption of multisample sphericity. When this assumption is not met, that is, when the $\binom{B}{2}$ treatment-difference variances are not equal and/or the between-subjects covariance matrices are not the same for all levels of A, the statistic in (4.35) will be biased.

One approach to circumvent this problem is to use a test statistic which estimates error variance on the basis of the data at a fixed level of A, that is,

$$F_{B \, \text{at} \, A_j} = \frac{MS_{B \, \text{at} \, A_j}}{MS_{B \times S/A_j}} \sim F[\alpha; (B - 1), (n_j - 1)(B - 1)]. \tag{4.36}$$

Essentially, this approach is equivalent to conducting a simple within-subjects analysis of the repeated factor B at a particular level of A. As such, the assumption of equality of the between-subjects covariance matrices is trivially satisfied because there is only one between-subjects covariance matrix. The use of the statistic given in (4.36), however, still requires that the data are spherical, i.e., the $\binom{B}{2}$ treatment-difference variances are homogeneous. Given that this assumption is seldom satisfied in educational and psychological research, the statistic in (4.36) cannot be recommended.

A robust univariate within-subjects simple effect test can be obtained by adopting one of the adjusted univariate approaches previously described. Using such an approach, the test statistic in (4.36) would be compared to a critical value based on $(B - 1)\hat{\varepsilon}$ and $(n_j - 1)(B - 1)\hat{\varepsilon}$ or $(B - 1)\tilde{\varepsilon}$ and $(n_j - 1)(B - 1)\tilde{\varepsilon}$ degrees of freedom, where $\hat{\varepsilon}$ and $\tilde{\varepsilon}$ are estimated on the basis of S_j, the sample covariance matrix of group j. Because of the reduction in the degrees of freedom, this approach will result in less powerful tests of the simple within-subjects effects. We nonetheless recommend it because it does not depend on the assumption of multisample sphericity.

As with the univariate approach, either one of two error terms (in this case, error matrices) can be used in calculating multivariate tests of the simple within-subjects effects. The first of these matrices is formed by pooling across the levels of the between-subjects factor A and, therefore, its use depends on the homogeneity of covariance matrices assumption. The use of the second of these matrices does not rest on this assumption because it is based only on the observations at a particular level of A. Because violation of the equality of covariance matrices assumption can be problematic, particularly in unbalanced designs, we prefer the use of a separate error matrix. Using this approach, the B repeated measurements for the n_j subjects at a fixed level of A are transformed into $B - 1$ D variables and a test of the simple within-subjects effect is performed using Hotelling's (1931) T^2 statistic. Upper $100(1 - \alpha)$ percentage points of Hotelling's T^2 distribution can be obtained from the relationship

$$F = \frac{n_j - B + 1}{(n_j - 1)(B - 1)} T^2 \sim F[\alpha; B - 1, n_j - B + 1]. \qquad (4.37)$$

Finally, with either approach and in order to limit the MFWER to α for the set of simple effect tests, each simple effect test should be conducted using a reduced significance level (Kirk, 1982, pp. 365–371; Maxwell and Delaney, 1990, pp. 265–266).

4.3.4.3 Interaction Contrasts

A second method of assessing the interaction effect is to perform a series of interaction contrasts. As in the case of simple effects, the choice of a interaction contrast test statistic depends on the assumptions one is willing to make about one's data.

Suppose, for example, that you are interested in testing whether the difference between B_1 and B_2 is the same for A_1 and A_2. Adopting a univariate approach to testing interaction contrasts and assuming that the data meet the multisample sphericity assumption, then $MS_{B \times S/A}$ can be used to estimate the standard error of the contrast. In this case, the statistic is

$$t = \frac{(\hat{\psi} - \Psi)}{\sqrt{MS_{B \times S/A}(\Sigma_j \, c_j^2/n_j)}} \sim t[\alpha; (N - A)(B - 1)], \tag{4.38}$$

where $\hat{\psi} = (\overline{X}_{11} - \overline{X}_{12}) - (\overline{X}_{21} - \overline{X}_{22})$.

If one cannot assume that the $\binom{B}{2}$ treatment-difference variances are equal, that is, that the data are spherical, then only the data associated with the specific levels of B that are involved in the contrast (in this case, B_1 and B_2) should be used in computing the standard error of the contrast. In this case, the test statistic is

$$t = \frac{(\hat{\psi} - \Psi)}{\sqrt{MS_{B(\hat{\psi}) \times S/A}(\Sigma_j \, c_j^2/n_j)}} \sim t[\alpha; (N - A)], \tag{4.39}$$

where $MS_{B(\hat{\psi}) \times S/A}$ is an estimate of error variance based on the data in levels B_1 and B_2.

Finally, when it further cannot be assumed that the variance of the B_1 vs. B_2 contrast is the same for all levels of A, that is, when the data do not satisfy either part of the multisample sphericity assumption, the standard error of the contrast should be estimated only from the data associated with the particular cells (in this case, the four cells) involved in the interaction contrast. That is, the standard error of the contrast must be estimated in a way that does not involve pooling over either the between- or within-subjects factors. In other words, the standard error of the contrast and, accordingly, the degrees of freedom must be estimated using the KKS procedure previously described [see Eqs. (4.28), (4.30), and (4.32)]. Applying this procedure to our example, the test statistic and degrees of freedom are

$$t = \frac{(\overline{X}_{11} - \overline{X}_{12}) - (\overline{X}_{21} - \overline{X}_{22})}{\sqrt{c_1^T S_1 c_1/n_1 + c_2^T S_2 c_2/n_2}} \tag{4.40}$$

and

$$v_s = \frac{\left(\sum_j \frac{\mathbf{c}^T\mathbf{S}_j\mathbf{c}}{n_j}\right)^2}{\sum_j \frac{(\mathbf{c}^T\mathbf{S}_j\mathbf{c}/n_j)^2}{n_j - 1}}, \tag{4.41}$$

respectively, where $\mathbf{c}_1^T = \mathbf{c}_2^T = [1 \; -1]$.

Adopting a multivariate approach and letting $D_{ij} = X_{ij1} - X_{ij2}$, our interaction contrast can be conceptualized as a between-group contrast between A_1 and A_2 on the dependent variable, D_{ij}. In this conceptualization, the test statistic is

$$t = \frac{(\hat{\psi} - \Psi)}{\sqrt{MS_{S/A(D)} \left(\sum_j c_j^2/n_j\right)}} \sim t[\alpha; N - A], \tag{4.42}$$

where $\hat{\psi} = \sum_j c_j \overline{D}_j$, and $MS_{S/A(D)}$ is the mean square error for the D variable. The test statistic in (4.42) rests on the assumption that the variances of the D variable at all levels of A are equal. If this assumption is not tenable, one can adopt the Welch (1947) statistic,

$$t = \frac{(\hat{\psi} - \Psi)}{\sqrt{\sum_j \frac{c_j^2 s_{j(D)}^2}{n_j}}}, \tag{4.43}$$

where $s_{j(D)}^2$ is the variance of the D variable at level j. This statistic, while not distributed as t, can be approximated by Student's t distribution with estimated Welch (1947) d.f. given by

$$v_w = \frac{(\sum_j c_j^2 s_{j(D)}^2/n_j)^2}{\sum_j \frac{(c_j^2 s_{j(D)}^2/n_j)^2}{n_j - 1}}. \tag{4.44}$$

It should be noted that the statistic and d.f. in (4.43) and (4.44) are equivalent to the test statistic and d.f. in (4.40) and (4.41), respectively, illustrating the equivalence of the multivariate and univariate approaches. [The use of Eqs. (4.43) and (4.44) is preferred on the basis of computational simplicity.] In either approach, MFWER control can be achieved using a Bonferroni critical value, $t[\alpha/(2c); v_w]$, where $c = \binom{A}{2}\binom{B}{2}$.

4.4   A COMPARISON OF ANALYSIS STRATEGIES

In summary, the data from repeated measures designs can be analyzed using either a univariate or a multivariate approach. For tests of between-subjects effects, the two approaches are equivalent and their validity depends on the same set of statistical assumptions, i.e., independence of observations, normality, and homogeneity of variance. For tests of within-subjects effects, these two approaches are not equivalent, and each approach has advantages and disadvantages. In general, we favor the multivariate approach for several reasons. First, this approach depends on a less restrictive set of assumptions than the univariate approach. Specifically, although both approaches depend on the assumptions of independence of observations, normality, and homogeneity of the between-subjects covariance matrices, the validity of the multivariate approach does not rest on the seldom-satisfied assumption of sphericity, as does the univariate approach. While adjusted univariate procedures have been shown to be relatively robust to violations of the sphericity assumption, these procedures are still only approximate, unlike the multivariate approach, which remains exact in the presence of non-sphericity provided its own derivational assumptions are satisfied.

Power considerations form the basis of a second reason for preferring the multivariate approach. When the assumptions of the univariate approach are met, this procedure is uniformly more powerful than the multivariate approach. In the more typical situation in which the sphericity assumption is not met, however, neither procedure is uniformly more powerful; rather, the relative power of the two approaches depends on the nature of the alternative hypothesis, the covariance structure of the data, and the relationship between these two variables. Nonetheless, Davidson indicated that, provided that the sample size is not too small, the power of the multivariate approach ranges from somewhat less than to much greater than that of the univariate approach. For a discussion of minimum sample size, the reader is referred to Maxwell and Delaney (1990, pp. 603, 676). Thus, only when sample size is too small to recommend the multivariate approach, or when it is so small that it precludes its use, do we recommend an adjusted univariate strategy.

The final and perhaps most important reason for our choice of the multivariate approach is based on our preference for the use of separate error terms in contrast testing and the consistency of the multivariate approach with this strategy. That is, unlike the univariate mixed-model approach, the multivariate approach treats sets of contrasts (e.g., a set of suitably defined contrasts representing an ANOVA effect) in such a way that each contrast remains linked with its specific error term. In short, the multivariate approach is the "natural generalization of the use of the specific type of error term for contrasts with one degree of freedom" (O'Brien and Kaiser, 1985, p. 319).

4.5 GENERALIZATIONS TO MORE COMPLEX REPEATED MEASURES DESIGNS

While all of the previously discussed procedures can be generalized to more complex repeated measures designs, we will illustrate this generalization for only our recommended data-analytic strategy, that being the use of multivariate omnibus tests and/or contrast testing procedures that use separate (i.e., nonpooled) error terms.

Consider a three factor $A \times B \times C$ design, where A is a between-subjects factor and B and C are within-subjects factors, and all factors, other than the random subject variable, are fixed. Further, let us assume that the between-subjects group sizes (n_j) are unequal and that this inequality is the result of the random loss of subjects. Thus, tests of unweighted hypotheses are in order.

In this design, seven effects, one between-subjects and six within-subjects, can be isolated and tested for statistical significance. Adopting the multivariate approach, these significance tests are accomplished by creating sets of linearly independent transformed variables (or, in certain cases, a single transformed variable), each reflecting an effect of interest, and then subjecting each of these sets of transformed variables to a multivariate analysis. One should remember that in cases in which the analysis is performed on a single transformed variable, this approach yields results equivalent to those of a univariate approach.

In order to test the between-subjects effect of factor A, one first creates a single transformed variable, the value of which is the score for each subject averaged over the BC repeated measurements, that is, AVG $= \Sigma_k \Sigma_l X_{ijkl}/BC$. To arrive at tests of the between-subjects effect, these scores are then subjected to a one-way between-subjects analysis. When group variances can be assumed to be homogeneous, a between-subjects ANOVA is an appropriate analysis strategy. When this assumption is not tenable, a robust alternative to the ANOVA should be used (see Section 4.3.1.1).

To arrive at tests of within-subjects effects in our three-factor design, three sets of transformed variables must be created to represent (1) the within-subjects factor B, (2) the within-subjects factor C, and (3) the completely within-subjects $B \times C$ interaction. The variable set reflecting factor B consists of $B - 1$ transformed variables, where each variable represents a linearly independent contrast among the levels of factor B, collapsed over factor C. Similarly, a set of $C - 1$ transformed variables, each representing a linearly independent contrast among the levels of factor C, collapsed over factor B, constitutes the variable set associated with factor C. Finally, the $B \times C$ interaction is reflected by a set of $(B - 1)(C - 1)$ transformed variables, each representing a linearly independent $B \times C$

interaction contrast. The interaction contrast coefficients are formed by taking the product of coefficients associated with the $B - 1$ transformed variables and the $C - 1$ transformed variables.

To illustrate this procedure, suppose that each of the within-subjects factors in our three-factor design has three levels, i.e., $B = C = 3$. In such a design, the following three sets of transformed variables can be used to arrive at multivariate tests of the within-subjects effects:

Set 1 (Factor B)

$$D_1 = (1)X_{ij11} + (1)X_{ij12} + (1)X_{ij13} + (-1)X_{ij21} + (-1)X_{ij22}$$
$$+ (-1)X_{ij23} + (0)X_{ij31} + (0)X_{ij32} + (0)X_{ij33}$$
$$D_2 = (0)X_{ij11} + (0)X_{ij12} + (0)X_{ij13} + (1)X_{ij21} + (1)X_{ij22}$$
$$+ (1)X_{ij23} + (-1)X_{ij31} + (-1)X_{ij32} + (-1)X_{ij33}$$

Set 2 (Factor C)

$$D_3 = (1)X_{ij11} + (-1)X_{ij12} + (0)X_{ij13} + (1)X_{ij21} + (-1)X_{ij22}$$
$$+ (0)X_{ij23} + (1)X_{ij31} + (-1)X_{ij32} + (0)X_{ij33}$$
$$D_4 = (0)X_{ij11} + (1)X_{ij12} + (-1)X_{ij13} + (0)X_{ij21} + (1)X_{ij22}$$
$$+ (-1)X_{ij23} + (0)X_{ij31} + (1)X_{ij32} + (-1)X_{ij33}$$

Set 3 (Factor $B \times C$)

$$D_5 = (1)X_{ij11} + (-1)X_{ij12} + (0)X_{ij13} + (-1)X_{ij21} + (1)X_{ij22}$$
$$+ (0)X_{ij23} + (0)X_{ij31} + (0)X_{ij32} + (0)X_{ij33}$$
$$D_6 = (0)X_{ij11} + (1)X_{ij12} + (-1)X_{ij13} + (0)X_{ij21} + (-1)X_{ij22}$$
$$+ (1)X_{ij23} + (0)X_{ij31} + (0)X_{ij32} + (0)X_{ij33}$$
$$D_7 = (0)X_{ij11} + (0)X_{ij12} + (0)X_{ij13} + (1)X_{ij21} + (-1)X_{ij22}$$
$$+ (0)X_{ij23} + (-1)X_{ij31} + (1)X_{ij32} + (0)X_{ij33}$$
$$D_8 = (0)X_{ij11} + (0)X_{ij12} + (0)X_{ij13} + (0)X_{ij21} + (1)X_{ij22}$$
$$+ (-1)X_{ij23} + (0)X_{ij31} + (-1)X_{ij32} + (1)X_{ij33}$$

Tests of the within-subjects main effects due to B and C and the completely within-subjects $B \times C$ interaction are accomplished by testing the hypothesis that the grand means of the corresponding set of transformed variables all equal zero, using Hotelling's (1931) T^2 statistic. (Remember that to arrive at unweighted hypothesis tests, the grand means of the transformed variables must be calculated so that the between-subjects groups are weighted equally.) Tests of the remaining within-subjects interaction effects are arrived at by performing a between-subjects analysis on each of the sets of transformed variables. In our three-factor design, one-way (A) between-subjects analyses on the sets of $B - 1$, $C - 1$, and $(B - 1)(C - 1)$ transformed variables result in tests of the $A \times B$ effect, the $A \times C$ effect, and the $A \times B \times C$ effect, respectively.

With respect to contrast testing, tests on between-subjects marginal means are computed using the transformed variable AVG as the dependent variable. As always, the choice of the method of estimating the standard error of a contrast, that is, using a pooled error term approach using $MS_{S/A}$ or a separate error term approach using Welch's (1947) procedure, depends on the tenability of the homogeneity of variance assumption (see Section 4.3.3.1). Our preference is for the separate error term approach.

Similarly, in order to circumvent the assumption of multisample sphericity, contrast tests on within-subjects marginal means should be performed using the KKS statistic and Satterthwaite's (1941, 1946) solution for degrees of freedom [see Eqs. (4.31) and (4.32) respectively]. In our three-factor design, the estimated value of a contrast of unweighted means is equal to

$$\hat{\psi} = \Sigma_j \, a_j(\Sigma_k \, \Sigma_l \, c_{kl}\overline{X}_{jkl}), \tag{4.45}$$

where $a_j = 1/A$ and c_{kl} is the contrast coefficient associated with mean $\overline{X}_{jkl}$ subject to the restriction that $\Sigma_k \, \Sigma_l \, c_{kl} = 0$. An estimate of the standard error of the contrast is given by

$$\hat{\sigma}(\hat{\psi}) = [\Sigma_j \, a_j^2(\mathbf{c}^T\mathbf{S}_j\mathbf{c})/n_j]^{1/2}, \tag{4.46}$$

where $\mathbf{S}_j$ is the $BC \times BC$ sample covariance matrix associated with group j and $\mathbf{c}$ is a vector of BC coefficients representing the contrast of interest.

To illustrate, assume again that $B = C = 3$ in our three-factor design. A test of the difference between the B_1 and B_2 marginal means can be accomplished by forming the contrast coefficient vector $\mathbf{c}$, where

$$\mathbf{c}^T = [1 \quad 1 \quad 1 \quad -1 \quad -1 \quad -1 \quad 0 \quad 0 \quad 0].$$

Simple between-subjects effects are conducted following the procedures described in Section 4.3.4.1. As described in this section, the main consideration in conducting these tests is the choice of error term, which depends on the tenability of the homogeneity of variance and multisample sphericity assumptions. For example, to test the effect of the between-subjects factor A at particular levels of the within-subjects factors B and C, one can adopt one of three error terms: (1) a pooled error term, MS_p, which rests on multisample sphericity, where $MS_p = [SS_{S/A} + SS_{B \times S/A} + SS_{C \times S/A} + SS_{BC \times S/A}]/[A(n_j - 1) + B(n_j - 1) + A(n_j - 1)(C - 1) + A(n_j - 1)(B - 1)(C - 1)]$; (2) $MS_{S/A}$ at BC_{kl}, an estimate of error variance which is based on only the data at the levels of k and l of interest, which does not rest on multisample sphericity but does depend on homogeneity of between-group variances; or (3) a heterogeneous variance procedure (Brown and Forsythe, 1974; Welch, 1951), computed on the data at the particular levels of k and l of interest, which depends on only the assumptions of independence of observations and normality. We prefer the last of these approaches.

Multivariate tests of simple within-subjects effects are performed using Hotelling's T^2 statistic, following the procedures outlined in Section 4.3.4.2. Depending on the tenability of the homogeneity of covariance assumption, we can use either an error matrix which pools across the levels of the between-subjects factors, and therefore depends on covariance homogeneity, or one which is based on only the observations at the level of the between-subjects factor of interest, which does not depend on this assumption. Thus, for our three-factor design, to test the simple within-subjects effect of factor B at a particular level of A, for example, one can use the pooled error matrix of the $B - 1$ transformed variables (based on all N scores in the design) or the separate error matrix (based on only the n_j scores at the level of A of interest). For reasons outlined previously, we prefer this latter approach.

Finally, interaction contrasts should be performed using the KKS statistic and Satterthwaite's solution for d.f. As mentioned previously, this statistic is our preferred choice because, in estimating the standard error of a contrast, it does not pool over either the between- or within-subjects factors and, therefore, does not depend on multisample sphericity.

For our three-factor design, the numerator and denominator of this statistic are given in (4.45) and (4.46), respectively; the Satterthwaite d.f. are given in (4.32). Assuming that $B = C = 3$, a test of whether the difference between C_1 and C_2 is the same at B_1 and B_2, for example, can be accomplished by forming the following contrast coefficient vector $\mathbf{c}$, where

$$\mathbf{c}^T = [1 \quad -1 \quad 0 \quad -1 \quad 1 \quad 0 \quad 0 \quad 0 \quad 0].$$

4.6 PRELIMINARY TESTING FOR MULTISAMPLE SPHERICITY

Before presenting a numerical example of several of the analysis procedures presented in this chapter, a final strategy should be mentioned, that of conducting a preliminary test of the multisample sphericity assumption and using the results of this test to choose between two analytical strategies.

For example, if the results of such a test suggested that the multisample sphericity assumption was tenable, univariate omnibus and contrast procedures based on a pooled error term approach could be adopted and would yield uniformly more powerful tests than the multivariate and contrast testing procedures based on a separate error term approach. On the other hand, if a preliminary test of multisample sphericity suggested that the data do not conform to this assumption, omnibus and contrast procedures based on a separate error term approach should be used.

There are both two-stage and simultaneous approaches to testing the multisample sphericity hypothesis. The two-stage approach involves Box's (1949) heteroscedasticity test and Mauchly's (1940) sphericity test. Keselman et al. (1980) found these tests to be so sensitive (both to departures from their respective null hypotheses and to their own derivational assumption of multivariate normality) that they virtually always signal violation of the multisample sphericity assumption. This two-stage approach, therefore, is not helpful in choosing between the univariate and multivariate procedures. The sensitivity of the simultaneous approach to testing multisample sphericity (Mendoza, 1980) has yet to be shown.

4.7 A NUMERICAL EXAMPLE

In this final section of the chapter, we provide a numerical example of our preferred repeated measures data-analytic strategy, that is, the use of multivariate tests of omnibus within-subjects effects and/or contrast testing procedures based on a separate error term approach. Based on the results of Olson (1974) and Rogan et al. (1979), we used the Pillai-Bartlett trace statistic to test omnibus within-subjects effects. For testing pairwise contrasts on marginal means, we used a studentized range critical value to limit the MFWER to α. Given that the design of our numerical example is unbalanced, we used Welch's (1951) statistic, which does not depend on homogeneous variances, to assess the between-subjects omnibus effect. Furthermore, in analyzing the data, we assumed that the unequal group sizes were the result of the random loss of subjects and, therefore, that interest was in tests of unweighted hypotheses. Finally, the significance level for all omnibus tests was .05; similarly, a .05 significance level was used for families of simple effect tests and of contrast tests.

To obtain the numerical results, we used the SAS (SAS Institute, 1990a,b,c) statistical package. In certain cases, the output of the BMDP (Dixon et al., 1988, 1990) statistical package was preferable to that of SAS; in these cases, BMDP procedures were assessed using the SAS PROC BMDP. For each of the analyses, we indicate the SAS programming statements that were required to produce the desired results.

4.7.1   Data Input and Variable Definitions

The data set for our numerical example is given in Table 4.1. As seen in this table, the data are from a repeated measures design containing a single between-subjects variable (A) with three levels and a single within-subjects factor (B) with four levels. Furthermore, the between-subjects group sizes are not equal $(n_1 = 13, n_2 = 10, n_3 = 7)$.

Our first task was to create a SAS data set to which we input our data and within which we defined any additional variables that were necessary for our analysis. In our example, the following SAS statements were used to create a SAS data set called ALL:

```
 1.  DATA ALL;
 2.  INPUT A B1 B2 B3 B4;
 3.  AVG = (B1 + B2 + B3 + B4)/4;
 4.  D1 = B1 − B2;
 5.  D2 = B1 − B3;
 6.  D3 = B1 − B4;
 7.  D4 = B2 − B3;
 8.  D5 = B2 − B4;
 9.  D6 = B3 − B4;
10.  CARDS;
11.  1 10 14 19 18
         ,
         ⋮
         ,
12.  2 8 10 10 17
         ,
         ⋮
         ,
13.  3 23 22 23 17
         ,
         ⋮
         ,
14.  ;
```

As defined in the input statement 2 and as illustrated in statements 11–13, for each subject, we input a group membership indicator A and the scores on each of the four repeated measurements B1, . . . , B4. In addition, our data set contained seven new variables, which, as will be seen later, were required for specific analyses. The variable AVG, which represents the average score for each subject, was used to test the between-subjects omnibus effect and to test contrasts among between-subjects marginal means.

Table 4.1 $A(3) \times B(4)$ Repeated Measures Data

Subject	B_1	B_2	B_3	B_4
			A_1	
1	10	14	19	18
2	15	16	18	23
3	17	19	20	21
4	13	16	16	21
5	17	17	20	18
6	14	16	17	21
7	11	14	18	19
8	13	16	18	21
9	12	15	18	22
10	18	18	20	20
11	17	19	20	21
12	17	17	20	19
13	17	17	20	21
			A_2	
14	8	10	10	17
15	12	11	17	16
16	8	12	14	16
17	8	8	10	15
18	16	16	16	18
19	13	18	16	20
20	15	16	16	16
21	8	7	12	12
22	11	17	15	20
23	14	13	16	15
			A_3	
24	23	22	23	17
25	24	20	28	20
26	30	23	27	26
27	23	23	22	23
28	24	19	25	24
29	18	19	20	22
30	29	23	19	20

The variables D1, . . . , D6 were used to conduct contrast tests on within-subjects marginal means and to test interaction contrasts.

4.7.2 Data Description

Tables 4.2 and 4.3 contain the cell means and sample covariance matrices, respectively, for our illustrative data set. These means and covariance matrices were generated by the following SAS statements:

Table 4.2 Cell Means

	B_1	B_2	B_3	B_4
A_1 ($n_1 = 13$)	14.69	16.46	18.77	20.38
A_2 ($n_2 = 10$)	11.30	12.80	14.20	16.50
A_3 ($n_3 = 7$)	24.43	21.28	23.43	21.71

15. PROC MEANS DATA = ALL MEAN MAXDEC = 2;
16. VAR B1 B2 B3 B4;
17. BY A;
18. PROC CORR DATA = ALL COV;
19. VAR B1 B2 B3 B4;
20. BY A;

4.7.3 Testing the Omnibus Between-Subjects Effect

In order to calculate Welch's (1951) statistic to test the between-subjects main effect, we used the BMDP (Dixon et al., 1990) one-way and two-

Table 4.3 Covariance Matrices

	B_1	B_2	B_3	B_4
		A_1		
B_1	7.2308	3.9038	2.4231	0.3782
B_2		2.6026	1.2820	0.4743
B_3			1.8590	−.7372
B_4				2.2564
		A_2		
B_1	10.0111	8.9556	6.7111	2.8333
B_2		14.8444	7.1556	7.6667
B_3			6.8444	2.3333
B_4				5.8333
		A_3		
B_1	16.2857	5.0238	3.6190	3.1429
B_2		3.5714	−.9762	−.0714
B_3			11.6190	2.9762
B_4				8.9048

way ANOVA program, 7D, which we accessed via the SAS PROC BMDP. The following SAS statements were used to generate the desired results:

21. PROC BMDP PROG = BMDP7D DATA = ALL UNIT = 3;
22. PARMCARDS;
23. /PROB TITLE = 'WELCH TEST OF THE OMNIBUS BETWEEN SUBJECTS EFFECT'.
24. /INPUT UNIT = 3. CODE = 'ALL'.
25. /HISTOGRAM GROUPING IS A.
26. VARIABLE IS AVG.
27. /END
28. /FINISH
29. ;

Note that the dependent variable for this single-factor between-subjects analysis was the average score for each subject, i.e., AVG. The results of this analysis indicated a significant main effect due to factor A, i.e., Welch $F_A = 30.87$ with d.f. $= (2, 13)$ at $p < .0001$. It should be noted that the statements above also generate the traditional ANOVA F test, which may be used when the homogeneity of variance assumption is tenable. For our data, $F_A = 41.58$ with d.f. $= (2, 27)$ at $p < .0001$.

The results of this analysis and all subsequent analyses are presented in Table 4.4.

4.7.4 Testing Omnibus Within-Subjects Effects

The SAS General Linear Model (GLM) procedure was used to arrive at multivariate tests of the within-subjects main and interaction effects. The following SAS statements were used to produce the desired results:

30. PROC GLM;
31. CLASS A;
32. MODEL B1 − B4 = A/NOUNI;
33. REPEATED B;

The SAS statements above produced an ANOVA F test of the between-subjects main effect due to A as well as both multivariate and adjusted univariate ($\hat{\varepsilon}$-adjusted and $\tilde{\varepsilon}$-adjusted) tests of the omnibus within-subjects effects. Note that, by default, SAS generates tests of unweighted hypotheses. Furthermore, for the within-subjects effects, four multivariate criteria and associate F approximations are reported. For our data, the F approximations associated with the Pillai-Bartlett statistic are $F_B = 11.37$ with d.f. $= (3, 25)$ at $p < .0001$ and $F_{A \times B} = 3.62$ with d.f. $= (6, 52)$ at $p < .0045$ (see Table 4.4).

Table 4.4 Summary of Numerical Example Results

Test	F	d.f.	Critical F	Result[a]
A Effect (BMDP7D)				
A	30.87	2, 13	3.81	S
B & $A \times B$ Effects (SAS GLM)				
B	11.37	3, 25	2.99	S
$A \times B$	3.62	6, 52	2.28	S
A Marginal Means' Contrasts (BMDP3D)				
1. A_1 vs. A_2	18.32	1, 12.7	7.00	S
2. A_1 vs. A_3	35.28	1, 8.9	7.83	S
3. A_2 vs. A_3	63.20	1, 14.7	6.77	S
B Marginal Means' Contrasts (SAS MEANS)				
1. B_1 vs. B_2	0.01	1, 13.24	8.57	NS
2. B_1 vs. B_3	9.86	1, 8.90	9.77	S
3. B_1 vs. B_4	15.23	1, 14.09	8.43	S
4. B_2 vs. B_3	10.27	1, 10.46	9.23	S
5. B_2 vs. B_4	24.58	1, 11.98	8.82	S
6. B_3 vs. B_4	1.46	1, 13.57	8.51	NS
Simple A Effects (BMDP7D)				
A at B_1	24.82	2, 14	6.09	S
A at B_2	23.00	2, 14	6.09	S
A at B_3	19.84	2, 12	6.45	S
A at B_4	11.13	2, 12	6.45	S
Simple B Effects (SAS GLM)				
B at A_1	23.55	3, 10	5.55	S
B at A_2	10.52	3, 7	6.94	S
B at A_3	1.76	3, 4	12.58	NS
Interaction Contrasts (BMDP3D)				
1. $(AB_{11} - AB_{12})$ vs. $(AB_{21} - AB_{22})$	0.08	1, 13	13.54	NS
2. $(AB_{11} - AB_{12})$ vs. $(AB_{31} - AB_{32})$	15.52	1, 7.4	19.32	NS
3. $(AB_{21} - AB_{22})$ vs. $(AB_{31} - AB_{32})$	10.30	1, 11.5	14.36	NS
4. $(AB_{11} - AB_{13})$ vs. $(AB_{21} - AB_{23})$	2.07	1, 20.4	11.56	NS
5. $(AB_{11} - AB_{13})$ vs. $(AB_{31} - AB_{33})$	7.84	1, 7.4	19.32	NS
6. $(AB_{21} - AB_{23})$ vs. $(AB_{31} - AB_{33})$	4.62	1, 7.4	19.32	NS
7. $(AB_{11} - AB_{14})$ vs. $(AB_{21} - AB_{24})$	0.14	1, 18.7	11.85	NS
8. $(AB_{11} - AB_{14})$ vs. $(AB_{31} - AB_{34})$	20.98	1, 9.1	16.48	S
9. $(AB_{21} - AB_{24})$ vs. $(AB_{31} - AB_{34})$	16.81	1, 10.4	15.16	S
10. $(AB_{12} - AB_{13})$ vs. $(AB_{22} - AB_{23})$	0.94	1, 12.5	13.78	NS
11. $(AB_{12} - AB_{13})$ vs. $(AB_{32} - AB_{33})$	0.01	1, 6.7	21.17	NS
12. $(AB_{22} - AB_{23})$ vs. $(AB_{32} - AB_{33})$	0.18	1, 9.6	15.92	NS
13. $(AB_{12} - AB_{14})$ vs. $(AB_{22} - AB_{24})$	0.06	1, 17.8	12.03	NS
14. $(AB_{12} - AB_{14})$ vs. $(AB_{32} - AB_{34})$	5.81	1, 8.1	17.94	NS
15. $(AB_{22} - AB_{24})$ vs. $(AB_{32} - AB_{34})$	4.58	1, 9.5	16.02	NS
16. $(AB_{13} - AB_{14})$ vs. $(AB_{23} - AB_{24})$	0.38	1, 17.5	12.10	NS
17. $(AB_{13} - AB_{14})$ vs. $(AB_{33} - AB_{34})$	4.41	1, 8.6	17.15	NS
18. $(AB_{23} - AB_{24})$ vs. $(AB_{33} - AB_{34})$	5.57	1, 10.5	15.08	NS

[a]S = significant; NS = nonsignificant.

4.7.5 Contrast Testing on Marginal Means

Although the presence of a significant $A \times B$ interaction effect would preclude any further analysis of the between- and within-subjects main effects, we present the SAS statements that would generate these contrast tests for illustrative purposes.

To conduct all possible pairwise contrasts on the between-subjects marginal means using a separate or nonpooled error term approach, we used the BMDP t test program, 3D, and the following SAS statements:

```
34.   PROC BMDP PROG = BMDP3D DATA = ALL UNIT = 3;
35.   PARMCARDS;
36.   /PROB TITLE = 'CONTRASTS ON BETWEEN SUBJECTS
      MARGINAL MEANS'.
37.   /INPUT UNIT = 3. CODE = 'ALL'.
38.   /TWOGROUP GROUPING = A.
39.   VARIABLE = AVG.
40.   /END
41.   /FINISH
42.   ;
```

Note that, like the test of the between-subjects main effect, the dependent variable is the average score for each subject, that is, AVG.

The 3D program conducts all possible pairwise contrasts among the levels of the grouping variable, in this case factor A, and for each pairwise contrast calculates two t statistics and associated d.f., one based on a pooled error term approach and the other on a separate error term approach. In the separate error term approach, the d.f. are calculated using Satterthwaite's solution. The values of the $\binom{4}{2} = 3$ nonpooled statistics and the associated Satterthwaite d.f. are presented in Table 4.4. For consistency with the omnibus results, these values are reported as F statistics, where $t^2 = F$. Referring these F values to a studentized range critical value, $(q[.05; 3, v_s])^2/2$, all contrasts were judged statistically significant.

To conduct contrast tests on the marginal within-subjects means while circumventing the assumption of multisample sphericity, we have recommended the KKS statistic and Satterthwaite's solution for d.f. given in (4.31) and (4.32), respectively. Unfortunately, neither this statistic nor its associated d.f. currently can be obtained directly from any of the popular statistical packages. The SAS PROC MEANS, however, can be used to obtain intermediate quantities that are necessary in the calculation of the KKS statistic and its associated d.f.

Recall that the KKS statistic is the ratio $\hat{\psi}/\hat{\sigma}(\hat{\psi})$, where, for pairwise contrasts and for tests of unweighted hypotheses,

$$\hat{\psi} = \frac{\Sigma_j(\overline{X}_{jk} - \overline{X}_{jk'})}{A} \quad \text{and} \quad \hat{\sigma}(\hat{\psi}) = \left\{\frac{\Sigma_j[(s_{jk}^2 + s_{jk'}^2 - 2s_{jkk'})/n_j]}{A^2}\right\}^{1/2}.$$

Letting

$$D_{ij} = X_{ijk} - X_{ijk'},$$

then

$$\overline{D}_j = \frac{\Sigma_j(X_{ijk} - X_{ijk'})}{n_{jk}} = \frac{\Sigma_j X_{ijk}}{n_{jk}} - \frac{\Sigma_j X_{ijk'}}{n_{jk'}} = \overline{X}_{jk} - \overline{X}_{jk'}.$$

The quantities to be averaged in the numerator of the KKS statistic, therefore, are mean differences. Similarly, $\hat{\sigma}_{D_{ij}}^2 = \hat{\sigma}_{X_{ijk} - X_{ijk'}}^2 = s_{jk}^2 + s_{jk'}^2 - 2s_{jkk'}$ and, therefore, the quantities to be averaged in the denominator of the KKS statistic are variances of difference scores. Accordingly, in order to arrive at these intermediate quantities, one can define a series of difference scores which reflect the contrasts of interest and then calculate the means and variances of these difference scores.

In our numerical example, a series of six D variables, representing all the possible pairwise contrasts among the levels of the within-subjects factor B, were defined when we created our original SAS data set. Thus, the following SAS statements produced the desired intermediate quantities:

43. PROC MEANS DATA = ALL MAXDEC = 2 MEAN VAR;
44. VAR D1 D2 D3 D4 D5 D6;
45. BY A;

It should be noted that a less direct way to arrive at these intermediate quantities is to calculate the cell means and covariance matrices of the original B1, . . . , B4 variables, using the SAS PROC MEANS and PROC CORR.

To illustrate the calculation of the KKS statistic, consider the pairwise contrast between the marginal means of levels B_1 and B_2, which is reflected in the D1 variable. The three mean differences for the D1 variable calculated by the SAS MEANS procedure were -1.77 (A_1), -1.50 (A_2) and 3.14 (A_3). Thus,

$$\hat{\psi} = \frac{(-1.77) + (-1.50) + 3.14}{3} = -0.043.$$

Similarly, the three variances of the D1 variable were 2.03 (A_1), 6.94 (A_2) and 9.81 (A_3). Thus,

$$\hat{\sigma}(\hat{\psi}) = \left[\frac{(2.03/13) + (6.94/10) + (9.81/7)}{3^2}\right]^{1/2} = .5002$$

and $t_{KKS} = -.043/.5002 = -.086.$

The intermediate quantities produced by the SAS PROC MEANS are also necessary for the calculation of the Satterthwaite d.f. given in (4.32). As $\mathbf{c}^T\mathbf{S}_j\mathbf{c} = s_{jk}^2 + s_{jk'}^2 - 2s_{jkk'}$, ν_s is equal to

$$\frac{(.5002)^4}{\dfrac{1}{3^4}\left(\dfrac{(2.03/13)^2}{12} + \dfrac{(6.94/10)^2}{9} + \dfrac{(9.81/7)^2}{6}\right)} = \frac{.06260006}{.00472677} = 13.24.$$

The values of the six test statistics (reported as F values) and the associated Satterthwaite d.f. are presented in Table 4.4. Referring each of these F values to a studentized range critical value, $(q[.05; 4, \nu_s])^2/2$, contrasts numbered two through five were declared significant.

4.7.6   Assessing the Interaction Effect

4.7.6.1   Simple Between-Subjects Effects

To arrive at tests of the simple between-subjects effects of factor A, we used the BMDP 7D program and the following SAS statements:

```
46.   PROC BMDP PROG = BMDP7D DATA = ALL UNIT = 3;
47.   PARMCARDS;
48.   /PROB TITLE = 'SIMPLE BETWEEN SUBJECTS EFFECTS'.
49.   /INPUT UNIT = 3. CODE = 'ALL'.
50.   /HISTOGRAM GROUPING IS A.
51.   VARIABLE IS B1, B2, B3, B4.
52.   /END
53.   /FINISH
54.   ;
```

These SAS statements generate Welch (1951) F and ANOVA F tests of the effects of the between-subjects factor A at each of the four levels of the within-subjects factor B (see Table 4.4). Considering the four simple between-subjects effect tests as a family of tests, each simple effect Welch F test was compared to a critical F value based on a reduced significance level equal to $\alpha/4 = .05/4 = .0125$. Applying this criterion, the simple effects of factor A at all levels of B were declared statistically significant.

4.7.6.2   Simple Within-Subjects Effects

To arrive at tests of the simple within-subjects effects of factor B, we used the SAS PROC GLM and the following SAS statements:

```
55.   PROC GLM DATA = ALL;
56.   MODEL B1 - B4 = /NOUNI;
```

57. REPEATED B;
58. BY A;

That is, by specifying a model which contained only a single within-subjects factor (statements 56 and 57) and by requesting that the analysis be performed at each level of the between-subjects factor A (statement 58), multivariate and adjusted univariate tests of the effect of the within-subjects factor B at each level of A were generated. Table 4.4 contains the F approximations associated with the multivariate tests of the three simple within-subjects effects. Considering the three simple within-subjects effects as a family of tests, each F value was compared to a critical F value based on a reduced significance level equal to $\alpha/3 = .05/3 = .01667$. Accordingly, the simple effects of factor B at A_1 and A_2 were declared significant.

4.7.6.3 Interaction Contrasts

As mentioned previously, a second way to assess interaction effects is to perform a series of interaction contrasts. To arrive at tests of interaction contrasts which use the data from only those cells involved in the contrast in estimating the standard error of the contrast (i.e., a separate error term approach), we used the BMDP 3D program and the following SAS statements:

59. PROC BMDP PROG = BMDP3D DATA = ALL UNIT = 3;
60. PARMCARDS;
61. /PROB TITLE = 'INTERACTION CONTRASTS'.
62. /INPUT UNIT = 3. CODE = 'ALL'.
63. /TWOGROUP GROUPING = A.
64. VARIABLE = D1, D2, D3, D4, D5, D6.
65. /END
66. /FINISH
67. ;

These statements generated all possible pairwise contrasts between the A levels of the between-subjects factor for each of the D variables indicated in statement 64. As each of these D variables represents a pairwise difference among the levels of the within-subjects factor B, these contrast tests are, in fact, interaction contrasts. For example, as D1 = B1 − B2, the contrast tests between the levels of A for the D1 variable are tests of whether the B1 − B2 difference is the same for all possible pairs of levels of the between-subjects factor A. The values of the test statistic and associated d.f. for the $\binom{A}{2}\binom{B}{2} = 18$ interaction contrasts are presented in Table 4.4. Considering these interaction contrasts as a family of tests, each test was referred to a critical value equal to $F[.05/18; 1, \nu_s]$. Adopting this procedure, contrasts numbered eight and nine were judged to be statistically significant.

REFERENCES

Alberton, Y. and Hochberg, Y. (1984). Approximations for the Distribution of a Maximal Pairwise t in Some Repeated Measures Designs. *Comm. Statist.— Theor. Meth.*, *13*: 2847–2854.

Appelbaum, M. I. and Cramer, E. M. (1974). Some Problems in the Nonorthogonal Analysis of Variance. *Psych Bull.*, *81*: 335–343.

Barcikowski, R. S. and Robey, R. R. (1984). Decisions in Single Group Repeated Measures Analysis: Statistical Tests and Three Computer Packages. *Amer. Statist.*, *38*: 148–150.

Bartlett, M. S. (1939). A Note on Tests of Significance in Multivariate Analysis. *Proc. Cambridge Phil. Soc.*, *35*: 180–185.

Best, D. J. and Rayner, J. C. W. (1987). Welch's Approximate Solution for the Behrens-Fisher Problem. *Technometrics*, *29*: 205–210.

Boik, R. J. (1981). A Priori Tests in Repeated Measures Designs: Effects of Nonsphericity. *Psychometrika*, *46*: 241–255.

Box, G. E. P. (1949). A General Distribution Theory for a Class of Likelihood Criteria. *Biometrika*, *36*: 317–346.

Box, G. E. P. (1954). Some Theorems on Quadratic Forms Applied in the Study of Analysis of Variance Problems, II. Effects of Inequality of Variance and Correlation Between Errors in the Two-Way Classification. *Ann. Math. Statist.*, *25*: 484–498.

Brown, M. B. and Forsythe, A. B. (1974). The ANOVA and Multiple Comparisons for Data with Heterogeneous Variances. *Biometrics*, *30*: 719–724.

Carlson, J. E. and Timm, N. H. (1974). Analysis of Nonorthogonal Fixed-Effects Designs. *Psych. Bull.*, *81*: 563–570.

Clinch, J. J. and Keselman, H. J. (1982). Parametric Alternatives to the Analysis of Variance. *J. Ed. Statist.*, *7*: 207–214.

Cole, J. W. L. and Grizzle, J. E. (1966). Applications of Multivariate Analysis to Repeated Measurement Experiments. *Biometrics*, *22*: 810–828.

Collier, R. O., Jr., Baker, F. B., Mandeville, G. K., and Hayes, T. F. (1967). Estimates of Test Size for Several Test Procedures Based on Conventional Variance Ratios in the Repeated Measures Design. *Psychometrika*, *32*: 339–353.

Davidson, M. L. (1972). Univariate Versus Multivariate Tests in Repeated Measures Experiments. *Psych. Bull.*, *77*: 446–452.

Dixon, W. J., Brown, M. B., Engelman, L., Hill, M. A., and Jennrich, R. I. (1988). *BMDP Statistical Software Manual* (Vol. 1). University of California Press, Los Angeles.

Dixon, W. J., Brown, M. B., Engelman, L., Hill, M. A., and Jennrich, R. I. (1990). *BMDP Statistical Software Manual* (Vol. 2). University of California Press, Los Angeles.

Dunn, O. J. (1961). Multiple Comparisons Among Means. *J. Amer. Statis. Assoc.*, *56*: 52–64.

Edgington, E. S. (1974). A New Tabulation of Statistical Procedures Used in APA Journals. *Amer. Psych.*, *29*: 25–26.

Games, P. A. and Howell, J. F. (1976). Pairwise Multiple Comparison Procedures with Unequal Ns and/or Variances: A Monte Carlo Study. *J. Ed. Statist.*, *1*: 113–125.

Greenhouse, S. W. and Geisser, S. (1959). On Methods in the Analysis of Profile Data. *Psychometrika*, *24*: 95–112.

Greenwald, A. G. (1976). Within-Subjects Designs: To Use or Not to Use? *Psych. Bull.*, *83*: 314–320.

Herr, D. G. and Gaebelein, J. (1978). Nonorthogonal Two-Way Analysis of Variance. *Psych. Bull.*, *85*: 207–216.

Hochberg, Y. (1988). A Sharper Bonferroni Procedure for Multiple Tests of Significance. *Biometrika*, *75*: 800–802.

Hochberg, Y. and Tamhane, A. C. (1987). *Multiple Comparison Procedures*. Wiley, New York.

Hotelling, H. (1931). The Generalization of Student's Ratio. *Ann. Math. Statist.*, *2*: 360–378.

Hotelling, H. (1951). A Generalized *t* Test and Measure of Multivariate Dispersion. *Proceedings of the Second Berkeley Symposium on Mathematical Statistics and Probability*. Neyman, J. (ed.), University of California Press, Berkeley, vol. 2, pp. 23–41.

Huynh, H. (1978). Some Approximate Tests for Repeated Measurement Designs. *Psychometrika*, *43*: 161–175.

Huynh, H. S. and Feldt, L. (1970). Conditions Under Which Mean Square Ratios in Repeated Measurements Designs Have Exact *F* Distributions. *J. Amer. Statist. Assoc.*, *65*: 1582–1589.

Huynh, H. and Feldt, L. S. (1976). Estimation of the Box Correction for Degrees of Freedom from Sample Data in Randomized Block and Split-Plot Designs. *J. Ed. Statist.*, *1*: 69–82.

Imhof, J. P. (1962). Testing the Hypothesis of No Fixed Main-Effects in Scheffé's Mixed Model. *Ann. Math. Statist.*, *33*: 1085–1095.

Keselman, H. J. (1982). Multiple Comparisons for Repeated Measures Means. *Multiv. Behav. Res.*, *17*: 87–92.

Keselman, H. J. (1991). Stepwise Multiple Comparisons of Repeated Measures Means Under Violations of Multisample Sphericity. Paper presented at a Symposium on Biostatistics and Statistics in Honour of Charles W. Dunnett (May, McMaster University, Hamilton, Ontario).

Keselman, H. J. and Keselman, J. C. (1988). Repeated Measures Multiple Comparison Procedures: Effects of Violating Multisample Sphericity in Unbalanced Designs. *J. Ed. Statist.*, *13*: 215–226.

Keselman, J. C. and Keselman, H. J. (1990). Analysing Unbalanced Repeated Measures Designs. *Brit. J. Math. Statist. Psych.*, *43*: 265–282.

Keselman, H. J., Keselman, J. C., and Shaffer, J. P. (1991). Multiple Pairwise Comparisons of Repeated Measures Means Under Violation of Multisample Sphericity. *Psych. Bull.*, *110*: 162–170.

Keselman, H. J., Rogan, J. C., Mendoza, J. L., and Breen, L. J. (1980). Testing the Validity Conditions of Repeated Measures *F* Tests. *Psych. Bull.*, *87*: 479–481.

Keuls, M. (1952). The Use of the 'Studentized Range' in Conjunction with an Analysis of Variance. *Euphytica, 1*: 112–122.

Kirk, R. E. (1982). *Experimental Design: Procedures for the Behavioral Sciences.* Brooks/Cole, Monterey, California.

Kogan, L. S. (1948). Analysis of Variance: Repeated Measurements. *Psych. Bull., 45*: 131–143.

Lawley, D. N. (1938). A Generalization of Fisher's z Test. *Biometrika, 30*: 180–187, 467–469.

Lewis, C. and van Knippenberg, C. (1984). Estimation and Model Comparisons for Repeated Measures Data. *Psych. Bull., 96*: 182–194.

Looney, S. W. and Stanley, W. B. (1989). Exploratory Repeated Measures Analysis for Two or More Groups: Review and Update. *Amer. Statist., 43*: 200–225.

Mauchly, J. W. (1940). Significance Test for Sphericity of a Normal n-Variate Distribution. *Ann. Math. Statist., 29*: 204–209.

Maxwell, S. E. (1980). Pairwise Multiple Comparisons in Repeated Measures Designs. *J. Ed. Statist., 5*: 269–287.

Maxwell, S. E. and Arvey, R. D. (1982). Small Sample Profile Analysis with Many Variables. *Psych. Bull., 92*: 778–785.

Maxwell, S. E. and Delaney, H. D. (1990). *Designing Experiments and Analyzing Data: A Model Comparison Perspective.* Wadsworth, Belmont, California.

McCall, R. B. and Appelbaum, M. I. (1973). Bias in the Analysis of Repeated Measures Designs: Some Alternative Approaches. *Child Develop., 44*: 401–415.

Mendoza, J. L. (1980). A Significance Test for Multisample Sphericity. *Psychometrika, 45*: 495–498.

Mendoza, J. L., Toothaker, L. E., and Nicewander, W. A. (1974). A Monte Carlo Comparison of the Univariate and Multivariate Methods for the Groups by Trials Repeated Measures Design. *Multiv. Behav. Res., 9*: 165–178.

Milligan, G. W., Wong, D. S., and Thompson, P.A. (1987). Robustness Properties of Nonorthogonal Analysis of Variance. *Psych. Bull., 101*: 464–470.

Newman, D. (1939). The Distribution of the Range in Samples from a Normal Population Expressed in Terms of an Independent Estimate of Standard Deviation. *Biometrika, 31*: 20–30.

Noe, M. J. (1976). A Monte Carlo Survey of Several Test Procedures in the Repeated Measures Design. Paper presented at the meeting of the American Educational Research Association, (April, San Francisco, California).

O'Brien, R. G. and Kaiser, M. K. (1985). MANOVA Method for Analyzing Repeated Measures Designs: An Extensive Primer. *Psych. Bull., 97*: 316–333.

Olson, C. L. (1974). Comparative Robustness of Six Tests in Multivariate Analysis of Variance. *J. Amer. Statist. Assoc., 69*: 894–908.

Pillai, K. C. S. (1955). Some New Test Criteria in Multivariate Analysis. *Ann. Math. Statist., 26*: 117–121.

Rogan, J. C., Keselman, H. J., and Mendoza, J. L. (1979). Analysis of Repeated Measurements. *Brit. J. Math. Statist. Psych., 32*: 269–286.

Rouanet, H. and Lépine, D. (1970). Comparison Between Treatments in a Repeated Measures Design: ANOVA and Multivariate Methods. *Brit. J. Math. Statist. Psych., 23*: 147–163.

Roy, S. N. (1953). On a Heuristic Method of Test Construction and Its Use in Multivariate Analysis. *Ann. Math. Statist.*, *24*: 220–238.

Ryan, T. A. (1960). Significance Tests for Multiple Comparison of Proportions, Variances and other Statistics. *Psych. Bull.*, *57*: 318–328.

SAS Institute, Inc. (1990a). *SAS/STAT User's Guide (Vol. 1) Version 6* (4th ed.). SAS Institute, Inc., Cary, North Carolina.

SAS Institute, Inc. (1990b). *SAS/STAT User's Guide (Vol. 2) Version 6* (4th ed.). SAS Institute, Inc., Cary, North Carolina.

SAS Institute, Inc. (1990c). *SAS Procedures Guide Version 6* (3rd ed.). SAS Institute, Inc., Cary, North Carolina.

SPSS, Inc. (1990). *SPSS Reference Guide*. SPSS, Inc., Chicago, Illinois.

Satterthwaite, F. E. (1941). Synthesis of Variance. *Psychometrika*, *6*: 309–316.

Satterthwaite, F. E. (1946). An Approximate Distribution of Estimates of Variance Components. *Biometrics*, *2*: 110–114.

Scheffé, H. (1959). *The Analysis of Variance*. Wiley, New York.

Seaman, M., Levin, J. R., and Serlin, R. C. (1991). New Developments in Pairwise Multiple Comparisons: Some Powerful and Practicable Procedures: *Psych. Bull.*, *110*: 577–586.

Searle, S. R. (1987). *Linear Models for Unbalanced Data*. Wiley, New York.

Stoloff, P. H. (1970). Correcting for Heterogeneity of Covariance for Repeated Measures Designs of the Analysis of Variance. *Ed. Psych. Meas.*, *30*: 909–924.

Timm, N. H. (1975). *Multivariate Analysis with Applications in Education and Psychology*. Wadsworth, Belmont, California.

Tomarkin, A. J. and Serlin, R. C. (1986). Comparison of ANOVA Alternatives under Variance Heterogeneity and Specific Noncentrality Structures. *Psych. Bull.*, *99*: 90–99.

Tukey, J. W. (1953). The Problem of Multiple Comparisons. Unpublished manuscript, Princeton University.

Wallenstein, S. and Fleiss, J. L. (1979). Repeated Measurements Analysis of Variance When the Correlations Have a Certain Pattern. *Psychometrika*, *44*: 229–233.

Welch, B. L. (1947). The Generalization of Student's Problem when Several Different Population Variances Are Involved. *Biometrika*, *34*: 28–35.

Welch, B. L. (1951). On the Comparison of Several Mean Values: An Alternative Approach. *Biometrika*, *38*: 330–336.

Welsch, R. E. (1977). Stepwise Multiple Comparison Procedures. *J. Amer. Statist. Assoc.*, *72*: 566–575.

Wilcox, R. R. (1987). *New Statistical Procedures for the Social Sciences*. Lawrence Erlbaum Associates, Hillsdale, New Jersey.

Wilks, S. S. (1932). Certain Generalizations in the Analysis of Variance. *Biometrika*, *24*: 471–494.

Wilson, K. (1975). The Sampling Distribution of Conventional, Conservative and Corrected F-ratios in Repeated Measurement Designs with Heterogeneity of Covariance. *J. Statist. Comput. Simul.*, *3*: 201–215.

Wilson, R. S. (1975). Analysis of Developmental Data: Comparison Among Alternative Methods. *Dev. Psych.*, *11*: 676–680.

Winer, B. J. (1971). *Statistical Principles of Experimental Design* (2nd ed.). McGraw-Hill, New York.

5

Latin Square Designs

JOHN W. COTTON University of California, Santa Barbara, California

5.1 INTRODUCTION. THE $p \times p$ LATIN SQUARE: BALANCING THREE FACTORS BUT CROSSING ONLY TWO AT A TIME

5.1.1 Example and Purpose of a Latin Square Experimental Design

If we were conducting an experiment on traffic safety and wanted to study the visibility of traffic signs as a function of shape, color, and illumination, we might use a so-called Latin square design such as the one below. Suppose that rows represent values of one experimental variable, such as shape of pattern, e.g., circle, square, triangle, and diamond. Let columns represent a second variable, such as color: red, green, blue, and yellow. Also let the interior symbols represent values of a third variable, such as illumination: 10, 100, 1000, and 10,000 in some reasonable units. Consider the square structure:

Row treatment	Column treatment			
	Red	Green	Blue	Yellow
Circle	10	100	1000	10,000
Square	100	10	10,000	1000
Triangle	1000	10,000	10	100
Diamond	10,000	1000	100	10

This square has a certain elegance, for we can see that no row or column emphasizes one illumination (the interior or so-called Latin variable) more than any other. More precisely, every row contains each of the first four illuminations exactly once, as does each column. This property is what is required for a 4 × 4 Latin square. We can, of course, substitute a different set of four symbols for our Latin variable, labeling its illumination values, I_1, I_2, I_3, and I_4, for example. Often an experimenter increases the amount of data available for analysis by replicating a single Latin square—having n observations in each cell of a design such as the one just described. Alternatively, independent Latin squares can be used for replications.

How might we interpret each of the 16 cells in our original square? In row 1, column 1 there is a 10. The row label implies that this treatment combination involves a circle, the column label implies that it is red, and the 10 inside the cell implies that this stimulus is illuminated at level 10. Suppose that, after appropriate instructions and pretraining, we ask 20 experimental subjects to identify this stimulus as quickly as possible. Then the average recognition time for it may be compared to the average recognition time for each of the other 15 cells, studied by a similar method and possibly with different 20 people in each case (making a replicated Latin square design with $n = 20$ replications). The reason for employing a Latin square or a replicated Latin square design for study of the effects of shape, color, and illumination is that the balancing of Latin (illumination) values across rows (shapes) and columns (colors) permits us in effect to perform three experiments with the effort ordinarily required for only one. Averaging data in one way suggests how shape affects reaction time for identifying stimuli, while averaging a second way suggests how color affects reaction time, and averaging a third way suggests how illumination level affects reaction time.

Unfortunately, there is also a very good reason to consider not using a Latin square design in this particular case. *The classical model for analyzing data from a Latin square design implicitly assumes that no interactions between row, column, and Latin variables are present.* In the present example, this assumption does not seem plausible. We will return to this question in Sections 5.1.4 and 5.2.1.3. First, however, we may ask where Latin squares were first constructed and how they were used.

5.1.2 A Little Historical Background

5.1.2.1 Magic Squares in Europe and Asia

Magic squares with constant sums of numbers in each row, column, and main diagonal possibly first appeared in China in about 2200 B.C., and a magic square appeared over a gate in India in A.D. 1100 (Fults, 1974).

Albrecht Durer's engraving "Melancholia I," dated 1514, included a magic 4 × 4 square with a constant sum of 34 for many sets of four entries; its bottom row also contained the date of the painting (Gardner, 1961; Panofsky, 1955).

5.1.2.2 Poetic Latin Squares in Medieval Europe

Arnault Daniel, an Italian troubadour in the eleventh century, popularized the poetic form called a sestina, later used by Dante, Spenser, Sidney, Swinburne, Pound, Auden, and Ashbery, among others (Koch and Farrell, 1982; Riesz, 1971). Cotton and Revlin (1987) have noted that the sestina's typical pattern of six 6-line stanzas (and one 3-line envoy with each line split) corresponds to a Latin square (plus envoy) with words as Latin entries, the same six words at the ends of lines (or half-lines in the envoy) appearing in each stanza in balanced order. We do not know whether any poet working with sestinas was aware of magic squares or of Latin squares. Nor do we know of any mathematical use of Latin squares inspired by the existence of sestinas.

5.1.2.3 Latin Squares Enter Modern Mathematics

Euler (1782) became the first prominent modern mathematician to study Latin squares (Dénes and Keedwell, 1974; Street and Street, 1987). In the process of discussing magic squares in general, Euler presented the "Officer's Problem" of arranging 36 officers of 6 different ranks and from 6 different regiments in a square formation, with each row and each column containing exactly one officer of each rank and exactly one officer from each regiment, but only one each of the 36 possible combinations of regiments and ranks appearing anywhere in the square. This would constitute a 6 × 6 Graeco-Latin square, i.e., a square with 6 different Greek letters and 6 different Latin letters, each combination of Greek and Latin letters occurring exactly once, and both Greek and Latin entries independently meeting the definition of a Latin square in relation to the row and column structure. While he knew of the existence of Graeco-Latin squares of any odd size (number of rows) or any size equal to a multiple of four, Euler hypothesized, and Tarry (1900–01) later showed, that no such 6 × 6 square exists.

5.1.2.4 Latin Squares Become Experimental Designs

Statistical interest in Latin squares stems from agricultural research in which a different fertilizer or a different type of seed is the Latin variable of a field in which the effects of row and column soil differences on crop yield can be removed from the estimated error variance. Yates (1933) and, in the 1925 version of a classic book, Fisher (1970) published early usages of Latin square designs where harvest yields were analyzed as a function of the fertilizer or other treatment used. Here, the Latin variable is the treatment of interest and the row and column variables are simply nuisance variables to be controlled.

5.1.3 Two Types of Latin Square Designs

5.1.3.1 Latin Squares That Are Also 1/p Fractional Factorial Designs

With three independent factors, such as the shape, color, and illumination of our original example, one can easily expand a 4 × 4 Latin square to be a 4 × 4 × 4 factorial design with all 64 possible combinations of values of these three factors studied once. Similarly, a 2 × 2 Latin square could be expanded into a 2 × 2 × 2 factorial design with the conditions combining values of A, B, and C factors as shown below:

A₁ B₁ C₁	$A_1 B_2 C_1$
$A_2 B_1 C_1$	**A₂ B₂ C₁**
$A_1 B_1 C_2$	**A₁ B₂ C₂**
A₂ B₁ C₂	$A_2 B_2 C_2$

Notice that four of the eight possibilities in this 2 × 2 × 2 design are shown in boldface form. They form a 2 × 2 Latin square as shown below:

	Period 1	Period 2
Subject 1	**A₁ B₁ C₁**	**A₁ B₂ C₂**
Subject 2	**A₂ B₁ C₂**	**A₂ B₂ C₁**

where A values are for the row variables, B values are for the column variable, and C values are for the interior or Latin variable. (The Latin labels A and B do not make them Latin variables here.) Notice that one could have made a slightly different Latin square design by using the italicized entries above.

With n subjects for each of the possible eight combinations, we would have a replicated three-way factorial design frequently employed in behavioral science. But with the n subjects for each of the four conditions just displayed, we not only have a replicated Latin square design but also have a ½ fractional factorial design of the sort discussed by Kirk in Chapter 6 of this volume. Coding each factor with 0 and 1 for lower and higher values of the factor labels, we have $A_1 B_1 C_1 = 000$, $A_1 B_2 C_2 = 011$, $A_2 B_1 C_2 = 101$, and $A_2 B_2 C_1 = 110$. Adding the entries such as 000 in modular arithmetic with a modulus of two shows that each of the four patterns in this Latin square sums to 0 (mod 2). This kind of calculation is illustrated in Kirk (1982, pp. 573–577, 683–684). A similar process performed with the four nonboldface patterns above shows that their alternate codings each sum to 1 (mod 2). A ½ fractional factorial design can be constructed by selecting all patterns whose codings sum to a constant in modulo 2 arithmetic. We selected the patterns with a sum of zero; these are half the possible patterns, and so our fraction is ½.

5.1.3.2 Repeated Measurements Latin Squares

Wilk and Kempthorne (1957) point out that some Latin squares cannot be considered fractional factorial designs because certain events are not repeatable or reproducible. These events are called experimental units. For example, in a repeated measurements experiment, one cannot present Subject 1 with Treatment 1 at (time) Period 1 and also present that subject with Treatment 2 at the same period. The experimental unit is the combination of subject and period, and it is nonreproducible. Accordingly, a repeated measurements experiment cannot be a fractional factorial design. Any future reference in this chapter to a Latin square design that is not a fractional factorial design will be a reference to a repeated measurements Latin square design, and it will have unreproducible experimental units.

Sometimes this kind of Latin square experiment is called a square with a treatment factor and two nuisance factors. Indeed, repeated measurements Latin square designs may have subjects and periods as nuisance factors with treatments the only factor of interest. But in other repeated measurements Latin square studies, such as learning experiments, the effects of periods (times or trials) may be important to the experimenter. In other cases the magnitude of difference among subjects may also be of interest. The number of nuisance variables in this design depends, then, on the goals of the experimenter.

Repeated Measurements Latin Squares as Crossover Designs
We may define a crossover design as one in which some or all subjects receive at least two different treatments, crossing over from one treatment to another on at least one period. Sometimes such a design is also called a changeover design. Almost all behavioral science use of a Latin square

design is also use of a repeated measurements Latin square design and thus of a crossover design. In this case p measures are taken from each individual subject, with a different treatment provided for each measurement, and successive measurements are categorized as occurring on different periods or trials. It is important to bring the theory of crossover designs to bear upon analysis of data obtained with repeated measurements Latin squares (Jones and Kenward, 1989).

5.1.4 General Theoretical Remarks

There is a partial match between the two types of Latin squares just defined and two classical Latin square models to be discussed in Section 5.2. These models are special cases of a general model almost identical to that of Wilk and Kempthorne (pp. 221–224). Those authors permit the possibility that one or more of the three Latin square variables may have more than the p levels allotted to it in such a square. Accordingly, their model assumes random sampling of those levels prior to random sampling of one square from the set of all possible squares of a given size. In view of this random sampling of squares, it is possible to invoke randomization theory (Kempthorne, 1973; Scheffé, 1959, pp. 157–158; Wilk and Kempthorne), either to justify permutation tests of relevant statistical hypotheses or to justify conventional F tests as approximations of those tests. Randomization theory has the advantages of allowing interaction effects to be interpreted as error and of eliminating need for a special assumption of equal variance of errors for each cell. However, possibly because conclusions apply to the domain of possible squares rather than to the specific Latin square selected at random, it is more common to use "normal theory" (Scheffé, pp. 151–156; Wilk and Kempthorne, p. 224) in which error consists of measurement error and random sampling error. With normal theory, bias due to interaction is controlled by random sampling of squares in the sense that this permits computation of expected mean squares so that one can assess the relative contribution of different kinds of effects to a given mean square and can tell whether a given F test is legitimate. The interpretation of data is therefore conditional upon the Latin square actually selected rather than upon the set of all possible Latin squares of a given size.

5.1.5 Constructing and Randomly Selecting Latin Squares

5.1.5.1 Unrestricted Latin Squares

In their Table XV, Fisher and Yates (1953) list all standard squares (transformation sets) of 4, 5, and 6 letters (treatments), plus examples for 7, 8, 9, 10, 11, and 12 letters. There are only one standard 2×2 Latin square and one standard 3×3 Latin square. In their Table XVI, Fisher and Yates present complete sets of orthogonal Latin squares for all sizes of square from 3×3 to 9×9, excluding the impossible 6×6 option.

Fisher and Yates point out that there are 576 different possible 4×4 Latin squares but 161,280 different 5×5 squares, with vast increases in the numbers of possibilities as the size, p, of the square increases further. This makes the selection of a random square of stated size potentially quite difficult. The following recommendations for relatively small squares are based on those of Fisher and Yates: first, select a standard square at random; second, for 3×3, 4×4, or 5×5 squares, select a random permutation of all rows but the first, keeping the first row plus that permutation; third, select a random permutation of all columns. The labeling of treatments with treatment numbers can be arbitrary. Or random assignment of numbers to treatments can be done in lieu of column randomization.

With larger squares, Fisher and Yates recommend randomly selecting by a prescribed method one of 9408 standard 6×6 squares and then performing random permutation of all rows and columns of 6×6 squares and randomly assigning treatments to letters. Because of the even larger numbers of standard squares for more than 6 treatments, Kempthorne recommends that when $p > 6$, one should select an arbitrary square before following the rest of the Fisher and Yates 6×6 square procedure.

5.1.5.2 Balanced Repeated Measurements Latin Squares

Sheehe and Bross (1961) provide a convenient method for constructing a balanced repeated measurements Latin square or pair of squares, i.e., square(s) satisfying Williams' criterion (Section 5.2.7.1) of having each treatment precede every other treatment an equal number of times. For any even-numbered p value, balance can be attained by generation of only one square. However, for Latin squares of sizes 3, 5, or 7, no single Latin square of that size will satisfy the criterion for balance; Sheehe and Bross show how to construct a pair of squares of a given size that together satisfy the requirements of balance. For a size of 9, one can find a balanced square example in Street and Street (1987). Note that Street and Street call any Williams' square row-complete (if rows constitute subjects) or column-complete (if columns constitute subjects).

Again we may wish to select squares at random, possibly more to prevent bias in the assignment of treatment sequences to different subjects than to facilitate calculation of expected mean squares across squares (apparently not discussed in the technical literature for balanced squares). However, the Sheehe and Bross procedure tells us almost nothing about the domain of possible balanced squares of a given size. If we wish to select treatment orders or balanced squares at random, we are in trouble except perhaps with the 3 × 3, 4 × 4, and 9 × 9 cases discussed next.

For the 3 × 3 situation, there is only one pair of orthogonal squares (Fisher and Yates), the one given in Section 5.2.7.1. Let the three treatments be called T_1, T_2, and T_3, using any convenient pairing of treatments and labels. We may think of the left-hand square in Section 5.2.7.1 as standard square 1 and the right-hand square as standard square 2. There are two ways to proceed, depending on whether one wishes to treat the experimental design as two distinct Latin squares or as six sequences of three treatments each, with one subject per sequence. For two squares, with subjects already assigned to each square, one could number the first three subjects in standard square 1 as subjects 1, 2, and 3, using any arbitrary rule desired (e.g., alphabetically). The three treatments could also be labeled arbitrarily. Then one could select one random permutation of rows (subject numbers to apply to standard square 1). This amounts to selecting one out of a possible six squares. One should keep the arbitrary assignment of treatment numbers when moving to standard square 2 but assign the remaining subjects arbitrarily as subjects 4 through 6. Selecting a random permutation of rows of standard square 2 yields one of a possible six squares. The two procedures mean that one has randomly selected one pair of 3 × 3 Latin squares out of a possible 6 × 6 = 36 such pairs.

Alternatively, the six subjects to be studied can be randomly assigned to the six treatment sequences in the two standard 3 × 3 squares, yielding 6! = 720 possible experimental designs.

For the 4 × 4 case, let a standard balanced square be one in which the first column has the sequence T_1 T_2 T_3 T_4. There seem to be six such standard balanced squares:

	I	II	III
Subject 1	$T_1 T_2 T_3 T_4$	$T_1 T_2 T_4 T_3$	$T_1 T_3 T_2 T_4$
Subject 2	$T_2 T_4 T_1 T_3$	$T_2 T_3 T_1 T_4$	$T_2 T_1 T_4 T_3$
Subject 3	$T_3 T_1 T_4 T_2$	$T_3 T_4 T_2 T_1$	$T_3 T_4 T_1 T_2$
Subject 4	$T_4 T_3 T_2 T_1$	$T_4 T_1 T_3 T_2$	$T_4 T_2 T_3 T_1$

	IV	V	VI
Subject 1	$T_1 T_3 T_4 T_2$	$T_1 T_4 T_2 T_3$	$T_1 T_4 T_3 T_2$
Subject 2	$T_2 T_4 T_3 T_1$	$T_2 T_1 T_3 T_4$	$T_2 T_3 T_4 T_1$
Subject 3	$T_3 T_2 T_1 T_4$	$T_3 T_2 T_4 T_1$	$T_3 T_1 T_2 T_4$
Subject 4	$T_4 T_1 T_2 T_3$	$T_4 T_3 T_1 T_2$	$T_4 T_2 T_1 T_3$

In every square of this example, rows refer to subjects and columns to periods. All six standard squares are balanced (row-complete). In addition, squares I and III satisfy Street and Street's definition of a complete square, meaning that they are both simultaneously row-complete and column-complete. Wagenaar (1969) has used the term diagram-balanced instead of complete for this type of square. Square I above has previously been displayed by Williams (1949) and by Wagenaar; squares II, III, and VI by Williams; and square V by Sheehe and Bross and by Jones and Kenward.

Since permutation of rows in a standard row-complete square does not change the frequency of adjacent treatments in the rows, each standard square can be row-permuted in each of its possible 4! = 24 ways without destroying its balance. So there are in total 6 × 24 = 144 balanced 4 × 4 squares from which one could select randomly, using a procedure analogous to that already described for unconstrained Latin squares. An alternative procedure is to start with any one standard square, assign one specific subject arbitrarily to its row 1, select a random assignment of treatments to letters (24 possibilities), and assign the other three subjects to the remaining rows at random (6 possibilities), making the probability $\frac{1}{144}$ of selecting any combination of treatment assignment and subject ordering, the same probability as results from random selection of a square by methods provided earlier.

The one known balanced 9 × 9 square may be row-permuted in 9! = 362,880 ways. This gives an excellent basis for randomization, even if no further balanced squares of this size can be found.

Will *every* row permutation of a complete Latin square also be complete? No, here are two such permutations from standard square 1 above and thus from each other as well. Although each is row-complete (balanced), the left square is complete and the right is not. So one can sometimes move from only row-complete to complete (right square to left square) or from complete to row-complete (left square to right square).

$T_1\ T_2\ T_3\ T_4$	$T_1\ T_2\ T_3\ T_4$
$T_3\ T_1\ T_4\ T_2$	$T_4\ T_3\ T_2\ T_1$
$T_2\ T_4\ T_1\ T_3$	$T_2\ T_4\ T_1\ T_3$
$T_4\ T_3\ T_2\ T_1$	$T_3\ T_1\ T_4\ T_2.$

Also, Street and Street report that some row-complete squares have at least one permutation which forms a complete square while others do not. Clearly, additional work in random selection of balanced squares for experimental use is needed in order to permit development of universal procedures.

5.1.6 Some Difficulties Associated with Use of Latin Square Designs

Although reducing the number of required observations is a powerful argument in favor of employing a Latin square design, there are a good many reasons to question its use in cases in which a replicated three-way factorial design is possible. Before they can be discussed, a specialization of our earlier concept of replicated Latin squares (Section 5.1) is necessary: A replicated *repeated measurements* Latin square is one in which there are multiple occurrences of each row with its distinctive treatment pattern such as $T_1 T_2 T_3$.

Five advantages of the replicated three-way factorial design over the unreplicated Latin square for the same three factors are: (1) ability to estimate error variance with greater precision because of more degrees of freedom; (2) ability to estimate error variance without bias because of exclusion of interaction components from that estimate; (3) ability to assess main effects separately from interaction effects (Section 5.2.1.3); (4) ability to assess first-order (e.g., *Row* × *Column* = *A* × *B*) and second-order (e.g., *Row* × *Column* × *Latin* = *A* × *B* × *C*) interactions separately; and (5) ability to employ any number of levels for each factor without constraining all three factors to have exactly p levels. Even a replicated Latin square lacks all but the first advantage of the factorial design.

A minor further concern is whether to employ more than three factors in an experiment. Graeco-Latin squares (mentioned in Section 5.1.2.3) are available for investigation of four factors at once; hyper-Graeco-Latin squares with even more factors could also be used (Federer, 1955, Ch. 15). However, these alternate designs have fewer degrees of freedom than Latin squares, and the difficulties just noted for Latin squares apply even more seriously there.

These issues or close counterparts also apply when a three-way factorial design is compared to a repeated measurements Latin square design. Most investigators wish to preserve the option of within-subject comparisons of different treatments; one probably cannot reject the use of repeated measurements Latin squares without rejecting all (or almost all) other such within-subject designs. Therefore, it seems appropriate to use Latin squares as needed, minimizing the difficulties just noted by taking into account recommendations by Wilk and Kempthorne and by Gaito (1958), as well as by taking advantage of new statistical techniques compatible with the use of Williams' balanced designs.

5.2 USING LATIN SQUARE DESIGNS IN
BEHAVIORAL SCIENCE

5.2.1 Using a Classical Fixed-Effects Model for an
Unreplicated $p \times p$ Latin Square

5.2.1.1 The Classical Fixed-Effects Model Itself

Let Y_{ijk} be the observation obtained for row condition i, column condition j, and Latin condition k. Let us assume that

$$Y_{ijk} = \mu + \alpha_i + \beta_j + \gamma_k + e_{ijk} \tag{5.1}$$

where μ is a population constant, α_i is a fixed effect for row i, β_j is a fixed effect for column j, γ_k is a fixed effect for Latin condition k, and e_{ijk}, an error for this observed score, Y_{ijk}, is normally distributed. The mean error is zero and the variance of errors is σ^2.

This model is appropriate either to a Latin square that is also a fractional factorial design or to a repeated measurements Latin square with fixed subject effects. In the former case, experimental procedure should ensure that every error score is uncorrelated with every other error score. In the latter case, for each subject on each period, the covariance between any subject's error scores on any two periods is assumed constant with value $\rho\sigma^2$, generating the so-called symmetric covariance or uniform covariance structure.

If a fixed model is desired, (5.1) applies, of course, to a Latin square of any p value. This equation follows from the general model implicit in Section 5.1.4 if the number of rows, columns, and Latins in the population is assumed to be p rather than some larger number.

This model is a version of the Wilk and Kempthorne model with all levels of row, column, and Latin variables sampled and without their assumption that each set of fixed effects sums to zero. For example, there is no assumption that $\Sigma_{i=1}^p \alpha_i = 0$.

5.2.1.2 Digression About the Problem of
Overparameterization

One reason for making an assumption such as $\Sigma_{i=1}^p \alpha_i = 0$ is as follows: If no constraints are placed on parameters, the number of parameters is likely to be larger than the rank of the design matrix, $\mathbf{X}$, used to summarize which parameter applies to each observation in the data set (Kirk, Ch. 5). Then the rank of $\mathbf{X}$ and also of the square product matrix, $\mathbf{X'X}$, is less than the number of columns of $\mathbf{X}$ and thus of the number of parameters in the model, preventing determination of an inverse, as needed in direct estimation of parameter values. This prevents unique solutions for estimates of the individual parameters, which otherwise would use the equation

$\hat{\beta} = (X'X)^{-1}X'Y$ to solve for the vector of estimated parameter values, $\hat{\beta}$. We call the inability to find a solution to this estimation problem the problem of overparameterization or of having a design matrix with less than full rank.

One approach to this problem is to reduce the number of linearly independent parameters in the model itself. Sufficient reduction by assumptions such as $\Sigma_{i=1}^{p} \alpha_i = 0$ gives the modified design matrix a rank equal to the number of linearly independent parameters in the original design matrix, X. Then the remaining parameters are estimable. This may be called the *sum-to-zero assumption for parameters* or use of a sum-to-zero model and was employed in the Wilk and Kempthorne model on which the implicit model of Section 5.1.4 is based.

An alternate approach, possibly stemming from Cornfield and Tukey (1956, p. 920), is to leave the model in a form that has no special assumptions such as the one just given. Instead, one accepts the fact that the original parameters may not be estimable, settling instead for methods of obtaining unique estimates of differences between pairs of effects such as $\alpha_1 - \alpha_p$. What is done is to employ a generalized inverse, changing the estimation equation to $\hat{\beta} = (X'X)^{-}X'Y$, where the exponent " $-$ " denotes a generalized inverse of the quantity to which it applies. There are infinitely many possible generalized inverses associated with any design matrix that is not of full rank. Each such possibility corresponds to one or more special assumptions about estimated values, not about parameter values. For example, one can either (1) assume $\Sigma_{i=1}^{p} \hat{\alpha}_i = 0$ (a *sum-to-zero assumption for parameter estimates*), (2) assume something like $\hat{\alpha}_1 = 0$ (a *set-to-zero assumption for estimates*, setting the estimate of the first member of a class of effects equal to zero), or (3) assume $\hat{\alpha}_p = 0$ (a set-to-zero assumption for estimates, setting the last member of a class of effects equal to zero).

Such different assumptions may yield wildly different estimated values of individual parameters. However, if a given parameter difference such as $\alpha_1 - \alpha_p$ is estimable, then employing reasonable methods for obtaining individual (and biased) parameter estimates such as $\hat{\alpha}_1$ and $\hat{\alpha}_p$, for example, ensures that their difference is indeed an unbiased estimate of $\alpha_1 - \alpha_p$, regardless of which assumption about parameter estimates was made (Searle, 1987, pp. 282–284). To check on estimability of such differences, one needs to examine the SAS® GLM output for Type II Estimable Functions for Periods, for example, possibly getting help from Milliken and Johnson (1984), SAS Institute (1988, pp. 96–100), and Searle (1987, pp. 465–468).

In many cases, a sum-to-zero or set-to-zero assumption results in the same conclusions. An example in which these assumptions produce discrepant results will be given in Section 5.4.4.

5.2.1.3   Phenomena Associated with Interaction Effects: Confounding, Aliasing, and Bias in Estimating Error Variance

Notice that no interaction term exists in (5.1), but such interaction is very commonly present in behavioral data. When one performs a Latin square experiment, the presence of all possible interactions would imply that estimation of one effect is always *confounded* or *aliased* with another effect. These two closely related expressions need to be defined. Scheffé (1959, pp. 154–155) discusses confounding in Latin squares, showing that for a sum-to-zero version of the model (5.1), expanded to include three two-way interactions and a three-way interaction, the expected value of the mean for Treatment 1 is equal to $\mu + \gamma_1$ in present notation plus a weighted sum of selected interaction terms involving rows and columns, plus interaction terms involving rows, columns, and treatments. Similarly, a contrast used to estimate a quantity such as $\gamma_1 - \gamma_2$ is shown to have an expected value equal to that quantity plus a function of interactions. Accordingly, one cannot know whether a certain value of $\hat{\gamma}_1 - \hat{\gamma}_2$ reflects a difference in treatment effects, a difference in weighted interaction effects, or some combination thereof. The combination of design properties and statistical model producing an inability to estimate one class of effects separately from another is called confounding. We will use the term aliasing for the exact pairing of completely confounded effects in a specific design and model.

We state without proof that the methods of Kirk (1982, pp. 666–667) imply that with a three-way interaction of A, B, and C as what is called the defining contrast, a 2×2 Latin square has its A effect aliased with the B $\times$ C interaction, its B effect with the A $\times$ C interaction, and its C effect with the A $\times$ B interaction. These relations are symmetric: the A $\times$ B effect is also aliased with the C effect, etc. Kirk (pp. 683–685) shows that aliasing patterns are more complex with larger Latin squares than with the size just discussed. For example, Kirk discusses a 3×3 square based on a specific component of three-way interaction, leading to aliasing of the A main effect with specific components of B $\times$ C and A $\times$ B $\times$ C interactions.

Confounding or aliasing may be useful as a means of reducing the amount of experimental effort expended by an investigator. However, it is often undesirable because of confusion about the source of estimated effects or significant differences associated with those effects. If we could escape the problem of confounding, we would prefer using Latin squares because this use would reduce the number of required observations. One method of escape is to provide supporting evidence for an assumption that the interactions confounded with main effects are not present in the data studied. A second method appropriate to the case of squares which are indeed fractional factorial designs is provided by Box, Hunter, and Hunter (1978, pp. 386–388) for a half-fraction of a 2^5 factorial design. They assume that no more than four of the five independent variables will exhibit any main effects and that no interaction involves the fifth independent variable. Hence the ultimate data analysis will ignore at least one original independent variable, reducing the design to a 2^4 factorial. This approach appears to apply directly to a replicated 2×2 Latin square, which reduces to a replicated 2^2 factorial if one can assume that no more than two independent variables have main effects and that no interaction involves the third independent variable. It deserves attention for higher-order fractional factorial Latin squares as well.

Now consider the problem of estimating error variance in the presence of interaction. If all three possible two-way interactions and the only possible three-way interaction exist, the expected value of a mean square for error will include contributions from each interaction, inflating that mean square, on the average. Corresponding mean squares for effects such as rows would also be inflated but not in a way that leads to an exact F test. These problems with fixed and mixed Latin square models lead one to ask whether Latin square designs should be employed in cases in which interactions may exist. The effects (on hypothesis test results) of inflating expected mean squares, $E(MS)$ values, may be seen by considering tables of these expected mean squares for fixed models (Wilk and Kempthorne, Table 3) and for random or mixed finite models or for random or mixed infinite models (Wilk and Kempthorne, Table 2).

Gaito (1958) presents a psychologically oriented discussion of all cases except the random or mixed finite models. His comparison of $E(MS)$ values for the numerators and denominators of various F tests shows negative bias (expectation of numerator over expectation of denominator is less than if all interactions were absent) in most cases. This suggests but does not absolutely imply that the expected value of F is too small in those cases, making significance tests conservative. However, Gaito also notes several instances of positive bias, as in the case of only one random independent variable, with a three-way interaction and a two-way interaction having certain constraints, suggesting that the reported F will be too large, on the average. In the cases discussed by Gaito, the presence of fixed-effect interactions implies that they appear in the expectations for the numerator and denominator of at least one F ratio for a Latin square, resulting in a doubly noncentral F distribution if other assumptions of analysis of variance are met. On the whole, cases of negative bias are less troublesome than cases of positive bias.

Wilk and Kempthorne (1957, pp. 228–229) are relatively sanguine about the use of Latin squares in some cases in which interactions may exist. On the one hand, they mention that main effects of treatments may be well estimated even though F is not as significant; a recommended corrective method is to transform the data to eliminate existing interactions. On the other hand, they suggest that even though a randomized block design allows unbiased estimation of treatment effects, the failure of the analysis method for that design to remove any existing period (column) effects from the mean squares for error and for treatments may lead to "lower real and apparent error" for the Latin square than for the randomized block design.

5.2.1.4 Univariate Analysis Methods for Data Conforming to the Fixed-Effects Model with Interactions Assumed Absent

Estimation with the Classical Fixed-Effects Model

First a system of estimated parameters based on the classical fixed-effects Latin square model (5.1) which are consistent with the SAS® PROC GLM statistical computing software will be presented. See the relevant SAS® manual (SAS Institute, Ch. 20) or Searle (Ch. 12) for further information on this approach. Let the symbol $\overline{Y}_{i..}$ stand for the sample row mean, also $\overline{Y}_{.j.}$, and $\overline{Y}_{..k}$ for corresponding column and Latin means, and $\overline{Y}_{...}$ for the grand mean of the sample. The least squares estimate of one row parameter difference (effect) is

$$\hat{\alpha}_1 - \hat{\alpha}_p = \overline{Y}_{1..} - \overline{Y}_{p..}, \tag{5.2}$$

i.e., the row 1 parameter minus the row p parameter is estimated by the difference between corresponding row means. More generally,

$$\hat{\alpha}_i - \hat{\alpha}_p = \overline{Y}_{i..} - \overline{Y}_{p..} \tag{5.3}$$

applies for any row i. Corresponding rules for estimating parameter differences in column and Latin effects are

$$\hat{\beta}_j - \hat{\beta}_p = \overline{Y}_{.j.} - \overline{Y}_{.p.} \tag{5.4}$$

and

$$\hat{\gamma}_k - \hat{\gamma}_p = \overline{Y}_{..k} - \overline{Y}_{..p}, \tag{5.5}$$

respectively. It should be emphasized that (5.2) through (5.5) are all obtainable from (5.1) with either a set-to-zero or sum-to-zero assumption for parameter estimates. They are also obtainable from a modified (5.1) with a sum-to-zero assumption for parameters. Using that additional assumption would also allow estimation of parameters (not just differences) as in the relation $\hat{\beta}_j^* = \overline{Y}_{.j.} - \overline{Y}_{...}$, where an additional tilde and (*) superscript are employed to emphasize the distinctiveness of the estimator.

Hypothesis Testing with the Classical Fixed-Effects Model

This procedure takes the customary form of computing sums of squares for each effect and for error, converting to mean squares, and forming F ratios. Appropriate formulas are given below. First, we have an often-present correction term for a sum of squares, the squared grand total of observed scores, divided by p^2 (the total number of observations) in this case:

$$\text{Correction} = \frac{(\Sigma_{i=1}^{p} \Sigma_{j=1}^{p} Y_{ijk})^2}{p^2}. \tag{5.6}$$

Next we have

$$SS_{\text{rows}} = \frac{\Sigma_{i=1}^{p} (\Sigma_{j=1}^{p} Y_{ijk})^2}{p} - \text{Correction}, \tag{5.7}$$

with its first term on the right-hand side being the sum of all squared row observation totals divided by the dimension of the Latin square.

$$SS_{\text{columns}} = \frac{\Sigma_{j=1}^{p} (\Sigma_{i=1}^{p} Y_{ijk})^2}{p} - \text{Correction} \tag{5.8}$$

is analogously defined, with its first right-hand term being the sum of all squared column observation totals, divided by the dimension of the square, and the same correction term as before. The sum of squares for the third kind of effects has a fully predictable form:

$$SS_{\text{Latins}} = \frac{\Sigma_{k=1}^{p} (\Sigma_{i=1}^{p} Y_{ijk})^2}{p} - \text{Correction}. \tag{5.9}$$

We call the next quantity the total sum of squares:

$$SS_{\text{total}} = \sum_{i=1}^{p} \sum_{j=1}^{p} Y_{ijk}^2 - \text{Correction}. \tag{5.10}$$

Finally the sum of squares for error is a residual sum of squares, the total sum of squares less the three sums of squares for main effects:

$$SS_{\text{error}} = SS_{\text{total}} - SS_{\text{rows}} - SS_{\text{columns}} - SS_{\text{Latins}}. \tag{5.11}$$

Degree of freedom values are

$$df_{\text{rows}} = df_{\text{columns}} = df_{\text{Latin}} = p - 1 \tag{5.12}$$

and

$$df_{\text{error}} = p^2 - 3p + 2, \tag{5.13}$$

making the total number of degrees of freedom equal to the total number of observations minus one, as desired. Since (5.13) implies $df_{\text{error}} = 0$ for $p = 2$, the unreplicated 2×2 Latin square design cannot be used in estimating error or performing significance tests. In order to increase power, investigators should also take care to use designs with relatively large df_{error} values. In particular, they should avoid the case $df_{\text{error}} = 1$, for which hypothesis tests are uninterpretable (Draper and Joiner, 1984).

The obvious mean square and F definitions apply, where *category* applies to any specific source of variation in a Latin square design:

$$MS_{\text{category}} = \frac{SS_{\text{category}}}{df_{\text{category}}} \tag{5.14}$$

and row, column, or Latin effects are tested with

$$F = \frac{MS_{\text{category}}}{MS_{\text{error}}}. \tag{5.15}$$

Evaluation of an F value follows the usual procedures. Expected mean squares for this analysis are

$$E(MS_{\text{rows}}) = \sigma^2 + \frac{p \, \Sigma_{i=1}^p (\alpha_j - \overline{\alpha})^2}{p - 1}, \tag{5.16}$$

$$E(MS_{\text{columns}}) = \sigma^2 + \frac{p \, \Sigma_1^p (\beta_j - \overline{\beta})^2}{p - 1}, \tag{5.17}$$

$$E(MS_{\text{Latins}}) = \sigma^2 + \frac{p \, \Sigma_1^p (\gamma_k - \overline{\gamma})^2}{p - 1}, \tag{5.18}$$

and

$$E(MS_{\text{error}}) = \sigma^2. \tag{5.19}$$

This shows that F's for rows, columns, and Latins do indeed test hypotheses that each of these kinds of effects is constant for all conditions studied.

Testing Assumptions Used to Justify the Analysis Methods for the Classical Fixed-Effects Model

What methods exist for evaluating assumptions for the F tests just recommended? First, although with a fixed-effects model there is no obvious population of observations for which a specific random sampling procedure can meaningfully be developed, it seems reasonable nonetheless to check for normality of errors with data from Latin square designs. Second, consider what is known about testing goodness of fit of normally distributed errors from other designs. It can be tested within a single treatment group by standard methods, or pooled tests across two or more independent treatment groups can be conducted (D'Agostino, 1986; Stephens, 1986). Cook and Weisberg (1982, pp. 55–58) discuss advantages and disadvantages of using studentized residuals in graphic investigations of normality. They consider a method of developing a probability envelope for normal data, to be compared with those residuals. Ideally, however, one would have a precise method for testing whether residuals, after fitting a classical fixed-effects Latin square model to data, are consistent with an assumption of normally distributed errors. Wood (1978) has such precision in her method of testing residuals remaining after estimating parameters for a randomized block design model.

Third, the Tukey (1949) test for transformable nonadditivity might be employed with Latin square data. This is a plausible recommendation and may indeed work well in practice. However, the mathematical justification for this test is based on the assumption of a randomized block design and model. It is not clear whether this test would have the F distribution for Latin square data when the additivity assumption plus other standard assumptions hold. Fourth, independence of Y_{ijk} values for different subjects is not tested but typically ensured by running subjects individually, with instructions not to discuss the experiment with other subjects, or by running subjects in groups with methods preventing any subject from knowing exactly what is occurring with other subjects.

Possibly, behavioral scientists should do more evaluation of statistical assumptions by graphic examination of original observations and their residuals after model fitting (Anscombe and Tukey, 1963; Box et al., 1978; Cook and Weisberg, 1982; Gentleman and Wilk, 1975; Goodall, 1983; Jones and Kenward, 1989, pp. 34–39, 234). In some cases, visual judgment about goodness of fit can substitute for formal statistical tests.

Instead of the Mauchly (1940) test, Jones and Kenward (pp. 301–303) provide a generally applicable full-information likelihood ratio (LR) test of uniform covariance structure applicable to repeated measurements Latin squares with either fixed- or mixed-effects models. They illustrate it in the case of a replicated within-subject design having the six sequences required for a Williams' 3 × 3 Latin square pair and also in the case of an unreplicated 14 sequence (random block) design with period and carryover effects included in the model. For the former data set, Jones and Kenward (p. 285) also report a different LR test based on within-subject information only, yielding the same degree of freedom and χ^2 values as with the LR test based on both between-subject and within-subject information. It may be necessary to employ the full-information LR test with most unreplicated within-subject designs. If the appropriate LR test is significant, multivariate analysis of the data is required; otherwise a univariate analysis is acceptable.

5.2.2 Using a Classical Mixed Model with Random Subject Effects

5.2.2.1 The Classical Mixed Model Itself

Let us organize our discussion of classical mixed-model Latin square designs by assigning subjects to rows (i values), periods to columns (j values), and treatments to Latins (k values). Ordinarily, period and treatment effects will be considered as fixed effects, with subjects considered as random effects because of tradition in the behavioral sciences. Now let us modify (5.1) to make subject effects random under the coding just described.

$$Y_{ijk} = \mu + a_i + \beta_j + \gamma_k + e_{ijk}, \qquad (5.20)$$

where a_i is a random effect for subject i and the other elements of (5.20) have the same properties as in (5.1). The model just presented may be viewed as a version of the Wilk and Kempthorne model (Section 5.1.4), with sampling of all possible levels of the row and column variables, sampling from an infinite number of rows (subjects) in the population, and without parameters assumed to sum to zero.

5.2.2.2 Univariate Analysis for Data Conforming to the Classical Mixed Model

Estimation

Given the definitions of row, column, and Latin variables as subjects, periods, and treatments, (5.4) and (5.5) now apply to period and treatment effects, but subject effects require new information. First, note that

$$E(MS_{subj}) = \sigma^2 + p\sigma_a^2. \qquad (5.21)$$

Combining (5.21) with our earlier equation for $E(MS_{error})$ leads to an equation for σ_a^2, which has the following counterpart for estimating that variance:

$$\hat{\sigma}_a^2 = \frac{MS_{subj} - MS_{error}}{p}. \qquad (5.22)$$

Hypothesis Testing

This part of the analysis uses the same sums of squares, mean squares, and related formulas, (5.6) to (5.15), as in the fixed-effects model. Letting $E(MS_{periods}) = E(MS_{columns})$ in (5.17), $E(MS_{treatments}) = E(MS_{Latins})$ in (5.18), and using $E(MS_{error})$ and $E(MS_{subj})$ from (5.19) and (5.21), respectively, permits verification that proper hypotheses are being tested with the F tests proposed.

Evaluating Assumptions of the Classical Mixed Model

The same assumptions of normally distributed errors and uniform covariance structure as made in the classical fixed-effects model (5.1) apply to the classical mixed model (5.20). Both equations are additive, as well, including no interaction terms. It may be noted that behavioral scientists usually prefer the mixed model because of a desire to generalize beyond the subjects actually used in their experiment. However, we almost never perform random sampling of subjects from a clearly defined population. (See Jones and Kenward, pp. 260–261 and 307–311, for mathematical reasons to treat subject effects as fixed.) Accordingly, our generalizations ordinarily should be only about properties of "some population that could have generated this sample," of which there are very many examples, of course. Therefore, we should exercise skepticism about any results obtained with a mixed model analysis which differ from those obtained with a fixed-effects model analysis of the same Latin square data.

5.2.3 Three Possible Advantages of Using Within-Subject (Crossover) Designs

One advantage of using a crossover design such as a repeated measurements Latin square design is the possibility that variability within subjects may be smaller than variability among subjects, with a resulting increase in power if the assumptions of the statistical tests involved are met. Suppose, for example, that p independent groups of p randomly selected subjects each receive different attributional instructions in a social psychological experiment, resulting in a set of treatment effects called γ_k as in (5.20). Kirk's (p. 142) formula for a noncentrality parameter for the F distribution of a one-way ANOVA for p treatment groups of size n simplifies to

$$\phi = \frac{\sqrt{\Sigma (\gamma_k - \bar{\gamma})^2}}{\sigma}, \tag{5.23}$$

if $n = p$ and we do not assume that the γ_k sum to zero. It is easy to see that (5.23) applies also to a Latin square design under either (5.1) or (5.20). But if σ^2 has a smaller value in a within-subject Latin square experiment than in a one-way between-subject experiment, ϕ will be larger with the Latin square design. Since power increases as ϕ increases, this implies that the within-subject design will have greater power than the between-subject design in this case.

A second reason to employ crossover designs is that correlated scores for two treatments may reduce the estimated standard error of an estimated treatment effect difference. Consider a direct-difference t-test comparing arithmetic scores of several subjects on two occasions. Consistent individual differences, reflected in a_i values in models such as (5.20), lead to a positive correlation between scores on two occasions,

$$\rho = \frac{\sigma_a^2}{\sigma^2 + \sigma_a^2}. \tag{5.24}$$

This in turn leads the direct-difference t value to be larger than if there were a zero or negative correlation between these paired scores. Increased power is present, and this increase also holds when the direct-difference t is compared to an independent groups t. Provided that correct analyses are performed, a similar advantage may be expected in Latin square experiments and with most other crossover designs. (Note that it is possible for the value of ρ to be too small to compensate for the halving of degrees of freedom which occurs when one moves from the independent groups t.)

A third possible advantage for crossover designs also stems from the use of each subject as his or her own control. We may be able to test for subject-by-treatment interactions or related effects with certain experimental designs. Such effects seem likely in experimental work and in real life. For example, we suspect that a Davis Cup tennis coach can obtain superior performance from Player 1 by applying Motivation Method 1 rather than Motivation Method 2. However, we also suspect that the coach can obtain superior performance from Player 2 by using Motivation Method 2 with him rather than Method 1. This possible interaction is a cousin of the aptitude-treatment interactions often investigated in educational research (Cronbach and Snow, 1977). A definitive crossover design and statistical test of this particular hypothesis may be difficult to conduct because of problems in assessing each individual's treatment effects independent of period and other effects. Nonetheless, it seems an appropriate goal for experimenters.

5.2.4 The Problem of Carryover Effects in Repeated Measurements Latin Square Designs

One of the related kinds of effects of concern in many situations is the carryover effect (also called a residual effect). *A carryover effect is an effect of a prior treatment upon performance under a current treatment.* The value of this effect may be (a) independent of the degree of similarity between prior and current treatments, making it simply the delayed effect of the prior treatment, or (b) dependent on the specific pair of treatments on the current and prior period. The first situation leads us to talk about the carryover from treatment A on one trial (λ_A), regardless of what treatment is presented on the next trial. The second leads us to talk about the carryover from treatment A on one trial to treatment B on the next trial (λ_{AB}), for example. In the case of an unreplicated Latin square there is always some treatment change from one period to the next. Therefore, it is possible to interpret any carryover effect either as of type (a) or type (b). There is no way to distinguish between types of carryover with this design. It may be noted that carryover effects may be viewed as transfer of training effects or as what Campbell and Stanley (1966) call "multiple treatment interference."

5.2.5 A Mixed Model with Carryover Effects Included

Now let a carryover effect, $\lambda_{k'}$, from treatment k' on the previous trial, be added to the random subjects model (5.20):

$$Y_{ijk} = \mu + a_i + \beta_j + \gamma_k + \lambda_{k'} + e_{ijk}, \qquad (5.25)$$

no other changes in assumptions being made. With this model, it is possible to assess carryover effects as well as the subject, period, and treatment effects in (5.20).

Note that a more complete notation would indicate a k' subscript on the left side of (5.25) as well as where it appears on the right. But crossover design model notation when complete can become cumbersome. A reasonable compromise in this example is, first, to include the i, j, and k subscripts for Y since they are necessary to identify the procedure when that specific observation was made; and, second, to add on the right side of the equation all further subscripts (k') that are also necessary to specify theoretical effects assumed to be operative for that observation. There is no ambiguity in identification of observations with this notation. Clearly, all cell entries in a Latin square design can be generated from the identification of subscripts for the different Y values. Then any prior treatment, k', for use on the right side of (5.25) can be identified by inspection of the resulting table.

5.2.6 Univariate and Multivariate Analyses of Repeated Measurements Latin Square Data, Using a Carryover Model

5.2.6.1 A Univariate Method

The most convenient univariate method of analyzing data conforming to (5.25) is to use the SAS® GLM statistical package, as illustrated in a numerical example in Section 5.4. A significant LR test of uniform covariance structure would indicate need for a multivariate analysis of such data.

5.2.6.2 A Multivariate Method

General

The best multivariate analysis methods for this situation may be those discussed in Jones and Kenward (1989, Ch. 7) for use with crossover designs in general. One of those methods, which uses ordinary least squares parameter estimation despite the presence of general covariance structure, is presented now. The only change from the univariate model of (5.25) is to replace subject effects, a_i, and error, e_{ijk}, by a $p \times p$ variance-covariance matrix (also called a dispersion matrix), Σ, of Y_{ijk} scores for pairs of periods j and j' without restrictions on the amount of correlation between those pairs (Jones and Kenward, p. 284). This matrix is assumed constant for each treatment sequence in the Latin square or replicated Latin square design that is employed.

The philosophy of this technique is that for cases where uniform variance structure is absent, the conventional estimates of parameters or their differences remain unbiased but their standard errors are badly estimated. If one can justify invoking large-sample theory of significance testing, one uses least squares parameter estimates as well as improved estimates of standard errors of those parameter estimates as the basis for such testing. Robust methods described by Jones and Kenward (pp. 270–273, 284–290) provide unbiased estimates of the dispersion matrix of squared standard errors and covariances, justifying tests of hypotheses about individual parameters or differences. These methods ordinarily require writing special computer programs in an approach described below, rather than using standard statistical packages in a routine manner.

Estimating the Dispersion Matrix and Standard Errors of Parameter Estimates

It may seem paradoxical to recommend estimating a variance-covariance matrix, Σ, from unreplicated data, where only one observation from each combination of experimental conditions exists. However, this is what Jones and Kenward advocate. For unreplicated Latin square designs, a between-subject estimate of the dispersion matrix, Σ, is estimable if one uses a weighted sum-of-cross-products matrix of residuals (i.e., observed period k means less their predicted values). For a reasonably sized unreplicated square (the required size is not yet known), a $\mathbf{Q}$ matrix associated with Jones and Kenward's (p. 288) Eq. 7.13 will have an inverse, leading to $\tilde{\sigma}_v = \mathbf{Q}^{-1}\mathbf{r}_v$, with $\tilde{\sigma}_v$ being a vector (symbolized by $\mathbf{v}$) constituting the main-diagonal and lower-off-diagonal entries of the estimated matrix, $\hat{\Sigma}$, and $\mathbf{r}_v$ being similar entries from the weighted sum-of-cross-products matrix of residuals just mentioned. The estimated covariance matrix $V[\hat{\xi}_{OLS}]$ of parameter estimates based on traditional analysis of variance procedures (the ordinary least squares or OLS estimates) is generated from Jones and Kenward's equations (p. 284).

Significance Testing

Now large-sample *t*-tests evaluate the individual parameter differences of interest. Note Jones and Kenward's warnings that asymptotic results are invoked in these tests. For large enough samples to employ the large-sample *t*-tests, Bonferroni adjustments would be appropriate when more than one hypothesis test is conducted within a statistical family such as that of treatment effects. More theoretical and Monte Carlo simulation work on appropriate sample sizes for use with these techniques would be desirable.

Special Techniques with a Replicated Latin Square

For a replicated Latin square design, one has two main options. First, use the within-subject sample dispersion matrix, S, as an unbiased estimate of Σ. Proceed to obtain an unbiased $V[\hat{\xi}_{OLS}]$ by the Jones and Kenward method (pp. 284–288), continuing with large sample significance tests as desired. Second, increase the efficiency of the procedure by including both between-subject and within-subject information. Let s_v be the vector of main- and lower-off-diagonal elements of S. For the current design, also compute $s_v + r_v$ in the process of solving for $\hat{\Sigma}$. Then solve for $V[\hat{\xi}_{OLS}]$ as required with other methods and perform large-sample tests as before.

5.2.7 Reasons to Balance Latin Square Designs for Prior Treatments

5.2.7.1 First-Order Balancing: Adjacent Pairs of Different Treatments are Equally Frequent

This section explains why the use of balanced Latin squares is desirable. Williams (1949) provided theory, together with a computational example, for constructing and using a more constrained type of repeated measurements Latin square. This type of square or pair of different squares has the additional property of each treatment preceding each other treatment the same number of times; i.e., for the subjects of the square(s), each pair of different treatments is equally frequent. For example, with $p = 3$, one must use two different squares, such as the following:

	Period				Period		
	1	2	3		1	2	3
Subject 1	T_1	T_2	T_3	Subject 4	T_1	T_3	T_2
Subject 2	T_2	T_3	T_1	Subject 5	T_2	T_1	T_3
Subject 3	T_3	T_1	T_2	Subject 6	T_3	T_2	T_1

with the result that T_1 precedes T_2 twice, and T_1 precedes T_3 twice, etc.

There are two reasons to use Williams' squares. First, in addition to the subject, period, and treatment effects of other repeated measurements Latin squares, carryover effects can be readily assessed. This makes the absence of tests for conventional interactions less serious. Carryover, e.g., being a property of a specific treatment pairing, is fully as interpretable psychologically as a treatment-by-period interaction. Accordingly, some method of assessing carryover seems desirable, and balanced squares meet the need. Second, although data from unconstrained Latin squares can also be analyzed for carryover effects (Cotton, 1989a), the efficiency of such squares is sometimes dramatically lower than that of balanced squares. (See Section 5.5 for an elaboration of this point.)

In view of the above, balanced repeated measurements Latin squares are almost always more advantageous than other Latin squares. If higher-order carryovers are thought to exist, one may even want to employ Williams' (1950) technique for constructing squares balanced for carryovers from two periods as well as from one period earlier. There is no known advantage or disadvantage associated with using digram-balanced (complete) designs rather than other balanced Latin square designs in repeated measurements studies. However, in agricultural field experiments where adjacency (carryover) effects might appear along both rows and columns, it would be advisable to consider use of digram-balanced designs.

Cochran and Cox (1957, p. 140) note that balanced squares have the disadvantage of larger standard errors for estimated carryover effects than for estimated treatment effects. If investigators wish to reduce this discrepancy, they can follow Cochran and Cox's demonstration of how to make a $p \times p$ square into a $p \times (p + 1)$ design with an extra final period duplicating the treatment pattern for the pth period of the balanced square.

5.2.7.2 Using Orthogonal Sets of Latin Squares

Two or more Latin squares are orthogonal if, when they are superimposed, every combination of treatments from the different squares appears in exactly one cell, with none in more than one cell. Look again at our two balanced 3×3 designs, now superimposed into one square with uppercase letters for treatments in the first original square and lowercase letters for those in the second:

	Period 1	Period 2	Period 3
Subject 1 or 4	T_1t_1	T_2t_3	T_3t_2
Subject 2 or 5	T_2t_2	T_3t_1	T_1t_3
Subject 3 or 6	T_3t_3	T_1t_2	T_2t_1

In the above diagram, Subjects 1 through 3 might be assigned to the square based on capital letters and Subjects 4 through 6 to the other square. Since every possible combination such as T_1t_1 appears exactly once, the two underlying 3 × 3 squares are orthogonal. In the present case, this orthogonal pair of squares also forms a balanced pair of squares. This is not true for larger squares, however, where more than two squares are needed for such a set. Except for 6 × 6, all sizes of squares from 3 × 3 to 11 × 11 can be constructed in orthogonal sets, with $p - 1$ squares being required to form such a set. Should an investigator want to work with orthogonal Latin square pairs of even larger size, helpful information may be found in Street and Street.

The advantage of using an orthogonal set of Latin squares is that it permits the highest possible efficiency of estimating treatment and carryover effects with a given number of observations, just as would the corresponding Williams' squares. For squares larger than 3 × 3, more subjects will be required with an orthogonal set of Latin squares than with Williams' squares. With a limited number of subjects to be used, one might therefore prefer Williams' squares. But, if additional replication of the basic study is intended anyway, use of orthogonal squares will have the advantage of increased diversity of treatment orders.

5.3 SPECIAL TOPICS

5.3.1 Using Data from a Replicated Latin Square

5.3.1.1 A Replicated Square Larger Than 2 × 2

Edwards (1985, pp. 372–381) provides a clear computational example of classical model analysis of data from a 5 × 5 repeated measurements Latin square replicated five times. Significance tests are reported for the usual effects of treatments and periods and for residual error, which is the discrepancy between the 25 cell means of the Latin square and predictions based on means for treatment sequences, treatments, and periods. Edwards' example generalizes easily to other sizes of squares and numbers of replications. Unfortunately, the specific square used by Edwards is a cyclic one with the same pattern used for every sequence, but with different starting points. It yields very poor efficiencies, as noted in Section 5.5.

Ordinarily, one would test only for the presence of first-order carryover effects. However, testing for first-order, second-order, and third-order carryover effects with Edwards' data would lead to separate evaluation of the same components on which Edwards' 12 degrees of freedom sum of squares for residual error was based. (Thompson, 1988, shows an alternate decomposition of this kind of residual sum of squares into $(p - 1) \times (p - 2)$ Treatment $\times$ Period contrasts with 1 df each.) Note that Morsbach, McCulloch, and Clark (1986), using a replicated 3 $\times$ 3 Latin square, computed what they called a Sound Conditions $\times$ Order sum of squares, which is a Treatments $\times$ Periods or Residual sum of squares in my terminology. Because there were only three periods, it can be shown that this sum of squares can also be interpreted as the within-subjects sum of squares for first-order carryover effects. It is relatively easy to use SAS® GLM to program a carryover effects model analysis for data sets such as those of Edwards and of Morsbach et al.

Again a lack of balance in the square employed reduces efficiency: a comparable design with two orthogonal Latin squares selected to provide balance, as in Section 5.2.7.1, would yield an efficiency of estimation of treatment effects of 80.00 compared to 20.00 for the present square.

Analysis of So-Called Sequence Effects

A definition of sequence effects is needed at this point. Suppose that, for a specific treatment sequence (row) in a repeated measurements Latin square, such as $T_1T_2T_3$ for a 3 by 3 square, a constant needs to be added to the prediction on every period for that sequence, different constants being required for other sequences. In that case the carryover parameter, λ_k in (5.25), might be replaced by a new parameter, v_q, where q would represent a specific treatment sequence. The v_q for a certain sequence q is called a sequence effect.

For the following reason, an investigator should not employ a sequence effects model like that just outlined: any specific sequence group member would have a constant sequence effect for every period, even a period prior to treatments differentiating the sequence from other sequences. To see the consequences of this fact, consider a slightly different design: if one treatment sequence group received a sequence of four treatments, $T_1T_1T_2T_1$, on successive periods and another received a different sequence, $T_1T_2T_2T_1$, a sequence effects model could predict different performance

on period 1 for members of the two groups even though they had been
treated identically (receiving treatment T_1) up to that point.

A better option is provided by Milliken and Johnson (1984, pp. 448–
450), considering a three-treatment design with all six possible treatment
sequences. They show estimation formulas for carryover parameter differences based on comparisons of sequence means, i.e., on between-subject
information. They use their F test for sequences to test for carryover effects
with between-subject data, as a supplement to the within-subjects test for
carryover effects.

Unlike the Edwards example, the Morsbach et al. (1986) study included
a significance test for treatment sequences (called group effects in their
paper), using the formula $F = MS_{\text{sequences}}/MS_{\text{Subjects}}$ in the present terminology and obtaining nonsignificant results.

5.3.1.2 Special Difficulties with Replicated 2 × 2 Latin Square Designs

Cotton (1989b) provides an introduction to special issues in the interpretation of data obtained with a replicated 2 × 2 Latin square design. On
the one hand, psychologists such as Poulton and Freeman (1966) have
expressed concern that asymmetrical transfer (differential carryover) may
occur from treatment T_1 to treatment T_2, as opposed to that from treatment
T_2 to treatment T_1, distorting inferences about direct treatment effects.
On the other hand, statisticians such as Grizzle (1965, 1974) and Hills and
Armitage (1979) have warned that period effects can neither be estimated
nor tested in the presence of carryover effects. (This is the analogue of
saying that larger Latin squares involve aliasing of period effects and carryover effects; subsequently, significance tests of differences can be made
among the last $p - 1$ periods but not among the entire set of periods.)

A simplified solution to this problem is to decide that Grizzle's analysis
procedure should be followed, with a test for carryover effects being conducted first. If that test is significant, one should use Period 1 data only
in conducting a t-test for treatment effects. No test for period effects should
be performed. Alternatively, if the carryover test is not significant, the
entire data set can be employed in separate t-tests for treatment and period
effects. Other views about appropriate analysis methods for use with this
design are summarized in Cotton (1989b).

It is also possible to be misled by graphic indications of interactions
between 2 × 2 Latin square variables, concluding falsely that differential
carryover has occurred. For example, the interaction equivalent to carryover is the interaction between periods and treatments, not between periods and treatment orders.

5.3.2 Analysis of Data from Two or More Different Squares of the Same Size

Sometimes an investigator may wish to approximate a three-way factorial design by including more than one Latin square pattern in an experiment. Edwards (pp. 364–372) illustrates the univariate analysis appropriate to a classical mixed model as applied to data from five different 4 × 4 Latin squares. This analysis separates total variation into sums of squares for replicates (squares), subjects within replications, treatments, periods, replications times treatments, replications times periods, and error. This experimental procedure seems most appropriate when the two squares differ in some respect other than treatment sequence (already claimed in Section 5.3.1.1 to be an unreasonable theoretical factor for analysis). For example, squares might differ in the dates on which their data were gathered or in the types of subjects (5-year olds, 10-year olds, etc.) who were observed in the experiment.

A carryover model may also be employed with data based on more than one Latin square. One example of data analysis appropriate to such a model appears in Cochran and Cox (pp. 135–139), who discuss alternate randomization procedures and analyses. (1) One may randomize treatments within each square separately and remove period effects separately from other effects in each square. Here one may want to follow Ratkowsky, Alldrege, and Cotton (1990) in using a modern type II SAS® GLM (SAS Institute, 1988) analysis method to obtain appropriate expected mean squares, justifying all significance tests independently of whether certain others of them proved significant, as is an issue in the variant of type I analysis performed by Cochran and Cox. (2) One may randomize treatments within all periods for all squares at one time and assess period effects pooled across sequences. An analysis of this type appears in Milliken and Johnson (1984, pp. 445–450).

5.3.3 Relation to Incomplete Latin Square and Randomized Block Designs

Sometimes there may be a reason to preserve some degree of balance of row, column, and Latin factors in an experiment without quite employing a Latin square design. It might be, for example, that setting up a certain row in a 4 × 4 Latin square design is more difficult or expensive than using the other three rows alone. It is entirely legitimate to perform an experiment with four levels for the columns and Latin variables but only three levels for the row variable, leaving out the fourth row entirely. This would be an example of an incomplete Latin square design. Hand computational procedures for incomplete Latin square designs analyzed with a classical model (Cochran and Cox, 1957; John, 1971) are not quite as simple as the computational methods given above (Section 5.2.1.4); it may be simpler for most investigators to use a general linear models program such as SAS® GLM. If enough observations are available in an incomplete Latin square design to allow for a reasonable number of degrees of freedom for error, a carryover model may also be used as the basis for analysis of data.

A greater departure from use of the Latin square design would be to employ a randomized block design. Any block (usually a subject) would receive each treatment exactly once in a random order. An arbitrary number of blocks would be employed as rows, but period is no longer the column variable. Period and carryover effects are both assumed nonexistent in the randomized block model. However, one could include such effects in the model and do univariate or multivariate analyses such as those shown for that design by Jones and Kenward (pp. 232–234, 288–290, 303–306).

5.4 A NUMERICAL EXAMPLE

5.4.1 Plan for This Example

In view of the advantages of balanced Latin squares and orthogonal sets of Latin squares (see Section 5.2.7), it seems appropriate to illustrate the analysis of data from one of these kinds of squares. First, a classical analysis based on (5.20) and the equations following it will be presented. Then the carryover model analysis based on (5.25) will be presented. The methods presented are appropriate for other Latin squares as well as for balanced ones. After some discussion of these two analyses, I report a carryover analysis that may be viewed as based on (5.25) either with a sum-to-zero assumption on relevant parameters or with a sum-to-zero constraint on parameter estimates. Comparison of the results of this analysis to earlier results is used to reflect further on the problem of overparameterization.

A fair number of recent examples of Latin squares are given in the behavioral science literature, such as Feeney and Noller (1990), but there are few cases of Williams' squares there. One older example of a balanced Latin square is that of Raymond, Lucas, Beesley, and O'Connell (1957), who analyzed the effects of five different tranquilizers and a placebo on psychoneurotic patients. Since raw data are not reported in that study, it seems appropriate here to make up data for a somewhat different hypothetical experiment. Consider a 4 × 4 balanced Latin square employing repeated measurements. Let the treatments in this experiment be four television episodes viewed by each subject. Observed scores are applause-

Table 5.1 Results of a Hypothetical Experiment on the Effects of Television
Episodes (T_1–T_4) on Subject Applause-Meter Scores Ratings

	\multicolumn{4}{c}{Period}			
	1	2	3	4	$\sum_{j=1}^{4} Y_{ijk}$
Subject 1	68 (T_1)	74 (T_2)	93 (T_4)	94 (T_3)	329
Subject 2	60 (T_2)	66 (T_3)	59 (T_1)	79 (T_4)	264
Subject 3	69 (T_3)	85 (T_4)	78 (T_2)	69 (T_1)	301
Subject 4	90 (T_4)	80 (T_1)	80 (T_3)	86 (T_2)	336
$\sum_{i=1}^{4} Y_{ijk}$	287	305	310	328	1230

$$\sum_{j=1}^{4} Y_{ij1} = 276 \qquad \sum_{j=1}^{4} Y_{ij2} = 298 \qquad \sum_{j=1}^{4} Y_{ij3} = 309 \qquad \sum_{j=1}^{4} Y_{ij4} = 347$$

meter readings, with high scores reflecting a great deal of applause. The
following fixed parameters were employed in generating the data to be
discussed below: $\mu = 70,00$, $\beta_1 = -2.70$, $\beta_2 = -1.40$, $\beta_3 = 0.60$,
$\beta_4 = 5.40$, $\gamma_1 = -4.55$, $\gamma_2 = 1.80$, $\gamma_3 = 5.80$, $\gamma_4 = 16.00$, $\lambda_1 = -2.50$,
$\lambda_2 = 0.85$, $\lambda_3 = 2.80$, and $\lambda_4 = 8.50$, with the subject variance being
$\sigma_a^2 = 25.00$ and the error variance being $\sigma^2 = 1.021$. Scores generated
with (5.25) and these parameters were truncated to yield no nonzero digits
after the decimal sign.

Table 5.1 presents the episodes (treatments) for each subject on each
period, together with hypothetical scores. Total scores indicate an increase
in average applause from Period 1 to Period 4 and a similar increase from
Treatment 1 to Treatment 4. In addition, subject differences appear, with
Subject 2 giving noticeably the least applause.

5.4.2 Classical Mixed-Model Analysis of Data from a Repeated Measurements Latin Square Design

Table 5.2 presents what are called "classical analysis of variance" results
based on (5.3) through (5.15) adapted for a mixed model because it uses
the classical model and methods first developed for analyzing data obtained
with Latin squares. Subject effects ($F_{3,6} = 10.22, p < 0.01$) and TV episode
(treatment) effects ($F_{3,6} = 8.42, p < 0.05$) are significant, but period effects
do not approach significance. Because the classical models for a Latin
square design have all factors orthogonal to one another, we would obtain
exactly the same sums of squares, degree of freedom values, mean squares,
and F values if we used a sum-to-zero assumption for parameters rather
than the current model.

Table 5.2 Analysis of Variance Summary for the Hypothetical Data of Table 5.1, Using the Classical Model of Eq. (5.20) for Repeated Measurements Latin Square Data

Source	Sum of squares	df	Mean square	F
Subjects (Rows)	802.25	3	267.42	10.22**
Periods (Cols)	213.25	3	71.08	2.72
Treatments (Latins)	661.25	3	220.42	8.42*
Error	157.00	6	26.17	
Total	1833.75	15		

$*p < 0.05.$
$**p < 0.01.$

Table 5.3 shows period and treatment (TV episode) means plus associated estimates of differences as well as actual differences in parameter values based on the simulated data of Table 5.1. Use of (5.22) with appropriate values from Table 5.2 yields an estimate for the variance of random subject effects, $\hat{\sigma}_a^2 = 60.31$.

Table 5.3 Parameter Estimates for the Data of Table 5.1, Using a Classical Model [Eq. (5.20)] and the Carryover Model [Eq. (5.25)]

Period mean	$\overline{Y}_{.1.} = 71.75$	$\overline{Y}_{.2.} = 76.25$	$\overline{Y}_{.3.} = 77.50$	$\overline{Y}_{.4.} = 82.00$
$\hat{\beta}_j - \hat{\beta}_4 = \overline{Y}_{.j.} - \overline{Y}_{.4.}$ (classical model)	-10.250	-5.750	-4.500	0
$\hat{\beta}_j - \hat{\beta}_4$ (carryover model)	Not estimable (-16.35 reported as biased est.)	-5.750	-4.500	0
$\beta_j - \beta_4$ (population values)	-7.100	-6.800	-4.800	0
Treatment mean	$\overline{Y}_{..1} = 69.00$	$\overline{Y}_{..2} = 74.50$	$\overline{Y}_{..3} = 77.25$	$\overline{Y}_{..4} = 86.75$
$\hat{\gamma}_k - \hat{\gamma}_4 = \overline{Y}_{..k} - \overline{Y}_{..4}$ (classical model)	-17.75	-12.25	-9.50	0
$\hat{\gamma}_k - \hat{\gamma}_4$ (carryover model)	-20.500	-14.075	-11.025	0
$\gamma_k - \gamma_4$ (population values)	-20.550	-14.200	-10.200	0
Lag 1 mean	$\overline{Y}_{..1'} = 77.67$	$\overline{Y}_{..2'} = 76.00$	$\overline{Y}_{..3'} = 76.67$	$\overline{Y}_{..4'} = 84.00$
$\hat{\lambda}_{k'} - \hat{\lambda}_4$	-11.000	-7.300	-6.100	0
$\lambda_{k'} - \lambda_4$ (population values)	-11.000	-7.650	-5.700	0

5.4.3 Carryover Analysis Paralleling the Classical Analysis Above

Table 5.4 presents the results of analysis of variance calculations performed by SAS® GLM programmed to invoke the model of (5.25). This is a so-called carryover analysis of Table 5.1 data. Subject effects and treatment (TV episode) effects reach the 0.0001 level of significance ($F_{3,3} = 1776.79$, $p < 0.0001$ and $F_{3,3} = 2286.40$, $p < 0.0001$); and tests for period and carryovers reach a slightly lower level ($F_{2,3} = 313.57$, $p < 0.001$ and $F_{3,3} = 447.57$, $p < 0.001$, respectively). A variation of the program used to generate Table 5.4 could be constructed to perform the classical model analysis reported in Tables 5.2 and 5.3.

Calling the mean for scores following treatment T_1 the lag 1 mean for T_1 (and similarly for treatment T_2), we can say that Table 5.3 also includes lag 1 means and associated estimates of differences in parameter values based on the carryover model (5.25) and generated by SAS® GLM.

5.4.3.1 Procedural Discussion of the Carryover Analysis

Programming Details

Appendix A is the SAS® GLM program for the carryover analysis just presented. Assignment of values to input variable columns should be reasonably clear after inspection of the program. One special procedure is to assign the carryover on Period 1 the value 0 because there is no prior treatment to consider. This is a well-known programming maneuver and is not intended to generate a new parameter for the model. Although five carryovers are listed in the Appendix A program, only three degrees of freedom appear for carryovers. Also $\hat{\lambda}_0 - \hat{\lambda}_4 = 0$ (not shown in Table

Table 5.4 Analysis of Variance Summary for the Hypothetical Data of Table 5.1, Using the Carryover Model of (5.25) for Repeated Measurements Latin Square Data[a]

Source	Sum of squares	df	Mean square	F
Subjects (Rows)	621.88	3	207.29	1776.79****
Periods (Cols)	73.17	2	36.58	313.57***
Treatments (Latins)	800.24	3	266.75	2286.40****
Carryovers	156.65	3	52.22	447.57***
Error	0.35	3	0.117	
Total	1833.75	14 (+1 not shown above)		

[a]Nonorthogonal contributions do not sum to SS_{total}
***$p < 0.001$.
****$p < 0.0001$.

5.3). This permits use of λ_4 rather than λ_0 for the reparameterization of Y_{ilk} to appear later in Appendix B (5.B.16). Accordingly, λ_0 functions as a pseudoparameter with no fundamental impact compared to an alternate analysis that does not include λ_0 (not convenient with SAS®).

Note the optional sequence E E1 E2 E3 E4 in the MODELS command, where E generates the general form of estimable functions and E1 generates estimable function information for different effects under type I analysis, etc. To reduce the amount of hard copy from this program, users of this program may replace E E1 E2 E3 E4 by SS2 once the program's implications are understood. These options for the MODEL command are discussed in a general way in the manual for SAS® statistics programs (SAS Institute). One aspect of their results is described below.

In Table 5.4 all sums of squares reported for fixed and random effects are so-called adjusted sums of squares taking into account prior estimation of all other effects and coming from a type II analysis. We can tell this procedure is acceptable from the following information: use of the E2 option in the program's MODEL statement and the separate 'RANDOM SUBJ' command has a by-product of reporting expected mean squares (in current notation) of effects as follows:

$$E(MS_{\text{Subjects}}) = \sigma^2 + 3.634\sigma_a^2,$$

$$E(MS_{\text{Periods}}) = \sigma^2 + Q(\text{Periods}),$$

$$E(MS_{\text{Treatments}}) = \sigma^2 + Q(\text{Treatments}),$$

and

$$E(MS_{\text{Carryovers}}) = \sigma^2 + Q(\text{Carryovers}),$$

where $Q(\textit{effects})$ is an abstract expression for a quadratic function only of the effects listed here. Since the expected value of the mean square for error is σ^2, each of the F tests reported in Table 5.4 is a ratio of quantities with the same expectation, under the null hypothesis for the test involved, partially justifying each type II test.

Careful inspection of the output from the E command (General Linear Models Procedure—General Form of Estimable Functions) and from the E2 (or other desired) command (General Linear Models Procedure—Type II Estimable Functions for: SUBJ, etc.) is also desirable in helping to understand what individual combinations of parameters are estimable with a specific type of analysis. This information is what tells us that $\beta_1 - \beta_4$ is not estimable for the present model.

Why Not Do a Carryover Model Latin Square Analysis Using Type I or Other Analysis?

A type I analysis is a sequential analysis in which extraction of each successive sum of squares is conditional upon prior extraction of all sums of squares preceding it in the MODEL statement or equivalent aspect of a computational procedure. Expected mean squares for type I depend upon the order of this extraction. The type I expected mean squares reported because of the E1 option in our MODEL statement are

$$E(MS_{\text{Subjects}}) = \sigma^2 + 4\sigma_a^2 + Q(\text{Carryovers}),$$

$$E(MS_{\text{Periods}}) = \sigma^2 + Q(\text{Periods,Carryovers}),$$

$$E(MS_{\text{Treatments}}) = \sigma^2 + Q(\text{Treatments,Carryovers}),$$

and

$$E(MS_{\text{Carryovers}}) = \sigma^2 + Q(\text{Carryovers}),$$

justifying only the F test for carryover from the four that type I analysis reports routinely with the commands above. Using a different order of calculation of type I sums of squares could lead to some other legitimate hypothesis test but would not make all four of the desired F tests legitimate. For the present design, type III and type IV analyses yield expected mean squares and sums of squares identical to those of type II. No matter how many different types of analyses are invoked, only one set of (biased) parameter estimates is computed and reported. The difference between types of analysis output shows up only in the sums of squares table for each analysis and in the information on estimability and expected mean squares.

Also, types I through IV always yield identical results for orthogonal model designs such as the classical model Latin square analysis of Table 5.2 or analyses of two-way crossed designs with equal cell numbers.

I prefer to focus on type II because of its interpretation as employing sums of squares conditional upon extraction of sums of squares for all other effects being studied. However, since another type of analysis might be technically superior in some other situation, one should always seek a type of analysis for which the expected mean squares lead to legitimate F tests when the assumptions of the model employed are satisfied. Checking expected mean squares and estimability of parametric functions involved in them is the surest way to accomplish this.

5.4.3.2 Substantive Discussion of the Classical and Carryover Effect Model Analyses

Comparison of parameter estimates with actual values used in generating the hypothetical data under analysis is instructive. Since only one data set was developed, this is not a Monte Carlo test of adequacy of alternate analyses. Rather, it is a single example of results of this kind. A striking feature of Tables 5.2 and 5.4 is the reduction in the mean square for estimated error from 26.17 to 0.117 as one moves from the classical mixed model to the carryover model. This compares to a population value of 1.02. Use of appropriate values from Table 5.4 plus the type II EMS values above yields an estimated variance of subject effects, $\hat{\sigma}_a^2$, equal to 56.97. This is slightly closer to the population value used in the generating this data set, 25.00, than the 60.31 reported with the classical model. Smaller errors result from carryover model estimation of treatment effect differences than from the classical analysis. Carryover differences are closely estimated with the carryover analysis.

The classical and carryover model analyses yield the same estimates of period effects, except that with the carryover model $\beta_1 - \beta_j$ can be estimated only in the trivial case, $j = 1$. An intuitively stated reason for this relies on the definition $\bar{\lambda} = .25(\lambda_A + \lambda_B + \lambda_C + \lambda_D)$. Then, under our assumption that there is no carryover on Period 1, the average period mean, $\bar{Y}_{j.}$, includes $\bar{\lambda}$ on every period except the first, making the expected value of the period difference, $E[\bar{Y}_{.1.} - \bar{Y}_{.4.}]$, be $\beta_1 - \beta_4 - \bar{\lambda}$, without a way to separate the period difference from the mean carryover effect. With the present model and design, it is not possible to test the more interesting hypothesis that $\beta_1 = \beta_2 = \beta_3 = \beta_4$, that is, that all four period effects are equal.

5.4.4 An Analysis Assuming That Each Set of Fixed Parameter Effects or of Parameter Estimates Sums to Zero

Now consider a carryover Latin square model analysis with SYSTAT's™ MGLH module, which appears to assume that the parameter values in an equation such as (5.25) sum to zero for period effects, for carryover effects, and also for treatment effects. Alternatively, it may be viewed as assuming (5.25) without zero sums of parameters but using a procedure requiring

Table 5.5 Parameter Estimates for the Data of Table 5.1, Using the Carryover Model (5.25) but with Each Set of Effect Estimates Summing to Zero

Period	1	2	3	4
$\tilde{\beta}_j$	-5.125	-0.625	0.625	5.125
$\tilde{\beta}_j - \tilde{\beta}_4$	-10.250	-5.750	-4.500	0
Treatment	1	2	3	4
$\tilde{\gamma}_k$	-9.100	-2.675	0.375	11.400
$\tilde{\gamma}_k - \tilde{\gamma}_4$	-20.500	-14.075	-11.025	0
Carryover from:	1	2	3	4
$\tilde{\lambda}_k,$	-4.900	-1.200	0	6.100
$\tilde{\lambda}_k - \tilde{\lambda}_4$	-11.000	-7.300	-6.100	0

parameter estimates in each family of effects to sum to zero. Although Wilkinson shows a new proposed analysis for data from the balanced combination of two 3 × 3 squares previously discussed by Cochran and Cox, it is more helpful here to compare SYSTAT™ output for Table 5.1 data to the carryover analysis reported in Tables 5.3 and 5.4. The purpose for this new analysis is to compare different statistical methods, not simply to compare the results of different statistical packages. Given an appropriate data file in the A drive, e.g., LATIN.SYS, one first enters SYSTAT on the computer and types MGLH. On receiving a ">" prompt, one types successive commands:

```
USE 'A:LATIN.SYS'
CATEGORY SUBJ = 4, PERIOD = 4, TREAT = 4,
   CARRY = 4
MODEL RESP = CONSTANT + PERIOD + TREAT
   + CARRY
PRINT = LONG
ESTIMATE
```

Table 5.5 shows parameter estimates from the new analysis. Since we know that the overparameterized model of (5.25) does not have individual effect parameters that are estimable, we do not expect SYSTAT™ output such as the treatment 1 SYSTAT™ parameter estimate, $\tilde{\gamma}_1$, to be the same as the comparable biased SAS® estimate, which will still be called $\hat{\gamma}_1$. However, Section 5.2.1.2 indicated that overparameterized models yield the same values for estimable functions, regardless of the simplifying assumptions made about estimators in order to permit estimation.

Interpretation of SYSTAT Results as Based on Requirement That Parameter Estimates Sum to Zero

If (5.25) is used in overparameterized form by both programs, this implies that these two programs will yield the same empirical values for the estimable functions of (5.25), making $\hat{\gamma}_1 - \hat{\gamma}_4 = \tilde{\gamma}_1 - \tilde{\gamma}_4$ for the two programs, as an example. A comparison of such entries in Tables 5.3 and 5.5 shows that treatment and carryover parameter difference estimates, as well as two of the period effect differences, do match for the two programs. However, the boldfaced Period 1 minus Period 4 effect difference estimate of Table 5.5 matches only the classical model estimate of Table 5.3. This is because that parameter difference is not estimable within (5.25) without a sum-to-zero assumption about *parameters*.

A further apparent discrepancy between results of analyses based on the two carryover models is revealed in their sums of squares. All the sums of squares in Table 5.4 are also obtained with SYSTAT® except that $SS_{periods} = 213.250$ for SYSTAT® (not shown in a separate table). However, the degrees of freedom for periods are now $p - 1 = 3$, not $p - 2 = 2$ as required by (5.25), with a related change in value for the corresponding mean square and F ratio.

The variant results just reported can be explained: The SAS GLM® program presented in Appendix A includes an ESTIMATE command and a CONTRAST command called PSEUDO BETA_1 − BETA_4 because of contamination by carryovers. This command reflects SAS® evidence that $\beta_1 + \lambda_0 - \beta_4 - \bar{\lambda}$ is estimable, making $\beta_1 - \beta_4 - \bar{\lambda}$ estimable when λ_0 is dropped or set to zero as with direct matrix calculations by GAUSS® (Aptech Systems, 1988). Output resulting from these commands gives an estimated value of -10.25 and $SS_{pseudoperiods} = 213.25$, just as reported for SYSTAT® above without reference to pseudoproperties of the estimates. The associated mean square and F statistic are also the same as before. Accordingly, Table 5.4 shows a test of pseudoperiod effects, testing H: $\beta_1 - \bar{\lambda} = \beta_2 = \beta_3 = \beta_4$ or, equivalently, that the expected Y values for the four periods are equal. With the present model (5.25) and design, it is not possible to test the more interesting hypothesis that $\beta_1 = \beta_2 = \beta_3 = \beta_4$, i.e., that all four period effects are equal unless one adds the arbitrary assumption that members of each parameter family sum to zero. (The specific ESTIMATE and CONTRAST commands just discussed would probably be omitted by a user unless pseudoperiod effects were of special interest in the study being analyzed.)

Continuing to view SYSTAT® as invoking a sum-to-zero assumption about parameter *estimates*, with the overparameterized model as the one of interest, one can also retrieve additional SAS® results from SYSTAT®. To obtain the $SS_{periods}$ value of 73.17 with two degrees of freedom for comparing periods 2 through 4, one first employs the SYSTAT® commands after performing the main SYSTAT® analysis for these data:

HYPOTHESIS
EFFECT = PERIOD
CONTRAST
0 1 −.5 −.5
TEST

to obtain a sum of squares of 32.667 for the comparison of $\hat{\beta}_2 - .5(\hat{\beta}_3 + \hat{\beta}_4)$. The same set of commands (except for a different numerical line of 0 0 .5 −.5) yields a sum of squares of 40.500 for the comparison of .5 ($\hat{\beta}_3 - \hat{\beta}_4$). These two orthogonal sums of squares sum to 73.167, the same value as in Table 5.4.

In summary, if SYSTAT™ is viewed as testing the original model, (5.25), then its use of a sum-to-zero assumption about parameter estimates leads initially to an incorrect test of period effects because proper estimability information is not provided. A correct test can be performed with SYSTAT™ only if appropriate estimability information is available from other sources. Appendix B shows further relations between (5.25) and sum-to-zero and set-to-zero solutions.

Interpreting SYSTAT™ Results Based on an Assumption That Parameter Values Sum to Zero

If SYSTAT™ is viewed as based on a sum-to-zero assumption about parameter families, different models are tested and slightly different, but otherwise legitimate hypothesis tests about period effects are performed by SAS® and SYSTAT™. However, despite the internal consistency of analysis methods using (5.25) with a sum-to-zero assumption about *parameter values*, that assumption is unnecessary and a bit unnatural—unnecessary because various kinds of assumptions about *parameter estimates* solve the problem of obtaining estimated values despite the singularity of the design matrix for the original model, and unnatural because, for example, there is no obvious reason why Periods 1 to 4 should have effects summing to zero in a 4-period experiment and Periods 1 to 5 should also have effects summing to zero in a comparable, 5-period experiment. Note, however, that in some cases without carryover effects, a cell means ANOVA model (e.g., by Searle, 1987, pp. 53–54, with a one-way design) can generate effects that sum to zero algebraically without special assumptions.

It must be noted, first, that there is still a substantial amount of technical literature making a sum-to-zero assumption for population parameters and, second, that some statistical packages appear to be based on this assumption. At least, one should be conscious which model one is testing.

5.5 MEASURING EFFICIENCIES OF DIFFERENT LATIN SQUARE DESIGNS

Jones and Kenward (1989, pp. 194, 216) provide formulas for measuring the efficiencies, E, of different experimental designs used while estimating parameter differences in comparison to the optimal design, using a t-test for mean differences between two independent groups (assuming equal error variances, σ^2, for all designs). These measures permit a judgment about which experimental design is preferable in a given situation under the assumption that the same number of observations is employed with each. A general expression for such an efficiency is

$$E_{\text{estimation of effect difference}} = 100 \, \frac{2\sigma^2/r}{V[\text{estimated effect difference}]} \qquad (5.26)$$

where r is the number of replications of each effect (not effect difference) in the design. The 100 multiplier allows (5.26) to express efficiency in percentage units. Provided that E is constant for all differences in a certain family of effects, such as treatments, (5.26) also yields the efficiency of the design for assessing that class of effects. Otherwise, a separate efficiency must be calculated for each pair of such effects, so that their average value can be reported.

Jones and Kenward consider three kinds of efficiencies: E_t for efficiency of treatment effect difference estimates when no adjustment is made for carryover effects; E_d for efficiency of treatment effect estimates when such adjustment has been made; and E_c for efficiency of carryover effects when adjustment for treatment effects has been made. Because of behavioral science interest in period effects for their own sake, a fourth measure, E_p for efficiency of period effects adjusted for treatment and carryover effects, will also be considered here.

If a classical fixed-effects model (5.1) or classical mixed model (5.20) is employed, we should act as if that model is true and report E_t and E_p, with the latter no longer requiring adjustment for effects other than periods. Jones and Kenward note that E_t is 100.00 for any Latin square design using a classical model. E_p is also equal to 100.00 in that situation. Now consider balanced squares and use of the carryover model (5.25). Jones and Kenward report that here $E_d = 90.91$ and $E_c = 62.50$ for the 4×4 balanced square labeled V in the set of six standard squares in Section 5.1.5.2 above. For all estimable period effect differences (those not involving Period 1), $E_p = 100.00$. These values also apply to square II, the one employed in the computational example associated with Table 5.1.

An algebraic method for estimating variances of parameter differences needed in determining efficiency stems from Searle (p. 349, Eq. 86; p. 474, Eq. 25). It is possible with this and other information associated with (5.26) to compute efficiency without knowing even a sample estimate of population error variance. However, sometimes a more convenient method to obtain values such as those just reported is to examine standard errors of estimates obtained with the SOLUTION command of the SAS® GLM program employed in our computational example. [To be certain that other parameter differences not estimated with SOLUTION do not give inconsistent results, one should also employ enough ESTIMATE commands to yield a total of $p(p - 1)/2$ standard errors of estimate for distinct treatment effect differences or for distinct carryover effect differences, or a total of $(p-1)(p-2)/2$ standard errors of estimate for distinct period effect differences.] Since what SAS® reports as a standard error of estimate for a Treatment 1 effect in the carryover model analysis associated with Tables 5.3 and 5.4 is really the standard error for Treatment 1 minus Treatment 4 effects, SAS® GLM yields 0.25331140 as the standard error for each estimated treatment effect difference. This implies a variance for each treatment effect difference of 0.064166665. But the mean square for error is 0.116667 carried to five significant figures, making

$$E_d = 100 \frac{(2)(.116667)/4}{.064166665}$$

or 90.91, just as Jones and Kenward reported for standard square V of Section 5.1.5.2. The same method shows efficiencies of 100.00 for Table 5.3 carryover model estimates of period effect differences, excluding differences involving Period 1, and of 62.50 for estimates of carryover effects.

The Jones and Kenward definition of efficiency for carryover effect estimation penalizes the experimenter because it acts as if each carryover is replicated p times rather than its actual $p-1$ times. This is a reasonable option, emphasizing the fact that carryover estimation uses less experimental data than other estimation. One is measuring efficiency based on the amount of information provided per observation. Given that definition, it is easy to extend Jones and Kenward's $E_c = 62.50$ result to the Table 5.1 design, using the same kind of evidence from SAS output for other effects. Some investigators may choose to use what Patterson and Lucas (1962) call efficiency in terms of the potential information available. For the carryover model (5.25) applied to a Latin square design, this amounts to defining

$$E_c^* = 100 \frac{2\sigma^2/(p-1)}{V[\hat{\lambda}_k - \hat{\lambda}_{k,}]}, \tag{5.27}$$

where the asterisk implies that efficiency is assessed on the basis of the potential information available and $p-1$ rather than p is used as the number of replications. With this definition, the Table 5.1 design has $E_c^* = 83.33$ rather than 62.50.

The foregoing remarks show that unconstrained Latin squares are optimal when the classical model applies, with no carryover effects or interactions present, and balanced Latin squares are excellent even when a carryover model holds. What can be said about use of an unconstrained Latin square when a carryover effects model is to be tested? If one wishes to employ such a square in that situation or to reanalyze existing data from such a square using a carryover model, one should certainly compute efficiencies and/or average efficiencies for that square for all effects of interest. In some cases the resulting efficiencies may be almost as good as with a balanced square. However, in others, a great reduction in efficiency may occur. For example, Jones and Kenward (p. 193) display an unbalanced 4×4 square which yields $E_d = 18.18$ and $E_c = 12.50$ even though the balanced 4×4 square of Table 5.1 has corresponding efficiencies of 90.91 and 62.50. With a one-order carryover model, the Edwards square of Section 5.3.1. has $\overline{E}_d = 15.79$, $\overline{E}_c = 12.00$, and $E_p = 100.00$ for Periods 2 through 5.

A consequence of this inefficiency for certain estimation with certain squares is that one cannot be sure what to do if adjustment of treatment effects for carryover effects yields a conclusion about the significance of treatment effects different from that obtained if adjustment is not performed. For balanced Latin square designs, Abeyasekera and Curnow (1984) provide evidence that the adjusted analysis should be retained; it is conventional with other designs to omit the adjustment in such cases (Jones and Kenward, p. 150).

5.6 CONCLUDING REMARKS

This chapter has provided information about the history, purpose, construction, and data analysis of Latin square experimental designs. Classical fixed-effects and mixed models have been discussed, together with carryover effects models for replicated and unreplicated designs. Estimation and hypothesis testing equations based on classical models have been provided and their use demonstrated with contrived data from an unreplicated 4×4 Latin square design. In addition, two different statistical packages were used to illustrate the analysis of the same data under a carryover model, noting special requirements if comparable results are to be obtained with the two packages.

The Latin square design permits simultaneous investigation of three independent variables with a minimum number of observations. However, problems of confounding limit its usefulness in situations in which interactions of two or more of those variables may be present. There is no way to provide estimates for interaction between row, column, and Latin variables or to test hypotheses about them; their presence also contaminates estimation and hypothesis testing of main effects of those variables. Often a negative bias exists, resulting in a conservative test procedure with less significant results than are appropriate. However, the type of bias should be checked for each situation used. Transformation of data to eliminate undesired interactions is appropriate if sufficient information about those interactions exists to guide that transformation. Alternatively, if possible, the design should be expanded to a full factorial design.

With a repeated measurements Latin square design, the inclusion of carryover effects in the statistical model permits assessment of effects that resemble interaction in addition to main effects. For example, in the replicated 2×2 Latin square design, the carryover sum of squares is equal to a sum of squares for period-by-treatment interaction and to a sum of squares for stimulus sequences. If no other interactions are included in the model, a legitimate test for treatment effects can be found. However, a test for period effects can be performed only if carryover effects are not significant. With larger repeated measurements Latin squares, one should think of carryover as approximating period-by-treatment interaction and then provide evidence that no other kind of interaction is distorting conclusions drawn from the data analysis.

This chapter also discusses the use of repeated measurements Latin squares balanced to make each pair of treatments adjacent to each other an equal number of times. This may substantially increase the efficiency of the experimental design for assessing treatment and carryover effects. Therefore, balanced squares are recommended highly.

APPENDIX A: AN SAS® GLM PROGRAM TO ANALYZE TABLE 5.1 DATA

```
DATA ONE;
INPUT SUBJ PERIOD TREAT CARRY Y;
CARDS;
1 1 1 0 68
1 2 2 1 74
1 3 4 2 93
1 4 3 4 94
2 1 2 0 60
2 2 3 2 66
2 3 1 3 59
2 4 4 1 79
3 1 3 0 69
3 2 4 3 85
3 3 2 4 78
3 4 1 2 69
4 1 4 0 90
4 2 1 4 80
4 3 3 1 80
4 4 2 3 86
RUN;
PROC GLM;
CLASS SUBJ PERIOD TREAT CARRY;
```

MODEL Y = SUBJ PERIOD TREAT CARRY/E E1 E2 E3 E4
    SOLUTION;
RANDOM SUBJ;
ESTIMATE 'PSEUDO BETA_1-BETA_4'
PERIOD 1 0 0 -1 CARRY 1-.25 -.25 -.25 -.25;
CONTRAST 'PSEUDO-PERIODS'
PERIOD 1 0 0 -1 CARRY 1 -.25 -.25 -.25 -.25, PERIOD 0 1 0 -1,
PERIOD 0 0 1 -1;
RUN;

APPENDIX B: INTERPRETATION OF PARAMETER VALUES UNDER SUM-TO-ZERO AND SET-TO-ZERO ASSUMPTIONS FOR ESTIMATED VALUES

This appendix should give further help in understanding differences between SAS® and SYSTAT™ output for Latin square data generated by
our carry-over model, (5.25). It shows connections between parameter
values in the original model and in two equivalent restatements of it, each
using only as many parameters as are estimable. Connections to hypothesis
testing procedures, though clear, are not discussed here. Following Grizzle
(1974) and in view of the fact that both SAS® and SYSTAT™ compute
fixed-effect estimators for subject effects in the process of treating them
as random, it is appropriate to replace a_i with its fixed counterpart, α_i,
below, with $\alpha_i - \bar{\alpha}$ summing to zero over i.

The Sum-To-Zero Model

This section primarily helps to understand SYSTAT™ estimation of parameter values, viewed for the moment as leaving the basic model unchanged but requiring that parameter estimates for a family such as periods
sum to zero. The new pair of model equations below each splits the value
of an observation into four or five parenthesized terms plus error:

$$Y_{i1k} = (\mu + \bar{\beta} + \bar{\gamma} + \bar{\alpha}) + (\alpha_i - \bar{\alpha})$$
$$+ (\beta_1 - \bar{\beta}) + (\gamma_k - \bar{\gamma}) + e_{i1k} \quad (5.B.1)$$

for Period 1 and

$$Y_{ijk} = (\mu + \bar{\beta} + \bar{\gamma} + \bar{\alpha}) + (\alpha_i - \bar{\alpha}) + (\beta_j - \bar{\beta} + \bar{\lambda})$$
$$+ (\gamma_k - \bar{\gamma}) + (\lambda_k, - \bar{\lambda}) + e_{ijk}, \quad j \neq 1 \quad (5.B.2)$$

for all other periods. It is easy to verify that these new equations are
algebraically consistent with a fixed effect form of (5.25), i.e., with (5.25)

modified by replacing a_i with α_i. The parenthesized sections of (5.B.1) and (5.B.2) are now used to suggest estimable parameters:

$$\mu^* = \mu + \bar{\beta} + \bar{\gamma} + \bar{\alpha}, \tag{5.B.3}$$

$$\alpha_i^* = \alpha_i - \bar{\alpha}, \tag{5.B.4}$$

$$\gamma_k^* = \gamma_k - \bar{\gamma}, \tag{5.B.5}$$

and

$$\lambda_{k,}^* = \lambda_{k,} - \bar{\lambda}. \tag{5.B.6}$$

Computational use of (5.B.3) and (5.B.4) requires knowledge of α_i values of 5.66, -9.05, -2.78, and 7.30 in order of subject labeling, generated randomly in the simulation producing the data of Table 5.1. Each of the above equations, except the one for μ^*, defines four parameters summing to zero. Hence there are three linearly independent parameters each for subjects, for treatments, and for carryovers. The four equations define a set of 10 independent parameters out of the 13 we know to be estimable with the 4×4 Latin square design. Obviously, the remaining three parameters must be associated with period effects. However, because they do not sum to zero across the four periods, it is not sufficient to use $\beta_1 - \bar{\beta}$ and $\beta_j - \bar{\beta} + \bar{\lambda}$ (for $j \neq 1$) as the transformed period effects. What is needed is to define

$$\beta_1^* = \beta_1 - \bar{\beta} - 0.75\bar{\lambda} \tag{5.B.7}$$

and

$$\beta_j^* = \beta_j - \bar{\beta} + 0.25\bar{\lambda}, \qquad j \neq 1, \tag{5.B.8}$$

making their four values sum to zero. The summary equation,

$$Y_{ijk} = \mu^* + \alpha_i^* + \beta_j^* + \gamma_k^* + \lambda_{k,}^* + e_{ijk} \qquad \text{for all } j, \tag{5.B.9}$$

makes Y_{ijk} algebraically equivalent to its value in the fixed-effect variant of (5.25). A unique organization of original parameters is used to express a model in sum-to-zero terms.

The estimators for the sum-to-zero constraint on estimators have expectations consistent with the reparameterization above:

$$E[\tilde{\mu}^*] = \mu + \bar{\alpha} + \bar{\beta} + \bar{\gamma}, \tag{5.B.10}$$

$$E[\tilde{\alpha}_i^*] = \alpha_i, \tag{5.B.11}$$

$$E[\tilde{\beta}_1^*] = \beta_1 - \bar{\beta} - 0.75\bar{\lambda}, \tag{5.B.12}$$

$$E[\tilde{\beta}_j^*] = \beta_j - \bar{\beta} + 0.25\bar{\lambda}, \qquad j \neq 1, \tag{5.B.13}$$

$$E[\tilde{\gamma}_k^*] = \gamma_k - \bar{\gamma}, \tag{5.B.14}$$

and

$$E[\tilde{\lambda}_{k,}{}^{*}] = \lambda_{k}, - \bar{\lambda}. \tag{5.B.15}$$

Now note that Table 5.5 uses a less complete notation than this appendix. Its $\tilde{\beta}_j$ values lack superscript $*$ symbols, but the $\tilde{\beta}_1 - \tilde{\beta}_4 = 10.25$ reported there is an estimate of the present $\beta_1^* - \beta_4^* = -10.5125$. A SYSTAT® value not previously reported, $\bar{\mu}^* = 76.875$, estimates $\mu^* = 75.52$.

Using the Set-to-Zero Constraint with Final Elements Set to Zero

This section primarily helps to understand SOLUTION command output of parameter estimates from SAS® GLM. We can rewrite the fixed-effect variant of (5.25) as

$$Y_{i1k} = (\mu + \alpha_4 + \beta_4 + \gamma_4 + \lambda_4) + (\alpha_i - \alpha_4)$$
$$+ (\beta_1 - \beta_4 - \lambda_4) + (\gamma_k - \gamma_4) + e_{i1k} \tag{5.B.16}$$

for Period 1 and

$$Y_{ijk} = (\mu + \alpha_4 + \beta_4 + \gamma_4 + \lambda_4) + (\alpha_i - \alpha_4) + (\beta_j - \beta_4)$$
$$+ (\gamma_k - \gamma_4) + (\lambda_k, - \lambda_4) + e_{ijk}, \quad j \neq 1 \tag{5.B.17}$$

for all other periods. Again these equations are consistent with the fixed-effect variant of (5.25). All the revised parameter values for this new format follow directly from parenthesized terms of (5.B.16) and (5.B.17):

$$\mu^+ = \mu + \alpha_4 + \beta_4 + \gamma_4 + \lambda_4, \tag{5.B.18}$$
$$\alpha_i^+ = \alpha_i - \alpha_4, \tag{5.B.19}$$
$$\beta_1^+ = \beta_1 - \beta_4 - \lambda_4, \tag{5.B.20}$$
$$\beta_j^+ = \beta_i - \beta_4, \quad j \neq 1, \tag{5.B.21}$$
$$\gamma_k^+ = \gamma_k - \gamma_4, \tag{5.B.22}$$

and

$$\lambda_k^+, = \lambda_k, - \lambda_4. \tag{5.B.23}$$

Equations (5.B.18) through (5.B.23) define for a set-to-zero model the 13 parameters we know to be definable with a 4×4 Latin square design. The summary equation is

$$Y_{ijk} = \mu^+ + \alpha_i^+ + \beta_j^+ + \gamma_k^+ + \lambda_k^+, + e_{ijk}. \tag{5.B.24}$$

Again, it is algebraically equivalent to the fixed-effect variant of (5.25).

The expectations of the revised estimators are again equal to the re-defined parameters above:

$$E[\hat{\mu}^+] = \mu + \alpha_4 + \beta_4 + \gamma_4 + \lambda_4, \tag{5.B.25}$$

$$E[\hat{a}_i^+] = \alpha_i - \alpha_4, \tag{5.B.26}$$

$$E[\hat{\beta}_1^+] = \beta_1 - \beta_4 - \lambda_4, \tag{5.B.27}$$

$$E[\hat{\beta}_j^+] = \beta_j - \beta_4, \qquad j \neq 1, \tag{5.B.28}$$

$$E[\hat{\gamma}_k^+] = \gamma_k - \gamma_4, \tag{5.B.29}$$

and

$$E[\hat{\lambda}_k{}^+] = \lambda_k{}, - \lambda_4. \tag{5.B.30}$$

Again note that Table 5.3 uses a less complete notation than this appendix. Its $\hat{\beta}_j - \hat{\beta}_4$ values lack the superscript $+$ symbols, but the $\hat{\beta}_1 - \hat{\beta}_4 = -16.35$ reported there is an estimate of the present β_1^+. Correspondingly, a value reported by SAS® GLM but not mentioned earlier, $\hat{\mu}^+ = 106.325$, estimates $\mu^+ = 107.2$, defined by (5.B.10) and the parameter values of Section 5.4.1.

One of many alternative specifications of the intercept and first period effect parameters and estimators uses

$$\mu^{++} = \mu + \beta_4 + \gamma_4 + \bar{\lambda} \tag{5.B.31}$$

and

$$\beta_1^+ = \beta_1 - \beta_4 - \bar{\lambda}, \tag{5.B.32}$$

leading to

$$E[\hat{\mu}^{++}] = \mu + a_4 + \beta_4 + \gamma_4 + \bar{\lambda} \tag{5.B.33}$$

and

$$E[\hat{\beta}_1^{++}] = \beta_1 - \beta_4 - \bar{\lambda}. \tag{5.B.34}$$

The value $\hat{\beta}_1 - \hat{\beta}_4 - \hat{\lambda} = -10.25$ in Section 5.4.4 is an estimate of $\beta_1^{++} = -10.5125$ in (5.B.34). Similar modifications to formulate $\hat{\gamma}_k^{++}$ and $\hat{\lambda}_k^{++}$, need not be shown here.

REFERENCES

Abeyasekera, S. and Curnow, R. N. (1984). The Desirability of Adjusting for Residual Effects in a Crossover Design. *Biometrics*, *40*: 1071–1078.

Anscombe, F. J. and Tukey, J. W. (1963). The Examination and Analysis of Residuals. *Technometrics*, *5*: 141–160.

Aptech Systems. (1988). *GAUSS™: The GAUSS System Version 2.0.* Aptech Systems, Inc., Kent, Washington.

Box, G. E. P., Hunter, W. G., and Hunter, J. S. (1978). *Statistics for Experimenters. An Introduction to Design, Data Analysis, and Model Building.* Wiley, New York.

Campbell, D. T. and Stanley, J. C. (1966). *Experimental and Quasi-Experimental Designs for Research on Teaching.* Rand McNally, Chicago.

Cochran, W. G. and Cox, G. M. (1957). *Experimental Designs* (2nd ed.). Wiley, New York.

Cook, R. D. and Weisberg, S. (1982). *Residuals and Influence in Regression.* Chapman and Hall, New York.

Cornfield, J. and Tukey, J. W. (1956). Average Values of Mean Squares in Factorials. *Ann. Math. Statist., 27*: 907–949.

Cotton, J. W. (1989a). Data Analysis for Repeated Measurements Latin Square Designs, Emphasizing Carry-Over Effects. Paper presented at Joint Statistical Meetings, Washington, D.C.

Cotton, J. W. (1989b). Interpreting Data from Two-Period Cross-Over Design (Also Termed the Replicated 2 × 2 Latin Square Design). *Psych. Bull., 106*: 503–515.

Cotton, J. W. and Revlin, R. (1987). Latin Squares and Sestinas: A Historical Puzzle. Paper presented at Joint Statistical Meetings, San Francisco, California.

Cronbach, L. J. and Snow, R. E. (1977). *Aptitudes and Instructional Methods: A Handbook for Research on Interactions.* Irvington/Naiburg, New York.

D'Agostino, R. B. (1986). Tests for the Normal Distribution. *Goodness of Fit Techniques.* D'Agostino, R. B. and Stephens, M. A. (eds.), Marcel Dekker, New York, pp. 367–419.

Dénes, J. and Keedwell, A. D. (1974). *Latin Squares and Their Applications.* Academic Press, New York.

Draper, N. R. and Joiner, B. L. (1984). Residuals with One Degree of Freedom. *Amer. Statist., 38*: 55–57.

Edwards, A. L. (1985). *Experimental Design in Psychological Research* (5th ed.). Harper and Row, New York.

Euler, L. (1782). Recherches sur Une Nouvelle Espèce de Quarrés Magiques. *Verh. Zeeuwsch., Genootsch. Wetensch. Vlissengen, 9*: 85–239. (Reference from Dénes and Kedwell and from Street and Street, original not seen.)

Federer, W. T. (1955). *Experimental Design. Theory and Application.* Macmillan, New York.

Feeney, J. A. and Noller, P. (1990). Attachment Style as a Predictor of Adult Romantic Relationships. *J. Pers. Soc. Psych., 58*: 281–291.

Fisher, R. A. (1970). *Statistical Methods for Research Workers* (14th ed.). Oliver and Boyd, Edinburgh.

Fisher, R. A. and Yates, F. (1953). *Statistical Tables for Biological, Agricultural and Medical Research* (4th ed.). Hafner Publishing Company, New York.

Fults, J. L. (1974). *Magic Squares.* Open Court, La Salle, Illinois.

Gaito, J. (1958). The Single Latin Square Design in Psychological Research. *Psychometrika, 23*: 369–378.

Gardner, M. (1961). *The 2nd Scientific American Book of Mathematical Puzzles and Diversions*. Simon and Schuster, New York.

Gentleman, J. F. and Wilk, M. B. (1975). Detecting Outliers in a Two-Way Table: I. Statistical Behavior of Residuals. *Technometrics*, *17*: 1–14.

Goodall, C. (1983). Examining Residuals. *Understanding Robust and Exploratory Data Analysis*. Hoaglin, D. C., Mosteller, F., and Tukey, J. W. (eds.). Wiley, New York, pp. 211–246.

Grizzle, J. E. (1965). The Two-Period Change-Over Design and Its Use in Clinical Trials. *Biometrics*, *21*: 467–480.

Grizzle, J. E. (1974). Correction. *Biometrics*, *30*: 727.

Hills, M. and Armitage, P. (1979). The Two-Period Cross-Over Clinical Trial. *Brit. J. Clin. Pharm.*, 8: 7–20.

John, P. W. M. (1971). *Statistical Design and Analysis of Experiments*. Macmillan, New York.

Jones, B. and Kenward, M. G. (1989). *Design and Analysis of Cross-Over Trials*. Chapman and Hall, London.

Kempthorne, O. (1973). *The Design and Analysis of Experiments*. Robert E. Krieger, Melbourne, Florida.

Kirk, R. E. (1982). *Experimental Design: Procedures for the Behavioral Sciences* (2nd ed.). Brooks/Cole, Belmont, California.

Koch, K. and Farrell, K. (1982). *Sleeping on the Wing: An Anthology of Modern Poetry with Essays on Reading and Writing*. Vintage Books, New York.

Mauchly, J. W. (1940). Significance Test for Sphericity of a Normal n-Variate Distribution. *Ann. Math. Statist.*, *11*: 204–209.

Milliken, G. A. and Johnson, D. E. (1984). *Analysis of Messy Data*, Vol. I, *Designed Experiments*. Lifetime Learning Publications, Belmont, California.

Morsbach, G., McCulloch, M., and Clark, A. (1986). Infant Crying as a Potential Stressor Concerning Mothers' Concentration Ability. *Psychologia*, *29*, 10–20.

Panofsky, E. (1955). *The Life and Art of Albrecht Dürer*. Princeton University Press, Princeton.

Patterson, H. D. and Lucas, H. L. (1962). *Change-Over Designs*. North Carolina Agricultural Experiment Station and U.S. Department of Agriculture, Raleigh, North Carolina.

Poulton, E. C. and Freeman, P. R. (1966). Unwanted Asymmetrical Effects with Balanced Experimental Designs. *Psych. Bull.*, *66*: 1–8.

Ratkowsky, D. A., Alldredge, J. R., and Cotton, J. W. (1990). Analyzing Balanced or Unbalanced Latin Squares and Other Repeated-Measures Designs for Carryover Effects Using the GLM Procedure, Proceedings of the Fifteenth Annual SAS® Users Group International Conference, Nashville, Tennessee, pp. 1353–1358.

Raymond, M. J., Lucas, C. J., Beesley, M. L., and O'Connell, B. A. (1957). A Trial of Five Tranquilizing Drugs in Psychoneurosis. *Brit. Med. J.*, July 13: 63–66.

Riesz, J. (1971). *Die Sestine. Ihre Stellung in der Literarischen Kritik und ihre Geschichte als Lyrische Genus*. Wilhelm Fink Verlag, Munich.

SAS Institute. (1988). *SAS/STAT®️ User's Guide. Release 6.03 Edition.* SAS Institute, Inc., Cary, North Carolina.

Scheffé, H. (1959). *The Analysis of Variance.* Wiley, New York.

Searle, S. R. (1987). *Linear Models for Unbalanced Data.* Wiley, New York.

Sheehe, P. R. and Bross, I. D. J. (1961). Latin Squares to Balance Immediate Residual, and Other Order, Effects. *Biometrics, 17*: 405–414.

Stephens, M. A. (1986). Tests Based on EDF Statistics. *Goodness of Fit Techniques.* D'Agostino, R. B. and Stephens, M. A. (eds.). Marcel Dekker, New York, pp. 97–193.

Street, A. P. and Street, D. J. (1987). *Combinatorics of Experimental Design.* Clarendon Press, Oxford.

Tarry, G. (1900–01). Le Problème des 36 Officiers. *C. R. Assoc. Fr. Avance. Sci., 1*: 113–120.

Thompson, P. A. (1988). Contrasts for the Residual Interaction in Latin Square Designs. *Educ. Psych. Meas., 48*, 83–88.

Tukey, J. W. (1949). One Degree of Freedom for Nonadditivity. *Biometrics, 5*: 232–242.

Wagenaar, W. (1969). Note on the Construction of Digram-Balanced Latin Squares. *Psych. Bull., 72*: 384–386.

Wilk, M. B. and Kempthorne, O. (1957). Non-Additivities in a Latin Square Design. *J. Amer. Statist. Assoc., 52*: 218–236.

Wilkinson, L. (1989). *SYSTAT: The System for Statistics.* SYSTAT Inc., Evanston, Illinois.

Williams, E. J. (1949). Experimental Designs Balanced for the Estimation of Residual Effects of Treatments. *Austral. J. Sci. Res., A2*: 149–168.

Williams, E. J. (1950). Experimental Designs Balanced for Pairs of Residual Effects. *Austral. J. Sci. Res., A3*: 351–363.

Wood, C. (1978). A Large Sample Kolmogorov-Smirnov Test for Normality of Experimental Error in a Randomized Block Design. *Biometrika, 65*: 673–676.

Yates, F. (1933). The Formation of Latin Squares for Use in Field Experiments. *Emp. J. Exp. Agr., 1*: 235–244.

6

Confounded Factorial Designs

ROGER E. KIRK   Baylor University, Waco, Texas

6.1   INTRODUCTION TO CONFOUNDING

Ronald A. Fisher's classic book *The Design of Experiments* (1935) helped to popularize three elements of good experimental design: randomization, replication, and local control. This chapter will focus on local control or blocking, in which experimental units or subjects are assigned to blocks so that the subjects within a block are more homogeneous with respect to the dependent variable than subjects in different blocks. Any variable that is correlated with the dependent variable other than the independent variable can be used to form homogeneous blocks. The following procedures are often used for this purpose.

1. Matching subjects with respect to a nuisance (blocking) variable that is correlated with the dependent variable.
2. Observing each subject under all or a portion of the treatment levels or treatment combinations.
3. Using identical twins or littermates.

In these examples, a block consists of, respectively, a set of matched subjects, a single subject who is observed repeatedly, or a set of identical twins or littermates. When researchers consider potential blocking variables, they often overlook characteristics of the environment or experimental setting. For example, a block could consist of all observations that

---

Portions of this chapter were reproduced from *Experimental Design: Procedures for the Behavioral Sciences* (2nd ed.) by Roger E. Kirk with the permission of the publisher, Brooks/Cole Publishing Co.

are made at the same time of day, day of the week, season, or of all observations that are made at the same location or with the same experimental apparatus.

Fisher's research in the 1920s at the Rothamsted Experimental Station demonstrated the benefits of blocking: more precise estimates of treatment effects and smaller error variance. Unfortunately, there is a practical problem associated with the three blocking procedures just described. As the block size increases, it becomes more and more difficult to match subjects precisely and achieve block homogeneity. Obtaining repeated observations on the same subjects is no solution because there is a limit to the number of times a subject can participate in an experiment, and the nature of the treatments often precludes obtaining more than one observation per subject. The use of identical twins limits the block size to two, and many litters are smaller than five. Even a relatively small design such as the randomized block factorial design shown in Figure 6.1 has $3 \times 3 = 9$ treatment combinations and requires blocks of size nine. Fisher's solution to the block-size problem was to assign only a portion of the treatment combinations to each block. His solution is called *confounding*. Confounding achieves a reduction in block size but, as we will see, always intermingles two or more sources of variation so that it is impossible to determine the unique contribution of each source. Confounding can take one of three forms: group-treatment confounding, group-interaction confounding, and treatment-interaction confounding. The three types of confounding are the basis for distinguishing among three types of factorial designs. Group-treatment confounding is used in a split-plot factorial design to reduce the block size. Group-interaction confounding accomplishes the same purpose for the con-

	Treat. Comb. a_jb_k	Treat. Comb. a_jb_k	Treat. Comb. a_jb_k	Treat. Comb. a_jb_k	Treat. Comb. a_jb_k	Treat. Comb. a_jb_k	Treat. Comb. a_jb_k	Treat. Comb. a_jb_k	Treat. Comb. a_jb_k
Block$_1$	11	12	13	21	22	23	31	32	33
Block$_2$	11	12	13	21	22	23	31	32	33
Block$_3$	11	12	13	21	22	23	31	32	33
.	.	.	.	.	.	.	.	.	.
Block$_n$	11	12	13	21	22	23	31	32	33

Figure 6.1 Layout for a randomized block factorial design with three levels of treatment A and three levels of treatment B (RBF-33 design). The numbers in the body of the table denote the levels of treatments A and B, respectively. Each block contains all nine treatment combinations.

founded factorial designs that are described in this chapter. Treatment-interaction confounding is used to reduce the number of treatment combinations that must be included in a fractional factorial design. All three forms of confounding achieve the same goal—stratification of subjects into more homogeneous subgroups.

One way to reduce the block size is to confound treatment A with groups. This confounding scheme produces the split-plot factorial design (SPF-3·3 design) shown in Figure 6.2. The subjects in Group$_1$, for example, receive the first level of treatment A (a_1) and all levels of treatment B (b_1, b_2, and b_3). Hence, each block contains only three of the nine treatment combinations. By comparison, each block in the randomized block factorial design (RBF-33 design) contains nine combinations. In the split-plot factorial design, differences among the three levels of treatment A are completely confounded with differences among the three groups. In other words, it is not possible to determine the unique contribution of either treatment A or Groups—the two sources of variation are intermingled. Treatment A is often referred to as a between-blocks comparison; treatment B and the AB interaction are within-blocks comparisons. It is well known that tests of between-blocks comparisons tend to be less powerful than tests of within-blocks comparisons (Kirk, 1982, pp. 253–256). If a researcher's primary interest is in tests of treatments A and B, the split-plot design in Figure 6.2 is not the best design choice. This follows because the test of treatment A is less powerful than the test of treatment B. A better design choice is

		Treat. Comb. a_jb_k	Treat. Comb. a_jb_k	Treat. Comb. a_jb_k
	Block$_1$	11	12	13
Group$_1$ (a_1)	Block$_2$	11	12	13
	Block$_3$	11	12	13
	Block$_4$	21	22	23
Group$_2$ (a_2)	Block$_5$	21	22	23
	Block$_6$	21	22	23
	Block$_7$	31	32	33
Group$_3$ (a_3)	Block$_8$	31	32	33
	Block$_9$	31	32	33

Figure 6.2 Layout for a split-plot factorial design (SPF-3·3 design). The levels of treatment A are confounded with groups. Each block contains three treatment combinations: one level of treatment A and all levels of treatment B.

a randomized block confounded factorial design (RBCF-3^2 design) that confounds an AB interaction component with Groups. This design, which is discussed in Section 6.6, is shown in Figure 6.3. The numbers 3^2 in the design nomenclature indicate that the design has two treatments with three levels each. This design, like the split-plot factorial design, has blocks of size three. In the RBCF-3^2 design, the tests of treatments A and B are within-blocks comparisons and are more powerful than the test of the AB interaction that is a between-blocks comparison. An important advantage of the RBCF-3^2 design over the SPF-3·3 design is that treatments A and B are evaluated with equal power. A disadvantage of the confounded factorial design, but not the split-plot factorial design, is that it must be possible to administer the levels of each treatment in every possible sequence. This requirement precludes the use of treatments whose levels consist of successive periods of time. To summarize, confounding achieves a reduction in block size but always results in some loss of power. In the RBCF-3^2 design, this loss occurs in testing the AB interaction. In the SPF-3·3 design, the loss occurs in testing treatment A. A researcher has to decide if the advantage of a smaller block size outweighs the disadvantage of loss of power. If it is possible to use blocks of size nine, the best design strategy is to use a RBF-33 design. Although the subjects in the larger blocks of an RBF-33 design tend to be less homogeneous, the design provides the maximum number of degrees of freedom for estimating the

		Treat. Comb. a_jb_k	Treat. Comb. a_jb_k	Treat. Comb. a_jb_k
	Block$_0$	00	12	21
Group$_0$ $(AB)_0$	Block$_1$	00	12	21
	Block$_2$	00	12	21
	Block$_3$	01	10	22
Group$_1$ $(AB)_1$	Block$_4$	01	10	22
	Block$_5$	01	10	22
	Block$_6$	02	11	20
Group$_2$ $(AB)_2$	Block$_7$	02	11	20
	Block$_8$	02	11	20

Figure 6.3 Layout for a randomized block confounded factorial design (RBCF-3^2 design). The use of 00, 12, and so on to denote the levels of treatments A and B is discussed in Section 6.2.2. A component of the AB interaction is confounded with groups. Each block contains three treatment combinations and each level of treatments A and B.

population error variance and evaluates treatments A and B and the AB interaction with equal power.

In this chapter it is not possible to describe all or even a majority of the confounded factorial designs. Coverage is restricted to those designs that are constructed from a randomized block design. For a discussion of designs constructed from a Latin square design, the reader is referred to Cochran and Cox (1957), Kempthorne (1952), Kirk (1982), Winer, Brown, and Michels (1991), and Yates (1937). The approach that has been adopted is to present selected designs that illustrate basic principles. Also presented are three computational algorithms: the sum-of-squares approach, the cell means model approach, and the regression model approach. The latter two approaches can be used when one or more observations is missing.

6.1.1 A Bit of History

The technique of confounding was first described by Fisher (1926) and used as early as 1927 in agricultural research at the Rothamsted Experimental Station. Several years later, Fisher and Wishart (1930) described the analysis of confounded factorial designs. In 1933, Frank Yates, drawing on his work with Fisher, described confounding schemes for designs with k treatments each having two or three levels (2^k designs and 3^k designs). At that time, confounding in the 3^k designs was accomplished by using a Latin square. In 1935, Yates provided more examples of confounded factorial designs and discussed their efficiency. Two years later, Yates (1937) described a more general confounding scheme and provided an extensive catalogue of designs involving treatments with two, three, or four levels. Research on the construction of confounded factorial designs continued with the work of Nair (1938, 1940) and Fisher (1942). Li's classic monograph in 1944 filled in many of the gaps in our knowledge of the analysis of designs involving two to four treatments with two, three, or four levels.

6.2 USE OF MODULAR ARITHMETIC IN CONSTRUCTING CONFOUNDED FACTORIAL DESIGNS

The construction of confounded factorial designs requires a scheme for assigning treatment combinations to groups of blocks so that variation between the groups is confounded with one or more interactions or interaction components. Several schemes have been devised for this purpose (Bailey, 1977; John and Dean, 1975; Kempthorne, 1947; Patterson and Bailey, 1978; Yates, 1937). One scheme that is applicable to designs of the form p^k where p is a prime number uses properties of finite fields. We will now describe those properties that are used to construct p^k designs.

Let I and m be any integers, with $m > 0$. If I is divided by m, we obtain a quotient q and a remainder z, because

$I = qm + z.$

For example, let $I = 14$ and $m = 3$. Then $q = 4$ and $z = 2$, because

$14 = 4(3) + 2.$

In modular arithmetic the remainder, z, is the term of interest. Consider now dividing $J = 5$ by $m = 3$. The remainder also is equal to 2, because

$5 = (1)(3) + 2.$

Note that 14 and 5 leave the same remainder when divided by 3. Two integers I and J that leave the same remainder when divided by a positive integer m are said to be *congruent* with respect to the modulus m. This relation—congruence—can be written

$I = J \,(\text{mod } m)$

and read "I is congruent to J modulo m."

Any integer I is always congruent to its remainder z; that is,

$I = z \,(\text{mod } m).$

For example, $I = 14$ and $z = 2$ are congruent modulo 3 because when 14 and 2 are reduced modulo 3 (divided by the modulus 3), they leave the same remainder,

$14 = 4(3) + 2 \quad \text{and} \quad 2 = 0(3) + 2.$

The possible values of the remainder z are $0, 1, 2, \ldots, m - 1$. Thus, an integer is always congruent to $z = 0, 1, 2, \ldots,$ or $m - 1$, where m is the modulus. Consider the following examples.

$$0 = 0 \,(\text{mod } 2) \qquad 0 = 0 \,(\text{mod } 3)$$
$$1 = 1 \,(\text{mod } 2) \qquad 1 = 1 \,(\text{mod } 3)$$
$$2 = 0 \,(\text{mod } 2) \qquad 2 = 2 \,(\text{mod } 3)$$
$$3 = 1 \,(\text{mod } 2) \qquad 3 = 0 \,(\text{mod } 3)$$
$$4 = 0 \,(\text{mod } 2) \qquad 4 = 1 \,(\text{mod } 3)$$
$$5 = 1 \,(\text{mod } 2) \qquad 5 = 2 \,(\text{mod } 3)$$

6.2.1 Modular Addition and Subtraction

Two operations of modular arithmetic are used in constructing confounded factorial designs: addition and multiplication. The operation of addition is illustrated by the following examples.

$$a_j + b_k = z \ (\text{mod } 2) \qquad a_j + b_k = z \ (\text{mod } 3)$$
$$0 + 0 = 0 \ (\text{mod } 2) \qquad 0 + 0 = 0 \ (\text{mod } 3)$$
$$1 + 0 = 1 \ (\text{mod } 2) \qquad 0 + 1 = 1 \ (\text{mod } 3)$$
$$0 + 1 = 1 \ (\text{mod } 2) \qquad 0 + 2 = 2 \ (\text{mod } 3)$$
$$1 + 1 = 0 \ (\text{mod } 2) \qquad 1 + 1 = 2 \ (\text{mod } 3)$$
$$1 + 2 = 0 \ (\text{mod } 3)$$
$$2 + 2 = 1 \ (\text{mod } 3)$$

To add two integers a_j and b_k, one obtains their sum and reduces it modulo m, that is, expresses it as a remainder with respect to the modulus m. This operation will be used later to confound an interaction with groups of blocks. We will let a_j, b_k, z, and m correspond to properties of an experimental design as follows:

a_j and b_k denote levels of treatments A and B, respectively.
z denotes a group of blocks.
m denotes the number of levels of each treatment.

The second operation of modular arithmetic that is used in constructing factorial designs is multiplication. This operation is illustrated by the following examples.

$$(1)(1) = 1 \ (\text{mod } 3)$$
$$(1)(2) = 2 \ (\text{mod } 3)$$
$$(2)(2) = 1 \ (\text{mod } 3)$$
$$(3)(2) = 0 \ (\text{mod } 3)$$

To multiply two integers, one obtains their product and expresses it as a remainder with respect to the modulus m.

6.2.2 Modified Notation Scheme for Treatment Levels

To use modular arithmetic in assigning treatment combinations to groups, we will let 0 denote the first level of a treatment, block, or group; 1, the second level; and so on. According to this scheme, the treatment levels of a RBCF-3^2 design are denoted by a_0, a_1, a_2, b_0, b_1, and b_2. The nine treatment combinations can be denoted by a_0b_0, a_0b_1, . . . , a_2b_2, or, more simply, by 00, 01, 02, 10, 11, 12, 20, 21, 22. The digit in the first position denotes the level of treatment A; the digit in the second position denotes the level of treatment B. A design with three treatments, say an RBCF-3^3 design, requires three digits to denote a treatment combination. For example, if treatments A, B, and C are all at the first level, the designation is $a_0 b_0 c_0$ or 000.

This notation scheme, carried to its logical conclusion, leads to an odd-looking summation notation. For example, the sum of $i = 0, \ldots, n - 1$ blocks would be written

$$Y_0 + Y_1 + \cdots + Y_{n-1} = \sum_{i=0}^{n-1} Y_i.$$

To keep the notation as simple as possible when summation is performed, we will use 1 to represent the first level of a treatment, block, or group. For example, i will range over $1, \ldots, n$ and not $0, \ldots, n - 1$. This dual notation scheme lets us write $\sum_{i=1}^{n} Y_i$ instead of $\sum_{i=0}^{n-1} Y_i$. In effect, we are adding 1 to the initial and terminal values of the index of summation, that is,

$$\sum_{i=0}^{n-1} Y_i \qquad \text{becomes} \qquad \sum_{i=0+1}^{n-1+1} Y_i.$$

6.2.3 Assignment of Treatment Combinations to Groups

A randomized block factorial design with two levels of treatments A and B has four treatment combinations—$a_0 b_0$, $a_0 b_1$, $a_1 b_0$, $a_1 b_1$ or, more simply, 00, 01, 10, 11—and requires blocks of size four. Suppose it is possible only to observe a subject twice and that the researcher's primary interest is in the two treatments rather than in the interaction. The block size can be reduced from four to two by confounding the AB interaction with Groups. Modular arithmetic is used to determine which treatment combinations are assigned to each group. Let a_j denote the jth level of treatment A, and b_k the kth level of treatment B. All treatment combinations satisfying the defining relation

$$a_j + b_k = z \ (\text{mod } 2)$$

where z is equal to 0, are assigned to group 0. Those satisfying the relation where z is equal to 1 are assigned to group 1. Modulus 2 is used because treatments A and B each have two levels. The range of z is 0 and 1, because all integers are congruent to $0, 1, \ldots,$ or $m - 1$, and m is equal to 2 in this example.

Solving for a_j and b_k, we obtain

$$\left. \begin{array}{l} 0 + 0 = 0 \ (\text{mod } 2) \\ 1 + 1 = 0 \ (\text{mod } 2) \end{array} \right\} \quad \text{group 0 or } (AB)_0,$$

$$\left. \begin{array}{l} 0 + 1 = 1 \ (\text{mod } 2) \\ 1 + 0 = 1 \ (\text{mod } 2) \end{array} \right\} \quad \text{group 1 or } (AB)_1.$$

Thus, treatment combinations 00 and 11 are assigned to group 0; combinations 01 and 10 are assigned to group 1. The notation $(AB)_z$ is an alternative way to denote the treatment combinations that are assigned to group $z = 0, 1$. A block diagram for this RBCF-2^2 design with $n = 4$ blocks in each group is shown in Figure 6.4.

We will now show that the layout in Figure 6.4 confounds the AB interaction with groups. Let μ_{ijkz} denote the population mean for block i, treatment combination $a_j b_k$, and group z. By definition, a two-treatment interaction effect has the form $\mu_{jk} - \mu_{jk'} - \mu_{j'k} + \mu_{j'k'}$. The AB interaction effect for the design in Figure 6.4 can be written as

$$\mu_{.000} - \mu_{.011} - \mu_{.101} + \mu_{.110} \quad \text{or} \quad (\mu_{.000} + \mu_{.110}) - (\mu_{.011} + \mu_{.101}).$$

Consider now the contrast, ψ, between the means for Group_0 and Group_1:

$$\psi = \mu_{...0} - \mu_{...1}.$$

		Treat. Comb. $a_j b_k$	Treat. Comb. $a_j b_k$
	Block$_0$	00	11
Group$_0$ $(AB)_0$	Block$_1$	00	11
	Block$_2$	00	11
	Block$_3$	00	11
	Block$_4$	01	10
Group$_1$ $(AB)_1$	Block$_5$	01	10
	Block$_6$	01	10
	Block$_7$	01	10

Figure 6.4 Layout for an RBCF-2^2 design. The AB interaction is confounded with groups.

This contrast can be written as

$$\psi = (\mu_{.000} + \mu_{.110})/2 - (\mu_{.011} + \mu_{.101})/2,$$

because $\mu_{...0} = (\mu_{.000} + \mu_{.110})/2$ and $\mu_{...1} = (\mu_{.011} + \mu_{.101})/2$. It is apparent that the contrast between $Group_0$ and $Group_1$ involves the same means as the AB interaction effect. Thus, the AB interaction and Groups are completely confounded, and it is impossible to determine the unique contribution of either source of variation.

The basic concepts underlying the construction of randomized block confounded factorial designs where each treatment has two levels have now been described. Before describing other confounded factorial designs, three procedures for computing sums of squares for an RBCF-2^2 design will be presented in detail because they generalize to other confounded designs.

6.3 RBCF-2^2 DESIGN

6.3.1 Computational Procedures: The Sum-of-Squares Approach

An experiment described by Kirk (1982, pp. 577–578) will be used to illustrate the main features of an RBCF-2^2 design. The experiment was designed to evaluate the relative effectiveness of several procedures for using computer-aided instructional material. The material was prepared to acquaint mechanics with servicing procedures for a new airplane engine. The criterion used to assess the effectiveness of the material was the number of simulated malfunctions in an engine that trainees were able to diagnose. The instruction material was presented to trainees by means of a computer terminal. Treatment A consisted of two presentation rates. Level a_0 was an unpaced rate in which trainees pressed the "return" key on the terminal when they were ready to view the next frame of information. Level a_1 was a paced presentation with 30 seconds between successive frames of information. The second variable was the type of response that trainees made to each frame of information. Two types of responses were investigated: one in which trainees responded to a frame of information by touching the appropriate area of the computer screen, b_0, and a second in which trainees typed a response using the computer keyboard, b_1.

The research hypotheses leading to the experiment can be evaluated by means of statistical tests of the following null hypotheses.

$H_0: \alpha_j = 0$ for all j
$H_1: \alpha_j \neq 0$ for some j

H_0: $\beta_k = 0$ for all k
H_1: $\beta_k \neq 0$ for some k

H_0: $(\alpha\beta)_{jk} = 0$ for all j, k
H_1: $(\alpha\beta)_{jk} \neq 0$ for some j, k

The researcher's primary interest was in evaluating hypotheses regarding the two treatments. The level of significance adopted for all tests was .05.

To evaluate these hypotheses, a random sample of 16 trainees was obtained. Aptitude-test data were used to assign the trainees to eight blocks of size two so that those in a block had similar aptitude-test scores. Following this, the matched trainees in each block were randomly assigned to the $a_j b_k$ treatment combinations appropriate for a group. The use of only 16 subjects simplifies the statistical presentation but provides too few degrees of freedom for experimental error. A minimum of 48 subjects should be used. The layout for this design with 16 subjects is the same as that shown in Figure 6.4.

The AB interaction was confounded with groups by using the defining relations

$$a_j + b_k = 0 \ (\text{mod } 2),$$

$$a_j + b_k = 1 \ (\text{mod } 2).$$

Treatment combinations 00 and 11 satisfy the first relation and were assigned to the blocks in group 0. Treatment combinations 01 and 10 satisfy the second relation and were assigned to the blocks in group 1. The sum of squares computational procedure for this RBCF-2^2 design is shown in Table 6.1. The analysis of variance is summarized in Table 6.2. The F ratios in Table 6.2 and throughout are appropriate for a mixed model (Model III) in which the treatments are fixed and blocks are random. According to the analysis, the null hypotheses for treatments A and B are rejected. On the basis of the information in Tables 6.1 and 6.2, we can conclude that the paced presentation, a_0, is superior to the unpaced presentation, a_1, and that typing a response using the computer keyboard, b_1, is better than touching the computer screen, b_0.

The computation of the sum of squares in Table 6.1 involves simple arithmetic and can be performed with a calculator. Performing the computations is a useful first step in understanding the main features of the design. However, the sum-of-squares approach has two serious disadvantages: it is not well suited for computers, and it cannot be used when one or more observations are missing. Two alternative approaches, the cell means model and the regression model that are described in Sections 6.3.3

Table 6.1 Sum of Squares Computational Procedure for a RBCF-2^2 Design

Data and notation [Y_{ijkz} denotes a score for the experimental unit in block i, treatment
combination a_jb_k, and group z; $i = 1, \ldots, n$ blocks (s_i); $j = 1, \ldots, p$ levels of
treatment $A(a_j)$; $k = 1, \ldots, q$ levels of treatment $B(b_k)$; $z = 1, \ldots, w$ groups (g_z);
$jk = 1, \ldots, v$ combinations of a_jb_k within a block]

ABGS Summary Table
Entry is Y_{ijkz}

	a_jb_k	a_jb_k	$\sum_{jk=1}^{v} Y_{ijkz}$	$\sum_{i=1}^{n}\sum_{jk=1}^{v} Y_{ijkz}$
	00	11		
s_0	3	16	19	
$g_0(AB)_0\ s_1$	5	14	19	81
s_2	6	17	23	
s_3	5	15	20	
	01	10		
s_4	14	7	21	
$g_1(AB)_1\ s_5$	14	6	20	91
s_6	16	7	23	
s_7	16	11	27	

AB Summary Table

Entry is $\sum_{i=1}^{n} Y_{ijkz}$

	b_0	b_1	$\sum_{i=1}^{n}\sum_{k=1}^{q} Y_{ijkz}$
	$n = 4$		
a_0	19	60	79
a_1	31	62	93
$\sum_{i=1}^{n}\sum_{j=1}^{p} Y_{ijkz} = 50$		122	

Computational symbols

$$\sum_{i=1}^{n} \sum_{jk=1}^{v} \sum_{z=1}^{w} Y_{ijkz} = 3 + 5 + \cdots + 11 = 172.000$$

$$\frac{(\sum_{i=1}^{n} \sum_{jk=1}^{v} \sum_{z=1}^{w} Y_{ijkz})^2}{nvw} = [Y] = \frac{(172.000)^2}{(4)(2)(2)} = 1849.000$$

$$\sum_{i=1}^{n} \sum_{jk=1}^{v} \sum_{z=1}^{w} Y_{ijkz}^2 = [ABGS] = (3)^2 + (5)^2 + \cdots + (11)^2 = 2220.000$$

$$\sum_{i=1}^{n} \sum_{z=1}^{w} \frac{(\sum_{jk=1}^{v} Y_{ijkz})^2}{v} = [GS] = \frac{(19)^2}{2} + \frac{(19)^2}{2} + \cdots + \frac{(27)^2}{2} = 1875.000$$

$$\sum_{z=1}^{w} \frac{(\sum_{i=1}^{n} \sum_{jk=1}^{v} Y_{ijkz})^2}{nv} = [G] = \frac{(81)^2}{(4)(2)} + \frac{(91)^2}{(4)(2)} = 1855.250$$

$$\sum_{j=1}^{p} \frac{(\sum_{i=1}^{n} \sum_{k=1}^{q} Y_{ijkz})^2}{nq} = [A] = \frac{(79)^2}{(4)(2)} + \frac{(93)^2}{(4)(2)} = 1861.250$$

$$\sum_{k=1}^{q} \frac{(\sum_{i=1}^{n} \sum_{j=1}^{p} Y_{ijkz})^2}{np} = [B] = \frac{(50)^2}{(4)(2)} + \frac{(122)^2}{(4)(2)} = 2173.000$$

$$\sum_{j=1}^{p} \sum_{k=1}^{q} \frac{(\sum_{i=1}^{n} Y_{ijkz})^2}{n} = [AB] = \frac{(19)^2}{4} + \frac{(60)^2}{4} + \cdots + \frac{(62)^2}{4} = 2191.500$$

Computational formulas

$SSTO = [ABGS] - [Y] = 371.000$

$SS\text{ w. }BL = [ABGS] - [GS]$
$\qquad\qquad = 345.000$

$SS\text{ b. }BL = [GS] - [Y] = 26.000$

$SSA = [A] - [Y] = 12.250$

$SSG \text{ or } SSAB = [G] - [Y] = 6.250$

$SSB = [B] - [Y] = 324.000$

$SSBL(G) = [GS] - [G] = 19.750$

$SSAB \times BL(G) = [ABGS] - [AB]$
$\qquad\qquad\qquad - [GS] + [G]$

$\qquad\qquad\qquad = 8.750$

Table 6.2 ANOVA Table for an RBCF-2^2 Design

Source	SS	df	MS	F	E(MS) Model III (A and B fixed, Blocks random)
1 Between blocks	26.000	$nw - 1 = 7$			
2 Groups or AB	6.250	$w - 1 = 1$	6.250	[2/3] 1.90	$\sigma_\epsilon^2 + v\sigma_\pi^2 + \dfrac{nv\,\Sigma_{z=1}^{q}\,\zeta_z^2}{w-1}$
3 Blocks w. G	19.750	$w(n-1) = 6$	3.292		$\sigma_\epsilon^2 + v\sigma_\pi^2$
4 Within blocks	345.000	$nw(v-1) = 8$			
5 A (presentation rate)	12.250	$p - 1 = 1$	12.250	[5/7] 8.40*	$\sigma_\epsilon^2 + \sigma_{\alpha\beta\pi}^2 + \dfrac{nw\,\Sigma_{j=1}^{p}\,\alpha_j^2}{p-1}$
6 B (response model)	324.000	$q - 1 = 1$	324.000	[6/7]222.22**	$\sigma_\epsilon^2 + \sigma_{\alpha\beta\pi}^2 + \dfrac{nw\,\Sigma_{k=1}^{q}\,\beta_k^2}{q-1}$
7 $AB \times BL(G)$	8.750	$w(n-1)(v-1) = 6$	1.458		$\sigma_\epsilon^2 + \sigma_{\alpha\beta\pi}^2$
8 $AB \times BL(g_0)$	4.375	$(n-1)(v-1) = 3$	1.458		
9 $AB \times BL(g_1)$	4.375	$(n-1)(v-1) = 3$	1.458		
10 Total	371.000	$nvw - 1 = 15$			

*$p < .03$.
**$p < .000006$.

and 6.3.4, respectively, do not have these disadvantages. Both approaches involve solving matrix equations and require a computer for their successful implementation. Unfortunately, neither approach provides the intuitive understanding of analysis of variance that comes from performing computations by hand. Before turning to these approaches, the experimental design model for the RBCF-2^2 design will be described.

6.3.2 Experimental Design Model

A score Y_{ijkz} in an RBCF-2^2 design is a composite that is equal to the following terms in the experimental design model.

$$Y_{ijkz} = \mu + \zeta_z + \pi_{i(z)} + \alpha_j + \beta_k + (\alpha\beta\pi)_{jki(z)} + \varepsilon_{ijkz}$$

$$(i = 1, \ldots, n; j = 1, \ldots, p; k = 1, \ldots, q; z = 1, \ldots, w)$$

where Y_{ijkz} is the score for the experimental unit in block i, treatment combination $a_j b_k$, and group z,

ζ_z is the effect of group z and is subject to the restriction $\sum_{z=1}^{w} \zeta_z = 0$; the effects of Groups and the AB interaction are completely confounded,

$\pi_{i(z)}$ is the effect of block i that is $NID(0, \sigma_\pi^2)$,

α_j is the effect of treatment level j and is subject to the restriction $\sum_{j=1}^{p} \alpha_j = 0$,

β_k is the effect of treatment level k and is subject to the restriction $\sum_{k=1}^{q} \beta_k = 0$,

$(\alpha\beta\pi)_{jki(z)}$ is the interaction of $(\alpha\beta)_{jk}$ and $\pi_{i(z)}$ and is $NID(0, \sigma_{\alpha\beta\pi}^2)$, and

ε_{ijkz} is the experimental error that is $NID(0, \sigma_\varepsilon^2)$; ε_{ijkz} is independent of $\pi_{i(z)}$. In this design, ε_{ijkz} cannot be estimated separately from $(\alpha\beta\pi)_{jki(z)}$.

Two sets of assumptions underlie the F tests for an RBCF-2^2 design: one set for the between-blocks test and a second set for the within-blocks tests. The assumptions underlying the between-blocks test are the same as those for a completely randomized design. The key assumption is that the population variances for Group$_0$ and Group$_1$ are homogeneous. If sufficiently many blocks are randomly assigned to the groups, it is reasonable to expect that the assumption is tenable.

The assumptions for the within-blocks tests are similar to those for the within-blocks tests for a split-plot factorial design. The mean square within-blocks error term, $MSAB \times BL(G)$, is a pooled term that is equal to $[SSAB \times BL(g_0) + SSAB \times BL(g_1)]/w(n-1)(v-1)$. Because there are only two treatment combinations in each group of blocks, $MSAB \times$

$BL(g_0)$ and $MSAB \times BL(g_1)$ are equal to the average of two variances minus a covariance. That is,

$$MSAB \times BL(g_0) = (\hat{\sigma}^2_{.000} + \hat{\sigma}^2_{.110})/2 - \hat{\sigma}^2_{(.000)(.110)},$$

$$MSAB \times BL(g_1) = (\hat{\sigma}^2_{.011} + \hat{\sigma}^2_{.101})/2 - \hat{\sigma}^2_{(.011)(.101)}.$$

The key assumption for the within-blocks tests is that the population variances estimated by $MSAB \times BL(g_0)$ and $MSAB \times BL(g_1)$ are equal. Procedures for testing this assumption are described in most experimental design textbooks. If the assumption is not tenable, the three-step testing strategy described by Kirk (1982, pp. 259–262) can be used to obtain approximate F tests of treatments A and B.

The advantage of an RBCF-2^2 design over an RBF-22 design is that it enables a researcher to reduce the block size from four to two. This reduction is accomplished without sacrificing power in testing treatments A and B because both are within-blocks treatments. The RBCF-2^2 design is a better choice than an SPF-2·2 design if a researcher's primary interest is in treatments A and B rather than the AB interaction. Recall that in a split-plot factorial design, the test of treatment A is less powerful than the test of treatment B.

6.3.3 Computational Procedures: The Cell Means Model Approach

The cell means model, unlike the sum-of-squares approach described in Section 6.3.1, can be used when one or more observations are missing. Another advantage, as we will see, is that a researcher can compute sums of squares for any interesting null hypothesis, and there is never any ambiguity about the hypothesis that is tested. For an introduction to the cell means model and other general linear model approaches, the reader is referred to Hocking (1985), Kirk (1982), Pedhazur (1982), Searle (1987), and Timm and Carlson (1975).

The model equation for the cell means model is quite different from that for the experimental design model described in Section 6.3.2. The latter model equation for an RBCF-2^2 design is

$$Y_{ijkz} = \mu + \zeta_z + \pi_{i(z)} + \alpha_j + \beta_k + (\alpha\beta\pi)_{jki(z)} + \varepsilon_{ijkz}$$

$$(i = 1, \ldots, n; j = 1, \ldots, p; k = 1, \ldots, q; z = 1, \ldots, w)$$

and contains the parameters μ, ζ_z, $\pi_{i(z)}$, and so on. The cell means model equation for the same design is

$$Y_{ijkz} = \mu_{ijkz} + \varepsilon_{ijkz}$$

$$(i = 1, \ldots, n; j = 1, \ldots, p; k = 1, \ldots, q; z = 1, \ldots, w)$$

and contains the parameters μ_{ijkz} and ε_{ijkz}. Except for the experimental error, ε_{ijkz}, all of the parameters are cell means—hence the name, cell means model. As discussed by Hocking (1985, pp. 38–51), Kirk (1982, pp. 281–292), Searle (1987, pp. 402–407), and Timm and Carlson (1975, pp. 27–34), restrictions can be imposed on the cell means. When this is done the model is called a *restricted cell means model*. Recall from Sections 3.1 and 3.2 that $MSAB \times BL(G)$ is the error term for testing treatments A and B. The RBCF-2^2 design does not provide a pure estimate of the population error variance, σ_ε^2. However, treatments A and B can be tested subject to restrictions on μ_{ijkz} that for each group

$$\mu_{i(jk)z} - \mu_{i'(jk)z} - \mu_{i(jk)'z} + \mu_{i'(jk)'z} = 0 \quad \text{for all } i, i', (jk), \text{ and } (jk)'.$$

These restrictions specify that the block $\times$ $a_j b_k$ treatment-combination effects equal zero for each group. It is assumed that ε_{ijkz} is $NID(0, \sigma_\varepsilon^2)$.

The null hypotheses for the cell means model are expressed in terms of means. Consider the following null hypothesis:

$$H_0: \mu_1 = \mu_2 = \mu_3 = \mu_4.$$

Two equivalent ways of expressing this hypothesis are

$$H_0: \mu_1 - \mu_4 = 0, \mu_2 - \mu_4 = 0, \mu_3 - \mu_4 = 0 \tag{6.1}$$

and

$$H_0: \mu_1 - \mu_2 = 0, \mu_2 - \mu_3 = 0, \mu_3 - \mu_4 = 0. \tag{6.2}$$

These null hypotheses can be written using matrix notation as

$$H_0: \begin{matrix} \mathbf{C'} \\ (p-1) \times h \end{matrix} \quad \begin{matrix} \boldsymbol{\mu} \\ h \times 1 \end{matrix} \quad \begin{matrix} \mathbf{0} \\ (p-1) \times 1 \end{matrix}$$

$$H_0: \begin{bmatrix} 1 & 0 & 0 & -1 \\ 0 & 1 & 0 & -1 \\ 0 & 0 & 1 & -1 \end{bmatrix} \begin{bmatrix} \mu_1 \\ \mu_2 \\ \mu_3 \\ \mu_4 \end{bmatrix} = \begin{bmatrix} 0 \\ 0 \\ 0 \end{bmatrix}$$

$$\begin{matrix} \mathbf{C'} \\ (p-1) \times h \end{matrix} \quad \begin{matrix} \boldsymbol{\mu} \\ h \times 1 \end{matrix} \quad \begin{matrix} \mathbf{0} \\ (p-1) \times 1 \end{matrix}$$

$$\text{and} \quad H_0: \begin{bmatrix} 1 & -1 & 0 & 0 \\ 0 & 1 & -1 & 0 \\ 0 & 0 & 1 & -1 \end{bmatrix} \begin{bmatrix} \mu_1 \\ \mu_2 \\ \mu_3 \\ \mu_4 \end{bmatrix} = \begin{bmatrix} 0 \\ 0 \\ 0 \end{bmatrix},$$

where $p - 1$ is the rank of $\mathbf{C'}$ and h is the number of parameters in $\boldsymbol{\mu}$. The coefficient matrix $\mathbf{C'}$ must be of full row rank. In other words, each row of $\mathbf{C'}$ must be linearly independent of the other rows. The maximum number of such rows is $p - 1$, which is why it is necessary to express the

Table 6.3 Estimators of the Cell Means for the RBCF-2^2 Design ($\hat{\mu}_{0000}$ Is Missing Because of an Equipment Malfunction)

ABGS Summary Table
Entry is μ_{ijkz}

		$a_j b_k$	$a_j b_k$	$\dfrac{\sum_{jk=1}^{v} \hat{\mu}_{ijkz}}{\sum_{jk}^{v} n_{ijkz}}$	$\dfrac{\sum_{i=1}^{njk} \sum_{jk=1}^{v} \hat{\mu}_{ijkz}}{\sum_{i=1}^{njk} \sum_{jk=1}^{v} n_{ijkz}}$
		00	11		
$g_0(AB)_0$	Block$_0$	Missing	$\hat{\mu}_{0110} = 16$	$\hat{\mu}_{0 \cdot \cdot 0} = 16.0$	
	Block$_1$	$\hat{\mu}_{1000} = 5$	$\hat{\mu}_{1110} = 14$	$\hat{\mu}_{1 \cdot \cdot 0} = 9.5$	$\hat{\hat{\mu}}_{\cdot \cdot \cdot 0} = 11.14$
	Block$_2$	$\hat{\mu}_{2000} = 6$	$\hat{\mu}_{2110} = 17$	$\hat{\mu}_{2 \cdot \cdot 0} = 11.5$	
	Block$_3$	$\hat{\mu}_{3000} = 5$	$\hat{\mu}_{3110} = 15$	$\hat{\mu}_{3 \cdot \cdot 0} = 10.0$	
		01	10		
$g_1(AB)_1$	Block$_4$	$\hat{\mu}_{4011} = 14$	$\hat{\mu}_{4101} = 7$	$\hat{\mu}_{4 \cdot \cdot 1} = 10.5$	
	Block$_5$	$\hat{\mu}_{5011} = 14$	$\hat{\mu}_{5101} = 6$	$\hat{\mu}_{5 \cdot \cdot 1} = 10.0$	$\hat{\hat{\mu}}_{\cdot \cdot \cdot 1} = 11.38$
	Block$_6$	$\hat{\mu}_{6011} = 16$	$\hat{\mu}_{6101} = 7$	$\hat{\mu}_{6 \cdot \cdot 1} = 11.5$	
	Block$_7$	$\hat{\mu}_{7011} = 16$	$\hat{\mu}_{7101} = 11$	$\hat{\mu}_{7 \cdot \cdot 1} = 13.5$	

ABG Summary Table
Entry is $\hat{\mu}_{\cdot jkz}$

	$a_j b_k$	$a_j b_k$	$\sum_{jk=1}^{v} \dfrac{\hat{\mu}_{\cdot jkz}}{v}$
g_0	$\hat{\mu}_{\cdot 000} = 5.33$	$\hat{\mu}_{\cdot 110} = 15.50$	$\hat{\mu}_{\cdot \cdot \cdot 0} = 10.42$
g_1	$\hat{\mu}_{\cdot 011} = 15.00$	$\hat{\mu}_{\cdot 101} = 7.75$	$\hat{\mu}_{\cdot \cdot \cdot 1} = 11.38$

AB Summary Table
Entry is $\hat{\mu}_{\cdot jk \cdot}$

	b_0	b_1	$\sum_{k=1}^{q} \dfrac{\hat{\mu}_{\cdot jk \cdot}}{q}$	$\dfrac{\sum_{k=1}^{q} n_{jk} \hat{\mu}_{\cdot jk \cdot}}{\sum_{k=1}^{q} n_{jk}}$
a_0	$n_{00} = 3$ $\hat{\mu}_{\cdot 00 \cdot} = 5.33$	$n_{11} = 4$ $\hat{\mu}_{\cdot 01 \cdot} = 15.00$	$\hat{\mu}_{\cdot 0 \cdot \cdot} = 10.17$	$\hat{\hat{\mu}}_{\cdot 0 \cdot \cdot} = 10.86$
a_1	$n_{01} = 4$ $\hat{\mu}_{\cdot 10 \cdot} = 7.75$	$n_{10} = 4$ $\hat{\mu}_{\cdot 11 \cdot} = 15.50$	$\hat{\mu}_{\cdot 0 \cdot \cdot} = 11.62$	$\hat{\hat{\mu}}_{\cdot 1 \cdot \cdot} = 11.62$

$\sum_{j=1}^{p} \dfrac{\hat{\mu}_{\cdot jk \cdot}}{p} = \quad \hat{\mu}_{\cdot \cdot 0 \cdot} = 6.54 \qquad \hat{\mu}_{\cdot \cdot 1 \cdot} = 15.25$

$\dfrac{\sum_{j=1}^{p} n_{jk} \hat{\mu}_{\cdot jk \cdot}}{\sum_{j=1}^{p} n_{jk}} = \quad \hat{\hat{\mu}}_{\cdot \cdot 0 \cdot} = 6.71 \qquad \hat{\hat{\mu}}_{\cdot \cdot 1 \cdot} = 15.25$

null hypothesis as (6.1) or (6.2). For convenience, the form shown for hypothesis (6.1) is used throughout.

Suppose that the computer-aided instruction experiment described in Section 6.3.1 has been performed and that observation $\overline{Y}_{0000} = 3$ is missing because of an equipment malfunction. The data for the experiment will be analyzed using the cell means model. Estimators of the cell means are given in Table 6.3.

An RBCF-2^2 design contains the following sources of variation.

Between Blocks	Within Blocks
Groups or AB	A
Blocks w. Groups, $BL(G)$	B
	$AB \times BL(G)$

Coefficient matrices can be formulated for testing hypotheses associated with each of these sources of variation. For three of the sources—Groups or AB, A, and B—two kinds of hypotheses may be of interest: hypotheses for unweighted means, μ, and hypotheses for weighted means, $\overline{\mu}$. The weights, n_{jk}, for the weighted means are the sample sizes used to compute the cell means. Estimators of the two kinds of means are given in Table 6.3.

The Between Blocks null hypothesis is

$$H_0: \mu_{0\cdot\cdot0} - \mu_{7\cdot\cdot1} = 0 \qquad \text{or} \qquad H_0: \mathbf{C}'_{BL}\, \mu = \mathbf{0}_{BL}$$

$$\mu_{1\cdot\cdot0} - \mu_{7\cdot\cdot1} = 0$$

$$\mu_{2\cdot\cdot0} - \mu_{7\cdot\cdot1} = 0$$

$$\mu_{3\cdot\cdot0} - \mu_{7\cdot\cdot1} = 0$$

$$\mu_{4\cdot\cdot1} - \mu_{7\cdot\cdot1} = 0$$

$$\mu_{5\cdot\cdot1} - \mu_{7\cdot\cdot1} = 0$$

$$\mu_{6\cdot\cdot1} - \mu_{7\cdot\cdot1} = 0$$

where

$$\mathbf{C}'_{BL} = \begin{bmatrix} 0 & 0 & 0 & 1 & 0 & 0 & 0 & 0 & 0 & 0 & -\frac{1}{2} & 0 & 0 & 0 & -\frac{1}{2} \\ \frac{1}{2} & 0 & 0 & 0 & \frac{1}{2} & 0 & 0 & 0 & 0 & 0 & -\frac{1}{2} & 0 & 0 & 0 & -\frac{1}{2} \\ 0 & \frac{1}{2} & 0 & 0 & 0 & \frac{1}{2} & 0 & 0 & 0 & 0 & -\frac{1}{2} & 0 & 0 & 0 & -\frac{1}{2} \\ 0 & 0 & \frac{1}{2} & 0 & 0 & 0 & \frac{1}{2} & 0 & 0 & 0 & -\frac{1}{2} & 0 & 0 & 0 & -\frac{1}{2} \\ 0 & 0 & 0 & 0 & 0 & 0 & 0 & \frac{1}{2} & 0 & 0 & -\frac{1}{2} & \frac{1}{2} & 0 & 0 & -\frac{1}{2} \\ 0 & 0 & 0 & 0 & 0 & 0 & 0 & 0 & \frac{1}{2} & 0 & -\frac{1}{2} & 0 & \frac{1}{2} & 0 & -\frac{1}{2} \\ 0 & 0 & 0 & 0 & 0 & 0 & 0 & 0 & 0 & \frac{1}{2} & -\frac{1}{2} & 0 & 0 & \frac{1}{2} & -\frac{1}{2} \end{bmatrix},$$

$$\mathbf{\mu}' = [\mu_{1000}\ \mu_{2000}\ \mu_{3000}\ \mu_{0110}\ \mu_{1110}\ \mu_{2110}\ \mu_{3110}\ \mu_{4011}\ \mu_{5011}\ \mu_{6011}\ \mu_{7011}$$
$$\qquad \mu_{4101}\ \mu_{5101}\ \mu_{6101}\ \mu_{7101}],$$
$$\mathbf{0}'_{BL} = [0\ \ 0\ \ 0\ \ 0\ \ 0\ \ 0\ \ 0].$$

The coefficients, $c_{ijkz} = \pm 1/\Sigma_{jk=1}^{v} n_{ijkz}$ or 0, in $\mathbf{C}'_{BL}$ can be multiplied by 2 to eliminate fractions. This does not affect the nature of the hypothesis that is tested. The sum of squares for Blocks is

$$SSBL = (\mathbf{C}'_{BL}\ \hat{\mathbf{\mu}})'(\mathbf{C}'_{BL}\ \mathbf{C}_{BL})^{-1}(\mathbf{C}'_{BL}\ \hat{\mathbf{\mu}}) = 46.4333,$$

where $\hat{\mathbf{\mu}}' = [5\quad 6\quad 5\quad 16\quad 14\quad 17\quad 15\quad 14\quad 14\quad 16\quad 16\quad 7\quad 6\quad 7\quad 11]$.
 The Groups or AB null hypothesis for unweighted means is

$$H_0: \mu_{\cdots 0} - \mu_{\cdots 1} = 0 \qquad \text{or} \qquad H_0: \mathbf{C}'_{1(G)}\ \mathbf{\mu} = \mathbf{0}_{1(G)}$$

where

$$\mathbf{C}'_{1(G)} = [\tfrac{1}{6}\ \tfrac{1}{6}\ \tfrac{1}{6}\ \tfrac{1}{8}\ \tfrac{1}{8}\ \tfrac{1}{8}\ \tfrac{1}{8}\ -\tfrac{1}{8}\ -\tfrac{1}{8}\ -\tfrac{1}{8}\ -\tfrac{1}{8}\ -\tfrac{1}{8}\ -\tfrac{1}{8}\ -\tfrac{1}{8}\ -\tfrac{1}{8}],$$
$$\mathbf{0}'_{1(G)} = [0].$$

The coefficients, $c_{ijkz} = \pm 1/n_{jk}v$ or 0, in $\mathbf{C}'_{1(G)}$ can be multiplied by 24 to eliminate fractions. The unweighted-means sum of squares for Groups is

$$SS1(G) = (\mathbf{C}'_{1(G)}\ \hat{\mathbf{\mu}})'(\mathbf{C}'_{1(G)}\ \mathbf{C}_{1(G)})^{-1}(\mathbf{C}'_{1(G)}\ \hat{\mathbf{\mu}}) = 3.3910.$$

If the sample sizes in the two groups reflect the sizes of the two respective populations, a researcher might want to test hypotheses for weighted means. For this case, the Groups or AB null hypothesis is

$$H_0: \overline{\mu}_{\cdots 0} - \overline{\mu}_{\cdots 1} = 0 \qquad \text{or} \qquad H_0: \mathbf{C}'_{2(G)}\ \mathbf{\mu} = \mathbf{0}_{2(G)}$$

where

$$\mathbf{C}'_{2(G)} = [\tfrac{1}{7}\ \tfrac{1}{7}\ \tfrac{1}{7}\ \tfrac{1}{7}\ \tfrac{1}{7}\ \tfrac{1}{7}\ \tfrac{1}{7}\ -\tfrac{1}{8}\ -\tfrac{1}{8}\ -\tfrac{1}{8}\ -\tfrac{1}{8}\ -\tfrac{1}{8}\ -\tfrac{1}{8}\ -\tfrac{1}{8}\ -\tfrac{1}{8}],$$
$$\mathbf{0}'_{2(G)} = [0].$$

The coefficients, $c_{ijkz} = \pm 1/\Sigma_{i=1}^{n}\Sigma_{jk=1}^{v} n_{ijkz}$ or 0, in $\mathbf{C}'_{2(G)}$ can be multiplied by 56 to eliminate fractions. The weighted-means sum of squares for Groups is

$$SS2(G) = (\mathbf{C}'_{2(G)}\ \hat{\mathbf{\mu}})'(\mathbf{C}'_{2(G)}\ \mathbf{C}_{2(G)})^{-1}(\mathbf{C}'_{2(G)}\ \hat{\mathbf{\mu}}) = 0.2012.$$

The Blocks within Groups, $BL(G)$, null hypothesis is

$$H_0: \mu_{0\cdot\cdot 0} - \mu_{3\cdot\cdot 0} = 0 \qquad \text{or} \qquad H_0: \mathbf{C}'_{BL(G)}\ \mathbf{\mu} = \mathbf{0}_{BL(G)}$$
$$\mu_{1\cdot\cdot 0} - \mu_{3\cdot\cdot 0} = 0$$
$$\mu_{2\cdot\cdot 0} - \mu_{3\cdot\cdot 0} = 0$$
$$\mu_{4\cdot\cdot 1} - \mu_{7\cdot\cdot 1} = 0$$
$$\mu_{5\cdot\cdot 1} - \mu_{7\cdot\cdot 1} = 0$$
$$\mu_{6\cdot\cdot 1} - \mu_{7\cdot\cdot 1} = 0$$

where

$$
\mathbf{C}'_{BL(G)} = \begin{bmatrix}
0 & 0 & -\frac{1}{2} & 1 & 0 & 0 & -\frac{1}{2} & 0 & 0 & 0 & 0 & 0 & 0 & 0 & 0 \\
\frac{1}{2} & 0 & -\frac{1}{2} & 0 & \frac{1}{2} & 0 & -\frac{1}{2} & 0 & 0 & 0 & 0 & 0 & 0 & 0 & 0 \\
0 & \frac{1}{2} & -\frac{1}{2} & 0 & 0 & \frac{1}{2} & -\frac{1}{2} & 0 & 0 & 0 & 0 & 0 & 0 & 0 & 0 \\
0 & 0 & 0 & 0 & 0 & 0 & 0 & \frac{1}{2} & 0 & 0 & -\frac{1}{2} & \frac{1}{2} & 0 & 0 & -\frac{1}{2} \\
0 & 0 & 0 & 0 & 0 & 0 & 0 & 0 & \frac{1}{2} & 0 & -\frac{1}{2} & 0 & \frac{1}{2} & 0 & -\frac{1}{2} \\
0 & 0 & 0 & 0 & 0 & 0 & 0 & 0 & 0 & \frac{1}{2} & -\frac{1}{2} & 0 & 0 & \frac{1}{2} & -\frac{1}{2}
\end{bmatrix},
$$

$\mathbf{0}'_{BL(G)} = [0 \quad 0 \quad 0 \quad 0 \quad 0 \quad 0]$.

The coefficients, $c_{ijkz} = \pm 1/\Sigma^{v}_{jk=1}\, n_{ijkz}$ or 0, in $\mathbf{C}'_{BL(G)}$ can be multiplied by 2 to eliminate fractions. The sum of squares for Blocks within Groups is

$$
SSBL(G) = (\mathbf{C}'_{BL(G)}\, \hat{\boldsymbol{\mu}})'(\mathbf{C}'_{BL(G)}\, \mathbf{C}_{BL(G)})^{-1}(\mathbf{C}'_{BL(G)}\, \hat{\boldsymbol{\mu}}) = 46.2321.
$$

The Within Blocks, w.BL, null hypothesis is

H_0: $\mu_{1000} - \mu_{1110} = 0$ or H_0: $\mathbf{C}'_{w.BL}\, \boldsymbol{\mu} = \mathbf{0}_{w.BL}$
    $\mu_{2000} - \mu_{2110} = 0$
    $\mu_{3000} - \mu_{3110} = 0$
    $\mu_{4011} - \mu_{4101} = 0$
    $\mu_{5011} - \mu_{5101} = 0$
    $\mu_{6011} - \mu_{6101} = 0$
    $\mu_{7011} - \mu_{7101} = 0$

where

$$
\mathbf{C}'_{w.BL} = \begin{bmatrix}
1 & 0 & 0 & 0 & -1 & 0 & 0 & 0 & 0 & 0 & 0 & 0 & 0 & 0 & 0 \\
0 & 1 & 0 & 0 & 0 & -1 & 0 & 0 & 0 & 0 & 0 & 0 & 0 & 0 & 0 \\
0 & 0 & 1 & 0 & 0 & 0 & -1 & 0 & 0 & 0 & 0 & 0 & 0 & 0 & 0 \\
0 & 0 & 0 & 0 & 0 & 0 & 0 & 1 & 0 & 0 & 0 & -1 & 0 & 0 & 0 \\
0 & 0 & 0 & 0 & 0 & 0 & 0 & 0 & 1 & 0 & 0 & 0 & -1 & 0 & 0 \\
0 & 0 & 0 & 0 & 0 & 0 & 0 & 0 & 0 & 1 & 0 & 0 & 0 & -1 & 0 \\
0 & 0 & 0 & 0 & 0 & 0 & 0 & 0 & 0 & 0 & 1 & 0 & 0 & 0 & -1
\end{bmatrix},
$$

$\mathbf{0}'_{w.BL} = [0 \quad 0 \quad 0 \quad 0 \quad 0 \quad 0 \quad 0]$.

The coefficients, c_{ijkz}, are equal to ± 1 or 0. The sum of squares for Within Blocks is

$$
SS\ w.\ BL = (\mathbf{C}'_{w.BL}\, \hat{\boldsymbol{\mu}})'(\mathbf{C}'_{w.BL}\, \mathbf{C}_{w.BL})^{-1}(\mathbf{C}'_{w.BL}\, \hat{\boldsymbol{\mu}}) = 260.5000.
$$

We want to test treatments A and B subject to the restrictions that all $AB \times BL(G)$ interaction effects equal zero. These restrictions can be expressed as follows.

H_0: $\mu_{1000} - \mu_{2000} - \mu_{1110} + \mu_{2110} = 0$ or H_0: $\mathbf{R'} \boldsymbol{\mu} = \boldsymbol{\theta}$

 $\mu_{2000} - \mu_{3000} - \mu_{2110} + \mu_{3110} = 0$

 $\mu_{4011} - \mu_{5011} - \mu_{4101} + \mu_{5101} = 0$

 $\mu_{5011} - \mu_{6011} - \mu_{5101} + \mu_{6101} = 0$

 $\mu_{6011} - \mu_{7011} - \mu_{6101} + \mu_{7101} = 0$

where

$$\mathbf{R'} = \begin{bmatrix} 1 & -1 & 0 & 0 & -1 & 1 & 0 & 0 & 0 & 0 & 0 & 0 & 0 & 0 & 0 \\ 0 & 1 & -1 & 0 & 0 & -1 & 1 & 0 & 0 & 0 & 0 & 0 & 0 & 0 & 0 \\ 0 & 0 & 0 & 0 & 0 & 0 & 0 & 1 & -1 & 0 & 0 & -1 & 1 & 0 & 0 \\ 0 & 0 & 0 & 0 & 0 & 0 & 0 & 0 & 1 & -1 & 0 & 0 & -1 & 1 & 0 \\ 0 & 0 & 0 & 0 & 0 & 0 & 0 & 0 & 0 & 1 & -1 & 0 & 0 & -1 & 1 \end{bmatrix},$$

$\boldsymbol{\theta'} = \begin{bmatrix} 0 & 0 & 0 & 0 & 0 \end{bmatrix}$.

The rows of $\mathbf{R'}$ are interaction terms of the form $\mu_{i(jk)z} - \mu_{i'(jk)z} - \mu_{i(jk)'z} + \mu_{i'(jk)'z}$. Formation of this matrix is facilitated by using the following sets of crossed lines. The means connected by a dashed line are

subtracted from the means connected by a solid line. Each row of $\mathbf{R'}$ represents one of the five sets of crossed lines. The coefficients, c_{ijkz}, in $\mathbf{R'}$ are equal to ± 1 or 0. The sum of squares for $AB \times BL(G)$ is

$$SSAB \times BL(G) = (\mathbf{R'} \hat{\boldsymbol{\mu}})'(\mathbf{R'} \mathbf{R})^{-1}(\mathbf{R'} \hat{\boldsymbol{\mu}}) = 5.3750.$$

To test treatment A, subject to the restrictions that $\mathbf{R'} \boldsymbol{\mu} = \boldsymbol{\theta}$, the unweighted-means coefficient matrix for A is adjoined to the coefficient matrix for $AB \times BL(G)$. The augmented matrix is denoted by $\mathbf{Q'}_{1(A)}$,

$$\mathbf{Q'}_{1(A)} = \begin{bmatrix} \mathbf{R'} \\ \mathbf{C'}_{1(A)} \end{bmatrix}.$$

The unweighted-means coefficient matrix for treatment A is

$$\mathbf{C'}_{1(A)} = \begin{bmatrix} \frac{1}{6} & \frac{1}{6} & \frac{1}{6} & -\frac{1}{8} & -\frac{1}{8} & -\frac{1}{8} & -\frac{1}{8} & \frac{1}{8} & \frac{1}{8} & \frac{1}{8} & \frac{1}{8} & -\frac{1}{8} & -\frac{1}{8} & -\frac{1}{8} & -\frac{1}{8} \end{bmatrix},$$

where c_{ijkz} is equal to $\pm n_{jk}v$ or 0. The coefficients in $\mathbf{C}'_{1(A)}$ can be multiplied by 24 to eliminate fractions. The $\mathbf{Q}'_{1(A)}$ augmented matrix is

$$\mathbf{Q}'_{1(A)} = \begin{bmatrix} \mathbf{R}' \\ \mathbf{C}'_{1(A)} \end{bmatrix}$$

$$= \begin{bmatrix} 1 & -1 & 0 & 0 & -1 & 1 & 0 & 0 & 0 & 0 & 0 & 0 & 0 & 0 & 0 \\ 0 & 1 & -1 & 0 & 0 & -1 & 1 & 0 & 0 & 0 & 0 & 0 & 0 & 0 & 0 \\ 0 & 0 & 0 & 0 & 0 & 0 & 0 & 1 & -1 & 0 & 0 & -1 & 1 & 0 & 0 \\ 0 & 0 & 0 & 0 & 0 & 0 & 0 & 0 & 1 & -1 & 0 & 0 & -1 & 1 & 0 \\ 0 & 0 & 0 & 0 & 0 & 0 & 0 & 0 & 0 & 1 & -1 & 0 & 0 & -1 & 1 \\ \tfrac{1}{6} & \tfrac{1}{6} & \tfrac{1}{6} & -\tfrac{1}{8} & -\tfrac{1}{8} & -\tfrac{1}{8} & -\tfrac{1}{8} & \tfrac{1}{8} & \tfrac{1}{8} & \tfrac{1}{8} & \tfrac{1}{8} & -\tfrac{1}{8} & -\tfrac{1}{8} & -\tfrac{1}{8} & -\tfrac{1}{8} \end{bmatrix}.$$

In a similar manner, the null vector $\mathbf{0}_{1(A)} = [0]$ is adjoined to the null vector, $\boldsymbol{\theta}$, which was defined earlier. The resulting vector is denoted by $\boldsymbol{\eta}_{1(A)}$,

$$\boldsymbol{\eta}_{1(A)} = \begin{bmatrix} \boldsymbol{\theta} \\ \mathbf{0}_{1(A)} \end{bmatrix}.$$

The restricted, unweighted-means null hypothesis for treatment A is

$$H_0: \mu_{\cdot 0 \cdot \cdot}{}^* - \mu_{\cdot 1 \cdot \cdot}{}^* = 0 \qquad \text{or} \qquad H_0: \mathbf{Q}'_{1(A)} \boldsymbol{\mu} = \boldsymbol{\eta}_{1(A)},$$

where "$*$" indicates that the population means are subject to the restrictions that for each group

$$\mu_{i(jk)z} - \mu_{i'(jk)z} - \mu_{i(jk)'z} + \mu_{i'(jk)'z} = 0 \qquad \text{for all } i, i', (jk), \text{ and } (jk)'.$$

The restricted, unweighted-means sum of squares for treatment A is

$$SS1(A) = (\mathbf{Q}'_{1(A)} \hat{\boldsymbol{\mu}})'(\mathbf{Q}'_{1(A)} \mathbf{Q}_{1(A)})^{-1}(\mathbf{Q}'_{1(A)} \hat{\boldsymbol{\mu}})$$
$$- (\mathbf{R}' \hat{\boldsymbol{\mu}})'(\mathbf{R}' \mathbf{R})^{-1}(\mathbf{R}' \hat{\boldsymbol{\mu}}) = 7.8526.$$

When, as in this example, the rows of $\mathbf{R}'$ are orthogonal to the row of $\mathbf{C}'_{1(A)}$, the formula for $SS1(A)$ simplifies to

$$SS1(A) = (\mathbf{C}'_{1(A)} \hat{\boldsymbol{\mu}})'(\mathbf{C}'_{1(A)} \mathbf{C}_{1(A)})^{-1}(\mathbf{C}'_{1(A)} \hat{\boldsymbol{\mu}}).$$

If the sample sizes for treatment levels a_0 and a_1 reflect the sizes of the respective populations, a researcher might want to test hypotheses for weighted means. For this case, the restricted, weighted-means null hypothesis is

$$H_0: \bar{\mu}_{\cdot 0 \cdot \cdot}{}^* - \bar{\mu}_{\cdot 1 \cdot \cdot}{}^* = 0 \qquad \text{or} \qquad H_0: \mathbf{Q}'_{2(A)} \boldsymbol{\mu} = \boldsymbol{\eta}_{2(A)}.$$

The weighted-means coefficient matrix for treatment A is

$$\mathbf{C}'_{2(A)} = [\tfrac{1}{7} \quad \tfrac{1}{7} \quad \tfrac{1}{7} \quad -\tfrac{1}{8} \quad -\tfrac{1}{8} \quad -\tfrac{1}{8} \quad -\tfrac{1}{8} \quad \tfrac{1}{7} \quad \tfrac{1}{7} \quad \tfrac{1}{7} \quad \tfrac{1}{7} \quad -\tfrac{1}{8} \quad -\tfrac{1}{8} \quad -\tfrac{1}{8} \quad -\tfrac{1}{8}],$$

where $c_{ijkz} = \pm 1/\Sigma_{i=1}^{n} \Sigma_{jk=1}^{v} n_{ijkz}$ or 0. The $\mathbf{Q}'_{2(A)}$ and $\mathbf{\eta}_{2(A)}$ augmented matrices are

$$\mathbf{Q}'_{2(A)}$$

$$= \begin{bmatrix} 1 & -1 & 0 & 0 & -1 & 1 & 0 & 0 & 0 & 0 & 0 & 0 & 0 & 0 & 0 \\ 0 & 1 & -1 & 0 & 0 & -1 & 1 & 0 & 0 & 0 & 0 & 0 & 0 & 0 & 0 \\ 0 & 0 & 0 & 0 & 0 & 0 & 0 & 1 & -1 & 0 & 0 & -1 & 1 & 0 & 0 \\ 0 & 0 & 0 & 0 & 0 & 0 & 0 & 0 & 1 & -1 & 0 & 0 & -1 & 1 & 0 \\ 0 & 0 & 0 & 0 & 0 & 0 & 0 & 0 & 0 & 1 & -1 & 0 & 0 & -1 & 1 \\ \frac{1}{7} & \frac{1}{7} & \frac{1}{7} & -\frac{1}{8} & -\frac{1}{8} & -\frac{1}{8} & -\frac{1}{8} & \frac{1}{7} & \frac{1}{7} & \frac{1}{7} & \frac{1}{7} & -\frac{1}{8} & -\frac{1}{8} & -\frac{1}{8} & -\frac{1}{8} \end{bmatrix},$$

$$\mathbf{\eta}_{2(A)} = \begin{bmatrix} 0 \\ 0 \\ 0 \\ 0 \\ 0 \\ 0 \end{bmatrix}.$$

The coefficients in $\mathbf{Q}'_{2(A)}$ can be multiplied by 56 to eliminate fractions. The restricted, weighted-means sum of squares for treatment A is

$$SS2(A) = (\mathbf{Q}'_{2(A)} \hat{\mathbf{\mu}})'(\mathbf{Q}'_{2(A)} \mathbf{Q}_{2(A)})^{-1}(\mathbf{Q}'_{2(A)} \hat{\mathbf{\mu}})$$
$$- (\mathbf{R}' \hat{\mathbf{\mu}})'(\mathbf{R}' \mathbf{R})^{-1}(\mathbf{R}' \hat{\mathbf{\mu}}) = 2.2012.$$

Hypotheses can be constructed for treatment B that are analogous to those for treatment A. The following hypothesis is for restricted, unweighted-means.

$$H_0: \mu_{..0.}^* - \mu_{..1.}^* = 0 \qquad \text{or} \qquad H_0: \mathbf{Q}'_{1(B)} \mathbf{\mu} = \mathbf{\eta}_{1(B)}$$

where

$$\mathbf{Q}'_{1(B)}$$

$$= \begin{bmatrix} 1 & -1 & 0 & 0 & -1 & 1 & 0 & 0 & 0 & 0 & 0 & 0 & 0 & 0 & 0 \\ 0 & 1 & -1 & 0 & 0 & -1 & 1 & 0 & 0 & 0 & 0 & 0 & 0 & 0 & 0 \\ 0 & 0 & 0 & 0 & 0 & 0 & 0 & 1 & -1 & 0 & 0 & -1 & 1 & 0 & 0 \\ 0 & 0 & 0 & 0 & 0 & 0 & 0 & 0 & 1 & -1 & 0 & 0 & -1 & 1 & 0 \\ 0 & 0 & 0 & 0 & 0 & 0 & 0 & 0 & 0 & 1 & -1 & 0 & 0 & -1 & 1 \\ \frac{1}{6} & \frac{1}{6} & \frac{1}{6} & -\frac{1}{8} & -\frac{1}{8} & -\frac{1}{8} & -\frac{1}{8} & -\frac{1}{8} & -\frac{1}{8} & -\frac{1}{8} & -\frac{1}{8} & \frac{1}{8} & \frac{1}{8} & \frac{1}{8} & \frac{1}{8} \end{bmatrix},$$

$$\mathbf{\eta}_{1(B)} = \begin{bmatrix} 0 \\ 0 \\ 0 \\ 0 \\ 0 \\ 0 \end{bmatrix}.$$

The restricted, unweighted-means sum of squares for treatment B is

$$SS1(B) = (\mathbf{Q}'_{1(B)}\,\hat{\mathbf{\mu}})'(\mathbf{Q}'_{1(B)}\,\mathbf{Q}_{1(B)})^{-1}(\mathbf{Q}'_{1(B)}\,\hat{\mathbf{\mu}})$$
$$- (\mathbf{R}'\,\hat{\mathbf{\mu}})'(\mathbf{R}'\,\mathbf{R})^{-1}(\mathbf{R}'\,\hat{\mathbf{\mu}}) = 280.0064.$$

A similar hypothesis for treatment B can be constructed for restricted, weighted means.

$$H_0: \overline{\mu}_{..0.}{}^* - \overline{\mu}_{..1.}{}^* = 0 \qquad \text{or} \qquad H_0: \mathbf{Q}'_{2(B)}\,\mathbf{\mu} = \mathbf{\eta}_{2(B)}$$

where

$$\mathbf{Q}'_{2(B)}$$

$$= \begin{bmatrix}
1 & -1 & 0 & 0 & -1 & 1 & 0 & 0 & 0 & 0 & 0 & 0 & 0 & 0 & 0 \\
0 & 1 & -1 & 0 & 0 & -1 & 1 & 0 & 0 & 0 & 0 & 0 & 0 & 0 & 0 \\
0 & 0 & 0 & 0 & 0 & 0 & 0 & 1 & -1 & 0 & 0 & -1 & 1 & 0 & 0 \\
0 & 0 & 0 & 0 & 0 & 0 & 0 & 0 & 1 & -1 & 0 & 0 & -1 & 1 & 0 \\
0 & 0 & 0 & 0 & 0 & 0 & 0 & 0 & 0 & 1 & -1 & 0 & 0 & -1 & 1 \\
\tfrac{1}{7} & \tfrac{1}{7} & \tfrac{1}{7} & -\tfrac{1}{8} & -\tfrac{1}{8} & -\tfrac{1}{8} & -\tfrac{1}{8} & -\tfrac{1}{8} & -\tfrac{1}{8} & -\tfrac{1}{8} & -\tfrac{1}{8} & \tfrac{1}{7} & \tfrac{1}{7} & \tfrac{1}{7} & \tfrac{1}{7}
\end{bmatrix},$$

$$\mathbf{\eta}_{2(B)} = \begin{bmatrix} 0 \\ 0 \\ 0 \\ 0 \\ 0 \\ 0 \end{bmatrix}.$$

The restricted, weighted-means sum of squares for treatment B is

$$SS2(B) = (\mathbf{Q}'_{2(B)}\,\hat{\mathbf{\mu}})'(\mathbf{Q}'_{2(B)}\,\mathbf{Q}_{2(B)})^{-1}(\mathbf{Q}'_{2(B)}\,\hat{\mathbf{\mu}})$$
$$- (\mathbf{R}'\,\hat{\mathbf{\mu}})'(\mathbf{R}'\,\mathbf{R})^{-1}(\mathbf{R}'\,\hat{\mathbf{\mu}}) = 272.0048.$$

The cell means model enables a researcher to test a variety of null hypotheses—any hypothesis for which the coefficient matrix is of full row rank. Furthermore, the hypotheses can be tested subject to restrictions on the cell means. Earlier, treatments A and B were tested subject to the restriction that all $AB \times BL(G)$ interaction effects equal zero. A researcher can test very specialized hypotheses by placing additional restrictions on the cell means. For example, the null hypothesis for unweighted means for treatment A can be tested subject to the restrictions that (1) all $AB \times BL(G)$ interaction effects equal zero and (2) the unweighted means for groups 0 and 1 are equal. Because the Groups source of variation is confounded with the AB interaction, this latter restriction is equivalent to the restriction that the AB interaction effect for the unweighted means is equal to zero. The computation of sum of squares for these more specialized hypotheses involves adjoining three matrices. To illustrate, sums of squares

222 **Kirk**

for unweighted-means hypotheses will be computed subject to the restrictions that $\mathbf{R}'\,\boldsymbol{\mu} = \boldsymbol{\theta}$ and $\mathbf{C}'_{1(G)}\,\boldsymbol{\mu} = \mathbf{0}_{1(G)}$. Similar sums of squares will be computed for weighted means subject to the restrictions that $\mathbf{R}'\,\boldsymbol{\mu} = \boldsymbol{\theta}$ and $\mathbf{C}'_{2(G)}\,\boldsymbol{\mu} = \mathbf{0}_{2(G)}$. The augmented matrices and sums of squares for treatments A and B are as follows.

$$\mathbf{Q}'_{3(A)} = \begin{bmatrix} \mathbf{R}' \\ \mathbf{C}'_{1(A)} \\ \mathbf{C}'_{1(G)} \end{bmatrix}$$

$$= \begin{bmatrix} 1 & -1 & 0 & 0 & -1 & 1 & 0 & 0 & 0 & 0 & 0 & 0 & 0 & 0 & 0 \\ 0 & 1 & -1 & 0 & 0 & -1 & 1 & 0 & 0 & 0 & 0 & 0 & 0 & 0 & 0 \\ 0 & 0 & 0 & 0 & 0 & 0 & 0 & 1 & -1 & 0 & 0 & -1 & 1 & 0 & 0 \\ 0 & 0 & 0 & 0 & 0 & 0 & 0 & 0 & 1 & -1 & 0 & 0 & -1 & 1 & 0 \\ 0 & 0 & 0 & 0 & 0 & 0 & 0 & 0 & 0 & 1 & -1 & 0 & 0 & -1 & 1 \\ \frac{1}{6} & \frac{1}{6} & \frac{1}{6} & -\frac{1}{8} & -\frac{1}{8} & -\frac{1}{8} & -\frac{1}{8} & \frac{1}{8} & \frac{1}{8} & \frac{1}{8} & \frac{1}{8} & -\frac{1}{8} & -\frac{1}{8} & -\frac{1}{8} & -\frac{1}{8} \\ \frac{1}{6} & \frac{1}{6} & \frac{1}{6} & \frac{1}{8} & \frac{1}{8} & \frac{1}{8} & \frac{1}{8} & -\frac{1}{8} & -\frac{1}{8} & -\frac{1}{8} & -\frac{1}{8} & -\frac{1}{8} & -\frac{1}{8} & -\frac{1}{8} & -\frac{1}{8} \end{bmatrix},$$

$$\boldsymbol{\eta}_{3(A)} = \begin{bmatrix} \boldsymbol{\theta} \\ \mathbf{0}_{1(A)} \\ \mathbf{0}_{1(G)} \end{bmatrix} = \begin{bmatrix} 0 \\ 0 \\ 0 \\ 0 \\ 0 \\ 0 \\ 0 \end{bmatrix},$$

$$SS3(A) = (\mathbf{Q}'_{3(A)}\,\hat{\boldsymbol{\mu}})'(\mathbf{Q}'_{3(A)}\,\mathbf{Q}_{3(A)})^{-1}(\mathbf{Q}'_{3(A)}\,\hat{\boldsymbol{\mu}}) - (\mathbf{R}'\,\hat{\boldsymbol{\mu}})'(\mathbf{R}'\,\mathbf{R})^{-1}(\mathbf{R}'\,\hat{\boldsymbol{\mu}})$$
$$- (\mathbf{C}'_{1(G)}\,\hat{\boldsymbol{\mu}})'(\mathbf{C}'_{1(G)}\,\mathbf{C}_{1(G)})^{-1}(\mathbf{C}'_{1(G)}\,\hat{\boldsymbol{\mu}}) = 7.1209,$$

$$\mathbf{Q}'_{4(A)} = \begin{bmatrix} \mathbf{R}' \\ \mathbf{C}'_{2(A)} \\ \mathbf{C}'_{2(G)} \end{bmatrix}$$

$$= \begin{bmatrix} 1 & -1 & 0 & 0 & -1 & 1 & 0 & 0 & 0 & 0 & 0 & 0 & 0 & 0 & 0 \\ 0 & 1 & -1 & 0 & 0 & -1 & 1 & 0 & 0 & 0 & 0 & 0 & 0 & 0 & 0 \\ 0 & 0 & 0 & 0 & 0 & 0 & 0 & 1 & -1 & 0 & 0 & -1 & 1 & 0 & 0 \\ 0 & 0 & 0 & 0 & 0 & 0 & 0 & 0 & 1 & -1 & 0 & 0 & -1 & 1 & 0 \\ 0 & 0 & 0 & 0 & 0 & 0 & 0 & 0 & 0 & 1 & -1 & 0 & 0 & -1 & 1 \\ \frac{1}{7} & \frac{1}{7} & \frac{1}{7} & -\frac{1}{8} & -\frac{1}{8} & -\frac{1}{8} & -\frac{1}{8} & \frac{1}{7} & \frac{1}{7} & \frac{1}{7} & \frac{1}{7} & -\frac{1}{8} & -\frac{1}{8} & -\frac{1}{8} & -\frac{1}{8} \\ \frac{1}{7} & \frac{1}{7} & \frac{1}{7} & \frac{1}{7} & \frac{1}{7} & \frac{1}{7} & \frac{1}{7} & -\frac{1}{8} & -\frac{1}{8} & -\frac{1}{8} & -\frac{1}{8} & -\frac{1}{8} & -\frac{1}{8} & -\frac{1}{8} & -\frac{1}{8} \end{bmatrix},$$

$$\boldsymbol{\eta}_{4(A)} = \begin{bmatrix} \boldsymbol{\theta} \\ \mathbf{0}_{2(A)} \\ \mathbf{0}_{2(G)} \end{bmatrix} = \begin{bmatrix} 0 \\ 0 \\ 0 \\ 0 \\ 0 \\ 0 \\ 0 \end{bmatrix},$$

$$SS4(A) = (\mathbf{Q}'_{4(A)}\,\hat{\boldsymbol{\mu}})'(\mathbf{Q}'_{4(A)}\,\mathbf{Q}_{4(A)})^{-1}(\mathbf{Q}'_{4(A)}\,\hat{\boldsymbol{\mu}}) - (\mathbf{R}'\,\hat{\boldsymbol{\mu}})'(\mathbf{R}'\,\mathbf{R})^{-1}(\mathbf{R}'\,\hat{\boldsymbol{\mu}})$$

$$- (\mathbf{C}'_{2(G)}\,\hat{\boldsymbol{\mu}})'(\mathbf{C}'_{2(G)}\,\mathbf{C}_{2(G)})^{-1}(\mathbf{C}'_{2(G)}\,\hat{\boldsymbol{\mu}}) = 2.3091,$$

$$\mathbf{Q}'_{3(B)} = \begin{bmatrix} \mathbf{R}' \\ \mathbf{C}'_{1(B)} \\ \mathbf{C}'_{1(G)} \end{bmatrix}$$

$$= \begin{bmatrix}
1 & -1 & 0 & 0 & -1 & 1 & 0 & 0 & 0 & 0 & 0 & 0 & 0 & 0 & 0 \\
0 & 1 & -1 & 0 & 0 & -1 & 1 & 0 & 0 & 0 & 0 & 0 & 0 & 0 & 0 \\
0 & 0 & 0 & 0 & 0 & 0 & 0 & 1 & -1 & 0 & 0 & -1 & 1 & 0 & 0 \\
0 & 0 & 0 & 0 & 0 & 0 & 0 & 0 & 1 & -1 & 0 & 0 & -1 & 1 & 0 \\
0 & 0 & 0 & 0 & 0 & 0 & 0 & 0 & 0 & 1 & -1 & 0 & 0 & -1 & 1 \\
\frac{1}{6} & \frac{1}{6} & \frac{1}{6} & -\frac{1}{8} & -\frac{1}{8} & -\frac{1}{8} & -\frac{1}{8} & -\frac{1}{8} & -\frac{1}{8} & -\frac{1}{8} & -\frac{1}{8} & \frac{1}{8} & \frac{1}{8} & \frac{1}{8} & \frac{1}{8} \\
\frac{1}{6} & \frac{1}{6} & \frac{1}{6} & \frac{1}{6} & \frac{1}{6} & \frac{1}{6} & \frac{1}{6} & -\frac{1}{8} & -\frac{1}{8} & -\frac{1}{8} & -\frac{1}{8} & -\frac{1}{8} & -\frac{1}{8} & -\frac{1}{8} & -\frac{1}{8}
\end{bmatrix},$$

$$\boldsymbol{\eta}_{3(B)} = \begin{bmatrix} \boldsymbol{\theta} \\ \mathbf{0}_{1(B)} \\ \mathbf{0}_{1(G)} \end{bmatrix} = \begin{bmatrix} 0 \\ 0 \\ 0 \\ 0 \\ 0 \\ 0 \\ 0 \end{bmatrix},$$

$$SS3(B) = (\mathbf{Q}'_{3(B)}\,\hat{\boldsymbol{\mu}})'(\mathbf{Q}'_{3(B)}\,\mathbf{Q}_{3(B)})^{-1}(\mathbf{Q}'_{3(B)}\,\hat{\boldsymbol{\mu}}) - (\mathbf{R}'\,\hat{\boldsymbol{\mu}})'(\mathbf{R}'\,\mathbf{R})^{-1}(\mathbf{R}'\,\hat{\boldsymbol{\mu}})$$

$$- (\mathbf{C}'_{1(G)}\,\hat{\boldsymbol{\mu}})'(\mathbf{C}'_{1(G)}\,\mathbf{C}_{1(G)})^{-1}(\mathbf{C}'_{1(G)}\,\hat{\boldsymbol{\mu}}) = 276.9245,$$

$$\mathbf{Q}'_{4(B)} = \begin{bmatrix} \mathbf{R}' \\ \mathbf{C}'_{2(B)} \\ \mathbf{C}'_{2(G)} \end{bmatrix}$$

$$= \begin{bmatrix}
1 & -1 & 0 & 0 & -1 & 1 & 0 & 0 & 0 & 0 & 0 & 0 & 0 & 0 & 0 \\
0 & 1 & -1 & 0 & 0 & -1 & 1 & 0 & 0 & 0 & 0 & 0 & 0 & 0 & 0 \\
0 & 0 & 0 & 0 & 0 & 0 & 0 & 1 & -1 & 0 & 0 & -1 & 1 & 0 & 0 \\
0 & 0 & 0 & 0 & 0 & 0 & 0 & 0 & 1 & -1 & 0 & 0 & -1 & 1 & 0 \\
0 & 0 & 0 & 0 & 0 & 0 & 0 & 0 & 0 & 1 & -1 & 0 & 0 & -1 & 1 \\
\frac{1}{7} & \frac{1}{7} & \frac{1}{7} & -\frac{1}{8} & -\frac{1}{8} & -\frac{1}{8} & -\frac{1}{8} & -\frac{1}{8} & -\frac{1}{8} & -\frac{1}{8} & -\frac{1}{8} & \frac{1}{7} & \frac{1}{7} & \frac{1}{7} & \frac{1}{7} \\
\frac{1}{7} & \frac{1}{7} & \frac{1}{7} & \frac{1}{7} & \frac{1}{7} & \frac{1}{7} & \frac{1}{7} & -\frac{1}{8} & -\frac{1}{8} & -\frac{1}{8} & -\frac{1}{8} & -\frac{1}{8} & -\frac{1}{8} & -\frac{1}{8} & -\frac{1}{8}
\end{bmatrix},$$

$$\boldsymbol{\eta}_{4(B)} = \begin{bmatrix} \boldsymbol{\theta} \\ \mathbf{0}_{2(B)} \\ \mathbf{0}_{2(G)} \end{bmatrix} = \begin{bmatrix} 0 \\ 0 \\ 0 \\ 0 \\ 0 \\ 0 \\ 0 \end{bmatrix},$$

$$SS4(B) = (\mathbf{Q}'_{4(B)}\,\hat{\boldsymbol{\mu}})'(\mathbf{Q}'_{4(B)}\,\mathbf{Q}_{4(B)})^{-1}(\mathbf{Q}'_{4(B)}\,\hat{\boldsymbol{\mu}}) - (\mathbf{R}'\,\hat{\boldsymbol{\mu}})'(\mathbf{R}'\,\mathbf{R})^{-1}(\mathbf{R}'\,\hat{\boldsymbol{\mu}})$$

$$- (\mathbf{C}'_{2(G)}\,\hat{\boldsymbol{\mu}})'(\mathbf{C}'_{2(G)}\,\mathbf{C}_{2(G)})^{-1}(\mathbf{C}'_{2(G)}\,\hat{\boldsymbol{\mu}}) = 274.4629.$$

In this section, sums of squares appropriate for testing a variety of null hypotheses have been described. As we have seen, the cell means model is extremely versatile. It enables a researcher to compute sums of squares for any interesting hypothesis, and there is never any ambiguity about the hypothesis that is being tested because the coefficient matrix for the null hypothesis is a part of the computational procedure. Before completing the analysis by computing mean squares and F statistics, we will describe another model, the regression model, that can be used to perform an analysis of variance when there are missing observations. Then we will compare the results of the two approaches.

6.3.4 Computational Procedures: The Regression Model Approach

The data from the computer-aided instruction experiment described in Section 6.3.1 will be used to illustrate the computational procedures for the regression model. We will assume that observation $\overline{Y}_{0000} = 3$ is missing because of an equipment malfunction. A qualitative regression model with $h - 1 = (w - 1) + w(n - 1) + (p - 1) + (q - 1) = 9$ independent variables $(X_{i1}, \ldots, X_{i9})$ can be formulated for this design as follows.

$$Y_i = \beta_0 X_0 + \overbrace{\beta_1 X_{i1}}^{\text{Groups}} + \overbrace{\beta_2 X_{i2} + \cdots + \beta_7 X_{i7}}^{BL(G)}$$

$$+ \overbrace{\beta_8 X_{i8}}^{A} + \overbrace{\beta_9 X_{i9}}^{B} + \varepsilon_i \qquad (i = 1, \ldots, N),$$

where ε_i is $NID(0, \sigma_\varepsilon^2)$. Effect coding will be used to code the independent variables because this coding scheme leads to a simple correspondence between the parameters of the regression model and the parameters of the experimental design model. Using this coding scheme, an independent variable is coded 1 if an observation is in the first level of the source of variation, -1 if an observation is in the last level of the source of variation, and 0 otherwise. The coding scheme for the computer-aided instruction experiment is shown in Table 6.4. A model equation that includes independent variables corresponding to vectors $x_0, \ldots, x_9$ is called a full model equation. Model equations that contain some but not all vectors $x_0, \ldots, x_9$ are called reduced model equations. A number of useful model equations can be formulated for the data in Table 6.4. These equations

Table 6.4 Coding Scheme for the Computer-Aided Instruction Experiment (Y_{0000} Is Missing Because of an Equipment Malfunction)

Y_{ijkz}	y	x_0	G x_1	BL(g_0) x_2	x_3	x_4	BL(g_1) x_5	x_6	x_7	A x_8	B x_9
Y_{1000}	5	1	1	0	1	0	0	0	0	1	1
Y_{2000}	6	1	1	0	0	1	0	0	0	1	1
Y_{3000}	5	1	1	-1	-1	-1	0	0	0	1	1
Y_{4100}	16	1	1	1	0	0	0	0	0	-1	-1
Y_{5100}	14	1	1	0	1	0	0	0	0	-1	-1
Y_{6100}	17	1	1	0	0	1	0	0	0	-1	-1
Y_{7100}	15	1	1	-1	-1	-1	0	0	0	-1	-1
Y_{8011}	14	1	-1	0	0	0	1	0	0	1	-1
Y_{9011}	14	1	-1	0	0	0	0	1	0	1	-1
$Y_{10,011}$	16	1	-1	0	0	0	0	0	1	1	-1
$Y_{11,011}$	16	1	-1	0	0	0	-1	-1	-1	1	-1
$Y_{12,101}$	7	1	-1	0	0	0	1	0	0	-1	1
$Y_{13,101}$	6	1	-1	0	0	0	0	1	0	-1	1
$Y_{14,101}$	7	1	-1	0	0	0	0	0	1	-1	1
$Y_{15,101}$	11	1	-1	0	0	0	-1	-1	-1	-1	1

and a simplified designation for the corresponding error sum of squares are as follows. The first equation is for the full model; the remaining equations are for reduced models.

Model equation	Designation for error sum of squares

$$\overset{\text{Groups}}{\overbrace{}}\quad\overset{BL(G)}{\overbrace{}}$$

$Y_i = \beta_0 X_0 + \beta_1 X_{i1} + \beta_2 X_{i2} + \cdots + \beta_7 X_{i7}$

$$\overset{A}{\overbrace{}}\quad\overset{B}{\overbrace{}}$$

$+\ \beta_8 X_{i8} + \beta_9 X_{i9} + \varepsilon_i$ .... $SSE(G, BL(G), A, B)$

$$\overset{\text{Groups}}{\overbrace{}}\ \overset{A}{\overbrace{}}\ \overset{B}{\overbrace{}}$$

$Y_i = \beta_0 X_0 + \beta_1 X_{i1} + \beta_8 X_{i8} + \beta_9 X_{i9} + \varepsilon_i$ .... $SSE(G, A, B)$

$$\overset{\text{Groups}}{\overbrace{}}\quad\overset{BL(G)}{\overbrace{}}$$

$Y_i = \beta_0 X_0 + \beta_1 X_{i1} + \beta_2 X_{i2} + \cdots + \beta_7 X_{i7} + \varepsilon_i$ .... $SSE(G, BL(G))$

$$\overset{\text{Groups}}{\overbrace{}}\ \overset{A}{\overbrace{}}$$

$Y_i = \beta_0 X_0 + \beta_1 X_{i1} + \beta_8 X_{i8} + \varepsilon_i$ .... $SSE(G, A)$

$$\overset{\text{Groups}}{\overbrace{}}\ \overset{B}{\overbrace{}}$$

$Y_i = \beta_0 X_0 + \beta_1 X_{i1} + \beta_9 X_{i9} + \varepsilon_i$ .... $SSE(G, B)$

Model equation	Designation for error sum of squares
$\overbrace{}^{A}$ $\overbrace{}^{B}$ $Y_i = \beta_0 X_0 + \beta_8 X_{i8} + \beta_9 X_{i9} + \varepsilon_i$	$SSE(A, B)$
$\overbrace{}^{\text{Groups}}$ $Y_i = \beta_0 X_0 + \beta_1 X_{i1} + \varepsilon_i$	$SSE(G)$
$\overbrace{}^{BL(G)}$ $Y_i = \beta_0 X_0 + \beta_2 X_{i2} + \cdots + \beta_7 X_{i7} + \varepsilon_i$	$SSE(BL(G))$
$\overbrace{}^{A}$ $Y_i = \beta_0 X_0 + \beta_8 X_{i8} + \varepsilon_i$	$SSE(A)$
$\overbrace{}^{B}$ $Y_i = \beta_0 X_0 + \beta_9 X_{i9} + \varepsilon_i$	$SSE(B)$
$Y_i = \beta_0 X_0 + \varepsilon_i$	$SSE(0)$

The matrix equation for computing an error sum of squares is

$$\mathbf{y}' \, \mathbf{y} - ((\mathbf{X}' \, \mathbf{X})^{-1} (\mathbf{X}' \, \mathbf{y}))' (\mathbf{X}' \, \mathbf{y}),$$

where $\mathbf{y}$ is defined in Table 6.4 and $\mathbf{X}$ consists of all or a subset of the column vectors $\mathbf{x}_0, \mathbf{x}_1, \ldots, \mathbf{x}_9$ as specified in a model equation. Consider, for example, the model equation associated with $SSE(G, A, B)$,

$$Y_i = \beta_0 X_0 + \beta_1 X_{i1} + \beta_8 X_{i8} + \beta_9 X_{i9} + \varepsilon_i.$$

The $\mathbf{X}$ matrix for computing $SSE(G, A, B) = \mathbf{y}' \, \mathbf{y} - ((\mathbf{X}' \, \mathbf{X})^{-1}(\mathbf{X}' \, \mathbf{y}))'$ $(\mathbf{X}' \, \mathbf{y})$ consists of column vectors $\mathbf{x}_0, \mathbf{x}_1, \mathbf{x}_8,$ and $\mathbf{x}_9$ from Table 6.4, that is,

$$\mathbf{X} = \begin{array}{cccc} \mathbf{x}_0 & \mathbf{x}_1 & \mathbf{x}_8 & \mathbf{x}_9 \\ \begin{bmatrix} 1 & 1 & 1 & 1 \\ 1 & 1 & 1 & 1 \\ 1 & 1 & 1 & 1 \\ 1 & 1 & -1 & -1 \\ 1 & 1 & -1 & -1 \\ 1 & 1 & -1 & -1 \\ 1 & 1 & -1 & -1 \\ 1 & -1 & 1 & -1 \\ 1 & -1 & 1 & -1 \\ 1 & -1 & 1 & -1 \\ 1 & -1 & 1 & -1 \\ 1 & -1 & -1 & 1 \\ 1 & -1 & -1 & 1 \\ 1 & -1 & -1 & 1 \\ 1 & -1 & -1 & 1 \end{bmatrix} \end{array}$$

Other error sums of squares are obtained by the appropriate choice of **x** column vectors.

The sum of squares for $AB \times BL(G)$ is equal to $SSE(G, BL(G), A, B)$. Other analysis of variance sums of squares are obtained by subtracting one error sum of squares from another error sum of squares. For designs in which there are no missing observations, several different model comparisons may give the same sum of squares. For example, the following model comparisons for computing the sums of squares for Groups, A, and B give the same results.

$$SSG = SSE(A, B) - SSE(G, A, B)$$
$$= SSE(0) - SSE(G)$$
$$SSA = SSE(G, B) - SSE(G, A, B)$$
$$= SSE(G) - SSE(G, A)$$
$$= SSE(B) - SSE(A, B)$$
$$= SSE(0) - SSE(A)$$
$$SSB = SSE(G, A) - SSE(G, A, B)$$
$$= SSE(G) - SSE(G, B)$$
$$= SSE(A) - SSE(A, B)$$
$$= SSE(0) - SSE(B)$$

When one or more observations are missing, these model comparisons produce different sums of squares and provide tests of different null hypotheses. Various null hypotheses were described in Section 6.3.3 in connection with the cell means model. Regression model comparisons for computing each of the sum of squares described for the cell means model are presented in Table 6.5.

In general, researchers are interested in hypotheses involving unweighted means, the hypotheses tested in Table 6.5. The null hypotheses for treatments A and B can be rejected. Although the regression model can be used to test the same hypotheses that were tested using the cell means model, the regression model provides few clues about the hypothesis being tested. The advantage of the cell means model is that the specific hypothesis being tested is clearly identified in terms of cell means.

The following sections present a number of confounded factorial designs that can be analyzed using the computational procedures just described.

6.4 RBCF-2^3 DESIGN

An experiment with three treatments, each having two levels, contains $2 \times 2 \times 2 = 8$ treatment combinations. If a randomized block factorial design (RBF-222 design) is used, each block contains either eight matched

Table 6.5 Cell Means Model Null Hypotheses and Corresponding Regression Model Comparisons

Cell means model null hypothesis[a]	Regression model comparison	SS	df	MS	F	p
$C'_{BL}\ \mu = 0_{BL}$	$SSE(0) - SSE(G, BL(G))$	46.4333	7			
$C'_{1(G)}\ \mu = 0_{1(G)}$	$SSE(A, B) - SSE(G, A, B)$	3.3910	1	3.3910	0.44	.53648
$C'_{2(G)}\ \mu = 0_{2(G)}$	$SSE(0) - SSE(G)$	0.2021	1			
$C'_{BL(G)}\ \mu = 0_{BL(G)}$	$SSE(G) - SSE(G, BL(G))$	46.2321	6	7.7054		
$C'_{w.BL}\ \mu = 0_{w.BL}$	$SSE(G, BL(G))$	260.5000	7			
$Q'_{1(A)}\ \mu = \eta_{1(A)}$	$SSE(G, B) - SSE(G, A, B)$	7.8526	1	7.8526	7.30	.04269
$Q'_{2(A)}\ \mu = \eta_{2(A)}$	$SSE(0) - SSE(A)$	2.2012	1			
$Q'_{3(A)}\ \mu = \eta_{3(A)}$	$SSE(B) - SSE(A, B)$	7.1209	1			
$Q'_{4(A)}\ \mu = \eta_{4(A)}$	$SSE(G) - SSE(G, A)$	2.3091	1			
$Q'_{1(B)}\ \mu = \eta_{1(B)}$	$SSE(G, A) - SSE(G, A, B)$	280.0064	1	280.0064	260.47	.00002
$Q'_{2(B)}\ \mu = \eta_{2(B)}$	$SSE(0) - SSE(B)$	272.0048	1			
$Q'_{3(B)}\ \mu = \eta_{3(B)}$	$SSE(A) - SSE(A, B)$	276.9245	1			
$Q'_{4(B)}\ \mu = \eta_{4(B)}$	$SSE(G) - SSE(G, B)$	274.4629	1			
$R'\ \mu = \theta$	$SSE(G, BL(G), A, B)$	5.3750	5	1.0750		

[a]The hypotheses are defined in Section 6.3.3.

subjects or subjects who are observed eight times. The size of a block can be reduced from eight to four by confounding an interaction with groups using a randomized block confounded factorial design (RBCF-2^3 design). This design has four interactions: AB, AC, BC, and ABC. The interaction that is confounded with groups is usually the highest-order interaction, in this example the ABC interaction. As we will now see, the confounding procedures described earlier for an RBCF-2^2 design can easily be extended to this RBCF-2^3 design.

6.4.1 Confounding the *ABC* Interaction with Groups

Let a_j denote the jth level of treatment A; b_k, the kth level of treatment B; and c_l, the lth level of treatment C. The ABC interaction can be confounded with groups by assigning the treatment combinations satisfying the defining relation $a_j + b_k + c_l = 0$ (mod 2) to group 0 and those satisfying $a_j + b_k + c_l = 1$ (mod 2) to group 1. Solving for a_j, b_k, and c_l, we obtain

$$
\left.\begin{array}{l}
000 = 0 \ (\text{mod } 2) \\
011 = 0 \ (\text{mod } 2) \\
101 = 0 \ (\text{mod } 2) \\
110 = 0 \ (\text{mod } 2)
\end{array}\right\} \begin{array}{l} \text{group 0 or} \\ (ABC)_0, \end{array}
\qquad
\left.\begin{array}{l}
001 = 1 \ (\text{mod } 2) \\
010 = 1 \ (\text{mod } 2) \\
100 = 1 \ (\text{mod } 2) \\
111 = 1 \ (\text{mod } 2)
\end{array}\right\} \begin{array}{l} \text{group 1 or} \\ (ABC)_1. \end{array}
$$

The layout for this design is shown in Figure 6.5. To have an adequate number of degrees of freedom for experimental error, each group should contain at least five blocks. The sum-of-squares computational formulas for this design are given in Table 6.6. If one or more observations are missing, the cell means model or the regression model can be used to obtain the sums of squares.

The confounded factorial design shown in Figure 6.5 is a good design choice if a researcher's primary interest is in treatments A, B, and C and the two-treatment interactions. A split-plot factorial design in which A is a between-blocks treatment and B and C are within-blocks treatments (SPF-2·22 design) would be a good choice if a researcher was interested primarily in treatments B and C and all interactions. Recall that in an SPF-2·22 design, the test of treatment A is less powerful than the tests of the within-blocks sources of variation: B, C, AB, and so on. In most research, it is desirable to test all treatments with equal power. Hence, the confounded factorial design is usually a better choice than the split-plot factorial design.

The design shown in Figure 6.5 confounds the ABC interaction with groups. Alternatively, any of the two-treatment interactions can be confounded with groups. Procedures for accomplishing this are described next.

		Treat. Comb. $a_j b_k c_l$	Treat. Comb. $a_j b_k c_l$	Treat. Comb. $a_j b_k c_l$	Treat. Comb. $a_j b_k c_l$
	Block$_0$	000	011	101	110
Group$_0$ $(ABC)_0$	Block$_1$	000	011	101	110
	Block$_2$	000	011	101	110
	Block$_3$	001	010	100	111
Group$_1$ $(ABC)_1$	Block$_4$	001	010	100	111
	Block$_5$	001	010	100	111

Figure 6.5 Layout for an RBCF-2^3 design. The ABC interaction is confounded with groups.

Table 6.6 Computational Formulas for an RBCF-2^3 Design

Source	Computational formula for sum of squares (SS)	df	F ratio
1 Between blocks	[a][GS] $-$ [Y]	$nw - 1$	
2 Groups or ABC	[G] $-$ [Y]	$w - 1$	[2/3][b]
3 Blocks w. G	[GS] $-$ [G]	$w(n - 1)$	
4 Within blocks	[$ABCGS$] $-$ [GS]	$nw(v - 1)$	
5 A	[A] $-$ [Y]	$p - 1$	[5/11]
6 B	[B] $-$ [Y]	$q - 1$	[6/11]
7 C	[C] $-$ [Y]	$r - 1$	[7/11]
8 AB	[AB] $-$ [A] $-$ [B] $+$ [Y]	$(p - 1)(q - 1)$	[8/11]
9 AC	[AC] $-$ [A] $-$ [C] $+$ [Y]	$(p - 1)(r - 1)$	[9/11]
10 BC	[BC] $-$ [B] $-$ [C] $+$ [Y]	$(q - 1)(r - 1)$	[10/11]
11 $ABC \times BL(G)$	[$ABCGS$] $-$ [ABC] $-$ [GS] $+$ [G]	$w(n - 1)(v - 1)$	
12 Total	[$ABCGS$] $-$ [Y]	$nvz - 1$	

[a]The terms [GS], [Y], and so on are a simplified notation for functions of observations. The meaning of the terms can be deduced from Table 6.1.
[b]The F ratio is equal to (SS row 2/df row 2)/(SS row 3/df row 3).

6.4.2 Confounding a Two-Treatment Interaction with Groups

A researcher can confound one of the two-treatment interactions, say AB, with groups by first determining the treatment combinations that satisfy the defining relations $a_j + b_k = z$ (mod 2), $z = 0, 1$. These combinations are

$$\left.\begin{matrix} 00 = 0 \ (\text{mod } 2) \\ 11 = 0 \ (\text{mod } 2) \end{matrix}\right\} \quad \text{group 0 } (AB)_0,$$

$$\left.\begin{matrix} 01 = 1 \ (\text{mod } 2) \\ 10 = 1 \ (\text{mod } 2) \end{matrix}\right\} \quad \text{group 1 } (AB)_1.$$

Next, the levels 0 and 1 of treatment C are added to these combinations in a balanced manner.

add 0 ⟶   ⟵ add 1

$$\left.\begin{matrix} 000 & 001 \\ 110 & 111 \end{matrix}\right\} \quad \text{group 0 } (AB)_0,$$

$$\left.\begin{matrix} 010 & 011 \\ 100 & 101 \end{matrix}\right\} \quad \text{group 1 } (AB)_1.$$

The treatment combinations assigned to groups 0 and 1 are as follows.

Group 0 $(AB)_0$ 000 001 110 111

Group 1 $(AB)_1$ 010 011 100 101

The procedures just described can be used to confound the AC interaction with groups. We begin with the treatment combinations that satisfy the defining relations $a_j + c_l = z$ (mod 2), $z = 0, 1$. They are

$$
\left.\begin{array}{l}
00 = 0 \text{ (mod 2)} \\
11 = 0 \text{ (mod 2)}
\end{array}\right\} \quad \text{group 0 } (AC)_0,
$$

$$
\left.\begin{array}{l}
01 = 1 \text{ (mod 2)} \\
10 = 1 \text{ (mod 2)}
\end{array}\right\} \quad \text{group 1 } (AC)_1.
$$

Next, the levels of 0 and 1 of treatment B are added to these combinations in a balanced manner. For example,

add 0 ⌐   ⌐——— add 1

$$
\left.\begin{array}{ll}
000 & 010 \\
101 & 111
\end{array}\right\} \quad \text{group 0 } (AC)_0,
$$

$$
\left.\begin{array}{ll}
001 & 011 \\
100 & 110
\end{array}\right\} \quad \text{group 1 } (AC)_1.
$$

The treatment combinations assigned to groups 0 and 1 are as follows.

Group 0 $(AC)_0$ 000 010 101 111

Group 1 $(AC)_1$ 001 011 100 110

6.5 COMPLETE VERSUS PARTIAL CONFOUNDING

The designs described so far are examples of completely confounded designs. In these designs, one of the interactions is confounded with groups. The confounded interaction is usually evaluated with less precision than the unconfounded, within-blocks effects. In designs having more than two treatments, each with two levels, it is possible to confound one interaction in one group of blocks, a second interaction in a second group of blocks, and so on. This confounding scheme has the advantage of providing some within-blocks information about an interaction from the blocks in which the interaction is not confounded. The procedure is called *partial confounding*, and the design is called a randomized block partially confounded factorial design (RBPF-p^k design). Consider the RBPF-2^3 design in Figure 6.6. The AB interaction is confounded with the blocks in group 0, the AC

interaction is confounded with the block in group 1, the BC interaction is confounded with the blocks in group 2, and the ABC interaction with the blocks in group 3. Because the AB interaction is confounded only in group 0, within-blocks information on this interaction is available from groups 1, 2, and 3. It should be apparent from an examination of Figure 6.6 that within-blocks information from three of the four groups is also available for the AC, BC, and ABC interactions. The advantage of partial confounding is that the block size can be reduced and still obtain partial within-blocks information for each of the confounded interactions. If previous research or a pilot study indicates that an interaction is insignificant, complete confounding in which the interaction is confounded in all of the groups is preferable to partial confounding. This follows because complete confounding provides full rather than partial within-blocks information for all of the nonconfounded interactions.

Federer (1955, p. 230) distinguishes between *balanced partial confounding* and *unbalanced partial confounding*. The former designation refers to designs in which all effects of a particular order, for example, all two-treatment interactions, are confounded with blocks an equal number of times. Balanced partial confounding is illustrated by the RBPF-2^3 design

			Treat. Comb. $a_j b_k c_l$	Treat. Comb. $a_j b_k c_l$	Treat. Comb. $a_j b_k c_l$	Treat. Comb. $a_j b_k c_l$
Group$_0$	$(AB)_0$	Block$_0$	000	001	110	111
	$(AB)_1$	Block$_1$	010	011	100	101
Group$_1$	$(AC)_0$	Block$_2$	000	010	101	111
	$(AC)_1$	Block$_3$	001	011	100	110
Group$_2$	$(BC)_0$	Block$_4$	000	011	100	111
	$(BC)_1$	Block$_5$	001	010	101	110
Group$_3$	$(ABC)_0$	Block$_6$	000	011	101	110
	$(ABC)_1$	Block$_7$	001	010	100	111

Figure 6.6 Layout for an RBPF-2^3 design. Each of the four interactions is confounded with the blocks in one group and unconfounded in the other three groups. This partially confounded design provided within-blocks information for each interaction from three of the four groups.

in which AB, AC, and BC are each confounded in one group of blocks. If all effects of a particular order are confounded with blocks an unequal number of times, the arrangement is called unbalanced partial confounding. For example, the AB and AC interactions could be confounded in groups 0 and 1, respectively. Because AB and AC are each confounded once but the BC interaction is not confounded, the design uses unbalanced partial confounding.

6.5.1 RBPF-2^3 Design

The sum-of-squares formulas for a randomized block partially confounded factorial design with balanced partial confounding are given in Table 6.7. If one or more observations are missing, the cell means model or the regression model can be used to obtain the sums of squares.

The advantage of balanced partial confounding is that a within-blocks estimate of all of the interactions can be computed. This advantage is gained

Table 6.7 Computational Formulas for an RBPF-2^3 Design with Balanced Partial Confounding

	Source	Computational formula for sum of squares (SS)	df	F ratio
1	Between blocks	$[GS] - [Y]$	$nw - 1$	
2	Groups	$[G] - [Y]$	$w - 1$	$[2/3]$[b]
3	Blocks w. G	$[GS] - [G]$	$w(n - 1)$	
4	Within blocks	$[ABCGS] - [GS]$	$nw(v - 1)$	
5	A	$[A] - [Y]$	$p - 1$	$[5/12]$
6	B	$[B] - [Y]$	$q - 1$	$[6/12]$
7	C	$[C] - [Y]$	$r - 1$	$[7/12]$
8	AB (w)	[a]$[AB]' - [A]' - [B]' + [Y]'$	$(p - 1)(q - 1)$	$[8/12]$
9	AC (w)	$[AC]' - [A]' - [C]' + [Y]'$	$(p - 1)(r - 1)$	$[9/12]$
10	BC (w)	$[BC]' - [B]' - [C]' + [Y]'$	$(q - 1)(r - 1)$	$[10/12]$
11	ABC (w)	$[ABC]' - [AB]' - [AC]' - [BC]'$ $+ [A]' + [B]' + [C]' - [Y]'$	$(p - 1)(q - 1)$ $(r - 1)$	$[11/12]$
12	Residual	$SSWITHIN\ BL - SSA - SSB$ $- SSC - SSAB - SSAC$ $- SSBC - SSABC$	$nvw - pqr -$ $nw + 1$	
13	Total	$[ABCGS] - [Y]$	$nvw - 1$	

[a]A prime following [ ] indicates that the source of variation is computed from the replications in which the sum of squares is not confounded with groups. For example, in computing $SSAB$, $[A]'$ is computed from replications 1–3. In computing $SSAC$, $[A]'$ is computed from replications 0, 2, and 3.
[b]The F ratio is equal to (SS row 2/df row 2)/(SS row 3/df row 3).

at the price of greater complexity in the analysis. Furthermore, the AB, AC, BC, and ABC interactions are computed from only three-fourths of the groups. The ratio 3/4 is called the *relative information* on the confounded effects (Yates, 1937). The choice between a completely confounded or a partially confounded design rests on the researcher's expectations about the interactions. If one interaction is known to be insignificant, a completely confounded design is the better choice.

Numerous design possibilities are inherent in partial confounding for researchers who have a good knowledge of a research area. For example, a researcher could choose to confound the AB interaction in groups 0 and 1 and the ABC interaction in groups 2 and 3. If sufficient subjects are available for five groups, a researcher could choose to confound AB, AC, and BC in groups 0 through 2, respectively, and ABC in groups 3 and 4. This design provides as much information on the AB, AC, and BC interactions as the completely confounded design in Table 6.6. In addition, it provides three-fifths relative information on the ABC interaction.

6.6 RBCF-3^2 AND RBPF-3^2 DESIGNS

6.6.1 Assignment of Treatment Combinations to Groups When p Equals Three

So far the discussion has focused on designs in which each treatment has two levels. We will now apply the principles developed for these designs to designs in which each treatment has three levels. The AB interaction in an RBCF-3^2 design can be partitioned into two orthogonal components as follows.

SS	df
AB	$(3 - 1)(3 - 1) = 4$
(AB)	2
(AB^2)	2

The two orthogonal components (AB) and (AB^2) have no special significance apart from their use in partitioning the interaction. Treatment combinations that satisfy the defining relations

$$a_j + b_k = 0 \ (\text{mod } 3) \quad (AB)_0$$

$$a_j + b_k = 1 \ (\text{mod } 3) \quad (AB)_1$$

$$a_j + b_k = 2 \ (\text{mod } 3) \quad (AB)_2$$

constitute the (AB) component of the interaction. These treatment combinations are

00	12	21	$(AB)_0$
01	10	22	$(AB)_1$
02	11	20	$(AB)_2$.

The (AB^2) interaction component is composed of the treatment combinations that satisfy the defining relations

$$a_j + 2b_k = 0 \text{ (mod 3)} \qquad (AB^2)_0$$
$$a_j + 2b_k = 1 \text{ (mod 3)} \qquad (AB^2)_1$$
$$a_j + 2b_k = 2 \text{ (mod 3)} \qquad (AB^2)_2.$$

where the powers of A and B are used as the coefficients of a_j and b_k. These treatment combinations are

00	11	22	$(AB^2)_0$
02	10	21	$(AB^2)_1$
01	12	20	$(AB^2)_2$.

By convention, the power of A is always equal to one. This is necessary to define the interaction components uniquely because (A^2B) and (AB^2) define the same treatment combinations. That is,

$$2a_j + b_k = z \text{ (mod 3)} \qquad (z = 0, 1, 2)$$

defines the same nine combinations as

$$a_j + 2b_k = z \text{ (mod 3)} \qquad (z = 0, 1, 2).$$

6.6.2 Construction of RBCF-3^2 and RBPF-3^2 Designs

An RBCF-3^2 design is constructed by confounding either the (AB) or (AB^2) interaction component with groups. A block diagram in which the (AB) interaction is confounded with groups is shown in Figure 6.3. This design provides within-blocks information on the (AB^2) component of the AB interaction. An alternative and preferable design strategy is to confound (AB) in one group of blocks and (AB^2) in a second group of blocks as shown in Figure 6.7. This confounding scheme provides one-half relative information on each component of the AB interaction. To provide sufficient degrees of freedom for testing the within-blocks null hypotheses, the number of blocks assigned to each $(AB)_0$, $(AB)_1$, . . . , $(AB^2)_2$ must be

			Treat. Comb. a_jb_k	Treat. Comb. a_jb_k	Treat. Comb. a_jb_k
	$(AB)_0$	Block$_0$	00	12	21
	$(AB)_0$	Block$_1$	00	12	21
Group$_0$	$(AB)_1$	Block$_2$	01	10	22
	$(AB)_1$	Block$_3$	01	10	22
	$(AB)_2$	Block$_4$	02	11	20
	$(AB)_2$	Block$_5$	02	11	20
	$(AB^2)_0$	Block$_6$	00	11	22
	$(AB^2)_0$	Block$_7$	00	11	22
Group$_1$	$(AB^2)_1$	Block$_8$	02	10	21
	$(AB^2)_1$	Block$_9$	02	10	21
	$(AB^2)_2$	Block$_{10}$	01	12	20
	$(AB^2)_2$	Block$_{11}$	01	12	20

Figure 6.7 Layout for an RBPF-3^2 design. The (AB) component is confounded in Group$_0$; the (AB^2) component is confounded in Group$_1$. To provide sufficient degrees of freedom, two blocks have been assigned to each $(AB)_0$, $(AB)_1$, . . . , $(AB^2)_2$ component.

equal to or greater than two. The sum of squares formulas for the RBPF-3^2 design in Figure 6.7 are given in Table 6.8. If one or more observations are missing, the cell means model or the regression model can be used to obtain the sums of squares.

6.7   HIGHER-ORDER CONFOUNDED DESIGNS

So far we have described 2×2, $2 \times 2 \times 2$, and 3×3 confounded factorial designs. The principles discussed in connection with these designs can be readily extended to any design of the form p^k, where p is a prime number. Experiments involving mixed primes, such as $3 \times 2 \times 2$ and $3 \times 3 \times 2$, are more difficult to lay out and analyze. A discussion of these designs is deferred to Section 6.8. In this section we will see how to extend the principles described earlier in connection with 2×2, $2 \times 2 \times 2$, and 3×3 designs to other unmixed designs. In the process we will see how to achieve further reductions in block size by confounding two or more interactions with groups of blocks.

Table 6.8 Computational Formulas for an RBPF-3^2 Design with Balanced Partial Confounding

	Source	Computational formula for sum of squares (SS)	df	F ratio
1	Between blocks	$^a[EGS] - [Y]$	$nuw - 1$	
2	Groups	$[G] - [Y]$	$w - 1$	$[2/4]^b$
3	AB (between)	$[EG] - [G]$	$w(u - 1)$	$[3/4]$
4	Blocks w. AB (between)	$[EGS] - [EG]$	$uw(n - 1)$	
5	Within blocks	$[ABEGS] - [EGS]$	$nuw(v - 1)$	
6	A	$[A] - [Y]$	$p - 1$	$[6/9]$
7	B	$[B] - [Y]$	$q - 1$	$[7/9]$
8	AB (within)	$[ABEG] - [AG] - [BG] - [EG]$ $+ 2[G]$	$(p - 1)(q - 1)$	$[8/9]$
9	Residual	$SSWITHIN\ BL - SSA - SSB$ $- SSAB$ (within)	$nuvw - nuw$ $- pq + 1$	
10	Total	$[ABEGS] - [Y]$	$nuvw - 1$	

$^a E$ denotes the components $(AB)_z$ or $(AB^2)_z$; $o = 1, \ldots, u$ levels of E. The meaning of the other terms can be determined from Table 6.1.
^{b}The F ratio is equal to (SS row 2/df row 2)/(SS row 4/df row 4).

6.7.1 RBCF-2^4 Designs with Blocks of Size Eight, Four, or Two

A randomized block factorial design with two levels of treatments A, B, C, and D has 16 treatment combinations and requires blocks of size 16. The block size can be reduced from 16 to 8 by confounding the $ABCD$ interaction with groups of blocks. An RBCF-2^4 design can be constructed using the defining relations

$$a_j + b_k + c_l + d_m = z \ (\text{mod } 2) \qquad (z = 0, 1).$$

The treatment combinations that are assigned to group 0 or $(ABCD)_0$ and group 1 or $(ABCD)_1$ are

Group 0 $(ABCD)_0 = 0000 \quad 0011 \quad 0101 \quad 0110 \quad 1001 \quad 1010 \quad 1100 \quad 1111$

Group 1 $(ABCD)_1 = 0001 \quad 0010 \quad 0100 \quad 0111 \quad 1000 \quad 1011 \quad 1101 \quad 1110.$

The block size of an RBCF-2^4 design can be further reduced by using two defining relations instead of one to assign treatment combinations to groups.

For example, a researcher could use the defining relations

$$a_j + c_l + d_m = z \pmod 2 \qquad (z = 0, 1),$$
$$a_j + b_k = z \pmod 2 \qquad (z = 0, 1).$$

Eight treatment combinations satisfy each of these defining relations. For example,

$$(ACD)_0 = 0000 \quad 0011 \quad 1101 \quad 1110 \quad 0100 \quad 0111 \quad 1001 \quad 1010$$
$$(AB)_0 = 0000 \quad 0011 \quad 1101 \quad 1110 \quad 0001 \quad 0010 \quad 1100 \quad 1111.$$

If we examine these combinations closely, we find that four of them satisfy both $(ACD)_0$ and $(AB)_0$. These combinations are

$$(ACD)_0(AB)_0 = 0000 \quad 0011 \quad 1101 \quad 1110.$$

A different set of four treatment combinations satisfies both $(ACD)_0$ and $(AB)_1$, and so on. Thus, we can assign four treatment combinations to blocks so that they satisfy, simultaneously, two defining relations. This procedure produces the RBCF-2^4 design with blocks of size four shown in Figure 6.8.

		Treat. Comb. $a_j b_k c_l d_m$	Treat. Comb. $a_j b_k c_l d_m$	Treat. Comb. $a_j b_k c_l d_m$	Treat. Comb. $a_j b_k c_l d_m$
	Block$_0$	0000	0011	1101	1110
Group$_0$ [$(ACD)_0$ $(AB)_0$]	Block$_1$	0000	0011	1101	1110
	Block$_2$	0000	0011	1101	1110
	Block$_3$	0100	0111	1001	1010
Group$_1$ [$(ACD)_0$ $(AB)_1$]	Block$_4$	0100	0111	1001	1010
	Block$_5$	0100	0111	1001	1010
	Block$_6$	0001	0010	1100	1111
Group$_2$ [$(ACD)_1$ $(AB)_0$]	Block$_7$	0001	0010	1100	1111
	Block$_8$	0001	0010	1100	1111
	Block$_9$	0101	0110	1000	1011
Group$_3$ [$(ACD)_1$ $(AB)_1$]	Block$_{10}$	0101	0110	1000	1011
	Block$_{11}$	0101	0110	1000	1011

Figure 6.8 Layout for an RBCF-2^4 design. The $(ACD)_z$ and $(AB)_z$ interaction components are confounded with groups. The generalized interaction component, $(BCD)_z$, is also confounded with groups.

Defining relations must be chosen with care because when two or more interactions are simultaneously confounded with groups, their *generalized interaction(s)* or *treatment(s)* is (are) also confounded. The generalized interaction(s) for any two interactions, symbolized by (X) and (Y), is (are) given by the product $(X)(Y)^{m-z}$, reduced modulo m, where m is the modulus and z assumes values $m - 1, m - 2, \ldots, m - (m - 1)$. For treatments with two levels each, $m - z = 2 - 1$ and $(X)(Y)^{2-1} = (X)(Y)$. For example, the generalized interaction for (ACD) and (AB) is

$$(ACD)(AB) = (A^2BCD) \text{ reduced modulo } 2 = (BCD).$$

Hence, if (ACD) and (AB) are simultaneously confounded with groups, another interaction, (BCD), is also confounded with groups. Defining relations should be chosen so that no treatment is confounded with groups. Such confounding would have occurred if we had chosen $(ABCD)$ and (ACD) because then treatment B would have been the generalized treatment,

$$(ABCD)(ACD) = (A^2BC^2D^2) \text{ reduced modulo } 2 = B.$$

For any confounded design of the form p^k, i interactions can be completely confounded in p^i blocks, where each block contains p^{k-i} treatment combinations. The resulting design will have $[p^i - (p - 1)i - 1]/(p - 1)$ generalized interactions.

Designs in which the block size must equal two occur frequently in research. Examples include research with identical twins, the use of pretest and posttests measures, and vision research in which one eye serves as a control. The block size of an RBCF-2^4 design can be reduced to two by using three defining relations. For example, a researcher could use the defining relations

$$a_j + b_k + c_l + d_m = z \ (\text{mod } 2) \qquad (z = 0, 1),$$
$$a_j + b_k = z \ (\text{mod } 2) \qquad (z = 0, 1),$$
$$a_j + c_l = z \ (\text{mod } 2) \qquad (z = 0, 1).$$

Eight treatment combinations satisfy each of these defining relations. For example,

$$
\begin{aligned}
(ABCD)_0 &= 0000 \quad 0011 \quad 0110 \quad 0101 \quad 1010 \quad 1001 \quad 1100 \quad 1111 \\
(AB)_0 &= 0000 \quad 0001 \quad 0010 \quad 0011 \quad 1100 \quad 1101 \quad 1110 \quad 1111 \\
(AC)_0 &= 0000 \quad 0001 \quad 0100 \quad 0101 \quad 1010 \quad 1011 \quad 1110 \quad 1111.
\end{aligned}
$$

If we examine these combinations closely we find that two of them simultaneously satisfy $(ABCD)_0$, $(AB)_0$, and $(AC)_0$. These combinations are

$$(ACD)_0(AB)_0(AC)_0 = 0000 \quad 1111.$$

Two other treatment combinations satisfy $(ABCD)_0$, $(AB)_0$, and $(AC)_1$, and so on. The RBCF-2^4 design that results from confounding $ABCD$, AB, and AC with Groups is shown in Figure 6.9. The $[p^i - (p - 1)i -$

		Treat. Comb. $a_j b_k c_l d_m$	Treat. Comb. $a_j b_k c_l d_m$
	Block$_0$	0000	1111
Group$_0$ [$(ABCD)_0$ $(AB)_0$ $(AC)_0$]	Block$_1$	0000	1111
	Block$_2$	0000	1111
	Block$_3$	0011	1100
Group$_1$ [$(ABCD)_0$ $(AB)_0$ $(AC)_1$]	Block$_4$	0011	1100
	Block$_5$	0011	1100
	Block$_6$	0101	1010
Group$_2$ [$(ABCD)_0$ $(AB)_1$ $(AC)_0$]	Block$_7$	0101	1010
	Block$_8$	0101	1010
	Block$_9$	0110	1001
Group$_3$ [$(ABCD)_0$ $(AB)_1$ $(AC)_1$]	Block$_{10}$	0110	1001
	Block$_{11}$	0110	1001
	Block$_{12}$	0001	1110
Group$_4$ [$(ABCD)_1$ $(AB)_0$ $(AC)_0$]	Block$_{13}$	0001	1110
	Block$_{14}$	0001	1110
	Block$_{15}$	0010	1101
Group$_5$ [$(ABCD)_1$ $(AB)_0$ $(AC)_1$]	Block$_{16}$	0010	1101
	Block$_{17}$	0010	1101
	Block$_{18}$	0100	1011
Group$_6$ [$(ABCD)_1$ $(AB)_1$ $(AC)_0$]	Block$_{19}$	0100	1011
	Block$_{20}$	0100	1011
	Block$_{21}$	0111	1000
Group$_7$ [$(ABCD)_1$ $(AB)_1$ $(AC)_1$]	Block$_{22}$	0111	1000
	Block$_{23}$	0111	1000

Figure 6.9 Layout for an RBCF-2^4 design. The $(ABCD)_z$, $(AB)_z$, and $(AC)_z$ interaction components are confounded with groups. The generalized interaction components, $(AD)_z$, $(BC)_z$, $(BD)_z$, and $(CD)_z$, are also confounded with groups.

$1]/(p - 1) = [2^3 - (2 - 1)3 - 1]/(2 - 1) = 4$ generalized interactions for this design are

$$(ABCD)(AB) = (A^2B^2CD) \text{ reduced modulo } 2 = (CD),$$
$$(ABCD)(AC) = (A^2BC^2D) \text{ reduced modulo } 2 = (BD),$$
$$(AB)(AC) = (A^2BC) \text{ reduced modulo } 2 = (BC),$$
$$(ABCD)(AB)(AC) = (A^3B^2C^2D) \text{ reduced modulo } 2 = (AD).$$

The sources of variation, degrees of freedom, and F ratios for the RBCF-2^4 design in Figure 6.9 are as follows.

Source	df	F ratio
1 Between blocks	$nw - 1 = 23$	
2 Groups, AB, AC, AD,	$w - 1 = 7$	[2/3]
BC, BD, CD, $ABCD$		
3 Blocks w. groups	$w(n - 1) = 16$	
4 Within blocks	$nw(v - 1) = 24$	
5 A	$p - 1 = 1$	[5/13]
6 B	$q - 1 = 1$	[6/13]
7 C	$r - 1 = 1$	[7/13]
8 D	$l - 1 = 1$	[8/13]
9 ABC	$(p - 1)(q - 1)(r - 1) = 1$	[9/13]
10 ABD	$(p - 1)(q - 1)(l - 1) = 1$	[10/13]
11 ACD	$(p - 1)(r - 1)(l - 1) = 1$	[11/13]
12 BCD	$(q - 1)(r - 1)(l - 1) = 1$	[12/13]
13 $ABCD \times BL(G)$	$w(n - 1)(v - 1) = 16$	
14 Total	$nvw - 1 = 47$	

6.7.2   RBPF-3^3 Designs with Blocks of Size Nine or Three

A factorial design with $3 \times 3 \times 3 = 27$ treatment combinations can be laid out in blocks of size nine by confounding an interaction component with groups. The block size can be reduced to three by confounding two interactions with groups. The layout using blocks of size nine will be described first.

We saw in Section 6.6 that the AB interaction in an RBPF-3^2 design can be partitioned into two orthogonal components as follows.

SS	df
AB	$(3 - 1)(3 - 1) = 4$
(AB)	2
(AB^2)	2

Similarly, the ABC interaction in an RBPF-3^3 design can be partitioned into four orthogonal components.

SS	df
ABC	$(3 - 1)(3 - 1)(3 - 1) = 8$
(ABC)	2
(ABC^2)	2
(AB^2C)	2
(AB^2C^2)	2

The ABC interaction can be partially confounded by assigning treatment combinations to groups using each of the following defining relations.

Group 0 $(ABC)_z = a_j + b_k + c_l = z \pmod 3$ $(z = 0, 1, 2)$

Group 1 $(ABC^2)_z = a_j + b_k + 2c_l = z \pmod 3$ $(z = 0, 1, 2)$

Group 2 $(AB^2C)_z = a_j + 2b_k + c_l = z \pmod 3$ $(z = 0, 1, 2)$

Group 3 $(AB^2C^2)_z = a_j + 2b_k + 2c_l = z \pmod 3$ $(z = 0, 1, 2)$

The layout is shown in Figure 6.10. This design provides within-blocks information on treatments A, B, and C and the AB, AC, and BC interactions from all four groups. It provides three-fourths relative information on each component of the ABC interaction.

The design shown in Figure 6.10 requires blocks of size nine. If this block size is considered too large, further confounding can be used to reduce the block size to three. This is accomplished by assigning treatment combinations to groups so that they simultaneously satisfy two defining relations instead of one. For example, the two defining relations might be

$(ABC)_z = a_j + b_k + c_l = z \pmod 3$ $(z = 0, 1, 2)$

$(AB^2)_z = a_j + 2b_k = z \pmod 3$ $(z = 0, 1, 2).$

If $i = 2$ interactions are confounded in groups of blocks, $[p^i - (p - 1)i - 1]/(p - 1) = 2$ generalized interactions are also confounded. These

		Treat. Comb. $a_j b_k c_l$	Treat. Comb. $a_j b_k c_l$	Treat. Comb. $a_j b_k c_l$	Treat. Comb. $a_j b_k c_l$	Treat. Comb. $a_j b_k c_l$	Treat. Comb. $a_j b_k c_l$	Treat. Comb. $a_j b_k c_l$	Treat. Comb. $a_j b_k c_l$	Treat. Comb. $a_j b_k c_l$
	$(ABC)_0$ Block$_0$	000	012	021	102	111	120	201	210	222
Group$_0$	$(ABC)_1$ Block$_1$	001	010	022	100	112	121	202	211	220
	$(ABC)_3$ Block$_2$	002	011	020	101	110	122	200	212	221
	$(ABC^2)_0$ Block$_3$	000	011	022	101	112	120	202	210	221
Group$_1$	$(ABC^2)_1$ Block$_4$	002	010	021	100	111	122	201	212	220
	$(ABC^2)_2$ Block$_5$	001	012	020	102	110	121	200	211	222
	$(AB^2C)_0$ Block$_6$	000	011	022	102	110	121	201	212	220
Group$_2$	$(AB^2C)_1$ Block$_7$	001	012	020	100	111	122	202	210	221
	$(AB^2C)_2$ Block$_8$	002	010	021	101	112	120	200	211	222
	$(AB^2C^2)_0$ Block$_9$	000	012	021	101	110	122	202	211	220
Group$_3$	$(AB^2C^2)_1$ Block$_{10}$	002	011	020	100	112	121	201	210	222
	$(AB^2C^2)_2$ Block$_{11}$	001	010	022	102	111	120	200	212	221

Figure 6.10 Layout for an RBPF-3^3 design in blocks of size nine. The four components of the ABC interaction are confounded in the four groups.

generalized interactions are

$$(ABC)(AB^2)^{3-1} = (ABC)(AB^2)^2 = (A^3B^5C) = (BC^2) \text{ reduced modulo } 3$$
$$(ABC)(AB^2)^{3-2} = (ABC)(AB^2) = (A^2B^3C) = (AC^2) \text{ reduced modulo } 3.$$

Actually, $(A^3B^5C) = (B^2C)$ and $(A^2B^3C) = (A^2C)$ when reduced modulo 3. As discussed earlier, the power of the first term should always equal 1 to define the interaction components uniquely. This is achieved by squaring (B^2C) and (A^2C) and then reducing them modulo 3. As we have seen, when (ABC) and (AB^2) are confounded with groups of blocks, their generalized interactions, (BC^2) and (AC^2), are also confounded. Care must be exercised in choosing defining relations so that one or more treatments are not confounded with groups. Kempthorne (1952, p. 299) lists 13 ways of confounding a $3 \times 3 \times 3$ design in blocks of size three; only four of these systems do not confound a treatment with groups. One layout for an RBPF-3^3 design in blocks of size three that provides within-blocks information on all main effects and interactions is shown in Figure 6.11.

6.7.3 RBPF-4^2 Design

The schemes described so far for assigning treatment combinations to groups can be used when the number of treatment levels is a prime number. Other

			Treat. Comb. $a_jb_kc_l$	Treat. Comb. $a_jb_kc_l$	Treat. Comb. $a_jb_kc_l$
	$(ABC)_0 (AB^2)_0$	Block$_0$	000	111	222
	$(ABC)_0 (AB^2)_1$	Block$_1$	021	102	210
	$(ABC)_0 (AB^2)_2$	Block$_2$	012	120	201
	$(ABC)_1 (AB^2)_0$	Block$_3$	001	112	220
Group$_0$	$(ABC)_1 (AB^2)_1$	Block$_4$	022	100	211
	$(ABC)_1 (AB^2)_2$	Block$_5$	010	121	202
	$(ABC)_2 (AB^2)_0$	Block$_6$	002	110	221
	$(ABC)_2 (AB^2)_1$	Block$_7$	020	101	212
	$(ABC)_2 (AB^2)_2$	Block$_8$	011	122	200
	$(ABC^2)_0 (AB^2)_0$	Block$_9$	000	112	221
	$(ABC^2)_0 (AB^2)_1$	Block$_{10}$	022	101	210
	$(ABC^2)_0 (AB^2)_2$	Block$_{11}$	011	120	202
	$(ABC^2)_1 (AB^2)_0$	Block$_{12}$	002	111	220
Group$_1$	$(ABC^2)_1 (AB^2)_1$	Block$_{13}$	021	100	212
	$(ABC^2)_1 (AB^2)_2$	Block$_{14}$	010	122	201
	$(ABC^2)_2 (AB^2)_0$	Block$_{15}$	001	110	222
	$(ABC^2)_2 (AB^2)_1$	Block$_{16}$	020	102	211
	$(ABC^2)_2 (AB^2)_2$	Block$_{17}$	012	121	200
	$\cdots$	$\cdots$	$\cdots$	$\cdots$	$\cdots$
	$(AB^2C^2)_0 (AB)_0$	Block$_{27}$	000	122	211
	$(AB^2C^2)_0 (AB)_1$	Block$_{28}$	012	101	220
	$(AB^2C^2)_0 (AB)_2$	Block$_{29}$	021	110	202
	$(AB^2C^2)_1 (AB)_0$	Block$_{30}$	002	121	210
Group$_3$	$(AB^2C^2)_1 (AB)_1$	Block$_{31}$	011	100	222
	$(AB^2C^2)_1 (AB)_2$	Block$_{32}$	020	112	201
	$(AB^2C^2)_2 (AB)_0$	Block$_{33}$	001	120	212
	$(AB^2C^2)_2 (AB)_1$	Block$_{34}$	010	102	221
	$(AB^2C^2)_2 (AB)_2$	Block$_{35}$	022	111	200

Figure 6.11 Layout for an RBPF-3^3 design in blocks of size three. The confounding scheme is based on the four components of the ABC interaction and the two components of the AB interaction.

		Treat. Comb. a_jb_k	Treat. Comb. a_jb_k	Treat. Comb. a_jb_k	Treat. Comb. a_jb_k
	Block$_0$	00	11	23	32
Group$_0$	Block$_1$	01	10	22	33
	Block$_2$	02	13	21	30
	Block$_3$	03	12	20	31
	Block$_4$	00	12	21	33
Group$_1$	Block$_5$	01	13	20	32
	Block$_6$	02	10	23	31
	Block$_7$	03	11	22	30
	Block$_8$	00	13	22	31
Group$_2$	Block$_9$	01	12	23	30
	Block$_{10}$	02	11	20	33
	Block$_{11}$	03	10	21	32

Figure 6.12 Layout for an RBPF-4^2 design.

schemes must be used when the number of treatment levels is not a prime number. The RBPC-4^2 design shown in Figure 6.12 was suggested by Yates (1937). The sources of variation, degrees of freedom, and F ratios are as follows.

Source	df	F ratio
1 Between blocks	$nw - 1 = 11$	
2 Groups	$w - 1 = 2$	[2/3]
3 Blocks w. groups	$w(n - 1) = 9$	
4 Within blocks	$nw(v - 1) = 36$	
5 A	$p - 1 = 3$	[5/8]
6 B	$q - 1 = 3$	[6/8]
7 AB (within)	$(p - 1)(q - 1) = 9$	[7/8]
8 Residual	$nvw - nw - pq + 1 = 21$	
9 Total	$nvw - 1 = 47$	

The design provides two-thirds relative information on the AB interaction.

Table 6.9 Guide to Confounded Factorial Designs

Design	Number of treatment combinations	Number of observations per block	Number of groups for balanced design[a]	Interaction(s) confounded and relative information[b]	Reference[c]
RBCF-2²	4	2	x	AB	(4) p. 61, (7) p. 577, (8) p. 319, (11), p. 608
RBCF-2³	8	4	x	Any interaction	(1) p. 220, (2) p. 427, (3) p. 233, (7) p. 583, (8) p. 321, (11) p. 609
RBPF-2³	8	4	4	AB (¾), AC (¾), BC (¾), ABC (¾)	(1) p. 220, (3) p. 244, (4) p. 62, (5) p. 201, (6) p. 275, (7) p. 591, (8) p. 329, (11) p. 609
RBCF-2⁴	16	8	x	Any interaction	(1) p. 220, (2) p. 429, (5) p. 193, (7) p. 614
RBCF-2⁴	16	4	x	AB, ACD, BCD; or AB, CD, ABCD	(1) p. 220, (2) p. 428, (5) p. 194, (7) p. 614
RBPF-2⁴	16	4	6	All two-factor interactions (⅝), all three-factor interactions (½)	(1) p. 220
RBCF-2⁴	16	2	x	AB, AC, BC, AD, BD, CD, ABCD	(6) p. 261
RBPF-2⁴	16	2	4	Main effects (¾), two-factor interactions (½), three-factor interactions (¼)	(6) p. 278
RBCF-2⁵	32	16	x	Any interaction	(2) p. 430
RBCF-2⁵	32	8	x	ABC, ADE, BCDE; or AB, CDE, ABCDE	(1) p. 220, (2) p. 429, (4) p. 72, (8) p. 222

Design					
RBPF-2^5	32	8	5	All three-factor interactions ($\frac{4}{5}$), all four-factor interactions ($\frac{4}{5}$)	(1) p. 220
RBCF-2^5	32	4	x	$AB, AC, BC, DE, ABDE, ACDE, BCDE$; or $AB, CD, ACE, BDE, ADE, BCE, ABCD$	(2) p. 428
RBCF-3^2	9	3	x	AB or AB^2	(2) p. 435, (3) p. 239, (7) p. 601, (8) p. 399
RBPF-3^2	9	3	2	AB components ($\frac{1}{2}$)	(3) p. 251, (4) p. 99, (6) p. 300, (7) p. 601, (11) p. 625
RBCF-3^3	27	9	x	Any interaction component	(4) p. 102, (8) p. 401
RBPF-3^3	27	9	4	ABC components ($\frac{3}{4}$)	(1) p. 222, (2) p. 438, (3) p. 251, (6) p. 302, (7) p. 617, (11) p. 636
RBPF-3^3	27	3	4	All two-factor interactions ($\frac{1}{2}$), ABC ($\frac{3}{4}$)	(2) p. 436, (3) p. 252, (7) p. 617, (11) p. 655
RBCF-3^4	81	9	x	$ABC, AB^2D^2, AC^2D, BC^2D^2$	(8) p. 403
RBPF-3^4	81	9	4	All three-factor interactions ($\frac{3}{4}$)	(1) p. 223, (6) p. 306
RBPF-4^2	16	4	3	AB components ($\frac{2}{3}$)	(1) p. 225
RBCF-4^3	64	16	x	ABC interaction	(9) p. 121
RBCF-5^2	25	5	x	$AB, AB^2, AB^3,$ or AB^4	(7) p. 621
RBPF-5^2	25	5	4	AB components ($\frac{3}{4}$)	(7) p. 621, (10) p. 57

[a]The symbol x indicates that any number of groups (replicates) can be used; for a balanced design, groups (replicates) should be a multiple of the number that appears in the column.

[b]If a fraction does not follow the interaction, it is completely confounded.

[c]Information about a design can be found in the following references: (1) Cochran and Cox (1957), (2) Davies (1956), (3) Federer (1955), (4) Gill (1978, Vol. 2), (5) Johnson and Leone (1964), (6) Kempthorne (1952), (7) Kirk (1982), (8) Montgomery (1991), (9) Nair (1938), (10) Nair (1940), and (11) Winer, Brown, and Michels (1991).

247

6.7.4   RBCF-5^2 and RBPF-5^2 Designs

A 5 × 5 factorial design has 25 treatment combinations. The block size can be reduced from 25 to 5 by confounding the AB interaction with groups. This interaction can be partitioned as follows.

SS	df
AB	$(5 - 1)(5 - 1) = 16$
(AB)	4
(AB^2)	4
(AB^3)	4
(AB^4)	4

A completely confounded design is constructed by confounding any one of the four interaction components with groups. A better design choice is to confound each of the interaction components with blocks within a group. A balanced partially confounded design can be laid out in four groups or a multiple of four groups. Treatment combinations are assigned to blocks within each group by means of the following defining relations.

Group 0 $(AB)_z = a_j + b_k = z$ (mod 5) $(z = 0, \ldots, 4)$

Group 1 $(AB^2)_z = a_j + 2b_k = z$ (mod 5) $(z = 0, \ldots, 4)$

Group 2 $(AB^3)_z = a_j + 3b_k = z$ (mod 5) $(z = 0, \ldots, 4)$

Group 3 $(AB^4)_z = a_j + 4b_k = z$ (mod 5) $(z = 0, \ldots, 4)$

Each group contains five blocks corresponding to $z = 0, 1, 2, 3$, and 4. The design provides three-fourths relative information on the confounded interaction.

A variety of confounded factorial designs of the form p^k where p is equal to two, three, four, or five has been described. The approach that has been adopted is to present selected examples that illustrate basic principles. A detailed coverage of each of the designs is beyond the scope of this chapter. Table 6.9 gives references that will be useful to the researcher who needs details about the analysis of these confounded factorial designs.

6.8   MIXED CONFOUNDED FACTORIAL DESIGNS

Confounded designs in which the number of levels of each treatment is not equal are called *mixed designs*. Examples of mixed designs are 3 × 2 × 2, 3 × 3 × 2, and 4 × 3 × 2 designs. These designs generally entail a more complex analysis than unmixed designs, and the choice of block size is much more restricted for the mixed designs. To avoid confounding a treatment, the block size must be an integral multiple of the number of levels of each treatment. Thus, for a $3 \times 2 \times 2$ design, the block size must equal six. This permits a_0, a_1, and a_2 to occur twice each in a block and b_0, b_1 and c_0, c_1 to occur three times. All 12 treatment combinations of a $3 \times 2 \times 2$ design can be assigned to blocks of size six as follows.

Block 0   000   011   101   110   201   210

Block 1   001   010   100   111   200   211

Each level of a_j, b_k, and c_l as well as each of the six combinations of $a_j b_k$ and $a_j c_l$ occurs equally often in blocks 0 and 1. This cannot be true for $b_k c_l$ because the interaction has four treatment combinations. As a result, the BC interaction is confounded with blocks. Furthermore, the ABC interaction that has 12 treatment combinations is also confounded with blocks.

A balanced RBPF-32^2 design can be laid out in three groups of two blocks each. The layout is shown in Figure 6.13. The sources of variation and degrees of freedom are as follows.

Source	df	F ratio
1 Between blocks	$nw - 1 = 5$	
2 Groups	$w - 1 = 2$	[2/3]
3 Blocks w. groups	$w(n - 1) = 3$	
4 Within blocks	$nw(v - 1) = 30$	
5 A	$p - 1 = 2$	[5/12]
6 B	$q - 1 = 1$	[6/12]
7 C	$r - 1 = 1$	[7/12]
8 AB	$(p - 1)(q - 1) = 2$	[8/12]
9 AC	$(p - 1)(r - 1) = 2$	[9/12]
10 BC(adj)	$(q - 1)(r - 1) = 1$	[10/12]
11 ABC(adj)	$(p - 1)(q - 1)(r - 1) = 2$	[11/12]
12 Residual	$nvw - nw - pqr + 1 = 19$	
13 Total	$nvw - 1 = 35$	

The designations BC(adj) and ABC(adj) indicate that these sources of variation must be adjusted for block effects. Although the adjustment is not difficult, the rationale underlying the adjustment is beyond the scope of this chapter. The interested reader can consult Federer (1955), Kempthorne (1952), Li (1944), Nair (1938), or Yates (1937). Mixed designs present special computational problems. Table 6.10 provides a summary

		Treat. Comb. $a_j b_k c_l$	Treat. Comb. $a_j b_k c_l$	Treat. Comb. $a_j b_k c_l$	Treat. Comb. $a_j b_k c_l$	Treat. Comb. $a_j b_k c_l$	Treat. Comb. $a_j b_k c_l$
Group$_0$	Block$_0$	000	011	101	110	201	210
	Block$_1$	001	010	100	111	200	211
Group$_1$	Block$_2$	001	010	100	111	201	210
	Block$_3$	000	011	101	110	200	211
Group$_2$	Block$_4$	001	010	101	110	200	211
	Block$_5$	000	011	100	111 ·	201	210

Figure 6.13 Layout for an RBPF-32^2 design. This design confounds the BC and ABC interactions with groups.

of the most useful mixed designs and references that describe the analysis procedures.

6.9 SUMMARY OF ADVANTAGES AND DISADVANTAGES OF CONFOUNDED FACTORIAL DESIGNS

Researchers today are well aware of the potential advantage of a randomized block factorial design in reducing error variation by isolating variation attributable to blocks. However, a relatively small RBF-pq design may require a prohibitively large block size. For example, an RBF-33 design requires blocks of size nine. A split-plot factorial design provides one solution to this problem by confounding treatment A with Groups and thereby reducing the block size to three. An alternative solution is provided by RBCF-3^2 and RBPF-3^2 designs. These two designs confound the AB interaction with Groups and also reduce the block size from nine to three. However, the advantages of reduced block size and better local control in split-plot and confounded factorial designs are achieved at a price. In the case of an SPF-3·3 design, the price a researcher pays is loss of power in evaluating a treatment. In the cases of RBCF-3^2 and RBPF-3^2 designs, the price is loss of power in evaluating an interaction. The particular research application will determine which kind of information can be sacrificed and, indeed, if either form of confounding is appropriate.

An examination of the current literature in the behavioral and social sciences reveals an almost total absence of experiments using the confounded factorial designs described in this chapter. This observation is in sharp contrast to the observation that a split-plot factorial design, which involves a different form of confounding, is one of the most popular designs.

Table 6.10 Guide to Mixed Confounded Factorial Designs

Design	Number of treatment combinations	Number of observations per block	Number of groups for balanced design[a]	Interaction(s) confounded and relative information[b]	Reference[c]
RBPF-32^2	12	6	3	BC ($8/9$), ABC ($5/9$)	(1) p. 224, (2) p. 253, (3) p. 348, (4), p. 627, (6) p. 647
RBPF-32^3	24	6	3	BC ($8/9$), BD ($8/9$), CD ($8/9$), ABC ($5/9$), ABD ($5/9$), ACD ($5/9$)	(1) p. 225
RBPF-3^22	18	6	4	AB ($7/8$), ABC ($5/8$)	(1) p. 224, (2) p. 254, (3) p. 355
RBPF-3^32	54	6	4	AB ($7/8$), AC ($7/8$), AD ($7/8$), ABD ($5/8$), ACD ($5/8$), BCD ($5/8$), ABC ($3/4$)	(6) p. 656, (7) p. 63
RBPF-42^2	16	8	3	ABC ($2/3$)	(1) p. 226, (5) p. 460
RBPF-432	24	12	9	AC ($26/27$), AB ($23/27$)	(1) p. 227
RBPF-43^2	36	12	2	BC ($7/8$), ABC ($7/8$)	(5) p. 475
RBCF-4^22	32	16	x	ABC	(5) p. 465
RBPF-4^23	48	12	3	AB ($26/27$), ABC ($23/27$)	(5) p. 486
RBPF-42^3	32	8	3	ABC ($2/3$), ABD ($2/3$), ACD ($2/3$)	(5) p. 467
RBPF-52^2	20	10	5	BC ($24/25$), ABC ($19/25$)	(5) p. 469

[a]The symbol x indicates that any number of groups (replicates) can be used; for a balanced design, groups (replicates) should be a multiple of the number that appears in the column.

[b]If a fraction does not follow the interaction, it is completely confounded.

[c]Information about a design can be found in the following references: (1) Cochran and Cox (1957), (2) Federer (1955), (3) Kempthorne (1952), (4) Kirk (1982), (5) Li (1944), (6) Winer, Brown, and Michels (1991), (7) Yates (1937).

251

The difference in the popularity of the two designs, both of which accomplish the same objective of reducing block size, can be attributed to several factors: (1) most computer packages can analyze data for split-plot factorial designs but cannot analyze data for confounded factorial designs, and (2) many researchers are not familiar with the advantages of confounded factorial designs. It should be abundantly clear that the selection of an analysis of variance design requires an intimate knowledge of a research area as well as a knowledge of the advantages and disadvantages of alternative designs.

The advantages of a confounded design are as follows.

1. The block size can be reduced by a factor of 1/2, 1/3, and so on, thereby achieving better local control.
2. All treatments are tested with equal power.
3. Differences among blocks are removed from the estimate of the within-blocks error.
4. An experiment can be designed so that only interactions believed to be insignificant are partially or completely sacrificed.
5. For designs in which each treatment has two levels, the block size can always be reduced to two by confounding a sufficient number of interactions.

The disadvantages of a confounded design are as follows.

1. Some information is lost with respect to one or more interactions.
2. The layout and analysis of confounded designs are more complex than the layout and analysis of other designs that use blocking to achieve local control.
3. It must be possible to administer the levels of each treatment in every possible sequence. This requirement precludes the use of treatments whose levels consist of successive periods of time.

REFERENCES

Bailey, R. A. (1977). Patterns of Confounding in Factorial Designs. *Biometrika*, *64*: 579–603.

Cochran, W. G. and Cox, G. M. (1957). *Experimental Designs*. Wiley, New York.

Davies, O. L. (ed.). (1956). *The Design and Analysis of Industrial Experiments*. Hafner, New York.

Federer, W. T. (1955). *Experimental Design: Theory and Application*. Macmillan, New York.

Fisher, R. A. (1926). The Arrangement of Field Experiments. *J. Min. Agric.*, *33*: 503–513.

Fisher, R. A. (1935). *The Design of Experiments*. Oliver and Boyd, Edinburgh.

Fisher, R. A. (1942). The Theory of Confounding in Factorial Experiments in Relation to the Theory of Groups. *Ann. Eugenics*, *11*: 341–353.

Fisher, R. A. and Wishart, J. (1930). *The Arrangement of Field Experiments and the Statistical Reduction of the Results*. Imperial Bureau of Soil Science, Harpenden, England.

Gill, J. L. (1978). *Design and Analysis of Experiments in the Animal and Medical Sciences*. Iowa State University Press, Ames, Iowa.

Hocking, R. R. (1985). *The Analysis of Linear Models*. Brooks/Cole, Pacific Grove, California.

John, J. A. and Dean, A. M. (1975). Single Replicate Factorial Experiments in Generalized Cyclic Designs: I. Symmetrical Arrangements. *J. Roy. Statist. Soc.*, *B37*: 63–71.

Johnson, N. L. and Leone, F. C. (1964). *Statistics and Experimental Design in Engineering and the Physical Sciences*, Vol. II. Wiley, New York.

Kempthorne, O. (1947). A Simple Approach to Confounding and Fractional Replication in Factorial Experiments. *Biometrika*, *34*: 255–272.

Kempthorne, O. (1952). *The Design and Analysis of Experiments*. Wiley, New York.

Kirk, R. E. (1982). *Experimental Design: Procedures for the Behavioral Sciences* (2nd ed.) Brooks/Cole, Pacific Grove, California.

Li, J. C. R. (1944). Design and Statistical Analysis of Some Confounded Factorial Experiments. *Iowa Agricultural Experiment Station Research Bulletin*, No. 333, Iowa Agricultural Experiment Station, Ames, Iowa, pp. 449–492.

Montgomery, D. C. (1991). *Design and Analysis of Experiments*. Wiley, New York.

Nair, K. R. (1938). On a Method of Getting Confounded Arrangements in the General Symmetrical Type of Experiments. *Sankhyā*, *4*: 121–138.

Nair, K. R. (1940). Balanced Confounded Arrangements for the 5^n Type of Experiment. *Sankhyā*, *5*: 57–70.

Patterson, H. D. and Bailey, R. A. (1978). Design Keys for Factorial Experiments. *Appl. Statist.*, *27*: 335–343.

Pedhazur, E. J. (1982). *Multiple Regression in Behavioral Research* (2nd ed.). Holt, Rinehart and Winston, Fort Worth, Texas.

Searle, S. R. (1987). *Linear Models for Unbalanced Data*. Wiley, New York.

Timm, N. H. and Carlson, J. E. (1975). Analysis of Variance Through Full Rank Models. *Multivariate Behavioral Research Monographs*, No. 75-1.

Winer, B. J., Brown, D. R., and Michels, K. M. (1991). *Statistical Principles in Experimental Design*. McGraw-Hill, New York.

Yates, F. (1933). The Principles of Orthogonality and Confounding in Replicated Experiments. *J. Agric. Sci.*, *23*: 108–145.

Yates, F. (1935). Complex Experiments. *J. Roy. Statist. Soc. (Suppl.)*, *2*: 181–233.

Yates, F. (1937). *The Design and Analysis of Factorial Experiments*. Imperial Bureau of Soil Science, Harpenden, England.

7
Multivariate Analysis of Variance

RICHARD J. HARRIS University of New Mexico, Albuquerque, New Mexico

7.1 INTRODUCTION

Multivariate analysis of variance (Manova) is situated between univariate analysis of variance (Anova) and canonical correlation analysis (Canona) in the panoply of statistical techniques. Like Anova, Manova is used to assess the grand mean and/or differences among independent group means on one or more dependent variables. It generalizes Anova by examining not only each one of the p original dependent variables separately but also linear combinations of these measures. As part of its examination of such linear combinations it tells us which particular linear combination provides the strongest evidence against the null hypothesis of no differences among corresponding population means.

On the other hand, Manova can be seen as a special case of Canona. Canona examines the relationships between linear combinations of the variables in each of two sets of measures. Manova is that special case of Canona in which the variables in one set of measures consist of *level-membership variables* determining the level of each factor into which each subject is assigned or classified. This chapter will focus on Manova as a generalization of Anova, as well as a special case of Canona.

Given that each dependent variable is usually selected because of its direct relevance to the independent variable distinguishing the independent groups from each other and given that it is less than intuitive to think of linear combinations of dependent variables as having meaning in their own right, we will find that the *possibility* of conducting a Manova need not imply the *desirability* of doing so and that multiple Anovas carried out with

Bonferroni-adjusted alpha levels are often an attractive alternative. This distinction between possibility and desirability of full-scale Manova is the focus of the next four sections.

7.1.1 Situations in Which Manova *Can* Be Applied

There are essentially four conditions under which Manova may be the analysis of choice:

1. Where there are $k \geq 2$ independent groups (possibly organized into a factorial or hierarchical structure) with scores on $p \geq 2$ dependent variables available for each sampling unit. This is the classic prescription for Manova and is a straightforward generalization of completely between-subjects Anova.
2. Where there is a single group ($k = 1$) with scores on $p \geq 3$ measures for each sampling unit, and the p measures constitute the levels of an independent variable or the combinations of the levels of two or more factors. This is the classic prescription for completely within-subjects Anova.
3. Where you have some combination of between- and within-subjects factors. That is, there are $k \geq 2$ independent groups and $p \geq 3$ measures having a factorial structure. This is the prescription for a mixed-design Anova.
4. Where, regardless of the number of independent groups, the $p \geq 4$ measures are generated from two or more dependent variables scored for each combination of the levels of one or more within-subject factors. This is what the SPSS manual's (SPSS, 1990) description of the MANOVA program refers to as a doubly multivariate design.

7.1.1.1 Advantages of Manova in Above Situations

Manova as applied to each of the above situations involves examining not only each of the p measures on each sampling unit separately but also linear combinations of these measures, with each such combination being analyzed as a new, emergent dependent variable in its own right. Where a within-subjects factor is involved, the combining weights involved in the linear combination are constrained to sum to zero (i.e., to constitute contrast coefficients). Control over the inflated familywise error rate (that analyzing each of the p measures with a separate Anova would entail) is accomplished by testing each Anova effect (including specific contrasts) against a critical value derived from the sampling distribution of the largest F for that effect obtainable from any linear combination of the measures. However, this control of familywise Type I error rate could be accomplished more simply and more powerfully by carrying out univariate An-ovas on each of the p measures, but using Bonferroni-adjusted critical values (such that the error rates for F's on individual dependent variables add to the desired familywise error rate) for each of the resultant F's. (One may also carry out Bonferroni-adjusted univariate Anovas on a priori linear combinations of the measures, but if the number of such combinations considered gets much larger than p you may find that the fully multivariate procedure provides more powerful tests.) The principal advantage of applying Manova must, then, lie elsewhere. In particular, the unique advantage of applying multivariate techniques is the identification of linear combinations of measures that (potentially) yield much larger F's than any single measure. This advantage carries with it the responsibility or opportunity of interpreting the discriminant function(s) or optimal within-subjects contrast(s) identified by Manova. This is both as exciting and as problematic as interpreting the factors identified by an exploratory factor analysis.

7.1.1.2   Conditions Under Which Multiple Anovas Should Be Preferred

As implied above, Manova should be employed only if one is interested in examining linear combinations of the p measures. If one is interested only in examining the effects of between-subjects manipulations or categorizations on each dependent variable separately, then conducting p separate univariate Anovas each at a Bonferroni-adjusted individual Type I error rate of $.05/p$ will provide much more powerful tests of these univariate effects than would a Manova. The Manova overall test must protect against the inflation of familywise alpha that comes from the possibility of computing an F for any one or more of the infinite number of linear combinations of the measures—rather useless protection if one has no interest in such combined variables. Any purported Manova should report and interpret at least one discriminant function or optimal within-subjects contrast to follow up each significant overall multivariate test. This is directly analogous to the relationship between the traditional overall F ratio and Scheffé (1953) contrasts versus the host of pairwise-comparison techniques. The overall F is a constant times the largest possible F_{contr} for *any* contrast among the k group means and is thus designed to control familywise alpha in the case where the researcher wishes to examine an unlimited number of contrasts of the form $\Sigma\, c_j \overline{Y}_j$ where $\Sigma\, c_j = 0$. If, rather, one is interested only in pairwise comparisons, a studentized range test of the overall null hypothesis followed by Tukey HSD tests of other pairwise differences provides much more powerful and internally consistent tests of these comparisons.

7.1.2 Combining Multiple Anovas with Full-Scale Manova

Use of Bonferroni-adjusted univariate Anovas on particular dependent variables provides very powerful tests of a priori hypotheses but forgoes the possibility, provided by a Manova, of uncovering unanticipated linear combinations of measures that differentiate among k groups (or p levels of a within-subjects factor) much more powerfully than any single dependent variable or any of the within-subjects contrasts selected a priori. Use of the overall Manova test (and follow-up tests based on the critical value of this overall test) provides this intriguing exploratory possibility, but at the cost of a rather stringent critical value for all tests, whether anticipated (a priori) or not. To circumvent this problem, one can carry out all a priori tests with individual error rates that sum to some value ($\alpha_{planned}$) less than the desired familywise error rate (α_{target}). Then one can carry out a Manova and associated post hoc follow-up tests, using the remaining portion of the desired familywise error rate ($\alpha_{target} - \alpha_{planned}$) as the alpha level. The post hoc linear combinations of variables you choose may be suggested by examination of the discriminant functions produced by Manova, by examination of the means, or by any other information, and need not be limited to any finite number.

Henceforth we will examine linear combinations of measures, either as a fully post hoc approach or as a part of the combined a priori/post hoc approach. The relevant familywise error rate for the fully multivariate approaches may either be the total α_{target} or the ($\alpha_{target} - \alpha_{planned}$).

7.2 THEORETICAL UNDERPINNINGS OF MANOVA

7.2.1 Basic Logic

To understand the way in which Manova generalizes Anova, consider the large variety of between-subjects univariate Anova designs. For any given effect to be tested in a Manova, one must compute the sum of squares and mean square for that effect (SS_{eff} and MS_{eff}), using in turn each of the p measures as the dependent variable. Similarly, one must compute SS_{err} and MS_{err} corresponding to the appropriate error term for the effect being tested. In a totally between-subjects Anova with all factors fixed, this error term will be MS_w, the within-cells mean square. For more complex designs, such as those involving one or more random factors, readers are referred to Kirk (1982), Winer, Brown, and Michels (1991), or Harris (in press).

The Manova overall test consists of finding the maximum possible value of $F_{eff} = MS_{eff}/MS_{err}$ computed on any *linear combination* of the p measures and comparing this maximized F ratio to the percentiles of the sampling distribution of such a maximized F. Follow-up tests consist of comparing F_{eff} computed on any linear combinations of the measures (such as one or more simplified versions of the *discriminant function*, which actually yielded the maximized F_{eff}) to the same critical value against which max F_{eff} was compared. The F computed on any particular linear combination of the measures clearly does not involve any more capitalization on chance than does max F_{eff}, so if its value exceeds that required to convince us that max F_{eff} represents a nonzero population effect, this particular linear combination must also be considered to have yielded a statistically significant effect.

There are mathematical shortcuts to implementing this maximization with respect to choice of combining weights of F_{eff}. Certain computer programs incorporate these shortcuts. Keep in mind that the properties of the resulting max F_{eff} and discriminant function(s) are exactly the same as if they had been arrived at via trial-and-error search of the p-variable parameter space. For instance, it is difficult to prove via algebraic manipulation of the computational formulas that adding any new, $(p + 1)$st measure on the same subjects (including a column of random numbers) must inevitably increase the max F_{eff}. Consideration of the properties of a trial-and-error search, however, makes this proof quite simple, since a successful search could always begin by using the same combining weights that had maximized F_{eff} for the original p measures, plus zero times each subject's score on the $(p + 1)$st measure. Moreover, since this increase (or, at worst, nondecrease) in max F_{eff} holds for every value in its null distribution, the α-level critical value for max F_{eff} based on $p + 1$ measures must be larger than that for max F_{eff} based on p measures, all other parameters such as sample size being held constant.

7.2.2 Matrix-Algebraic Shortcut to Maximizing F_{eff}

Now, armed with the ability to compute SS_{eff} and SS_{err} for your effect, using any one of the original measures as the dependent variable, proceed with the following mathematical shortcut:

(1) Construct a matrix **H**, the hypothesis cross-product matrix, putting in its ith main diagonal entry SS_i, the SS_{eff} computed for measure i. In the off-diagonal position (i, j), put $SP_{i,j}$, obtained by applying the formula for SS_{eff} to the cross-product of scores on measures i and j, rather than to the square of either. Thus, for instance, $\Sigma\,(T_j^2/n_j)$ in the computational formula for SS_j becomes $\Sigma\,(T_iT_j/n_j)$ for SP_{ij}.

(2) Construct a similar matrix, **E**, the error cross-product matrix, putting in its ith main diagonal entry SS_i, the SS_{err} computed for measure i. In its off-diagonal position (i, j) put $SP_{i,j}$, obtained by applying the formula

for SS_{err} to the cross-product of scores on measures i and j, rather than to the square of either.

The F_{eff} for any linear combination of the measures, $\mathbf{Y}\,\mathbf{a}$, where $\mathbf{a}$ is a vector containing the combining weights, $(a_1, a_2, \ldots, a_p)$, can then be computed as

$$F_{eff} \text{ on linear combination} = (df_{err}/df_{eff})\cdot\mathbf{a}'\mathbf{H}\mathbf{a}/\mathbf{a}'\mathbf{E}\mathbf{a}.$$

[See Harris (1985, Derivation 4.1) for a formal proof.] This will, in fact, be a convenient formula to use when exploring particular linear combinations of the measures post hoc. For the moment, however, the goal is to find those values of a_1 through a_p that maximize F_{eff}, for which purpose we must turn to computation of characteristic roots and vectors.

(3) Compute the characteristic roots and vectors of the matrix product $\mathbf{E}^{-1}\mathbf{H}$. In other words, find the s pairs of characteristic roots λ_i and corresponding characteristic vectors $\mathbf{a}_i$ that solve the set of simultaneous equations,

$$[\mathbf{H} - \lambda\mathbf{E}]\mathbf{a} = \mathbf{0}.$$

A number of texts, including Horst (1961) and Harris (1985, Digression 2) can provide detailed methods for finding characteristic roots and vectors. The largest of the characteristic roots (eigenvalues), λ_1, provides the maximum possible value of SS_{eff}/SS_{err} obtainable for any linear combination of the measures, while the characteristic vector (eigenvector) corresponding to λ_1 gives the combining weights that define the linear combination that yields this maximized "stripped F ratio" (F_{eff} sans degrees of freedom) and thus also maximizes F_{eff} itself. The linear combination is the *discriminant function* for this effect. If $s = \min(df_{eff}, p) > 1$, there will be additional pairs of characteristic roots and corresponding characteristic vectors, λ_i being the ith largest of the roots. λ_i gives the largest possible value of SS_{eff}/SS_{err} obtainable from any linear combination of the p measures that is uncorrelated with any of the first $i - 1$ discriminant functions. Each subsequent discriminant function, then, gives another dimension, nonredundant with those already identified, along which the specified effect manifests itself.

7.2.3 Statistical Significance

The test of the overall null hypothesis (that the effect being tested is zero in the population with respect both to each of the p measures and to any and all linear combinations thereof) consists of comparing λ_1 to the $100(1 - \alpha)$th percentile of its null sampling distribution. Available tables and corresponding computer subroutines are actually based on the sampling distribution of θ_1, which is the maximum (with respect to **a**) of $SS_{eff}/(SS_{eff} + SS_{err})$. θ_1 has the advantages of being bounded between 0.0 and 1.0 and of being interpretable as the squared canonical R between the p measures and the set of level-membership variables (i.e., essentially contrast-coded dummy variables used in carrying out Anova via multiple regression) that represent the between-subjects effect being tested. It is readily computed from λ_1 via $\theta_1 = \lambda_1/(1 + \lambda_1)$. Since both θ_i and λ_i are referred to in computer printouts as eigenvalues (or as Roy's criterion) printouts, one must carefully examine which one is being reported. [In their most recent editions, SAS's (1990) PROC GLM reports λ_1 and SPSS's (1990) MANOVA reports θ_1.] The $100(1 - \alpha)$th percentile of the sampling distribution of θ_1, that is, the critical value of the g.c.r. (greatest characteristic root) distribution is symbolized as $\theta_a(s, m, n)$ and has degree-of-freedom parameters

$$s = \min(df_{eff}, p),$$

$$m = (|df_{eff} - p| - 1)/2,$$

and

$$n = (df_{err} - p - 1)/2.$$

The most complete set of tables, covering $s = 1$ through $s = 20$, is provided by Harris (1985), whose text also provides a printout of a computer program to compute g.c.r. critical values interactively. The BMDP4V program (Dixon, 1985) for Manova prints the p value associated with the θ_1, thus obviating the need for tables or a separate computer program. On the other hand, SAS's PROC GLM reports an upper bound of an F approximation to the g.c.r. distribution that is too liberal to be useful, while SPSS leaves determination of the statistical significance of the θ_1 to the user. However, if $s = 1$, then the g.c.r. distribution is identical to that of the three multiple-root tests: Wilks' lambda (Wilks, 1932) and the Pillai (1955) and Hotelling (1951) trace statistics, for which an F-based critical value (Jones, 1966) is provided by all three programs. [Hotelling's trace statistic, which is the sum of all s values of λ_1 through λ_s, should not be confused with Hotelling's (1931) T^2 statistic, which is the maximum squared t ratio obtainable for the difference between two group means with respect to any linear combination of a set of dependent variables. When $s = 1$, Hotelling's $T^2 = df_{err}$ (Hotelling's trace).] The critical value against which to compare max F_{eff} is easily computed by hand, aided only by a table of ordinary F critical values, as

$$\text{c.v. for max } F_{eff} = \frac{p(df_{err})}{df_{err} - p + 1} F_\alpha(p, df_{err} - p + 1),$$

which approaches $p \cdot F_\alpha(p, df_{err})$ and is thus highly reminiscent of the Scheffé fully post hoc critical value (Scheffé, 1953) for Anova.

Follow-up tests of F_{eff} as computed on any single measure or on any linear combination of measures follow Roy and Bose's (1953) union-intersection principle (which is also the basis of the familiar Scheffé post hoc critical value in Anova) by taking as the critical value for any such post hoc F_{eff} the same critical value that would be applied to the deliberately maximized F_{eff} on the first discriminant function, namely

fully post hoc c.v. for $F_{eff} = (df_{err}/df_{eff}) \cdot \theta_\alpha(s, m, n)/[1 - \theta_\alpha(s, m, n)]$.

In words, we consider F_{eff} computed on any particular linear combination of the measures statistically significant if and only if it exceeds the value that would have convinced us that F_{eff} computed on the discriminant function (i.e., the absolutely optimal linear combination of the measures) was statistically significant. This provides total consistency between the overall test and the post hoc tests, in that the test of θ_1 yields statistical significance if and only if there is at least one linear combination of the measures (specifically, the discriminant function) that yields an F_{eff} that is statistically significant by the fully post hoc critical value.

The job of interpreting a statistically significant overall test is not over when we have identified one or more linear combinations of the measures for which F_{eff} is statistically significant. As in Anova, we still need to specify for each such dependent variable one or more specific contrasts identifying particular ways in which the means on that dependent variable differ from each other. The fact that we have accepted maximization of F_{eff} as a goal implies that we wish to consider general contrasts, rather than only pairwise differences. F_{eff} is a constant times the maximized F_{contr} and thus considers an infinitely larger set of contrasts than just pairwise comparisons. Were our interest confined to pairwise differences among the means, we would employ the studentized range statistic as our overall test for any given measure or linear combination thereof. Thus our critical value for any and all post hoc contrasts on any and all linear combinations of the measures is obtained by the Scheffé adjustment of the critical value for F_{eff} on such linear combinations of the measures, namely

fully post hoc contrast on differences among means on
post hoc linear combination of measures

$= df_{eff} \cdot$ (post hoc c.v. for F_{eff})

$= df_{eff} \cdot (df_{err}/df_{eff})[\theta_\alpha(s, m, n)/(1 - \theta_\alpha(s, m, n))]$

$= df_{err} \cdot \theta_\alpha(s, m, n)/[1 - \theta_\alpha(s, m, n)]$.

7.2.4 Application to Within-Subjects Factors

We have restricted our attention thus far to situations in which the Anova we would conduct on any one of the measures involves only between-subjects factors. Given the power of within-subjects designs, this might appear to be a major limitation. However, it is readily transcended by incorporating the measures at all levels of the within-subjects factor(s) as part of the vector of p outcome measures, then restricting the linear combinations of measures considered in maximizing F to those that represent contrasts among the within-subjects factor(s). The test of the grand mean with respect to the various linear combinations that represent a particular within-subjects effect then provides a test of that effect, while the test of any other between-subjects effect with respect to this same restricted set of linear combinations of the outcome measures provides a test of the *interaction* between the between-subjects and the within-subjects effects.

As applied to the simplest case, that of a single within-subjects effect assessed at each level by a single dependent variable, the number of levels of the within-subjects factor will equal the total number of measures for each sampling unit, p. The overall test of the within-subjects factor is accomplished by finding the linear combination of the p measures that yields the maximum value of $F_{gm} = N \cdot \overline{Y}^2 MS_{err}$, subject to the additional constraint that the combining weights sum to zero (so that they constitute a contrast among the levels of the within-subjects factor).

Actually, putting such an additional constraint on the maximization procedure is much less efficient than an alternative approach in which the set of p measures are transformed into $p - 1$ contrasts among those original measures. We may call this set of $p - 1$ contrasts the *basis contrasts* for the analysis. They need not be mutually orthogonal, but no one of them can be a perfect linear combination of the others. Then, since any linear combination of a set of contrasts among the p original measures will itself be a contrast among those measures, we can apply the characteristic-root-and-vector procedure to find the linear combination of the basis contrasts that maximizes F_{gm}. The optimal contrast among the p levels of the within-subjects factor is then computed by applying the combining weights for the $p - 1$ basis contrasts to the definition of each contrast as a linear combination of the p measures. If the ith row of the $p \times (p - 1)$ matrix **C** contains the contrast coefficients for the ith basis contrast and the $(p - 1)$-element column vector **a** contains the coefficients that define the optimal linear combination of the basis contrasts, then the matrix product **C a** yields the contrast coefficients defining the optimal contrast among the p levels of the within-subjects factor.

Since $df_{\text{eff}} = 1$ (and thus $s = 1$) here, max F_{gm} can be compared directly to

$$p^*(df_{\text{err}})F_\alpha(p^*, df_{\text{err}} - p^* + 1)/(df_{\text{err}} - p^* + 1),$$

where $p^* = p - 1$ and $df_{\text{err}} = N - 1$. Thus the critical value for this particular case of a single within-subjects factor and no between-subjects factors is given by

$$[(p - 1)(N - 1)F_\alpha(p - 1, N - p)]/(N - p).$$

This is also the fully post hoc critical value for follow-up tests of any particular contrast among the levels of the within-subjects factor. The F_{contr} for any particular contrast using this approach is very closely related to the F_{contr} as computed in the univariate approach because F_{contr} under the multivariate approach equals the ratio,

$$\frac{SS_{\text{contr}}}{\text{normalized variance of that contrast across subjects}}.$$

The *normalized variance* of the contrast is the variance of the contrast after its coefficients have been divided by $\sqrt{\Sigma c_j^2}$ so as to have unit length. Thus the normalized variance of the contrast is $(1/\Sigma c_j^2)$ times the variance across subjects of its numerical value. By comparison, F_{contr} under the univariate approach equals the ratio, $SS_{\text{contr}}/MS_{\text{Subjects} \times \text{Treatments}}$, where it can be shown that $MS_{\text{Subjects} \times \text{Treatments}}$ is the simple average of the normalized variances of any $p - 1$ mutually orthogonal contrasts among the levels of the within-subjects factor. So far as computation of F_{contr} for particular contrasts is concerned, then the only and very crucial difference between the univariate and multivariate approaches is that the Manova approach tests each contrast against its particular variance across subjects, while the Anova approach tests it against an average, pooled variance of all contrasts.

The generalization to multiple within-subjects factors is straightforward. Each within-subjects effect (main effect or interaction) is represented by a set of non-linearly-dependent contrasts equal in number to the degrees of freedom for that effect. Sets of contrasts to represent interactions are most easily computed as all possible products of the main-effect contrasts for the within-subjects factors involved in that interaction. The overall test of the within-subjects effect involves comparing max F_{gm}, where the maximization is across all possible linear combinations of the set of contrasts representing that effect, to the appropriate percentile of the sampling distribution of such a maximized F. The critical value for this maximized F is the same as for the single-within-subjects-factor case, except that p^* is replaced by df_w, the degrees of freedom for the within-subjects effect being tested.

The generalization to mixed between- and within-subjects designs is also straightforward. Each within-subjects effect is once again represented by a set of non-linearly-dependent contrasts, and that effect is tested by comparing max F_{gm} to the same critical value as above. Each between-subjects effect is tested by computing F_{eff}, using the simple sum or average of all measures for a given subject as the dependent variable. Finally, each interaction between a within-subjects effect and a between-subjects effect is tested by comparing max F_{eff}, maximized with respect to all possible linear combinations of the set of within-subject basis contrasts representing the within-subjects effect, to

$$(df_{err}/df_{eff})[\theta_\alpha(s, m, n)/(1 - \theta_\alpha(s, m, n))],$$

where $s = \min(df_{eff}, df_w)$, $m = (|df_{eff} - df_w| - 1)/2$, $n = (df_{err} - df_w - 1)/2$, and θ_1 is, the greatest characteristic root of $\mathbf{E}^{-1}(\mathbf{E} + \mathbf{H})$. If $s = 1$ (in this context, if $df_{eff} = 1$), this critical value can be computed from F c.v.s as

$$[(df_{eff})(df_w)/(df_{err} - df_w + 1)] \cdot F_\alpha(df_w, df_{err} - df_w + 1).$$

For instance, the test of the $A \times B \times C \times D$ interaction, where C and D are within-subject effects, involves maximizing F_{AB} with respect to all possible linear combinations of a set of $(df_C)(df_D)$ contrasts where scores on each interaction contrast are obtained as the product of scores on one of the C-effect contrasts with scores on one of the D-effect contrasts.

The generalization to doubly multivariate designs is similarly straightforward. Now the test of any given within-subjects effect or its interaction with a between-subjects effect requires consideration of a set of within-subjects contrasts for *each* dependent variable in the battery of measures administered to subjects at each combination of levels of the within-subjects factor. The effective value of p for significance tests is now $p_{battery}(df_w)$, where $p_{battery}$ is the number of (non-linearly-dependent) measures in the battery of measures collected from each subject at each combination of levels of the within-subjects factors. Thus we arrive at the most general of the critical values for max F_{eff}, maximized with respect to all $p_{battery}(df_w)$ basis contrasts computed on the various measures in the battery:

$$\left(\frac{df_{err}}{df_{eff}}\right)\left[\frac{\theta_\alpha(s, m, n)}{1 - \theta_\alpha(s, m, n)}\right],$$

where

$$s = \min(df_{eff}, p_{battery}df_w),$$
$$m = (|df_{eff} - p_{battery}df_w| - 1)/2,$$

and

$$n = (df_{err} - p_{battery}df_w - 1)/2.$$

This simplifies, for single-df between-subjects effects, to

$$\frac{(df_{err})(df_w)(p_{battery})F_\alpha(p_{battery}df_w, \, df_{err} - p_{battery}df_w + 1)}{(df_{err} - p_{battery}df_w + 1)}.$$

The follow-up to any significant overall multivariate test will include testing specific contrasts across the levels of the between-subjects factor for specific linear combinations of the basis contrasts on the various dependent variables in the battery against the above critical value multiplied by df_{eff}.

7.2.5 Computerized Manova

The amount of hand calculation involved in carrying out a Manova could be prodigious.

The first level of assistance in carrying out Manovas is provided by matrix-manipulation packages such as Pro-Matlab (Moler, Little, Banger, and Kleiman, 1990) or PROC IML within SAS (1990). (PROC MATRIX was used in SAS editions through Version 5.) Such packages allow the user to carry out matrix multiplication, matrix inversion, and computation of characteristic roots and vectors via simple commands that are very similar in structure to the matrix-algebraic equations corresponding to those operations. For example, the Pro-Matlab command for finding the characteristic roots and vectors needed for the overall test of a given effect in a Manova is

⟨roots,vecs⟩ = eig(inv(E)*H).

Detailed application of the matrix algebra was described in Section 7.2.2. However, these matrix-manipulation systems cannot compete with the fully canned procedures incorporated in the major nationally distributed statistical packages for ease of data entry, simplicity of commands to request a wide variety of subanalyses within the overall Manova, and integration of the Manova input and output with other analyses of the same data in the same computer run.

An important benefit of learning to use a matrix-manipulation program is its use in converting the discriminant functions reported by Manova programs for analyses involving within-subjects factors [which invariably are expressed as linear combinations of the basis contrasts specified (or defaulted to) for the analysis] to the optimal contrast among the levels of the repeated-measures factor. Even though this would be a trivial additional step to incorporate into these programs, to date no statistical package has taken this step.

Readers are warned that some matrix-manipulation systems (*not* including Pro-Matlab, but true of SAS PROC IML through Version 6) restrict their computation of characteristic roots and vectors to *symmetric* matrices, which $E^{-1}H$ almost never is (though each of H and E is). Moreover, such systems do not always warn about this limitation on line, but instead proceed to provide erroneous characteristic roots and vectors if one requests same for an asymmetric matrix. Carefully check the user's manual for the system, or run a test problem to determine whether the system being considered can handle the general eigenvalue problem.

The emphasis will be on SPSS's MANOVA program, with supplementary discussion of SAS's PROC GLM and BMDP's BMDP4V program. Sample setups have been checked by running on Version 4.0 of SPSS and Version 6 of SAS.

7.2.5.1 Basic Setup for SPSS MANOVA

The basic setup to accomplish a Manova via SPSS's MANOVA program is

```
MANOVA measure list BY bsf1 (bl1a,bl1b) bsf2 (bl2a,bl2b)/
   WSFACTORS = wsf1 (wl1) wsf2 (wl2)/
   CONTRAST (wsf2) = SPECIAL (1 1 1, 1 0 −1, 1 −2 1)/
   WSDESIGN  = wsf1, wsf2, wsf1 BY wsf2/
   CONTRAST (bsf2) = POLYNOMIAL (1,2,4,8)/
   PRINT = TRANSFORM CELLINFO (MEANS) DISCRIM (RAW
      STAN) SIGNIF (UNIV HYPOTH) ERROR (COV)/
   DESIGN/DESIGN = bsf1, bsf2(1), bsf2(2), bsf2(3),
      bsf1 by bsf2(1), bsf1 by bsf2(2), bsf1 by bsf2(3)/
```

In the above, uppercase terms are MANOVA subcommands and specifications and lowercase terms are inputs that must be supplied by the user. In particular:

measure list is a list of all $p_{\text{battery}} \cdot wl_1 \cdot wl_2$ measures obtained from each sampling unit that are involved in this analysis.

*bsf*1, *bsf*2, etc. are the names of variables (already read in) that indicate what level of one of the between-subjects factors was assigned to or measured for that subject; and (*bl1a*, *bl1b*) provides the lower and upper bounds of the levels of this factor. For example, "ANX (2,5)" indicates that between-subjects factor ANX takes on the values 2, 3, 4, and 5. These values must be consecutive integers and must have been read in for each subject, recoded (via the RECODE command) from another variable that was read in, or constructed in some other way from the variables that were read in.

*wsf*1 is the name given within-subjects factor 1 and is used only within the MANOVA program.

(wl1) gives the number of levels of within-subjects factor 1. Unlike the between-subjects factors, the within-subjects factor names are *not* names of variables read in for each subject but are internal to the MANOVA program. The levels of each within-subjects factor are inferred from the structure of the *measure list* by counting, with dependent variable moving most slowly, then wsf1, then wsf2, etc. Thus, for instance, if our set of 24 measures on each subject comes from having measured the same battery of 4 dependent variables at each of the 6 combinations of one of the 3 levels of ANX with one of the 2 levels of task difficulty, the "WSFACTORS = ANX (3) TD (2)" specification tells SPSS that the measure list has the structure

V1A1TD1, V1A1TD2, V1A2TD1, . . . , V4A3TD1, V4A3TD2.

This is the portion of the MANOVA setup you are most likely to get wrong the first time around, so Rule Number 1 when using MANOVA with within-subjects factors is to *always* (without fail) include a PRINT = TRANSFORM request. This will produce a list of the basis contrasts actually employed by MANOVA, from which any lack of fit between the ordering of your *measures list* and the order in which the WSFACTORS are listed will be immediately apparent.

CONTRAST(*wsfi*) must come between WSFACTORS and WSDESIGN. Since the SIGNIF (UNIV) request prints the univariate F_{contr} for each of the basis contrasts and for its interaction with each of the between-subjects effects, it is important to spell out the particular contrasts in which one is most interested. However, control over the basis contrasts one wishes to employ is lost if the contrasts specified are not mutually orthogonal, since SPSS MANOVA will, in that case, insist on "orthogonalizing" them— usually by retaining one of the contrasts which has been specified (not necessarily the first) and selecting additional contrasts orthogonal to that one. In earlier editions of MANOVA one could get around the totally unnecessary restriction by selecting one of MANOVA's keyword-specified sets of contrasts, such as

CONTRAST (wsfactor) = SIMPLE,

which tested every other level of the within-subjects factor against the first level. In the version available at this writing, however, MANOVA even "corrects" its own internally generated sets of contrasts. Again, *always* be sure to request PRINT = TRANSFORM to be able to know to what the UNIVARIATE Fs and discriminant function coefficients refer.

No such restriction applies to contrasts among levels of between-subjects factors, although such a restriction would make some sense there, since mutually orthogonal contrasts do additively partition SS_{eff} for between-subjects effects but not for within-subjects effects.

Finally, the first, unadorned DESIGN subcommand requests a standard factorial breakdown into main effects and interactions, while the second, more involved DESIGN subcommand requests tests of the specific $bsf2$ contrasts. [ANX(2) for example, refers to the second single-df contrast specified in the CONTRAST subcommand for ANX.]

7.2.5.2 Supplementary Computations

Although SPSS's MANOVA program is detailed, it does *not* directly provide any of the following:

1. The maximized F_{eff}
2. The critical value to which max F_{eff} is to be compared (and thus also the fully post hoc critical value for F_{eff} computed on any linear combination of the measures)
3. F_{eff} for various post hoc linear combinations of the measures
4. In the case of within-subjects factors, the optimal contrast among the levels of that within-subjects factor
5. For doubly multivariate designs, the optimal linear combination of the measures when we restrict attention to contrasts among the levels of the within-subjects factor(s) with respect to a particular linear combination of the dependent-variable battery.

These additional bits of information are easily, if sometimes tediously, computed by hand calculations and/or supplementary computer runs.

First, max F_{eff}, i.e., the value of F_{eff} computed on the discriminant function (the optimal linear combination of the measures): MANOVA reports ROYS CRITERION, which is max $SS_{eff}/(SS_{eff} + SS_{err}) = \theta_1$. This is easily converted to max $F_{eff} = (df_{err}/df_{eff})\theta_1/(1 - \theta_1)$. Some programs (at this writing, SAS's PROC GLM among them) report an EIGENVALUE that is max $SS_{eff}/SS_{err} = \lambda_1$, from which max $F_{eff} = (df_{err}/df_{eff})\lambda_1$. Comparison of max F_{eff} to the largest of the UNIVARIATE F's (i.e., the largest F_{eff} yielded by any one of the measures) tells one whether there has been any *multivariate gain*, i.e., whether the optimal combination of the measures yields an F_{eff} substantially larger than that attainable by using any single measure.

Second, the critical value against which max F_{eff} is compared to test the overall H_0: this is computed from $(df_{err}/df_{eff})\theta_{crit}/(1 - \theta_{crit})$, where $\theta_{crit} = \theta_a(s, m, n)$ is obtained from tables of the greatest characteristic root distribution (Harris, 1985) or by running a separate computer program, such as the one listed in Harris (1985, Appendix B). The BMDP4V program for carrying out Manova reports the p value associated with θ_1. Even with BMDP4V, however, the explicit critical value for max F_{eff} is needed for use as the fully post hoc critical value for follow-up F_{eff}'s computed on particular linear combinations of the measures.

Third, just as a significant F_{eff} in a univariate Anova should be followed up by specific contrasts to determine the sources of rejection of the overall H_0, a statistically significant θ_1 must be followed up by tests of F_{eff} on particular linear combinations of the measures. Each of these tests of the overall F_{eff} for a particular linear combination of the measures will then, if significant, be followed up by F_{contr}'s on the significant dimension of difference thus identified. The follow-up F_{eff}'s should include a test of each of the original measures and tests of one or more simplified versions of the discriminant function. By simplified version we mean the linear combination of the measures implied by one's interpretation (verbal description) of the absolutely optimal linear combination (the discriminant function). For instance, a discriminant function of $.179Y_1 - 3.216Y_2 + 4.713Y_3$ might be interpreted as essentially the difference between Y_3 and Y_2 and thus be followed up with a test of F_{eff} computed on the simplified discriminant function (single new or emergent dependent variable) $Y_3 - Y_2$. If one is not interested in exploring linear combinations but, e.g., is interested only in testing each of the individual measures and/or a relatively few a priori linear combinations thereof, the test of the overall multivariate null hypothesis should be omitted in favor of Bonferroni-adjusted univariate F_{eff}'s. (See Section 7.1.3.)

The F_{eff}'s on the original measures are provided by the UNIVARIATE F's requested by the

PRINT = . . . SIGNIF (UNIV . . .)

subcommand. The value of SS_{eff} for each of these F_{eff}'s also appears in the corresponding main diagonal entry of the HYPOTHESIS SUM OF SQUARES AND CROSS PRODUCTS matrix printed because of the

PRINT = . . . SIGNIF (. . . HYPOTH)

request, while the value of MS_{err} appears in the corresponding main diagonal entry of the ERROR COVARIANCE MATRIX printed because of the

PRINT = . . . ERROR (COV)

request. More generally, F_{eff} computed on the linear combination $\mathbf{Ya} = a_1 Y_1 + a_2 Y_2 + \cdots + a_p Y_p$ (where Y_i is the ith measure specified in the *measures list*) can be computed as $(\mathbf{a'Ha}/df_{\text{eff}})/(\mathbf{a'Ea}/df_{\text{err}})$, where $\mathbf{H}$ is the HYPOTHESIS SSCP matrix and $\mathbf{E}/df_{\text{err}}$ is the ERROR COVARIANCE MATRIX. Alternatively, one can force the computation of F_{eff} by carrying out a second SPSS run in which one computes scores on the desired linear combination via, e.g.,

COMPUTE DFSIMP = Y3 − Y2

and then names this newly created variable as one of the measures in the measures list of a new run through the MANOVA program.

Fourth (relevant only when the measures represent the levels of a within-subjects factor) is computation of the optimal contrast among the levels of that within-subjects factor. Remember that we handle each within-subjects effect by converting the original measures to a set of *basic contrasts* representing that within-subjects effect. Unfortunately, the DISCRIMINANT FUNCTION COEFFICIENTS reported by SPSS (and, presently, all Manova programs) give the combining weights for the optimal linear combination of these basis contrasts, *not* for the optimal contrast among the levels of the within-subjects factor, trivial though such a computation would be as a program add-on. To find the contrast that yields the maximum possible F_{eff} for that within-subjects effect (remembering that F_{gm} computed on a within-subjects contrast yields the overall test of that within-subjects contrast, while any other F_{eff} computed on the within-subjects contrast yields a test of the interaction between the between-subjects effect and that within-subjects contrast), one must apply the discriminant function coefficients provided by the printout to the definitions of the various basis contrasts (pulled from the transformation matrix requested in the "PRINT = TRANSFORM" specification) as linear combinations of the levels of the within-subjects factor(s) and then collect terms. The description is actually more complex than the process. If your basis contrasts are

$$T1 = .707Y_1 + 0Y_2 - .707Y_3 \quad \text{and}$$

$$T2 = .408Y_1 - .816Y_2 + .408Y_3,$$

then a discriminant function of

$$.306T1 - .159T2$$

translates into an optimal contrast of

$$[.306(.707) - .159(.408)]Y_1 + (-.159)(-.816)Y_2$$

$$+ [.306(-.707) - .159(.408)]Y_3 = .151Y_1 + .130Y_2 - .281Y_3.$$

Once one has computed the optimal contrast, one can simplify it to a readily interpretable form—probably in this example to a $(1,1,-2)$ contrast among the levels of this within-subjects factor. One can then obtain the F_{eff} for the simplified contrast in one of the two ways discussed above. Computation of the optimal contrast can also be accomplished by a bit of matrix algebra, namely

$$\mathbf{c}' = \mathbf{a}'T,$$

where $\mathbf{c}'$ is a p-element row vector containing the optimal contrast coefficients, $\mathbf{a}'$ is a $(p - 1)$-element row vector of discriminant function coefficients, and **T** is a $(p-1) \times p$ matrix with each row providing the definition of one of the basis contrasts as a linear combination of the original measures.

Fifth, when the measures list is doubly multivariate, one must first apply the procedure just outlined to find the optimal linear combination of the original measures. This optimal linear combination, however, will almost certainly *not* have what Harris (1985, Section 5.4.2) refers to as a repeated-battery structure; i.e., it will not be perfectly reproducible as a contrast across the levels of the within-subjects factor(s) with respect to a single linear combination of the dependent variables. For example, if our six basis contrasts had been

$$T1 = .707Anx_1 + 0Anx_2 - .707Anx_3,$$

$$T2 = .408Anx_1 - .816Anx_2 + .408Anx_3,$$

$$T3 = .707Dep_1 + 0Dep_2 - .707Dep_3,$$

$$T4 = .408Dep_1 - .816Dep_2 + .408Dep_3,$$

$$T5 = .707Euph_1 + 0Euph_2 - .707Euph_3, \text{ and}$$

$$T6 = .408Euph_1 - .816Euph_2 + .408Euph_3,$$

where the subscripts refer to the three levels of our "Measurement Occasion" within-subjects factor, then a discriminant function of

$$.714T1 - .013T2 - .492T3 + .219T4 + .064T5 + .443T6$$

translates into the combination

$$.4995Anx_1 + .0106Anx_2 - .5101Anx_3 - .2585Dep_1 - .1787Dep_2$$
$$+ .4372Dep_3 + .2260Euph_1 - .3615Euph_2 + .1355Euph_3.$$

This involves a different contrast across measurement occasions for each dependent measure [for example, $(.50, .01, -.51)$ for *Anx* but $(.23, -.36, .14)$ for *Euph*] and is thus much more difficult to interpret than would be a contrast across the three measurement occasions with respect to a single linear combination of the measure in the dependent-variable battery. Although there are a number of ways to approach finding a repeated-battery breakdown that approximates this optimal linear combination of the nine measures, the most systematic approach begins with a reordering of the nine coefficients into a 3×3, dependent variable $\times$ occasion table, **X** =

	Anx	Dep	Euph
1	.4995	−.2585	.2260
2	.0106	−.1787	−.3615
3	−.5101	.4372	.1355

We now select two three-element column vectors, c and a, so that $X^* = c \, a'$ closely approximates X. c provides the coefficients for a contrast representing change across measurement occasions, while a provides combining weights for a linear combination of the dependent measures. Strong candidates are c = the characteristic vector associated with the largest eigenvalue of XX' and a = the characteristic vector associated with the largest eigenvalue of $X'X$. In the current example the characteristic vectors are $c = (-.6296, -.1355, .7651)$ and $a = (.8044, -.5940, -.0118)$, i.e., a $(-.63, -.14, .77)$ contrast across occasions in the magnitude of $.8Anx - .59Dep - .01Euph$. This is interpretable as essentially the change from Occasion 1 to Occasion 3 in the tendency for subjects' reported anxiety to be greater than their reported depression.

This provides a close approximation to the optimal linear combination of the three basis contrasts on the three dependent variables in that it maximizes the sum of the three $SS_{Occasion}$'s (one for each dependent variable) computed as if the coefficients in X were means based on identical cell sizes. This is, however, not as important as the use of the technique to suggest a repeated-battery interpretation of the optimal combination of the measures to which θ_1 refers and should by no means preclude examining other, less formally derived interpretations. This procedure is a straight-forward extension of a corresponding technique for finding that contrast of contrasts that most closely approximates the optimal interaction contrast in factorial univariate Anova (Boik, 1981).

7.2.5.3 Setups for SAS PROC GLM and BMDP4V

The SAS Version 5 Edition of PROC GLM has all the tools one needs to apply Manova to any of Conditions 1–4 of Section 7.1.3. For Condition 1 (no within-subjects factors), an example of the required setup is

```
PROC GLM;
    CLASS vetgen agegrp ;
    MODEL a1 a2 a3 a4 a5 a6 a7
        = vetgen agegrp vetgen*agegrp ;
    MEANS vetgen agegrp vetgen*agegrp ;
    MANOVA H = _ALL_ ;
    TITLE 'vetgen, agegrp effects on after-movie mood' ;
```

where *vetgen* and *agegrp* are equivalent to SPSS MANOVA's *bsfi* factor names and must, like them, be read in or otherwise created as SAS variables before the PROC GLM command. There is no provision for establishing the range of values on each between-subjects factor within GLM, so before entering GLM one must declare as missing all values of the variables that are not to be included in the Manova. Note, too, that SAS indicates interaction effects with "*", rather than BY. The MEANS statement requests tables of means relevant to the specified main effects and interactions; $a1$ through $a7$ are the dependent variables. The output includes a univariate Anova summary table for each dependent variable as well as the discriminant functions and the usual four multivariate overall tests for each between-subjects effect specified on the MODEL statement.

For Conditions 2 and 3 (where measures for each subject consist entirely of scores on a single dependent variable at every combination of the levels of one or more within-subjects factors), an example setup is

```
PROC GLM;
    CLASS vetgen agegrp ;
    MODEL a1 b1 a2 b2 a3 b3 a4 b4 a5 b5 a6 b6 a7 b7
        = vetgen agegrp vetgen*agegrp / NOUNI ;
    REPEATED moodmeas 7 PROFILE , befaft 2 / NOU PRINTM
    PRINTRV
                        SHORT SUMMARY ;
    TITLE '2 rep-measures factors';
```

In this example, $a1$ through $a7$ are mood measures obtained after a movie, $b1$ through $b7$ are premovie measures, and the battery of seven mood measures are treated as the levels of a within-subjects factor called *moodmeas*. Note that, as in SPSS MANOVA, the first-named within-subjects factor varies most slowly. The NOUNI option on the MODEL statement tells GLM not to print univariate Anova summary tables for each of the 14 original measures. If these summary tables *are* wanted, NOUNI should be omitted. *PROFILE* on the REPEATED statement requests that adjacent-difference contrasts (level i − level $i + 1$) be used as the basis contrasts for the *moodmeas* factor. (Various other options, including POLYNOMIAL contrasts, are available.) The options appearing after the slash on the REPEATED statement and their meanings follow.

NOU: Do not print the univariate-approach Anovas on the within-subjects effects.

PRINTM: Print the contrast coefficients that define the basis contrasts (an important check, though GLM appears to give the contrasts requested without orthogonalizing them).

PRINTRV: Print the eigenvalues and associated discriminant function coefficients. Without this specification one gets only the overall tests, with no indication of what optimal contrast got you there.

SHORT: Print relatively short messages as to how the various multivariate overall tests were computed.

SUMMARY: Provide a univariate Anova summary table for each of the within-subjects basis contrasts.

For Condition 4 (doubly multivariate design), one must explicitly construct within-subjects contrasts with respect to each measure in the dependent-variable battery. The SPSS MANOVA "kluge" of specifying a within-subjects factor with a number of levels that is a multiple of the number of measures in the *measures list* does not work here. An example setup is

```
PROC GLM ;
   CLASS vetgen agegrp ;
   MODEL b1 b2 b3 b4 b5 b6 b7 a1 a2 a3 a4 a5 a6 a7
      = vetgen agegrp vetgen*agegrp / INT ;
   MANOVA H = _ALL_ M = b1 − a1, b2 − a2, b3 − a3, b4 − a4,
      b5 − a5, b6 − a6, b7 − a7 PREFIX = befaft / SHORT
      SUMMARY;
   MANOVA H = _ALL_ M = b1 + a1, b2 + a2, b3 + a3, b4 + a4,
      b5 + a5, b6 + a6, b7 + a7 PREFIX = sum / SHORT SUMMARY;
   TITLE 'Doubly-multiv analysis, own befaft contrasts' ;
```

In this setup the order in which the measures is named on the MODEL statement is irrelevant. The INT option is highly relevant in that it requests that, in addition to the between-subjects factors and their interactions, the grand mean be tested. Two MANOVA statements are necessary to be able to use the PREFIX option (which labels the basis contrasts as BEFAFT1, BEFAFT2, etc., and the transformed variables used for testing purely between-subjects effects as SUM1, SUM2, etc.). The SUMMARY option on each MANOVA statement requests a univariate Anova summary table for each transformed variable.

Crucial in all of the above are the semicolons, which are statement separators in SAS. Their omission is the most common and most subtle error made in setting up SAS analyses.

The setup for a BMDP4V analysis of a study involving two between-subjects and two within-subjects factors is illustrated in Appendix B. In that listing, CODES are the values of variables read into the program that define the levels of the between-subjects factors; NAMES are alphabetic labels for the levels of either kind of factor; $a1$ through $a7$ are the answers to questions assessing after-movie mood; $b1$ through $b7$ are the answers to those same questions before viewing the movie; and the DESIGN FACTOR sets of statements establish various within-subjects contrasts as the basis contrasts for the *question* factor.

While the setup for the BMDP4V program is quite different from the other two (using BMDP's paragraph format), the output is very similar to that from SPSS MANOVA, the major exception being that BMDP4V provides a *p* value for the g.c.r. statistic.

7.3 NUMERICAL EXAMPLES

It is not possible to demonstrate each design to which Manova is applicable. We will first examine a relatively straightforward factorial Manova with no within-subjects factors, then analyze a study involving two within-subjects factors and a multiplicity of between-subject error terms.

7.3.1 A Study of Person Perception

Harris, Harris, and Bochner (1982) had separate groups of subjects evaluate eight stimulus persons (all named "Chris Martin") having several characteristics in common but differing in whether the said person was male or female, of average weight or overweight, and a wearer of glasses or not. The evaluation was in terms of 11 adjective pairs, with 7 different ratings available to indicate how close "Chris" was to one or the other end of the continuum defined by each adjective pair. We thus have a 2 × 2 × 2 factorial Manova with an outcome vector consisting of 11 dependent variables. A portion of the SPSS MANOVA run that was used for the initial analysis of these data is reproduced in Appendix A. Preceding the portion listed there, a number of data preparation commands (explained in the *SPSS Reference Guide*) defined the format for reading the data in, provided labels for variables and the values thereof, and converted all adjective-pair responses so that a high score represented a check near the more socially desirable adjective. Then came the essence of the analysis, the MANOVA command, which followed the format defined in Section 7.2.5 rather closely.

Since all responses had the same range, it was decided to base interpretation of the discriminant functions on the raw-score discriminant-function coefficients [whence the "DISCRIM (RAW)" specification.] The "OMEANS" specification requests the printing of marginal means for each of the three between-subject factors.

Next in the listing comes a portion of the output from this initial analysis, with a focus on the results relevant to the main effect of OBESity. The portion reproduced in Appendix A was preceded by a listing of both weighted and unweighted means for the main-effect tables requested by "OMEANS." Since we have by default adopted the full-model (regression, Type III) approach to this unequal-n design, the unweighted means are actually being analyzed by Manova and should therefore be the means examined when interpreting our results. The unweighted mean for the Overweight level of OBES, for example, is the simple (unweighted) average of the four cell means that are relevant to that level, i.e., of the (M,Ov,Gl), (M,Ov,NoGl), (F,Ov,Gl), and (F,Ov,NoGl) cell means.

The "ERROR (COV)" specification yielded the "WITHIN-CELLS variances and covariances" matrix; only the first four rows and columns of this 12×12 matrix are reproduced in Appendix A. This matrix is ($1/df_{err}$) times the **E** matrix common to all effects in this all-fixed-factors design. Each main-diagonal entry is MS_{err} for the corresponding dependent variable.

SPSS MANOVA prints the tests of the highest-order interactions first, then works its way down to the main effects. The printout for the OBES main effect begins with the **H** matrix ("Adjusted Hypothesis Sum-of-Squares and Cross-Products"), whose main-diagonal entries are SS_{eff} for the OBES factor as computed for a single dependent variable. (Only the first four rows and columns of this 12×12 matrix are reproduced in Appendix A.) Reading this matrix and the **E** matrix into a matrix-manipulation package provides an efficient way of exploring various simplifications (interpretations) of the OBES discriminant function. We will shortly demonstrate an alternative approach.

Next come the "Multivariate Tests of Significance." In this situation, where all between-subjects effects are single-df effects and thus $s = 1$, all four multivariate criteria lead to the same conclusion and can be converted to exactly the same F-distributed variable. No need, then, to go in search of a g.c.r. table or program.

Directly below the multivariate tests come the "Univariate F-tests with (1,146) D.F.". These univariate Fs are self-explanatory—and too often they are also the sole follow-up to the overall multivariate test. Note that the p values listed there are for fully a priori tests. If these 11 univariate Fs encompass the sum total of one's interest in the effect of the average weight/overweight manipulation, one can control familywise alpha by using $F_{\alpha/11}(11,46)$, which equals 8.307, 11.470, and 16.207 at the .05, .01, and .001 levels, respectively, as the critical value for each of the Fs. In that case, one should ignore the overall multivariate test (since that provides familywise protection for F_{OBES}'s on all possible linear combinations of the 11 dependent variables) and consider the overall, multivariate hypothesis of no OBES main effect rejected if and only if one or more univariate Fs exceeds the Bonferroni-adjusted critical value. Assuming that one *is* interested in exploring linear combinations, the appropriate critical value for the above Fs is $[11(146)/136]F_{\alpha}(11,136)$, which equals 21.96, 28.11, and 36.42 at the .05, .01, and .001 levels, respectively. This is the same critical value we employ for F_{OBES}'s computed on various linear combinations of the dependent variables, such as the discriminant function defined by the "Raw discriminant function coefficients" and simplifications of that discriminant function. The discriminant function coefficients for OBES tell us that $.129X1 - .665X2 + .153X3 + \cdots - .365X11 - .028SEXAPP$

yields the highest F_{OBES} of any linear combination of the dependent variables. The numerical value of that maximized F_{OBES} (i.e., of F_{OBES} calculated on the discriminant function) is given by $(146/1)(.55097/.44903) =$ 179.145. This is considerably larger than the F_{OBES} of 89.58 yielded by our most effective single dependent variable, X2. It seems reasonable to begin our search for linear combinations of the measures that are especially effective in differentiating between average-weight and overweight stimulus persons by composing a verbal description of the new, emergent variable represented by the discriminant function, computing F_{OBES} for the linear combination of the measures implied by that "simplified discriminant function," and comparing that F_{OBES} both to our fully post hoc c.v. and to max F_{eff}. One is free to explore several interpretations of the discriminant function and to test any other linear combination of the measures that seems interesting. A procedure that often yields promising interpretations is to assign a weight of $+1$ or -1 to each dependent variable (sign matching the sign of that variable's discriminant function coefficient) that has one of the highest discriminant function coefficients (in absolute value) and a weight of 0 to each of the other dependent variables. Where to draw the line between high and negligible discriminant function coefficients is somewhat subjective, akin to applying the scree test to eigenvalues in a principal components or factor analysis. Starting with the highest absolute value of any discriminant function coefficient, one keeps assigning $+1$'s and -1's to variables having lower and lower absolute discriminant function coefficients until there is a sharp drop in absolute value and/or the ratio between the absolute value of this coefficient and the largest absolute value gets to be larger than, say, 2 or 3. In the present case this procedure leads to considering

X5 + X9 − X2 − X8 − X11

or, if we "drop down" to absolute value of .139,

X3 + X5 + X9 − X2 − X6 − X8 − X11.

Once we learn that the overweight stimulus persons are rated higher on this combination than are average-weight stimulus persons, the first of these is readily interpretable as a tendency for the overweight to be seen as more outgoing and popular than one would expect, given how inactive, physically unattractive, and unathletic they are perceived to be. The second simplified discriminant function adds "Intelligence" to the positive side of this description and "Happy" to the negative (predictor) side.

This use of $+1$'s and -1's makes sense where, as in the present case, the dependent variables are essentially bipolar in nature, so it is simply a matter of which end of the scale is being referenced. When the variables are essentially unipolar, it is often better to take the additional step of dividing each +1 weight by the number of such weights and each −1 weight by their number, i.e., to interpret the discriminant function as the difference between the averages of two subsets of the dependent variables.

To test the adequacy of these two interpretations, make a second run through SPSS, preceding the MANOVA program with COMPUTE statements as follows:

COMPUTE DFOBES = X5 + X9 − X2 − X8 − X11
COMPUTE DFOBES2 = X3 + X5 + X9 − X2 − X6 − X8 − X11
. . . .
COMPUTE DFOBES7 = .129*X1 − .665*X2 + .153*X3 + .037*X4 + .312*X5 − .139*X6 − .656*X8 + .440*X9 − .059*X10 − .365*X11 − .028*SEXAPP
COMPUTE DFSPECS = X5 + X9 − X3 − X4 − X11
. . . .
COMPUTE LOADOB3 = X1 + X2 + X4 + X6 + X8 + X9 + X10 + X11 + SEXAPP
MANOVA DFOBES TO LOADOB3 BY SEXT (1,2) OBES (3,4) SPECS (4,5) /
    PRINT = DISCRIM (RAW) SIGNIF (UNIV HYPOTH)
       ERROR (COV) OMEANS(TABLES(SEXT OBES SPECS))/
DESIGN

This run included several other versions of the simplified discriminant function for OBESity, as well as interpretations of the SEXT and SPECS discriminant functions and (in LOADOBES, LOADOB2, and LOADOB3) linear combinations that would arise from interpreting the OBES discriminant function on the basis of the *loadings* of each variable on (correlation of each variable with) the discriminant function. (More about this in Section 7.4.4.) DFOBES yielded a univariate F_{OBES} of 143.29, while DFOBES2 yielded an F_{OBES} of 121.55. One should not make too much of this difference in univariate F's, but focus on whichever interpretation makes more practical or theoretical sense—or use both.

The SPECS main effect yielded no single dependent variable that was statistically significant by the fully post hoc criterion (though glasses wearers' tendencies to be rated as less outgoing and popular than nonwearers yielded F_{SPECS}'s of 14.55 and 13.66, respectively, significant at the .01 familywise level by the Bonferroni-adjusted criterion), but glasses wearers showed a significantly greater tendency than did nonwearers to be rated

Table 7.1 Summary of Multivariate Tests, FFEAF Study

Source	df	$\theta(1,4.5,67)$	$F(11,136)$	max F_{eff}
Sex of Target (SEXT)	1	.1499	2.18*	25.75
Obesity (OBES)	1	.5510	15.17***	179.14
Glasses (SPECS)	1	.2154	3.39***	40.08
SEXT × OBES	1	.0651	<1	10.16
SEXT × SPECS	1	.0748	1.00	11.81
OBES × SPECS	1	.1194	1.68	19.80
SEXT × OBES × SPECS	1	.0571	<1	8.84

*$p < .05$.
***$p < .001$.

as assertive, intelligent, hardworking, athletic, and sex-role-consistent but inactive, reserved, unattractive, and unpopular ($F_{SPECS} = 32.98$).

Despite the statistically significant overall multivariate test for SEXT, no readily interpretable linear combination of the dependent variables quite reached fully post hoc statistical significance, although the tendency for males, relative to females, to be rated as more active, hardworking, and

Table 7.2 Means, Univariate F's, and Discriminant Function Coefficients for Obesity Main Effect, FFEAF Study[a]

Variable	Average weight	Over-weight	Univ. F	Discr. funct. coefficient
X1 (Assertive)	3.82	3.62	1.24	.129
X2 (Active)	4.69	3.10	89.58***	−.665←
X3 (Intelligent)	4.86	4.75	.68	.153
X4 (Hardworking)	4.52	4.01	9.59	.037
X5 (Outgoing)	4.37	4.19	.84	.312←
X6 (Happy)	4.95	4.59	3.36	−.139
X8 (Attractive)	4.85	3.79	50.66***	−.656←
X9 (Popular)	4.89	4.64	2.88	.440←
X10 (Successful)	4.71	4.48	2.31	−.059
X11 (Athletic)	4.29	2.78	59.57***	−.365←
SEXAPP	1.73	.98	5.31	−.026

[a]Note: F_{Obes} for discriminant function = 179.145; for the linear combination X5 + X9 − X2 − X8 − X11 it's 143.289.
***$p < .001$ by fully post hoc (g.c.r.-based) criterion. (F of 21.96 required for significance at .05 level.)

attractive but less intelligent, popular, and sex-role-consistent yielded an F_{SEXT} of 20.48. Bird and Hadzi-Pavlovic (1983) suggest that in such cases it is best to treat the statistical significance of the effect as unestablished because doing so very nearly eliminates the nonrobustness of the g.c.r. statistic under violation of multivariate normality (see Section 7.4.1).

Conventions for presenting the results of a Manova are not nearly as well established as those for Anova summary tables. Tables 7.1 and 7.2, however, represent one approach to presenting the overall multivariate tests and the followup tests for the Obesity factor, respectively.

7.3.2 Effects of Exercise on Mood

We will now show how to carry out a Manova when there is a mix of fixed, random, crossed, and nested effects in the between-subjects portion of a design.

Gilliland (1986) examined changes on the POMS (Profile of Mood States) after three kinds of aerobic exercise and three kinds of nonaerobic exercise. This examination yielded a very mixed design involving two within-subjects factors (POMS scale and time of measurement) and three between-subjects factors (gender, aerobic vs. nonaerobic exercise, and particular kind of aerobic or nonaerobic exercise), with the third factor nested within the second. Due to differential participation by the two genders in the various kinds of exercise, it is also an unequal-n design. Gilliland did not want to restrict her statements about results to the three particular examples of aerobic and the three particular nonaerobic exercises employed in this study, so we should think of type of exercise as a random factor. This leads to the following design for the between-subjects effects:

Source	df	Sources of variance	Error term
G: Gender	1	W, G, GX(T)	$MS_{GX(T)}$
T: Type of Exercise	1	W, T, X(T)	$MS_{X(T)}$
G × T	1	W, GT, GX(T)	$MS_{GX(T)}$
X(T): Particular Exercise	4	W, X(T)	MS_w
G × X(T)	4	W, GX(T) × D	MS_w
Subjs(GX(T))	$N - 12$	W	

Note: The above table is the basis for the DESIGN subcommand within SPSS-X MANOVA.

Each subject gets a score on each of the six subscales of the POMS before exercising (PRE), immediately after exercising (POST) and about 20 minutes later (FOLLOWUP). Assuming that the POMS subscales have a common unit of measurement (all 6 are scored as % of max possible socially desirable score), think of these 18 scores as having been generated by all possible combinations of three levels of "Time of Measurement" (TIME) and six levels of "Particular Subscale" (SCALE). Then break the 18 scores down into 18 linear combinations: the sum or average of all 18 scores (used on computing main effects of between-subjects factors), two contrasts representing the TIME main effect, five representing the SCALE main effect, and $5 \times 2 = 10$ contrasts representing SCALE $\times$ TIME. Putting this within-subjects structure together with the between-subjects portion of the design leads to the following SPSS MANOVA setup:

```
TITLE CHANGES IN POMS IN AEROBIC VS. NONAEROB
   EXERCISE
DATA LIST FREE / ID EXCAT AEROB GENDER TA1 TA2 TA3
   DEP1 DEP2 DEP3
ANG1 ANG2 ANG3 VIG1 VIG2 VIG3 FAT1 FAT2 FAT3 CONF1
   CONF2 CONF3
VAR LABELS EXCAT KIND OF EXERCISE /
TA1 TENSION ANXIETY,PRE / TA2 TENSION ANXIETY,
   POST /
TA3 TENSION ANXIETY,FOLLOWUP / DEP1 DEPRESSION,
   PRE /
DEP2 DEPRESSION, POST / DEP3 DEPRESSION, FOLLOWUP /
ANG1 ANGER, PRE / ANG2 ANGER, POST / ANG3 ANGER,
   FOLLOWUP /
VIG1 VIGOR, PRE / VIG2 VIGOR, POST / VIG3 VIGOR,
   FOLLOWUP /
FAT1 FATIGUE, PRE / FAT2 FATIGUE, POST / FAT3
   FATIGUE, FOLLOWUP /
CONF1 CONFUSION, PRE / CONF2 CONFUSION, POST /
CONF3 CONF, FOLLOWUP /
VALUE LABELS EXCAT (1) RUNNER (2) RACE WALKER (3)
   SWIMMER
(4) RACQUETBALL (5) WT LIFTERS (6) WALKERS /
AEROB (1) AEROBIC (2) NON-AEROBIC / GENDER (1)
   MALE (2) FEMALE
/* SINCE VIGOR IS THE ONLY POSITIVELY WORDED */
   /* SCALE, ITS SCORING */
/* IS REVERSED SO AS TO HAVE HIGH SCORES */
   /* REPRESENT NEGATIVE */
/* MOODS FOR ALL 6 SCALES.        */
DO REPEAT X1 = VIG1,VIG2,VIG3 /
```

```
X2 = VIGR1,VIGR2,VIGR3 /
COMPUTE X2 = 19 - X1
END REPEAT
COMPUTE XRC = EXCAT
IF (EXCAT GT 3) XRC = EXCAT - 3
VAR LABELS XRC EXCAT WITH NESTED LEVEL LABELS /
VALUE LABELS XRC (1), RUN, RACQUET (2) RACE WALK,
   WT LIFT (3) SWIM, WALK /
MANOVA TA1,TA2,TA3,DEP1,DEP2,DEP3,ANG1,ANG2,
   ANG3,VIGR1,VIGR2,VIGR3
FAT1,FAT2,FAT3,CONF1,CONF2,CONF3
  BY GENDER (1,2) AEROB (1,2) XRC (1,3) /
WSFACTORS = SCALE (6) TIME (3) /
CONTRAST (TIME) = SPECIAL (1 1 1
        1 -1 0
        0 1 -1) /
CONTRAST (SCALE) = SPECIAL (1 1 1 1 1 1
        1 1 1 -5 1 1
        -3 2 -3 0 2 2
        -1 0 1 0 0 0
        0 1 0 0 1 -2
        0 1 0 0 -1 0)/
WSDESIGN = SCALE TIME SCALE BY TIME /
RENAME = AVE VIGVOTH MPM0PP M0P000 ZP00PM ZP00M0,
   PREPOST,
POSTFOLL, SCTM1,SCTM2,SCTM3,SCTM4,SCTM5,SCTM6,
   SCTM8, SCTM9,SCTM10 /
PRINT = TRANSFORM DISCRIM (RAW STAN) ERROR (SSCP) /
NOPRINT = SIGNIF (DIMENR) /
METHOD = SSTYPE (UNIQUE) /
ANALYSIS (REPEATED) /
DESIGN = GENDER VS 2, AEROB VS 1, GENDER BY AEROB
   VS 2, XRC WITHIN AEROB = 1 VS W,
   GENDER BY XRC WITHIN AEROB = 2 VS W /
```

7.4 ISSUES AND DEVELOPMENTS

The reader has thus far been exposed to the author's approach to applications of Manova. In Sections 7.4.1 through 7.4.4 we present some ways in which other researchers' practices differ from this approach. Since each issue has been discussed in more detail elsewhere, we will provide only brief summaries in this chapter.

7.4.1 Greatest-Characteristic-Root Versus Multiple-Root Tests

Up to this point in our chapter, all tests of the overall null hypothesis have been based on the sampling distribution of the maximum F_{eff} attainable from any linear combination of the measures, or the monotonic transforms of F_{eff}, θ_1 (the maximized ratio of SS_{eff} to $SS_{eff} + SS_{err}$) and λ_1 (the maximized ratio of SS_{eff} to SS_{err}), with θ_1 being the statistic whose critical values are available in published tables. However, whenever $s = \min(p, q)$ is greater than 1, i.e., whenever the between-subjects effect involved has more than a single degree of freedom, we also get nonzero values of θ_2, $\theta_3, \ldots, \theta_s$. Each of these additional (and smaller) θ's represents a separate dimension of differences among the levels of the between-subjects effect. Manova's status as a special case of Canona is useful here in making clear that the contrasts that yield the maximum F_{contr} for each of the discriminant functions associated with the θ_i's are, in fact, mutually orthogonal. (Those optimal contrasts are canonical variates for the set of level-membership variables in the Canona representation of Manova.)

It has been extremely tempting to combine all s θ_i's into a single test statistic on which to base the test of the overall multivariate null hypothesis. Each of the major computer programs for Manova reports three *multiple-root* tests in addition to the (single-root) g.c.r. statistic: Pillai's trace, $\Sigma \lambda_i$; Hotelling's trace, $\Sigma \theta_i$; and Wilks' lambda, $\Pi(1 - \theta_i)$. Unlike the g.c.r. statistic, each of these multiple-root statistics can be readily transformed to a statistic that is distributed as the familiar F ratio and thus requires no special tables of critical values. Furthermore, the multiple-root tests provide more powerful tests of the overall null hypothesis than does θ_1 when the corresponding population roots are not greatly different (but less powerful tests when the first population root is much larger than the second population root). Finally, there is evidence that the multiple-root tests are more robust against violation of multivariate normality than is the g.c.r. test (Olson, 1974, 1976).

Why, then, has the present chapter ignored the multiple-root tests? The strong preference for the g.c.r. test flows from consideration of what an overall test provides. This author takes the position that the role of the overall test is *not* really to tell us whether or not the overall, multivariate null hypothesis of a precisely zero population effect on every one of the measures is true (we know a priori that it cannot be). If you doubt this, ask yourself whether you truly believe that *any* overall test would continue to be nonsignificant if you employed a sample size of 10 or 20 billion. The purpose of the overall test is to tell whether or not there are any particular linear combination(s) of the measures for which there is enough evidence to be able to tell the direction in which that (those) combination(s) departs from the null hypothesis. This logic leads inexorably to adoption of Roy and Bose's (1953) union-intersection criterion, of which the g.c.r. statistic and the Scheffé post hoc critical value for Anova contrasts are examples. It is useless to conclude that we have enough evidence to reject the overall multivariate H_0 if we cannot also specify one or more of the infinite number of linear combinations of the measures that should be considered statistically significant on a post hoc basis; and *that* requires concentrating on max F_{eff} (and θ_1), rather than on the suboptimal linear combinations indexed by the other θ_i's. One might consider using a linear combination of all s discriminant functions, but a linear combination of the discriminant functions is itself reducible to a linear combination of the original variables, and, by definition, the first discriminant function (the one associated with θ_1) yields the largest F_{eff} of any such combination. If we are interested in identifying particular ways in which particular linear combinations of the measures display the between-subjects effect, rather than solely in the meaningless exercise of demonstrating the obvious fact that no population effect is truly zero, we have little choice but to base our overall test and follow-ups thereto on the g.c.r. statistic, despite the need this presents for special tables of or computer subroutines for computing its critical values and despite the evidence of its lack of robustness to severe violations of multivariate normality. Bird and Hadzi-Pavlovic, however, show that this nonrobustness is very nearly eliminated if we withhold judgment as to statistical significance until and unless at least one simplified, interpretable version of the discriminant function has been shown to be statistically significant by the g.c.r.-based fully post hoc critical value.

Further details and references on this issue are provided in Section 4.5 of Harris (1985).

7.4.2 Univariate Versus Multivariate Approaches to Within-Subjects Anova

This chapter has presented only the multivariate approach to testing within-subjects effects. The core of the multivariate approach is that each within-subjects contrast is tested against its own variance across subjects. The univariate approach, on the other hand, ignores any differences among the variances (inversely related to the intersubject reliabilities) of different contrasts, using instead an average normalized variance as the error term for every contrast. This makes clear that the univariate approach is valid if and only if all within-subjects contrasts have identical population variances, a most unlikely circumstance. In repeated-measures designs, where the levels of the within-subjects factor represent successive measurements, it is very likely that recent behavior will be more predictive of (more highly correlated with) current behavior than will behavior more distant in time, so that the variance of the difference between pairs of measures will be an increasing function of their separation in time. We can also expect very heterogenous contrast variances when individual patterns of response fit a common functional form, but with individual differences in parameters. For instance, Harris (1980) and Harris and Joyce (1980) find that individual recommendations as to how much of a group's profits each participant should receive are very well fit by the assumption of a linear relationship between individual contributions and recommended outcomes, but with considerable variation from subject to subject in the steepness of the slope of this linear relationship (i.e., in how much that subject wishes to differentiate between the lowest-input and highest-input partners). As a result, the linear trend contrast usually has a variance across subjects that is over a *hundred* times larger than the variance of other trend contrasts. As a result the univariate-approach F for linear trend is close to four times larger than warranted by its intersubject reliability and four times larger than the F_{contr} computed under the multivariate approach. Conversely, the F_{contr} for, say, cubic trend computed under the univariate approach is more than 25 times smaller than its intersubject reliability warrants, while the multivariate-approach F_{contr} for cubic trend is a much more accurate reflection of the mean value of that contrast relative to its variability across subjects.

There is available a relatively simple method of correcting the univariate overall F test by adjusting the degrees of freedom of the critical value to which it is compared so as to yield an overall test that has very nearly the right alpha level. However, attempts to "patch up" the univariate approach in order to take into account degree of departure from the assumption of identical contrast variances overlook the following:

1. The *df*-correction approach just mentioned applies *only* to the overall test and is of no use in adjusting the F's for specific contrasts. In fact, it corrects in the wrong direction the F_{contr}'s for those contrasts having below-average variance, bringing the actual alphas for those contrasts even farther below their nominal alphas.
2. While the corrected overall F has the right alpha level, it is almost irrelevant to the subsequent specific F_{contr}'s. In particular, the corrected univariate overall F can be statistically significant even though *no* specific F_{contr} can be found that is statistically significant, and (worse) there could be one or more contrasts that would be highly statistically significant if tested against their own variance across subjects, even though the corrected univariate overall F does *not* achieve statistical significance. Some authors have suggested using the corrected univar-

iate overall F as the overall test but then switching to the multivariate approach to testing specific contrasts if and only if the corrected univariate overall F achieves significance. (See, e.g., Chapter 13 of Maxwell and Delaney, 1990, where this is discussed as a viable alternative when sample size is low.) As this point (2) makes clear, this procedure is often not much better than simply proceeding to correctly computed F_{contr}'s if and only if a uniformly distributed random number yields a value less than .05. It is a potentially enormous and unnecessary waste of power.

7.4.3 Full Model Versus Sequential Analyses

Within SPSS MANOVA, one can let the "METHOD = " specification default to SSTYPE(UNIQUE) or explicitly request "METHOD = SSTYPE(SEQUENTIAL)." Other Manova programs incorporate a similar distinction (for instance, Type I versus Type II sums of squares in SAS). What is addressed by this distinction is best seen by considering Manova as a special case of Canona and generalizing the testing of univariate Anova effects as increments to multiple R^2 to the testing of Manova effects as increments to canonical R^2. Much has been written on alternative ways of analyzing unequal-n factorial Anova designs. The distinctions among these approaches are primarily a matter of what sort of increment to R^2 is involved: the increment yielded (1) when the contrasts representing this between-subjects effect are the first predictors considered in the regression analysis, (2) when they are the last predictors added to the model, or (3) when each between-subjects effect is tested in terms of its increment to R^2 when added at a particular point in an a priori ordering of effects.

Remember that each alternative approach tests a different set of contrasts among the levels of the various between-subjects effects. Any effect tested in terms of its increment when added first involves contrasts among *weighted* means, each weighted by its cell size. Such a test is equivalent to carrying out a one-way Anova on that factor, disregarding confounds with other factors.

When a given effect is tested on the basis of its increment when added last, the contrasts being tested involve the *unweighted* means. These unweighted means are estimates of the level means that would have been obtained had a balanced, equal-n design been employed. The corresponding contrasts test the given effect free of confounds with other between-subjects effects.

When the sequential approach to testing effects in unbalanced factorial designs is employed, the contrasts being implicitly examined in the tests of all but the first and last effects entered into the regression equation involve comparisons of very complicated weighted averages of the cell means that are almost certainly not of any interest to the researcher (Kirk, 1982). SPSS MANOVA's advice in this respect is well put. In its current edition, whenever SSTYPE(SEQUENTIAL) is requested and the cell sizes are unequal, the program prints out a warning that the solution matrix should be consulted for the actual hypotheses being tested. This can be an eye-opening experience for those who have been using sequential analyses mistakenly believing they knew what effects they were testing.

Realistically, then, the choice among approaches reduces to testing each effect uncorrected for all others (i.e., when the contrasts representing that effect are the sole predictors) or corrected for all other effects (i.e., when added to the equation last, or equivalently in the case of single-df effects, when tested in terms of the statistical significance of the regression coefficient for that predictor). The first approach amounts to doing separate one-way Anovas, rather than a factorial analysis. The only appropriate use of sequential approaches is in trying to develop a simpler Anova model by determining whether higher-order interactions can be dropped from the model without serious loss of predictability. Even here, however, once the number of terms to be retained has been determined, tests of the effects that are retained should be based on the increment to R^2 each effect provides when added last.

7.4.4   Scoring Coefficients Versus Loadings as the Basis for Interpreting Discriminant Functions and Canonical Variates

Our discussion of the process of interpreting discriminant functions emphasized simplifying the discriminant function coefficients, since these define the discriminant function as a linear combination of the original variables. This, however, is inconsistent with the predominant practice in factor analysis of interpreting factors on the basis of the correlation of each original variable with a given factor (equivalent, for orthogonal factors, to the *loading* of the variable on the factor), rather than on the factor scoring coefficients that indicate how scores on a given factor are estimated from scores on the original variables. This practice has, moreover, been extended explicitly to Manova and Canona by Huberty (1984) and Thompson (1985).

The scoring coefficients (discriminant function coefficients, canonical variate coefficients) tell us how to compute scores on this new, emergent variable (discriminant function, canonical variate) as a function of their scores on the original variables. These scoring coefficients are thus the proper basis for trying to infer from the properties of the original variables what this new variable substantively represents. The correlation of a given measure with the discriminant function, on the other hand, tells us only how well we could predict scores on the discriminant function if we "tied our hands behind our backs" and used only scores on this single variable to estimate scores on the discriminant function. It does *not* tell us what role this measure plays in defining the profile of scores on all measures that defines the discriminant function. Each such correlation is, then, an answer to a *univariate* question that ignores the relationships among the various measures, rather than an answer to the multivariate question of what relationship among the measures yields the maximum evidence against the overall null hypothesis. In fact, when $s = 1$, the correlation of each measure with the discriminant function equals the square root of the F_{eff} computed on that measure and is completely independent of what other measures are included in the analysis.

Harris, Harris, and Bochner (see Section 7.3.1) found that subjects judged average-weight target persons more favorably than they did obese target persons on every adjective-pair, with several of the univariate F's being statistically significant by a post hoc criterion. Had they interpreted the discriminant function on the basis of loadings, they would have concluded that the major difference between ratings of average-weight and obese stimulus persons is a uniformly negative stereotype of the obese. The discriminant function as defined by the discriminant function coefficients, however, indicated that the stereotype of the obese that best accounts for the obese/average weight difference is of someone who is *more* outgoing and popular than one would expect on the basis of his or her low levels of physical attractiveness, activity, and athleticism. This simplified discriminant function yielded a much higher $F_{obesity}$ than did the simple sum of the measures implied by interpreting the correlations between single adjective pairs and the discriminant function. Similarly, Harris (1989) presents a dramatic (though hypothetical) example in which interpreting canonical variates on the basis of structure coefficients (correlations between measures and canonical variates) leads to a misguided attempt to match male and female Beefy-Breasted Bowery Birds on the basis of total head length (which correlates only .46 between the genders), whereas perfect matching is attainable if you use instead the canonical variate coefficients and thus match on the basis of (total head length − skull length) = beak length. A number of more direct arguments and counterarguments on this issue are discussed in the latter pages of this paper. See also Harris (1985, Example 6.5, pp. 282ff) for an example of how an attempt to interpret principal components can be seriously hindered by reliance on loadings, rather than scoring coefficients, as the basis of the interpretation.

7.4.5   When 4 > Infinity: Bonferroni-Adjusted Versus Fully Post Hoc Critical Values

As stated strongly in Section 7.1.3, one should *not* carry out a full Manova on a within-subject factor or factors unless interested in finding the optimal contrast among the levels of that within-subjects factor(s). That statement was correct but potentially misleading because one aspect of the Manova treatment (its test of each within-subjects contrast against its own variance across subjects, rather than against the average variance represented by $MS_{S \times T}$) should always be retained even if one is only going to test a number of a priori within-subjects contrasts. Moreover, one may also wish to retain the fully post hoc critical value from the Manova (even if one does not compute max F_{contr}), because it may be less stringent than using a Bonferroni-adjusted critical value based on $\alpha_i = \alpha_{target}/n_t$. [For the case in which you wish to examine only pairwise differences among the levels of your within-subjects factor Keselman, Keselman, and Shaffer (1991) recommend use of the lower critical values provided by the studentized range or the studentized maximum modulus critical value. However, those statistics were derived for the independent-means case and their application to the repeated-measures case leads to inflated Type I error rates if the sphericity assumption is violated. Kesselman et al. find this inflation to be slight for ε values as low as .40.] This is highly similar to the situation in univariate Anova where the fully post hoc Scheffé critical value for contrasts may be less stringent than a Bonferroni-adjusted critical value when n_t is large and, in fact, always is less stringent if you wish to consider all possible subset contrasts (the difference between the simple average of one subset of the means and the simple average of another subset of the means). Two aspects of the situation with respect to within-subjects contrasts bear emphasizing:

1.   The fully post hoc critical value for within-subjects contrasts can be lower than a Bonferroni-adjusted critical value for some surprisingly small values of n_t. For example, when $df_{err} = 20$ and there are $p = 3$ correlated means to be examined, a Bonferroni-adjusted critical value for $n_t = 4$ contrasts among these three means is given by $F_{.05/4}(1,20) = 7.532$, while the Manova-based fully post hoc critical value is given by $[2(20)/(20 - 2 + 1)]F_{.05}(2,19) = 7.415$. Thus, providing familywise protection for an infinite number of comparisons "costs less" than providing this protection for only four contrasts. This cannot hold for the true critical values for these two procedures, but is an artifact of the conservatism of Bonferroni-adjusted procedures when the tests employed are not mutually othogonal. One may use whichever of the two critical values is smaller.

2. Unlike the independent-means situation, the Manova-based, fully post hoc critical value may *not* be less stringent than a Bonferroni-adjusted critical value for examination of all possible subset contrasts. This is obviously true when $p > df_{err}$, since in this case the fully post hoc, Manova-based c.v. is undefined (or, if you prefer, infinite). It is also true for some situations where $p < df_{err}$. For example, if $df_{err} = 10$ and $p = 6$, the Manova-based critical value is given by $[5(10)/(10 - 5 + 1)]F_{.05(5,6)} = 36.56$. There are a total of 301 possible subset comparisons among 6 means, but we can carry out even 400 contrasts under the protection of Bonferroni adjustment with a less stringent c.v., since $F_{.05/400}(1,10) = 36.50$.

The moral is clear. Even if one is not interested in the full panoply of advantages conferred by a Manova of one's within-subjects factor(s), one should always compute both a Bonferroni-adjusted and the fully post hoc critical value and use whichever one is smaller. Furthermore, do *not* assume that in testing all possible subset comparisons among your p means you are better off using the fully post hoc critical value rather than a Bonferroni-adjusted one.

7.5 CONCLUDING REMARKS

1. Manova has two contributions to make:
    (a) It allows for control over the inflation of Type I error rate that could otherwise result from analyzing *sets* of measures, and
    (b) it identifies linear combinations of measures that reveal more about the differences among one's independent groups or among the levels of one's within-subjects factor(s) than does any single dependent variable or a priori contrast.
2. Contribution (1b) is exploratory. If you prefer to deal solely with single variables or a priori linear combinations thereof, control of familywise error rate is provided by Bonferroni-adjusted critical values at considerably less cost to the power of your tests.
3. The gain in discriminative power from employing Manova—that is, the difference between the F yielded by the discriminant function and the best single original variable is often dramatic.
4. Taking advantage of this gain, however, requires carrying out specific comparisons on at least one simplified, interpretable version of the discriminant function and demonstrating
    (a) that it is statistically significant when compared against the same critical value we used for the maximum possible F, and

(b) that it differentiates among the levels of the effect being tested almost as powerfully as does the multidecimal version of the discriminant function.

5. This exploration is often very rewarding in terms of suggesting new, emergent variables worthy of further study in their own right. However, we throw away this benefit if we attempt to interpret our discriminant function in terms of its correlations with single original measures. For single-*df* effects these correlations are directly proportional to the univariate F's on the original variables.

6. As points 4 and 5 imply, our assessment of the virtues of alternative approaches to designs involving multiple measures must focus on how naturally they lead to specific comparisons and on their performance as bases for specific comparisons. Relative power or robustness of the omnibus test of the overall null hypothesis is of minor concern if such comparisons would lead us to choose an omnibus test that bears no clear relationship to specific comparisons (such as the epsilon-adjusted univariate-approach overall F for repeated measures) or that provides low power for these specific comparisons (such as multiple-root tests of the overall multivariate null hypothesis).

APPENDIX A: OUTPUT FROM SPSS MANOVA ANALYSIS OF PERSON PERCEPTION STUDY

Ellipses (. . .) indicate omitted portions of the output.

MANOVA X1 TO X6, X8 TO X11, SEXAPP BY SEXT (1,2) OBES (3,4) SPECS (4,5)/

PRINT = CELLINFO (MEANS) DISCRIM (RAW) SIGNIF (UNIV HYPOTH)

ERROR (COV) OMEANS(TABLES(SEXT OBES SPECS))/

DESIGN

. . .

WITHIN CELLS Variances and CoVariances

	X1	X2	X3	X4
X1	1.25793			
X2	.42877	1.07783		
X3	.05580	.16553	.79470	
X4	.17615	.29706	.31961	1.07307

. . .
*********ANALYSIS OF VARIANCE -- DESIGN 1
EFFECT .. OBES
Adjusted Hypothesis Sum-of-Squares and Cross-Products

	X1	X2	X3	X4
X1	1.56443			
X2	12.29012	96.55097		
X3	.91789	7.21096	.53855	
X4	4.01314	31.52720	2.35463	10.29471
. . .				

Multivariate Tests of Significance (S = 1, M = 4 1/2, N = 67)

Test Name	Value	Exact F	Hypoth. DF	Error DF	Sig. of F
Pillais	.55097	15.17021	11.00	136.00	.000
Hotellings	1.22700	15.17021	11.00	136.00	.000
Wilks	.44903	15.17021	11.00	136.00	.000
Roys	.55097				

Note.. F statistics are exact.

. . .

EFFECT .. OBES (Cont.)
Univariate F-tests with (1,146) D.F.

Variable	Hypoth.	SS	Error SS	F	Sig. of F
X1	1.56443	183.65721	1.24366		.267
X2	96.55097	157.36287	89.57921		.000
X3	.53855	116.02646	.67768		.412
X4	10.29471	156.66775	9.59373		.002
X5	1.34741	235.08715	.83681		.362
X6	5.04236	218.94828	3.36237		.069
X8	43.40135	125.09256	50.65527		.000
X9	2.26299	114.64644	2.88187		.092
X10	2.03553	128.41282	2.31432		.130
X11	87.48823	214.41004	59.57408		.000
SEXAPP	21.54485	592.53410	5.30864		.023

. . .

EFFECT .. OBES (Cont.)
Raw discriminant function coefficients

Variable	Function No. 1
X1	.12919
X2	−.66540
X3	.15259
X4	.03673
X5	.31171
X6	−.13853
X8	−.65579
X9	.43954
X10	−.05911
X11	−.36531
SEXAPP	−.02758

APPENDIX B: BMDP4V SETUP TO ANALYZE MOVIE MOOD STUDY

VARIABLE NAMES ARE vetgen,agegrp,b1,b2,b3,b4,b5,b6,b7
         a1,a2,a3,a4,a5,a6,a7./
PRINT CELLS. MARGINALS = 1./
BETWEEN FACTORS = vetgen, agegrp.
  CODES(1) = 1,2,3.
  CODES(2) = 1,2.
  NAMES(1) = malevet,malenvet,femnvet.
  NAMES(2) = le35,gt35./
WITHIN FACTORS = occasn, question.
  USE = a1,a2,a3,a4,a5,a6,a7, b1,b2,b3,b4,b5,b6,b7.
  NAMES(1) = befmovies,allexps.
  NAMES(2) = quest1,quest2,quest3,quest4,quest5,quest6,quest7./
WEIGHTS BETWEEN ARE EQUAL. WITHIN ARE EQUAL./
END/
    Data go here.
END/
DESIGN FACTOR = question. TYPE = within, regression.
  CODE = READ.
   VALUES = −3, 2, 2, 2, 0, −3, 0.

```
NAME = agitvsdep./
DESIGN FACTOR = question.
VALUES = 1, 0, 0, 0, 0, -1, 0.
NAME = depvconf./
   . . .
ANALYSIS PROC = FACTORIAL. MULT.
EVAL. VECT.DISP./
FINISH/
```

REFERENCES

Bird, K. D. and Hadzi-Pavlovic, D. (1983). Simultaneous Test Procedures and the Choice of a Test Statistic in MANOVA. *Psych. Bull.*, *93*: 167–178.

Boik, R. J. (1981). Testing the Rank of a Matrix with Applications to the Analysis of Interaction in ANOVA. *J. Amer. Statist. Assoc.*, *81*: 243–248.

Dixon, W. J. (ed.). (1985). *BMDP Statistical Software, 1985 Printing*. University of California Press, Berkeley.

Gilliland, S. (1986). Effects of Exercise on Mood: Differences among Six Types of Exercise. Unpublished Senior Honors thesis, University of New Mexico.

Harris, M. B., Harris, R. J., and Bochner, S. (1982). Fat, Four-Eyed, and Female: Stereotypes of Obesity, Glasses, and Gender. *J. Appl. Soc. Psych.*, *12*: 503–516.

Harris, R. J. (1980). Equity Judgments in Hypothetical, 4-Person Partnerships. *J. Exp. Soc. Psych.*, *16*: 96–115.

Harris, R. J. (1985). *A Primer of Multivariate Statistics*. Harcourt Brace Jovanovich, San Diego.

Harris, R. J. (1989). A Canonical Cautionary. *Multi. Behav. Res.*, *24*: 17–39.

Harris, R. J. (In press). *An Analysis of Variance Primer*. Peacock, Itasca, Illinois.

Harris, R. J. and Joyce, M. A. (1980). What's Fair? It Depends on How You Phrase the Question. *J. Pers. Soc. Psych.*, *38*: 165–179.

Horst, P. (1961). *Matrix Algebra for Social Scientists*. Holt, New York.

Hotelling, H. (1931). The Generalization of Student's Ratio. *Ann. Math. Statist.*, *2*: 360–378.

Hotelling, H. (1951). A Generalized *T* Test and Measure of Multivariate Dispersion. *Proceedings of the Second Berkeley Symposium on Mathematical Statistics and Probability*. University of California Press, Berkeley, pp. 23–41.

Huberty, C. J. (1984). Issues in the Use and Interpretation of Discriminant Analysis. *Psych. Bull.*, *92*: 505–512.

Jones, L. V. (1966). Analysis of Variance in its Multivariate Developments. *Handbook of Multivariate Experimental Psychology*. Cattell, R. B. (ed.). Rand McNally, Chicago, pp. 244–266.

Keselman, H. J., Keselman, J. C., and Shaffer, J. P. (1991). Multiple Pairwise Comparisons of Repeated Measures Means Under Violation of Multisample Sphericity. *Psych. Bull.*, *110*: 162–170.

Kirk, R. L. (1982). *Experimental Design: Procedures for the Behavioral Sciences* (2nd ed.). Brooks/Cole, Belmont, California.

Maxwell, S. E. and Delaney, H. D. (1990). *Designing Experiments and Analyzing Data: A Model Comparison Perspective.* Wadsworth, Belmont, California.

Moler, C., Little, J., Banger, S., and Kleiman, S. (1990). *Pro-MATLAB for VAX Computers—User's Guide.* The Mathworks, Inc., Natick, Massachusetts.

Olson, C. L. (1974). Comparative Robustness of Six Tests in Multivariate Analysis of Variance. *J. Amer. Statist. Assoc., 69*: 894–908.

Olson, C. L. (1976). On Choosing a Test Statistic in Multivariate Analysis of Variance. *Psych. Bull., 83*: 579–586.

Pillai, K. C. S. (1955). Some New Test Criteria in Multivariate Analysis. *Ann. Math. Statist., 26*: 117–121.

Roy, S. N. and Bose, R. C. (1953). On a Heuristic Method of Test Construction and its Use in Multivariate Analysis. *Ann. Math. Statist., 24*: 220–238.

SAS. (1990). *SAS/STAT User's Guide*, Version 6 (4th ed.). SAS Institute, Inc., Cary, North Carolina.

SPSS. (1990). *SPSS Reference Guide.* SPSS, Inc., Chicago, Illinois.

Scheffé, H. (1953). A Method for Judging all Contrasts in the Analysis of Variance. *Biometrika, 43*: 227–231.

Thompson, B. (1985). *Canonical Correlation Analysis: Users and Interpretation.* Sage, Beverly Hills, California.

Wilks, S. S. (1932). Certain Generalizations in the Analysis of Variance. *Biometrika, 24*: 471–494.

Winer, B. J., Brown, D. R., and Michels, K. M. (1991). *Statistical Principles in Experimental Design* (3rd ed.). McGraw-Hill, New York.

8

Unified Power Analysis for *t*-Tests Through Multivariate Hypotheses

RALPH G. O'BRIEN   University of Florida, Gainesville, Florida

KEITH E. MULLER   University of North Carolina, Chapel Hill, North Carolina

8.1   INTRODUCTION

Determining adequate and efficient sample sizes is often critical in designing worthy studies. Yet too many studies have sample sizes that are too small to ensure enough statistical power to confirm meaningful effects. Freiman, Chalmers, Smith, and Kuebler (1979) concluded this about clinical trials in medicine. Sedlmeier and Gigerenzer (1989) reached a similar judgment about studies in psychology. The message in both articles is cogent to all fields that rely on statistical inference. Perhaps such articles are having positive effects, for we see signs that researchers are now paying more attention to power. For example, reviewers of research proposals now often require that sound power analyses be done before they will recommend funding or access to facilities and subject populations.

Going through the process of determining and justifying the sample size also has an important ancillary effect: it catalyzes the synergism between science and statistics at the study's conception. The statistician who performs a thorough power analysis is more likely to scrutinize the proposed design, assess issues regarding data management, and develop a sound plan for the data analysis. Such involvement can improve the proposal in a number of ways, thus increasing its chance for approval, funding, scientific success, and publication.

In this chapter, we present a strategy for performing power analyses that is applicable to the broad range of methods subsumed by the classical normal-theory univariate or multivariate general linear models. First, we introduce the requisite concepts of statistical power using concepts from the familiar t-test to compare two independent group means. Second, we proceed to the comparison of two correlated means (matched-pairs problem) and on to the one-way analysis of variance (ANOVA) with contrasts for a completely randomized design. Third, we develop power analysis for the univariate general linear model, thus providing a broad range of applications. We illustrate this with an analysis of covariance (ANCOVA) problem that has unequal distributions of the covariate's values among the groups as well as heterogeneous slopes. Fourth, we broaden the range still further by outlining an approximation for determining power under the multivariate general linear model. This is illustrated with a repeated-measures problem solved by using the multivariate analysis of variance (MANOVA) approach. Fifth and finally, we outline power analysis strategies developed for other types of methods, especially for tests to compare two independent proportions.

With writings on sample-size choice and power analysis for many methods now so plentiful, why do we offer yet another one? Like Kraemer and Thiemann (1987), we present an approach that unifies many seemingly diverse methods. We think our method is intuitive, because we develop *the strong parallels between ordinary data analysis and power analysis*. To make the methods easier to use, we distribute modules of statements to direct the popular SAS® System (SAS Institute, 1990) to perform and table (or graph) sets of power computations. Our ultimate goal is to show how a single approach covers a broad class of tests.

Rather than restricting attention to the power of the traditional tests (e.g., overall main effects and interactions), our methods allow one to easily examine statistical hypotheses that are more tailored to specific research questions. Departing from most writings on statistical power, we take unbalanced designs to be the norm rather than the exception. Many effective research designs use unequal sample sizes, as when certain types of subjects are easier than others to recruit or when certain treatments are more expensive per subject to apply. Thus, researchers and statisticians must decide how the total sample size will be allocated among the different groups of cases, with a balanced allocation being a special case.

We avoid oversimplifying the concept of effect size, as researchers often do when they employ rules of thumb, such as Cohen's (1988, 1992) "small," "medium," and "large" categorizations. A tiny effect size for one research question and study could be a huge effect size in another. Researchers often claim that their studies promise "medium" effect sizes, but they have no objective grounds to justify such a claim. Our scheme forces researchers to give specific conjectures or estimates for the relevant statistical parameters, such as the population means and standard deviations for an ANOVA problem. The conjectures are then used directly to calculate effect sizes, which determine statistical power for a proposed sample size. Our detailed examples illustrate how straightforward it is to do these things. Sample-size analysis is not harder to do than data analysis; as we shall see, the two problems are very similar.

The best hypothesis-driven research proposals include quite definite plans for data analyses, plans that merge the scientific hypotheses with the research design and the data. A good sample-size analysis must be congruent with a good plan for the data analysis. There are hundreds of other common statistical procedures besides normal-theory linear models, and there are thousands of uncommon methods and an unlimited number of "customized" ones that are developed for unique applications. While we can cover power for many of the common methods for statistical inference, one chapter cannot be exhaustive of all known or possible sample-size methods.

One final general point is in order. Many treatments on statistical power choose to address questions like "If I perform this particular hypothesis test and my conjectures are true about the expected data, what sample size do I need to achieve a power of .90?" But most studies have multiple hypotheses to test, and thus this kind of sample-size analysis determines separate sample sizes for the different hypothesis tests within a single study. This is confusing. Instead, our approach addresses questions like "If I study a total of N subjects, what are the powers for the different hypothesis tests I plan to perform?" This is the way of most studies—one total sample size, several hypotheses to be examined. In those limited cases in which there really is but one key hypothesis, one is still limited by resources on how large the sample size can be. The question here becomes: "I can study, at most, N subjects. Will this give me sufficient power?" If that value for N gives "too much" power, one can spend a few minutes investigating lower values to find something more efficient.

At the end of each major section, we explore the art of power analysis by presenting realistic, detailed examples inspired by actual studies conducted within the interface of behavioral and medical research. All of these case studies are fictionalized to some degree to make them more useful in this chapter. We use bogus names for the researchers and drugs, as well as for many of the scientific terms and measurements.

8.2 UNIVARIATE t-TESTS AND THE ONE-WAY ANOVA

8.2.1 Comparing Two Independent Means

Common t-tests probably are used more frequently than any other statistical method. In comparing the population means (μ) from two independent groups, we are formally comparing a null hypothesis, H_0: $\mu_1 = \mu_2$, with a research ("alternative") hypothesis that is either nondirectional, H_A: $\mu_1 \neq \mu_2$, or directional, H_A: $\mu_1 > \mu_2$ or H_A: $\mu_1 < \mu_2$. We "assess the improbability of H_0" by measuring the observed statistical difference between the sample means with

$$t = \frac{\hat{\mu}_1 - \hat{\mu}_2}{\hat{\sigma}(1/n_1 + 1/n_2)^{1/2}}, \tag{8.1}$$

where $\hat{\mu}_j$ is the sample mean for group j, n_j is the sample size for group j, and $\hat{\sigma}$ is the pooled sample standard deviation. Using $N = n_1 + n_2$, we can reexpress the t statistic as

$$t = N^{1/2}\{(w_1 w_2)^{1/2}[(\hat{\mu}_1 - \hat{\mu}_2)/\hat{\sigma}]\}, \tag{8.2}$$

where $w_j = n_j/N$ is the proportion of cases in group j. Obviously, $w_2 = 1 - w_1$. We have grouped the terms to show the distinct components of this t statistic. We call the term $(\hat{\mu}_1 - \hat{\mu}_2)/\hat{\sigma} = \hat{\psi}$ the estimate of *nature's effect size*. $\hat{\psi}$ is used often in meta analysis to measure the relative difference between the sample means. The structure of the design defines the term $(w_1 w_2)^{1/2}$; thus $(w_1 w_2)^{1/2}\hat{\psi}$ is an estimate of the effect size specific to that structure. Factoring in $N^{1/2}$ gives us t. We will soon see how the t and its components translate directly into their counterparts in power analysis.

If H_0 is true, and the observations are independent, and they are distributed as normal random variables having the same variance in both groups, then t is a *central t* random variable with $N - 2$ degrees of freedom for error (df_E), denoted here as $t(df_E)$. Let t_α be the upper-tail critical value, satisfying $\alpha = \Pr[t(df_E) \geq t_\alpha]$. Thus, α is the Type I error rate. Directional tests are based on t_α (or $-t_\alpha$, depending on how H_A is defined). Nondirectional tests can use $t_{\alpha/2}$ and $-t_{\alpha/2}$. This is identical to using $F = t^2$ and $F_\alpha = t_{\alpha/2}^2$, which is a form that better unifies our discussion throughout. While strict normality usually does not hold, the practical consequences of the central limit theorem allow us to suppose that t is still distributed as $t(df_E)$ for many nonnormal situations.

8.2.1.1 Directional Research Hypothesis

First we discuss the power of the directional test, and we define it using H_A: $\mu_1 > \mu_2$. The power is simply the rejection rate, $\Pr[t \geq t_\alpha]$, when H_0 is false. (When H_0 is true, the rejection rate is simply α.) Power is dependent on the *noncentrality* value,

$$\delta = N^{1/2}\{(w_1 w_2)^{1/2}[(\mu_1 - \mu_2)/\sigma]\}, \tag{8.3}$$

which is positive for H_A: $\mu_1 > \mu_2$. Note that δ is merely the t value that would result if one had "exemplary" data, that is, data in which $\hat{\mu}_1 \equiv \mu_1$,

$\hat{\mu}_2 \equiv \mu_2$, and $\hat{\sigma} \equiv \sigma$. Thus δ measures the statistical difference between two population means as realized with n_1 and n_2 observations.

It is helpful to see δ constructed as diagrammed in Figure 8.1. Begin with nature's effect size

$$\psi = (\mu_1 - \mu_2)/\sigma, \tag{8.4}$$

which depends on parameters related only to *what* is being studied, not *how* it is going to be studied. Because μ_1, μ_2, and σ are almost always unknown in practice, making conjectures or estimates for these values comes after considering the coordinating theory of the problem, reviewing

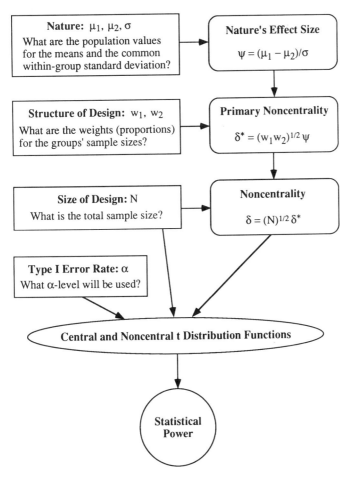

Figure 8.1 Building the noncentrality for the two-group t statistic.

the literature, and, sometimes, collecting pilot data. The *primary noncentrality* is

$$\delta^* = (w_1 w_2)^{1/2} \psi. \tag{8.5}$$

This $(w_1 w_2)^{1/2}$ factor depends only on the structure of design. Note that δ^* is maximized at $w_1 = w_2 = .5$. Factoring in the total sample size gives the noncentrality,

$$\delta = N^{1/2} \delta^*. \tag{8.6}$$

This description shows how noncentrality, which determines power, is a function of three distinct components: what nature hides from us (ψ), the structure of the design (w_1 and w_2), and the size of the sample (N).

Under the same distributional assumptions outlined above, t is distributed as a *noncentral t* random variable with $df_E = N - 2$ and noncentrality δ, which is denoted as $t(df_E, \delta)$. The central t is simply $t(df_E, 0)$. The power is

$$\Omega = \Pr[t(df_E, \delta) \geq t_\alpha]. \tag{8.7}$$

Power values can be found using a suitable computing routine, table, or graph. For example, one can use the SAS statements:

```
t_alpha = TINV(1 - alpha, dfE, 0);
power = 1 - PROBT(t_alpha, dfE, delta);
```

Note that TINV and PROBT are defined with respect to the cumulative distribution function, hence the need for "1 - alpha" and "1 - PROBT( )" to get values with respect to the upper tail.

The mean of a noncentral t random variable is approximated well by

$$\mathcal{E}[t(df_E, \delta)] \approx N^{1/2} \delta^* [(4df_E - 1)/(4df_E - 4)] \qquad \text{for } df_E > 1, \tag{8.8}$$

a result immediately deducible from the work on meta analysis by Hedges (1981). Note that the mean increases as either N or δ^* increase. This always increases the power of the test.

8.2.1.2 Nondirectional Research Hypothesis

When the research hypothesis is H_A: $\mu_1 \neq \mu_2$, it is more straightforward to use the statistic $F = t^2$, so that the split rejection region of the central t distribution is unified into the upper tail of the central F. In general, F statistics in ANOVA and linear models can be denoted as being central (if H_0 is true) or noncentral F random variables with noncentrality λ (with df_H and df_E degrees of freedom in the numerator and denominator, respectively). We denote this as $F(df_H, df_E, \lambda)$; $\lambda = 0$ defines the familiar central F. For the two-group problem, we test H_0 by taking $F = t^2$ to be

$F(1, N - 2, 0)$. In general, we define F_α to be the upper-tail critical value, satisfying: $\alpha = \Pr[F(df_H, df_E, 0) \geq F_\alpha]$.

For $F = t^2$ the noncentrality and primary noncentrality are

$$\lambda = \delta^2 \quad \text{and} \quad \lambda^* = \lambda/N = (\delta^*)^2. \tag{8.9}$$

The power for the nondirectional ("two-tailed") t-test is then

$$\Omega = \Pr[F(df_H, df_E, \lambda) \geq F_\alpha], \tag{8.10}$$

with $df_H = 1$ and $df_E = N - 2$. We can compute this using the SAS functions FINV and PROBF, as in:

F_alpha = FINV(1 - alpha, dfH, dfE, 0);

power = 1 - PROBF(F_alpha, dfH, dfE, lambda);

It helps in understanding the noncentral F to know that the expected value of any F random variable is

$$\mathcal{E}[F] = (1 + \lambda/df_H)[df_E/(df_E - 2)]. \tag{8.11}$$

As $df_E/(df_E - 2)$ is usually close to 1.0, we have

$$\mathcal{E}[F] \approx 1 + \lambda/df_H. \tag{8.12}$$

Thus with increasing λ, the distribution of F is shifted to the right, making it more likely that F will exceed F_α, thus increasing the power of the test.

Example 1: Two-Group t-Test

Dr. Seth Alalgia is planning a study as to whether yet another form of biofeedback therapy can reduce, at least in the short term, the frequency and severity of vascular headaches (see Blanchard et al., 1990). Specifically, he plans to conduct a double-blind, randomized trial in which patients will receive either an enhanced thermal (ET) biofeedback therapy or a sham placebo (SP) that simply gives noncontingent feedback to the patient. Each patient will be studied from a Monday to a Friday. On Monday morning, patients will be admitted to the university's General Clinical Research Center (GCRC) to begin identical, standard medical therapy consisting of 5 µg/kg/day of litosamine. On Monday and Tuesday evenings, patients will complete an extensive questionnaire that produces the pretreatment score on the Vascular Headache Index (PreVHI). On Wednesday morning, patients will be randomized to ET or SP groups, using the minimization scheme of Pocock and Simon (1975) to minimize group differences on PreVHI, gender, and age strata. The main behavioral therapy session will be done on the Wednesday afternoon, with a follow-up session early Thursday morning. Questionnaire data from Thursday and Friday will produce the PostVHI measure.

The main outcome measure is to be the relative change in the VHI values,

$$RC = (\text{PreVHI} - \text{PostVHI})/\text{PreVHI}.$$

PreVHI will serve as a covariate for RC, but this action is expected only to reduce error variance; the dynamic blocking minimizes group differences on the PreVHI, and no PreVHI $\times$ Group interaction is expected. Thus for our purposes here, we do not need to consider that PreVHI is to be a covariate, except when specifying σ^2. The costs for recruiting and studying ET and SP subjects are the same, so that a balanced design ($w_1 = w_2 = .50$) is optimal.

Dr. Alalgia has no pilot data. Based on his knowledge and experience, he conjectures that the ET group's true mean on RC is $\mu_{ET} = -.30$, while the SP group's is $\mu_{SP} = -.15$ (due to placebo effects). He postulates that the common within-group standard deviation is about $\sigma = .125$ because, if the data are normally distributed, about 95% of the patients' RC scores should lie within $.25 = 2\sigma$ of the group mean. (This logic helps researchers conjecture values for the standard deviation, a task difficult to do.) In continuing discussions with his statistician, Dr. Alalgia agrees that σ could possibly be 50% higher, at $\sigma = .1875$. The $\sigma = .125$ scenario is depicted in Figure 8.2.

Dr. Alalgia estimates that he has enough resources to study $N = 20$ patients. The GCRC's Scientific Review Committee hopes that a valid study can be conducted by using fewer than the 100 hospital bed-days requested (20 patients $\times$ 5 days/patient). This study is an early clinical trial of this new "enhanced" behavioral treatment, and research in biofeedback therapy has a history of conflicting findings. Thus Dr. Alalgia and his statistician feel that a directional test is unwarranted at this time; only the nondirectional test will be considered. Finally, he will concentrate

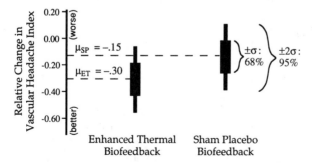

Figure 8.2 Dr. Alalgia's scenario for the two-group design.

on the powers for $\alpha = .05$ but is still curious to see the powers under $\alpha = .01$. Under these conditions, will $N = 20$ provide acceptable power? Would fewer suffice?

Listing 8.1 gives input and output related to executing *OneWyPow*, a module of standard SAS® data step statements. (See Appendix.) Here we use *OneWyPow* to get powers for Dr. Alalgia's scenario by all combinations of $N \in \{14, 20, 26, 32\}$ with $\sigma \in \{.1250, .1875\}$. The $\alpha = .05$ results for the nondirectional test (labeled "2-tailed *t*") indicate that $N = 20$ may be

Listing 8.1 Computations for Dr. Alalgia's two-group design.

Input

```
options ls=72 nosource2;
title1 "Dr. Alalgia: Enhanced Thermal vs Sham-Placebo Biofeedback";
%include OneWyPow;
cards;
mu -.30 -.15 .
weight .50 .50 .
sigma .125 .1875 .
alpha .05 .01 .
Ntotal 14 20 26 32 .
end
%include FPowTabl;
```

Selected Output

```
Effect: Two-Group Test,
DF Hypothesis: 1,
AND Primary SSHe: 0.005625
```

		Std Dev								
		0.125				0.1875				
		Total N				Total N				
		14	20	26	32	14	20	26	32	
		Pow-er	Pow-er	Pow-er	Pow-er	Pow-er	Pow-er	Pow-er	Pow-er	
Test Type	Alpha									
2-tailed t	0.05	.541	.718	.835	.907	.281	.395	.499	.591	
	0.01	.264	.445	.607	.735	.101	.172	.250	.331	
1-tailed t	0.05	.681	.825	.908	.953	.408	.530	.632	.714	
	0.01	.370	.561	.712	.819	.160	.251	.344	.434	

too few subjects to run: Ω = .72 for σ = .125, and Ω = .40 for σ = .1875. Even N = 26 gives powers of only .84 and .50 under the same conjectures. Dr. Alalgia is discouraged about these results and understands that the GCRC cannot afford to provide more than 100 bed-days for his project. He decides to restructure the design. See Example 2.

Even researchers with limited experience in statistical planning readily understand the rationale and implications of a power analysis like that just presented. The results in Listing 8.1 show concretely how power is increased by (1) increasing the α level, (2) decreasing the variance, (3) increasing the total sample size, and (4) using a directional test.

8.2.1.3 Using Data to Specify δ^*, λ^*, ψ, and ψ^2

General Comments and Directional Test

The population group means, μ_1 and μ_2, are the focus of this t-test. The population standard deviation of the observations within the groups, σ, is usually a nuisance parameter—in both a mathematical and an everyday sense. In practice, one may be forced to make a best guess or, preferably, a range of plausible guesses for them. Often it is prudent to set ($\mu_1 - \mu_2$) or ψ at some level that represents the lowest effect that would be of scientific or practical (clinical) interest. Although some researchers are uncomfortable with the subjectivity of making such choices, a well-done sample size analysis usually convinces them that statistical planning is valuable.

Often, just one part of ψ is unable to be fixed by a solid scientific conjecture or other argument. This unfixed part may be μ_1, μ_2, ($\mu_1 - \mu_2$), or σ. If one has good preliminary (pilot) data, one can get a point estimate or confidence interval for the unfixed part, then use the other, fixed parts to get an estimate or interval for δ^*. The standard methods can be used to get confidence intervals for μ_1, μ_2, or ($\mu_1 - \mu_2$). The problem comes when the unfixed part is σ^2. For example, μ_1 may be a control-group mean that is "well known" and thus fixed; μ_2 may be set by a "minimum practical (clinical) effect" argument at, say, $.80\mu_1$; and σ may be the lone unfixed parameter. If pilot data are available, an estimate for σ can be obtained using the usual pooled estimate, $\hat{\sigma}$. If those data are normally distributed, the traditional confidence limits for σ^2 can be obtained and used to define a range of σ^2 values for specifying ψ. This interval would be based on taking $(N - 2)\hat{\sigma}^2/\sigma^2$ to have a chi-square distribution with $N - 2$ degrees of freedom.

Unless the data are strictly normal, the traditional χ^2-based confidence intervals for σ^2 cannot be trusted to have their nominal confidence levels, even for infinitely large sample sizes. Somewhat better confidence intervals for σ^2 may be obtained by adapting the transform developed by O'Brien (1979, 1981) for testing variances. Simply convert the raw observations,

y_{ij}, to

$$r_{ij} = \frac{(n_j - 1.5)n_j(y_{ij} - \hat{\mu}_j)^2 - .5\hat{\sigma}_j^2(n_j - 1)}{(n_j - 1)(n_j - 2)}, \tag{8.13}$$

and use them to construct a one-mean confidence interval based on $\bar{r}$. This can be done in most statistics packages. The method works because $\bar{r} = \hat{\sigma}_j^2$, so the interval based on $\bar{r}$ is really an interval for σ^2. The method gives proper intervals when the sample size is large. By conducting Monte Carlo studies, Melton (1986) showed that while it is considerably more robust than the normal-theory method, it cannot be recommended for small N. We are not aware of acceptably robust methods for forming confidence limits on σ^2 when the sample size is small.

We can see from (8.8) that given an observed t value based on suitable pilot data, a nearly unbiased estimate of the primary noncentrality is

$$\hat{\delta}^* = (t/N^{1/2})[(4df_E - 4)/(4df_E - 1)]. \tag{8.14}$$

Thus for the two-group problem, we have the nearly unbiased estimator,

$$\hat{\psi} = \hat{\sigma}^*/(w_1 w_2)^{1/2} = [(\hat{\mu}_1 - \hat{\mu}_2)/\hat{\sigma}][(4df_E - 4)/(4df_E - 1)]. \tag{8.15}$$

Confidence limits for ψ can also be formed. It may be particularly worthwhile to find the lower limit, $\hat{\psi}_\gamma$, of a *one-tailed*, $1 - \gamma$ interval (no upper limit) by finding the noncentrality, $\hat{\delta}_\gamma$, that solves

$$\Pr[t(df_E, \hat{\delta}_\gamma) \geq t] = \gamma, \tag{8.16}$$

where t is the value observed with pilot data. This can be done using the SAS statement:

delhatgm = TNONCT(t, dfE, 1 − gamma);

$\hat{\delta}_\gamma$ may then be converted to the two forms of effect size,

$$\hat{\delta}_\gamma^* = \hat{\delta}_\gamma/N^{1/2} \tag{8.17}$$

and

$$\hat{\psi}_\gamma = \hat{\delta}_\gamma^*/(w_1 w_2)^{1/2}. \tag{8.18}$$

Using this logic allows one to say with $1 - \gamma$ confidence that the nature's effect size is *at least* $\hat{\psi}_\gamma$. Setting $\gamma = .50$ gives a median estimator that competes with the mean estimator, $\hat{\psi}$.

Either $\hat{\psi}$, $\hat{\psi}_{.50}$, or $\hat{\psi}_\gamma$ can be combined with particular N and $(w_1 w_2)^{1/2}$ values to form noncentralities and corresponding estimates of power. Suppose that a pilot study with $n_1 = 6$ and $n_2 = 4$ produces $t = 1.50$ ($p = .086$). The above formulas yield estimates for nature's effect size of $\hat{\psi} = .875$, $\hat{\psi}_{.50} = .937$, and $\hat{\psi}_{.20} = .359$. For a full study having $N = 50$ and a balanced design, the corresponding powers are $\hat{\Omega} = .920$, $\hat{\Omega}_{.50} = .948$, and $\hat{\Omega}_{.20} = .347$ for the .05-level, directional test. Thus, the researcher would see that although the mean-type and median-type estimates of ψ give powers greater than .90, there is a .20 chance that the power may be less than .35. This is hardly reassuring. If it seems erroneously low, realize that $\gamma = .20$ is close to $p = .086$, and $\hat{\psi}_{.086} = 0.00$, which corresponds to a "power" of .05. If $\gamma < p$ is used, then $\hat{\psi}_\gamma < 0$, and $\Omega < \alpha$, a result that would perplex most people.

Most power analyses are *prospective* in that they assess power for a study yet to be done, as we just discussed. *Retrospective power analyses* assess the power of a study already completed. It is often requested after a nonsignificant outcome: "What *was* my power in this (nonsignificant) study?" "How large should my *N* have been to have ensured acceptable power?" "What if I had had a more balanced design?" The estimation methods discussed here can help address these questions also.

Suppose that a study was completed with $n_1 = 17$, $n_2 = 15$, $\hat{\mu}_1 - \hat{\mu}_2 = 4.30$, and $\hat{\sigma} = 9.25$, which give $t = 1.31$ ($p = .10$). The researcher asks what the power was for a value of $\mu_1 - \mu_2 = 8.00$, which would have been the minimum difference worth seeing. Fixing $\mu_1 - \mu_2$ at 8.00 and using $\hat{\sigma} = 9.25$, the mean estimate of nature's effect size is $\hat{\psi} = 8/9.25 = .865$, which translates to a power of .771 using $\alpha = .05$ and the above sample sizes. A more conservative assessment begins by finding the lower .10 critical value for the $\chi^2(17 + 15 - 2 = 30)$ distribution, which is 20.6. Under normal theory, $\Pr[\sigma < 9.25(30/20.6)^{1/2} = 11.2] = .90$. This gives a lower 90% one-sided confidence limit of $\hat{\psi}_{.10} = 8/11.16 = .717$, which translates to a power of .630. Technically, these are powers for an exact replication of the study. If the study were to be replicated with twice the total sample size, these two values of $\hat{\psi}$ translate to powers of $\hat{\Omega} = .958$ and $\hat{\Omega}_{.10} = .875$.

If faced with the power results we have just seen, some researchers would decide that the "ongoing" study does not have enough subjects yet, so they would double the total sample size by *adding* another 32 subjects. This strategy will inflate the Type I error rate unless the whole sequence of *interim analyses* is planned at the outset using special statistical methods (see Fleming, Harrington, and O'Brien, 1984).

As the first example (with $n_1 = 6$ and $n_2 = 4$) shows, confidence intervals for ψ can be quite wide when based on small pilot studies, providing values that researchers find unreasonable when planning a new study. It is useful to use something like $\hat{\psi}_{.33}$ to estimate a lower limit for the effect size, understanding, of course, that these offer lower levels of confidence than do more standard values, such as $\hat{\psi}_{.05}$, $\hat{\psi}_{.10}$, and $\hat{\psi}_{.20}$. From the frequentist's perspective, using $\hat{\psi}_{.05}$ is more likely to lead to choosing an *N* that provides adequate power. But researchers' knowledge of the subject matter should not be discounted entirely in favor of pilot results. Bayesian strategies may one day be commonly available to mix one's subjective beliefs about ψ with estimates of it from pilot data.

Nondirectional Test

Based on (8.11), it can be shown that

$$\hat{\lambda}^* = \{[(df_E - 2)/df_E]df_H F - df_H\}/N \tag{8.19}$$

is an unbiased estimate of λ^*. Unfortunately, if $F < df_E/(df_E - 2)$, then $\hat{\lambda}^* < 0$. The adjusted estimator,

$$\hat{\lambda}^*_{adj} = \text{Max}[0, \hat{\lambda}^*] \tag{8.20}$$

can still be positively biased for small λ.

Adapting a method introduced by Venables (1975), the lower limit of a one-tailed $1 - \gamma$ interval for λ^* can be taken to be $\hat{\lambda}^*_\gamma = \hat{\lambda}_\gamma/N$, where $\hat{\lambda}_\gamma$ solves

$$\text{Pr}[F(df_H, df_E, \hat{\lambda}_\gamma) \geq F] = \gamma. \tag{8.21}$$

The SAS statement that will do this is:

lamhatgm = FNONCT(F, dfH, dfE, 1 − gamma);

A solution for $\hat{\lambda}_\gamma$ is possible only if γ exceeds the *p*-value for F, that is, $F > F_{\alpha = \gamma}$. As above, we may set $\gamma = .50$ for a median-type estimator or use $\gamma = .20$ or $.33$ to get some other useful lower bound on λ. We remark once again that with small N, setting γ too low often results in a lower bound for λ^* that is of little practical use.

When applied to the two-group problem, any of the estimators for λ^* can be converted to estimators for nature's effect size squared, $\psi^2 = (\mu_1 - \mu_2)^2/\sigma^2$, using

$$\hat{\psi}^2 = \lambda^*/(w_1 w_2). \tag{8.22}$$

This can then be the basis for estimating noncentralities for different $w_1 w_2$ values than were used in the pilot or previous study.

While these methods for estimating λ for a nondirectional test share the practical weaknesses of their counterparts for estimating δ for a directional test, they still may have utility when ample pilot or previous data are available. Unfortunately, there are no excellent methods for using small pilot studies to provide firm, objective estimates for effect sizes and non-centralities. Because there is usually little or no pilot data available for prospective power analyses, the methods suggested in this section rarely play a leading role in developing and justifying specific estimates. *In prac-*

tice, most good estimates for effect sizes are conjectures formed in the mind of thoughtful researchers, experienced in their science, and careful in their use of pilot data or related results found in the literature.

8.2.2 Comparing Two Correlated Means (Matched Pairs)

The common matched-pairs t-test applies to studies that have only one group of cases, but each case provides two measurements, y_1 and y_2, whose means will be compared to each other. The analysis focuses on the difference, $y_d = y_1 - y_2$, which we will take to have a mean of μ_d and a standard deviation of σ_d. Now we work with a null hypothesis, H_0: $\mu_d = 0$ (no difference), and a research ("alternative") hypothesis that is either non-directional (H_A: $\mu_d \neq 0$) or directional (say, H_A: $\mu_d > 0$). The statistic is

$$t = N^{1/2}(\hat{\mu}_d/\hat{\sigma}_d), \tag{8.23}$$

where N is the number of *pairs*. We take t to be distributed as a $t(df_E, \delta)$ random variable with $df_E = N - 1$ and

$$\delta = N^{1/2}(\mu_d/\sigma_d) \quad \text{and} \quad \delta^* = \psi = (\mu_d/\sigma_d). \tag{8.24}$$

Here again, t and δ have the same form. There is no way to vary the sample-size structure of a one-group design; thus the primary noncentrality and nature's effect size are identical, involving only μ_d and σ_d.

If it is difficult to specify σ_d directly, it may be specified indirectly by conjecturing values of σ_1 and σ_2, the standard deviations of y_1 and y_2, as well as ρ, the ordinary product-moment correlation between y_1 and y_2. Then,

$$\sigma_d = (\sigma_1^2 + \sigma_2^2 - 2\rho\sigma_1\sigma_2)^{1/2}. \tag{8.25}$$

Once δ^* is determined, power is computed in the same manner as given in the previous section, except that we have $df_E = N - 1$ pairs, instead of $df_E = N - 2$ individual subjects.

Example 2: Matched-Pairs t-Test
Dr. Seth Alalgia now considers a new design to study the efficacy of the enhanced thermal (ET) biofeedback therapy relative to the sham-placebo (SP) therapy in the relief of vascular headache. (See Example 1.) Patients will be run in pairs that are matched on several factors, including a screening version of the Vascular Headache Index (VHI). They will no longer be admitted to the hospital's research unit during their pretreatment phase, but will be seen as outpatients on Monday and Tuesday to complete the PreVHI measure and begin their standardized medical treatment with li-

tosamine. The pair will be admitted on Wednesday, already randomly split into the two treatment groups. The ET subject will be treated first, getting legitimate biofeedback. Then the SP subject will be treated as a "yoked control," getting the ET subject's session as noncontingent biofeedback. The same relative change score, RC, will serve the outcome measure, so that our paired differences are $d = RC_{SP} - RC_{ET}$. The paired difference on the PreVHI may serve as a covariate to lower the error variance in d. But, as in Example 1, covariates need not concern us here other than to influence our specification of σ_d. It is hoped that by matching and yoking pairs of patients, the error variance can be markedly reduced. Note also that now only three GCRC bed-days per subject will be used, rather than five, as before.

Dr. Alalgia believes that this study carries with it the same general treatment effects and variability as for the two-group study. Thus, he takes $\mu_d = .15$. As some of the reviewers consider this to be overly optimistic, he decides to look at $\mu_d = .10$ as well. Dr. Alalgia believes that the correlation between RC_{SP} and RC_{ET} is at least .40; thus it is decided to base the power analysis on both $\sigma_d = .125[2 - 2(.40)]^{1/2} \approx .137$ and $\sigma_d = .1875[2 - 2(.40)]^{1/2} \approx .205$. He is prepared to study up to 17 pairs of patients, which will require $17 \times 6 = 102$ bed-days. Is this enough? Too many?

Listing 8.2 gives the input to *OneWyPow* and some of its output. Under Dr. Alalgia's scenario ($\mu_d = .15$), the $\alpha = .05$, nondirectional test with $N = 17$ has power of .99 and .81 for the two standard deviations used. The corresponding powers under $\mu_d = .10$ (not shown) are .81 and .47. Dr. Alalgia decides to go forward with $N = 17$ and feels he can now easily defend this choice before the Scientific Review Committee.

8.2.3 One-Way Design: Overall Test and Planned Contrasts

The one-way analysis of variance is an extension of the two-group t-test. To compare the means of J independent groups, we work with two types of hypothesis tests, overall ("between-groups") tests and planned contrasts. The overall hypothesis is $H_0: \mu_1 = \mu_2 = \cdots = \mu_J$ (all populations means are equal), and the research ("alternative") hypothesis, $H_A: \mu_j \neq \mu_{j'}$, for some $j \neq j'$ (at least two means are different). But note that the overall research hypothesis is a nonspecific one. Often the researcher is more interested in one or more planned contrasts, which can be specified using $H_0: c_1\mu_1 + c_2\mu_2 + \cdots + c_J\mu_J = \theta_0 = 0$. The research hypothesis is either nondirectional, $H_A: \theta_0 \neq 0$, or it is directional, which we take to be $H_A: \theta_0 > 0$. The contrast weights, c_j, are usually defined such that $c_1 + c_2 +$

Listing 8.2 Computations for Dr. Alalgia's matched-pairs design.

Input

```
options ls=72 nosource2;
%include OneWyPow;
title1 "Dr. Alalgia: Yoked-Pairs Design for Biofeedback Study";
title2 "mu_diff = .15";
cards;
mu .15 .
sigma .137 .205 .
alpha .05 .01 .
Ntotal 10 14 17 20 .
end
%include FPowTab1;

%include OneWyPow;
title2 'mu_diff = .10';
cards;
mu .10 .
sigma .137 .205 .
alpha .05 .01 .
Ntotal 10 14 17 20 .
end
%include FPowTab1;
```

Selected Output

```
                                   mu_diff = .15
Effect: One-Group Test,
DF Hypothesis: 1,
AND Primary SSHe: 0.0225
-----------------------------------------------------------------------
|                         |                     Std Dev                |
|                         |--------------------------------------------|
|                         |      0.137        |        0.205          |
|                         |-------------------+----------------------- |
|                         |     Total N       |      Total N          |
|                         |-------------------+----------------------- |
|                         | 10 | 14 | 17 | 20 | 10 | 14 | 17 | 20 |
|                         |----+----+----+----+----+----+----+----|
|                         |Pow-|Pow-|Pow-|Pow-|Pow-|Pow-|Pow-|Pow-|
|                         | er | er | er | er | er | er | er | er |
|-------------------------+----+----+----+----+----+----+----+----|
|Test Type  |Alpha        |    |    |    |    |    |    |    |    |
|-----------+-----------  |    |    |    |    |    |    |    |    |
|2-tailed t |0.05         |.868|.966|.988|.996|.542|.716|.808|.873|
|           |-----------  |----+----+----+----+----+----+----+----|
|           |0.01         |.598|.838|.927|.970|.251|.427|.551|.659|
|-----------+-----------  |----+----+----+----+----+----+----+----|
|1-tailed t |0.05         |.938|.987|.996|.999|.688|.828|.893|.934|
|           |-----------  |----+----+----+----+----+----+----+----|
|           |0.01         |.727|.908|.963|.986|.362|.550|.667|.761|
-----------------------------------------------------------------------
```

$\cdots + c_J = 0$. Contrasts are formed to represent the specific hypotheses of interest. For example in a four-group design, the contrast H_0: $(.5)\mu_1 + (.5)\mu_2 + (-1)\mu_3 + (0)\mu_4 = 0$ compares the average of the first two means to the third mean. Example 3 illustrates several others. Planned contrasts are useful because they provide a more sharply focused analysis than do overall tests. This usually makes tests of planned contrasts easier to interpret and more powerful. In those rare cases in which the overall test captures the scientific question under study, the power of that test should naturally be of concern. On the other hand, if the main purpose is to test specific planned contrasts, one should choose a sample size based on those contrasts and pay little, if any, attention to the power of the overall test. For more on this, see Rosenthal and Rosnow (1985).

OneWyPow computes power for both overall tests and contrasts in the one-way design.

8.2.3.1 Overall Test in the One-Way ANOVA

The overall test of the equality of J independent means is performed using

$$F = N\left[\frac{\sum_{j=1}^{J} w_j(\hat{\mu}_j - \hat{\bar{\mu}})^2}{\hat{\sigma}^2(J - 1)}\right], \tag{8.26}$$

where $w_j = n_j/N$ is the proportion of cases in group j,

$$\hat{\bar{\mu}} = (w_1\hat{\mu}_1 + w_2\hat{\mu}_2 + \cdots + w_J\hat{\mu}_J) \tag{8.27}$$

is the weighted average of the sample means, and $\hat{\sigma}^2$ is the weighted (pooled) average of the sample variances. Under the classic normal-theory assumptions, the observed overall F statistic is an F random variable with $df_H = J - 1$ degrees of freedom for the hypothesis (numerator) and $df_E = N - J$ degrees of freedom for error (denominator). Its noncentrality is $\lambda = N\lambda^*$, where

$$\lambda^* = \frac{\sum_{j=1}^{J} w_j(\mu_j - \bar{\mu})^2}{\sigma^2}, \tag{8.28}$$

with $\bar{\mu} = w_1\mu_1 + w_2\mu_2 + \cdots + w_J\mu_J$. Note that $\lambda/(J - 1)$ is isomorphic to the overall F.

8.2.3.2 ANOVA Contrasts

It will simplify matters at times if we express H_0: $c_1\mu_1 + c_2\mu_2 + \cdots + c_J\mu_J = 0$ by only writing out the contrast coefficients in a matrix with one row, $\mathbf{C} = [c_1 c_2 \cdots c_J]$. Because $df_H = 1$, the choice must be made between a nondirectional or directional test. For the directional test, we form the statistic

$$t = N^{1/2} \frac{\sum_{j=1}^{J} c_j \hat{\mu}_j}{\hat{\sigma}(\sum_{j=1}^{J} c_j^2/w_j)^{1/2}}, \tag{8.29}$$

and use it as we used the two-group t-test described in Section 8.2.1.1 except that we now have $df_E = N - J$. The primary noncentrality is

$$\delta^* = \frac{\sum_{j=1}^{J} c_j \mu}{\sigma(\sum_{j=1}^{J} c_j^2/w_j)^{1/2}}, \tag{8.30}$$

and the noncentrality is $\delta = N^{1/2}\delta^*$. ψ cannot be defined, as the w_j are linked to the c_j and cannot be factored out of δ^*.

The material in Section 8.2.1.1 covers some methods to use pilot data to help estimate δ, and these extend to the t-tests for contrasts.

For the nondirectional case, the test involves $F = t^2$, which is taken to be $F(1, N - J, \lambda)$, where $\lambda = \delta^2$. Methods for using pilot data to help estimate λ follow from the material in Section 8.2.1.2.

Example 3: One-Way ANOVA with Contrasts

Dr. Mindy Bowdy studies relationships between personality characteristics and immune functioning (see Jemmott et al., 1990). For her next study, up to 100 (but preferably 60) male college students will give a sample of blood and complete a set of psychological instruments that are used to partition the sample into four types of people (D, O, L, and F):

Dominator: one who has a strong need to impact others; argumentative, assertive, overly competitive
Ordinary: one who is not a Dominator, a Loner, or a Friendly
Loner: one who has a minimal need for friendships; no desire to impact others
Friendly: one who has a very strong need to create and nurture friendships; accepts this as an ideal and an end in its own right

Other studies have found Friendlies to have better immune functioning than Ordinaries, who have better functioning than Dominators. Dr. Bowdy believes that the immune functioning of Loners may only be slightly better than that of Ordinaries. The specific question now is: "How do the DOLF groups differ with respect to Q-type killer cell activity (QKCA), as measured by the percentage of Q843 human leukemia cells lysed at a 50:1 effector-to-target ratio?"

Dr. Bowdy conjectures that the mean lysis rates are $\mu_D = .35$ (worst), $\mu_O = .50$, $\mu_L = .52$, $\mu_F = .60$ (best). Data from previous studies working with QKCA and Q843 suggest that the within-group standard deviation (σ) in this population is between .16 and .19. Furthermore, Dr. Bowdy expects that subjects will be partitioned according to the following proportions: $w_D = .20$, $w_O = .50$, $w_L = .10$, $w_F = .20$.

Many strategies are available to compare the groups' means in this study; good scientists and statisticians might disagree on what is best. Dr. Bowdy recognizes that the overall test of whether the four groups have equal lysis rates is weakened by the small difference between Ordinaries and Loners. She will assess it anyway, as peer reviewers and editors will expect to see it. Several contrasts are meaningful to Dr. Bowdy. The largest nature's effect size, ψ, for any two-group comparison is "Dominators vs. Friendlies," H_A: $\mu_F - \mu_D > 0$, i.e., the directional test of $\mathbf{C} = [-1\ 0\ 0\ 1]$. She also wants directional contrasts of "Friendlies vs. Ordinaries and Loners," $\mathbf{C} = [0\ -5/6\ -1/6\ 1]$, where the 5/6 and 1/6 are used because Ordinaries are expected to outnumber Loners by 5:1. Similarly, we have "Dominators vs. Ordinaries and Loners," $\mathbf{C} = [-1\ 5/6\ 1/6\ 0]$, directional. This family of three contrasts will be protected using a Bonferroni-adjusted level of $\alpha = .05/3$ for each. Though "Loners vs. Ordinaries" probably will have little power, Dr. Bowdy will test it with a nondirectional test of $\mathbf{C} = [0\ 1\ -1\ 0]$ at $\alpha = .05$.

Listing 8.3.1 gives the input and Listing 8.3.2 gives some of the key output related to using *OneWyPow*. These results demonstrate the benefits of looking at more than just overall tests when selecting a sample size. Here the overall test for the four-group design is fairly powerful; even if

Listing 8.3 Computations for Dr. Bowdy's four-group design.

Listing 8.3.1. Input

```
options ls=72 nosource2;
%include OneWyPow;
title1 "Mindy Bowdy : Four-Group Design";
*Order of groups: D O L F;
cards;
mu       .35 .50 .52 .60 .
weight   .20 .50 .10 .20 .
sigma    .16 .19 .
alpha    .05 .0167 .
Ntotal 60 80 100 .
contrasts
"Friendlies vs Ordin & Loners"  0 -.83 -.17  1 .
"Dominators vs Ordin & Loners" -1  .83  .17  0 .
"Friendlies vs Dominators"     -1  0  0  1 .
"Ordinaries vs Loners"          0  1 -1  0 .
"Almost Overall (2 DF)"         1 -.83 -.17  0 .
>                               0 -.83 -.17  1 .
end
%include FPowTab2;
run;
```

Listing 8.3.2. Selected output

```
ALPHA 0.05
```

		Std Dev					
		0.16			0.19		
		Total N			Total N		
		60	80	100	60	80	100
		Pow-er	Pow-er	Pow-er	Pow-er	Pow-er	Pow-er
Overall test	Regular F	.899	.970	.992	.763	.887	.951
Ordinaries vs Loners	2-tailed t	.059	.062	.065	.056	.058	.060
	1-tailed t	.086	.093	.099	.079	.084	.090
Almost Overall (2 DF)	Regular F	.933	.982	.996	.821	.923	.969

```
ALPHA 0.0167
```

		Std Dev					
		0.16			0.19		
		Total N			Total N		
		60	80	100	60	80	100
		Pow-er	Pow-er	Pow-er	Pow-er	Pow-er	Pow-er
Friendlies vs Ordin & Loners	2-tailed t	.265	.366	.464	.182	.253	.325
	1-tailed t	.362	.473	.573	.263	.347	.428
Dominators vs Ordin & Loners	2-tailed t	.659	.806	.897	.487	.637	.754
	1-tailed t	.755	.874	.938	.597	.735	.832
Friendlies vs Dominators	2-tailed t	.909	.974	.993	.772	.896	.956
	1-tailed t	.948	.987	.997	.849	.938	.976

σ is large (.19), $N = 80$ gives a .05-based power of .89. It can be shown that the noncentrality for the "Friendlies vs. Dominators" comparison captures over 95% of noncentrality for the overall test. Examining the powers for "Dominators vs. Ordinaries and Loners," we find that most of the power in this design is due to the conjectured immune inferiority of the Dominators. The immune superiority of the Friendlies is not likely to be supported by a significant "Friendlies vs. Ordinaries and Loners" test. Reasonable assurance that this will be significant seems to require far more than a total of 100 subjects, especially when we use Bonferroni protection.

Finally, we see almost no power for the .05-level comparison of the Ordinaries vs. Loners. It might be that we could create a more powerful overall test, call it "Almost Overall," by pooling the means over the Ordinaries and Loners. This can be done by aggregating the between-groups variance defined by $\mathbf{C} = [1 -5/6 -1/6 \ 0]$ and $\mathbf{C} = [0 -5/6 -1/6 \ 1]$ giving a two-degree-of-freedom contrast. The theory for this is reviewed in Section 8.3, but the programming of *OneWyPow* is straightforward, as shown in Listing 8.3.1. Because there is so little difference between the conjectured means for the Ordinaries and Loners, pooling them creates a test that is a little more powerful than the ordinary overall test.

8.3 UNIVARIATE GENERAL LINEAR (FIXED-EFFECTS) MODEL

All of the tests we have thus far considered, as well as many we have not considered, are special cases of tests within univariate general linear modeling (GLM). Understanding noncentrality in GLM testing, and knowing how to perform the computations easily, creates a wonderfully broad spectrum of power analyses.

Consider the standard model

$$\mathbf{y} = \mathbf{X}\boldsymbol{\beta} + \boldsymbol{\varepsilon}, \tag{8.31}$$

where $\mathbf{y}$ is the $N \times 1$ vector of the dependent variable, $\mathbf{X}$ is the $N \times r$ model matrix of nonrandom, known predictor values, $\boldsymbol{\beta}$ is the $r \times 1$ vector of the fixed, unknown coefficients, and $\boldsymbol{\varepsilon}$ is the $N \times 1$ vector of true residuals. Without loss of generality, we require the columns of $\mathbf{X}$ to be linearly independent, i.e., rank($\mathbf{X}$) = r. For tests on $\boldsymbol{\beta}$, we take the elements of $\boldsymbol{\varepsilon}$ to be independent $N(0, \sigma^2)$ random variables. The usual estimates are

$$\hat{\boldsymbol{\beta}} = (\mathbf{X}^T\mathbf{X})^{-1}\mathbf{X}^T\mathbf{y} \tag{8.32}$$

and

$$\hat{\sigma}^2 = (\mathbf{y} - \mathbf{X}\hat{\boldsymbol{\beta}})^T(\mathbf{y} - \mathbf{X}\hat{\boldsymbol{\beta}})/(N - r). \tag{8.33}$$

We focus here on the general linear hypothesis,

$$H_0: \mathbf{C\beta} = \mathbf{\theta}_0, \tag{8.34}$$

where $\mathbf{C}$ is $df_H \times r$ with rank($\mathbf{C}$) $= df_H \leq r$. $\mathbf{\theta}_0$ is a vector of constants appropriate to the research question. It is usually chosen to be $\mathbf{0}$. For $df_H > 1$, $H_A: \mathbf{C\beta} \neq \mathbf{\theta}_0$ is the only alternative we consider. For $df_H = 1$, the directional alternative might be appropriate, which is defined here to be $H_A: \mathbf{C\beta} > \mathbf{\theta}_0$. The test statistic is $F = SSH/(df_H \hat{\sigma}^2)$, where

$$SSH = (\mathbf{C}\hat{\mathbf{\beta}} - \mathbf{\theta}_0)^T [\mathbf{C}(\mathbf{X}^T\mathbf{X})^{-1}\mathbf{C}^T]^{-1}(\mathbf{C}\hat{\mathbf{\beta}} - \mathbf{\theta}_0) \tag{8.35}$$

is the sums of squares for the hypothesis. F is distributed $F(df_H, df_E, \lambda)$, where $df_E = (N - r)$ and

$$\lambda = (\mathbf{C\beta} - \mathbf{\theta}_0)^T [\mathbf{C}(\mathbf{X}^T\mathbf{X})^{-1}\mathbf{C}^T]^{-1}(\mathbf{C\beta} - \mathbf{\theta}_0)/\sigma^2. \tag{8.36}$$

This general structure encompasses t-tests, fixed-effects ANOVA and ANCOVA, and ordinary multiple regression.

λ may be expressed in a way that displays its distinct components. Let $\ddot{\mathbf{X}}$ be the $q \times r$ essence model matrix formed by assembling the q unique rows of $\mathbf{X}$. In other words, $\ddot{\mathbf{X}}$ is the collection of the unique design points (e.g., groups) for the proposed study. Let $\mathbf{W}$ be the $q \times q$ diagonal matrix having elements w_j, the proportion of the total sample size associated with the jth row of $\ddot{\mathbf{X}}$. Thus, $N\mathbf{W}$ holds the q sample sizes. It can be shown that

$$\lambda = N\{(\mathbf{C\beta} - \mathbf{\theta}_0)^T [\mathbf{C}(\ddot{\mathbf{X}}^T\mathbf{W}\ddot{\mathbf{X}})^{-1}\mathbf{C}^T]^{-1}(\mathbf{C\beta} - \mathbf{\theta}_0)/\sigma^2\}. \tag{8.37}$$

We see that the primary noncentrality, $\lambda^* = \lambda/N$, is based on the design points to be used ($\ddot{\mathbf{X}}$), the sample size weightings for those points ($\mathbf{W}$), the conjectured estimates for the unknown effects ($\mathbf{\beta}$) and the variance (σ^2), and the specification of the hypothesis ($\mathbf{C}, \mathbf{\theta}_0$).

One particularly straightforward and useful application of the GLM involves the use of the cell means model, $y_{ij} = \mu_j + e_{ij}$, for the J-group ANOVA. Here $\ddot{\mathbf{X}} = \mathbf{I}$, the $J \times J$ identity matrix, and $\mathbf{\beta} = \mathbf{\mu}$, the vector of population means. Taking $\mathbf{\theta}_0 = \mathbf{0}$, we have

$$\lambda = N(\mathbf{C\mu})^T [\mathbf{CW}^{-1}\mathbf{C}^T]^{-1}(\mathbf{C\mu})/\sigma^2. \tag{8.38}$$

For the two-group t-test, $\mathbf{I}$ is 2×2, $\mathbf{\mu}^T = [\mu_1 \ \mu_2]$, $\mathbf{W} = \text{diag}[w_1 \ w_2]$, and $\mathbf{C} = [1 \ -1]$. One can use (8.38) to match the square of (8.3). For the three-group problem, $\mathbf{I}$ is 3×3, $\mathbf{\mu}^T = [\mu_1 \ \mu_2 \ \mu_3]$, and $\mathbf{W} = \text{diag}[w_1 \ w_2 \ w_3]$. The overall test can then use

$$\mathbf{C} = \begin{bmatrix} 1 & -1 & 0 \\ 1 & 0 & -1 \end{bmatrix}.$$

This can be extended easily to J groups. To derive the λ for constrasts in a J-group ANOVA [see (8.30)], use $\mathbf{C} = [c_1 \, c_2 \cdots c_J]$. It can also be used to specify the effects for factorial designs. For example, a 2×3 factorial design can usually be handled by taking $\mathbf{I}$ to be 6×6, $\boldsymbol{\mu}^\mathrm{T} = [\mu_{11} \, \mu_{12} \, \mu_{13} \, \mu_{21} \, \mu_{22} \, \mu_{23}]$, and $\mathbf{W} = \mathrm{diag}[w_{11} \, w_{12} \, w_{13} \, w_{21} \, w_{22} \, w_{23}]$. The interaction test requires

$$\mathbf{C} = \begin{bmatrix} 1 & -1 & 0 & -1 & 1 & 0 \\ 1 & 0 & -1 & -1 & 0 & 1 \end{bmatrix},$$

and the main effects can be defined as is appropriate for the research design and question.

Other constructions for $\ddot{\mathbf{X}}$ will handle designs with missing cells, nested factors, fixed blocking factors, and multiple covariates. Therefore, power analysis can be done for any hypothesis that can be handled by the fixed-effects features of any linear models program. Thus, there is rarely a justification for limiting the power analysis to a minor part of a large design, such as the comparison of just two of the groups contained within a $2 \times 3 \times 3$ factorial.

Exemplary Data: SSH_e, SSH_e^*
Note that if we use an *exemplary data set* of N_e cases conforming to $\mathbf{y} \equiv \mathbf{X}\boldsymbol{\beta}$, we obtain $\hat{\boldsymbol{\beta}} \equiv \boldsymbol{\beta}$. This makes $SSH \equiv \lambda\sigma^2$, which is called the exemplary sums of squares hypothesis, or SSH_e. Defining $SSH_e^* = SSH_e/N_e$ to be the primary SSH_e, we have

$$\lambda = N \cdot SSH_e^*/\sigma^2. \tag{8.39}$$

In comparing means using t-tests or the ANOVA, the exemplary data must be designed to produce sample means that are identical to the conjectured population means: $\hat{\mu}_j \equiv \mu_j$. For the two-group case this yields

$$SSH_e = N_e w_1 w_2 (\mu_1 - \mu_2)^2 = N_e \sigma^2 \lambda^*. \tag{8.40}$$

SSH_e^* subsumes the complex parts of computing the noncentrality parameter. By first computing SSH_e using standard software for data analysis, an array of λ values can be formed by combining SSH_e^* with various values of N and σ^2. SSH_e^* is a common concept that unifies the notion of λ across all cases of the general linear model. As a computing scheme, it is very helpful in handling complex linear models, such as in the unbalanced AN-COVA design illustrated below. The module *PowSetUp* has been designed to create tables of power probabilities based on SSH_e, N_e, N, σ, and α values.

Random **X** *Variables*

As noted well by Gatsonis and Sampson (1989), if **X** involves random variables (the correlational model), then the noncentral *F* results given here do not hold strictly, unless one is willing to take the *conditional* viewpoint, that is, "given this **X**." Fortunately, the conditional results discussed in this chapter do provide reasonable approximations to methods developed under the assumption that the columns of **X** are multivariate normal. The practical discrepancy between the two approaches disappears as the sample size increases. Because the values for the population parameters are conjectures or estimates, strict numerical accuracy of the power computations is usually not critical. However, those who wish may use the excellent tables or software provided by Gatsonis and Sampson, which have their own limitation due to their dependence on multivariate normality.

Example 4: Univariate GLM for ANCOVA

Dr. Mindy Bowdy completed the four-group study (Example 3). She now wants to extend the work to focus on whether greater psychological stress is associated with poorer immune function, as Cohen, Tyrrell, and Smith (1991) reported by correlating higher stress with an increased risk of catching a cold. Dr. Bowdy's proposed design is similar to her D-O-L-F four-group design, except that the Ordinaries and Loners are now combined into a single group called Regulars. Thus we now have three groups, and we still expect the Dominators to have lower QKCA-Q843 lysis rates than the Regulars, who will have lower rates than the Friendlies. This study, however, will focus on the Life Events Stress Index (LESI), which takes on values:

- -2: low stress
- -1: below average stress
- 0: average stress
- $+1$: above average stress
- $+2$: high stress

Regulars and Friendlies are conjectured to have uniform distributions for the LESI, whereas the Dominators are conjectured to have LESI values higher than normal. The main research questions involve the relationships between stress and immunity, as measured by the LESI and the lysis rates. Dr. Bowdy conjectures that the Friendlies have no relationship between the LESI and lysis rates, the Regulars have a weak negative relationship, and the Dominators have a stronger negative relationship. She understands that addressing this question probably requires several hundred subjects.

An essence matrix for this design is

$$
\ddot{\mathbf{X}} =
\begin{bmatrix}
1 & -2 & 0 & 0 & 0 & 0 \\
1 & -1 & 0 & 0 & 0 & 0 \\
1 & 0 & 0 & 0 & 0 & 0 \\
1 & 1 & 0 & 0 & 0 & 0 \\
1 & 2 & 0 & 0 & 0 & 0 \\
0 & 0 & 1 & -2 & 0 & 0 \\
0 & 0 & 1 & -1 & 0 & 0 \\
0 & 0 & 1 & 0 & 0 & 0 \\
0 & 0 & 1 & 1 & 0 & 0 \\
0 & 0 & 1 & 2 & 0 & 0 \\
0 & 0 & 0 & 0 & 1 & -2 \\
0 & 0 & 0 & 0 & 1 & -1 \\
0 & 0 & 0 & 0 & 1 & 0 \\
0 & 0 & 0 & 0 & 1 & 1 \\
0 & 0 & 0 & 0 & 1 & 2
\end{bmatrix},
\begin{matrix}
(\text{``D''} \quad \text{LESI} = -2) \\
(\text{``D''} \quad \text{LESI} = -1) \\
(\text{``D''} \quad \text{LESI} = 0) \\
(\text{``D''} \quad \text{LESI} = +1) \\
(\text{``D''} \quad \text{LESI} = +2) \\
(\text{``R''} \quad \text{LESI} = -2) \\
(\text{``R''} \quad \text{LESI} = -1) \\
(\text{``R''} \quad \text{LESI} = 0) \\
(\text{``R''} \quad \text{LESI} = +1) \\
(\text{``R''} \quad \text{LESI} = +2) \\
(\text{``F''} \quad \text{LESI} = -2) \\
(\text{``F''} \quad \text{LESI} = -1) \\
(\text{``F''} \quad \text{LESI} = 0) \\
(\text{``F''} \quad \text{LESI} = +1) \\
(\text{``F''} \quad \text{LESI} = +2)
\end{matrix}
$$

which has corresponding regression coefficients

$$\boldsymbol{\beta}^{\mathrm{T}} = [\mu_D \ \beta_D \ \mu_R \ \beta_R \ \mu_F \ \beta_F].$$

This defines three distinct simple regression models of the form

$$\mathscr{E}[\text{lysis}_{ij}] = \mu_j + \beta_j \, \text{LESI}, \qquad \text{where } j \in \{\text{``D''} \ \text{``R''} \ \text{``F''}\}. \tag{8.41}$$

Dr. Bowdy specifies the regression coefficients to be

$$\boldsymbol{\beta}^{\mathrm{T}} = [.3350 \quad -.0300 \quad .5033 \quad -.0100 \quad .6000 \quad .0000].$$

She takes σ to be .12 or .15. Dr. Bowdy cannot control how the total sample size will disperse among the 15 combinations of DRF and LESI values, but this is conjectured to follow the weights

$$
\begin{aligned}
\mathbf{w}^{\mathrm{T}} &= \text{diag}(\mathbf{W}) \\
&= [.02 \ .03 \ .04 \ .05 \ .06 \ .12 \ .12 \ .12 \ .12 \ .12 \ .04 \ .04 \ .04 \ .04 \ .04].
\end{aligned}
$$

Listing 8.4.1 gives basic SAS statements showing how an exemplary data set can be constructed and then analyzed with a standard linear models routine, PROC GLM. Note how the "n" variable and the FREQ statement define the weights, which are 100**w**. Thus, the first data line effectively creates two observations, the second creates three, etc. Thus $N_e = 100$ cases are produced, all with lysis values equal to their respective expected values. We could have used PROC GLM's WEIGHT statement to take the noninteger weights of **w** (here; .02, .03, . . .), thus giving $N_e = 1.00$,

Listing 8.4 Computations for Dr. Bowdy's ANCOVA design.

Listing 8.4.1. **SAS statements to compute SSH$_e$ values using regular data analysis on exemplary data.**

```
options ls =72;
title "Mindy Bowdy: ANCOVA design";
data;
/*
Three groups, one covariate, unequal slopes;
* Order of exemplary data:
  D = Dominator
  R = Ordinary or Loner
  F = Friendly
*/
input lysis0 DRF $ LESI n;

*create exemplary data;
if DRF = 'D' then beta = -.03;
if DRF = 'R' then beta = -.01;
if DRF = 'F' then beta =  .00;
lysis = lysis0 + beta*LESI;

*    lysis at                      ;
*    LESI=0   DRF   LESI    n      ;      cards;
     .3350    D     -2     02
     .3350    D     -1     03
     .3350    D      0     04
     .3350    D      1     05
     .3350    D      2     06
     .5033    R     -2     12
     .5033    R     -1     12
     .5033    R      0     12
     .5033    R      1     12
     .5033    R      2     12
     .6000    F     -2     04
     .6000    F     -1     04
     .6000    F      0     04
     .6000    F      1     04
     .6000    F      2     04

* analyze exemplary data to get SSHe values;
proc glm order=data; class DRF; freq n;
model lysis = DRF DRF*LESI/noint solution;
contrast 'DRF main|LESI=0' DRF 1 -1 0, DRF 0 1 -1;
contrast 'Means:DvsR|LESI=0' DRF 1 -1 0;
contrast 'Means:FvsR|LESI=0' DRF 0 1 -1;
contrast 'LESI main | DRF' DRF*LESI 1 1 1;
contrast 'DFR*LESI' DRF*LESI 1 -1 0, DRF*LESI 0 1 -1;
contrast 'Slopes: D vs R' DRF*LESI 1 -1 0;
contrast 'Slopes: F vs R' DRF*LESI 0 1 -1;
```

but not all routines will handle this. If, as in some routines, PROC GLM had been unable to handle data giving a sum of squares residuals of 0.0, we could have simply replaced the last data line, which is

  .6000 F 2 04

with the two lines

  .5000 F 2 02
  .7000 F 2 02

This forces two residuals to be $+0.10$ and two to be -0.10.

Listing 8.4.2. Selected portions from SAS output giving SSH_e values.

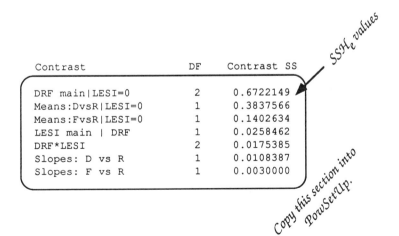

```
Dependent Variable: LYSIS
Frequency:          N

Source              DF

Uncorrected Total   100    ◄──── This shows Nₑ value.
```

This shows N_e value.

```
Contrast              DF    Contrast SS
DRF main|LESI=0        2      0.6722149
Means:DvsR|LESI=0      1      0.3837566
Means:FvsR|LESI=0      1      0.1402634
LESI main  | DRF       1      0.0258462
DRF*LESI               2      0.0175385
Slopes:  D vs R        1      0.0108387
Slopes:  F vs R        1      0.0030000
```

SSH_e values

Copy this section into PowSetUp.

Listing 8.4.3. SAS statements to obtain power tables.

```
options ls=72 nosource2;
title1 "Mindy Bowdy: DRF groups and LESI stress measure";
%include PowSetUp;
cards;
Ne 100
alpha .05 .01 .
sigma .12 .15 .
Ntotal  200 300 500 .
numparms 6
effects
  DRF main | LESI = 0           2     0.6722149
  Means: D vs R | LESI = 0      1     0.3837566
  Means: F vs R | LESI = 0      1     0.1402634
  LESI main | DRF               1     0.0258462
  DFR*LESI                      2     0.0175385
  LESI slopes: D vs R           1     0.0108387
  LESI slopes: F vs R           1     0.0030000
end
*;
%include FPowTab2;
```

*This section
was copied
from SAS GLM
output, then the
effects' titles were
modified slightly.*

The MODEL statement defines a model matrix corresponding to $\ddot{X}$, albeit with the first three columns defining the three intercepts and labeled DRF, and the last three defining the three slopes and labeled DRF*LESI. Knowing this, the CONTRAST statements shown here are quickly discernible, especially after reading the effect titles. The main point here is that power computations are practical for any test within a fixed-effects general linear model framework. Just begin by defining an exemplary data set and analyzing it as if it were the real data.

Listing 8.4.2 gives the key parts of output produced by PROC GLM's analysis of this exemplary data. The uncorrected total degrees of freedom simply gives the value for N_e; the Contrast SS "statistics" give the SSH_e values. Listing 8.4.3 shows how another SAS module, *PowSetUp* (see Appendix), turns these SSH values into power probabilities. Most input lines are self-explanatory; NUMPARMS gives the rank of **X** and the lines following EFFECTS are those copied verbatim from PROC GLM's output. Running the statements in Listing 8.4.3 produces the power tables, one of which is Listing 8.4.4.

These results show that to address whether stress is related to (QKCA) immune function will require at least 500 subjects, and even this number is likely to fail to show that the DRF groups have different LESI slopes (DRF*LESI effect). It is unclear at this point whether Dr. Bowdy should

Listing 8.4.4. Output from PowSetUp and FPowTab2 SAS modules.

```
ALPHA 0.05
```

			Std Dev					
			0.12			0.15		
			Total N			Total N		
			200	300	500	200	300	500
			Pow-er	Pow-er	Pow-er	Pow-er	Pow-er	Pow-er
DRF main LESI = 0	Regular F		.999	.999	.999	.999	.999	.999
Means: D vs R | LESI = 0	2-tailed t		.999	.999	.999	.999	.999	.999
	1-tailed t		.999	.999	.999	.999	.999	.999
Means: F vs R | LESI = 0	2-tailed t		.992	.999	.999	.940	.991	.999
	1-tailed t		.997	.999	.999	.970	.996	.999
LESI main DRF	2-tailed t		.470	.638	.848	.326	.456	.667
	1-tailed t		.596	.749	.911	.447	.582	.773
DRF*LESI	2-tailed t		.264	.380	.588	.182	.256	.404
LESI slopes: D vs R	2-tailed t		.231	.322	.491	.164	.224	.341
	1-tailed t		.336	.442	.615	.252	.328	.462
LESI slopes: F vs R	2-tailed t		.098	.124	.175	.081	.097	.129
	1-tailed t		.158	.196	.266	.129	.155	.203

go ahead with the study at $N = 500$, seek to increase that total sample size, or redesign the study to create something with more powerful tests of hypotheses.

8.4 THE MULTIVARIATE GENERAL LINEAR MODEL

Many of the common methods in multivariate analysis can be developed within the framework of the multivariate general linear model, including

Hotelling's T^2 test, multivariate analysis of variance and covariance (MANOVA and MANCOVA), linear discriminant analysis, profile analysis, and univariate and MANOVA-based repeated measures analyses. To our knowledge, a *general and practical* method for computing the power for such tests was not available until Muller and Peterson (1984) showed that the scheme just described for the univariate GLM extends to a sound method for the multivariate GLM. We briefly summarize the theory here and present a straightforward example. Note how our development here parallels that of the last section. In fact, all of univariate results given heretofore in this chapter can be shown to be special cases of the multivariate results given in this section. Although the conceptual jump from the univariate to the multivariate power problem is not a particularly difficult one, actually conducting a multivariate power analysis can be a most complex exercise due to the need to make conjectures about many parameters. For more on this, see Muller, LaVange, Ramey, and Ramey (1992).

Consider the standard model

$$\mathbf{Y} = \mathbf{XB} + \boldsymbol{\mathcal{E}}, \tag{8.42}$$

in which $\mathbf{Y}$ is the $N \times p$ matrix of the dependent variables, $\mathbf{X}$ is the $N \times r$ model matrix (full column rank, as in the univariate case), $\mathbf{B}$ is an $r \times p$ matrix of the fixed, unknown coefficients, and $\boldsymbol{\mathcal{E}}$ is the $N \times p$ matrix with independent row vectors of true residuals having covariance matrix $\boldsymbol{\Sigma}$. The usual estimates are

$$\hat{\mathbf{B}} = (\mathbf{X}^T\mathbf{X})^{-1}\mathbf{X}^T\mathbf{Y}, \tag{8.43}$$

and

$$\hat{\boldsymbol{\Sigma}} = (\mathbf{Y} - \mathbf{X}\hat{\mathbf{B}})^T(\mathbf{Y} - \mathbf{X}\hat{\mathbf{B}})/(N - r). \tag{8.44}$$

We focus here on the general linear hypothesis,

$$H_0: \mathbf{CBU} = \boldsymbol{\Theta}_0, \tag{8.45}$$

where $\mathbf{C} = df_C \times r$ with full rank$(\mathbf{C}) = df_C \leq r$, identical in form and function to that of the univariate case. The $\mathbf{U}$ matrix is $p \times df_U$ with full rank$(\mathbf{U}) = df_U \leq p$. The degrees of freedom for H_0 is $df_H = df_C df_U$. $\boldsymbol{\Theta}_0$ is almost always chosen to be $\mathbf{0}$. Just as $\mathbf{C}$ controls contrasts on the rows of $\mathbf{B}$, $\mathbf{U}$ controls contrasts on its columns. The net effect of $\mathbf{U}$ is to create a set of df_U "contrast variables," $\tilde{\mathbf{Y}} = \mathbf{YU}$, giving the model

$$\tilde{\mathbf{Y}} = \mathbf{XBU} + \boldsymbol{\mathcal{E}}\mathbf{U} = \mathbf{X}\tilde{\mathbf{B}} + \tilde{\boldsymbol{\mathcal{E}}}, \tag{8.46}$$

This now involves the covariance matrix $\mathbf{U}^T\boldsymbol{\Sigma}\mathbf{U} = \tilde{\boldsymbol{\Sigma}}$ and the hypothesis $H_0: \mathbf{C}\tilde{\mathbf{B}} = \boldsymbol{\Theta}_0$. Thus if $df_U = 1$, the problem becomes a univariate GLM on $\bar{\mathbf{y}} = \mathbf{YU}$ and the methods of the last section apply directly. Such a reduction in dimensionality is often done in data analysis. The matched-pairs t-test discussed in Section 8.2.2 is a basic example of this, using $p = 2$, $\mathbf{U}^T = [1 \ -1]$. Ordinary MANOVAs require $\mathbf{U} = \mathbf{I}$, along with the same $\mathbf{X}$ and $\mathbf{C}$ matrices used for their univariate counterparts. The cell means model, for example, becomes the cell mean-vectors model. For a three-group case, $\mathbf{C} = [1 \ -.5 \ -.5]$ tests the first mean vector versus the average of the other two mean vectors. For a full treatment on how forming contrast variables can handle repeated measures, see O'Brien and Kaiser (1985).

All of the common test statistics for H_0 are based on the assumption that the rows of $\mathcal{E}\mathbf{U}$ are independent, multivariate normal vectors. *SSH* from the univariate model now generalizes to

$$\begin{aligned}
\mathbf{H} &= (\mathbf{C}\hat{\mathbf{B}}\mathbf{U} - \boldsymbol{\Theta}_0)^T[\mathbf{C}(\mathbf{X}^T\mathbf{X})^{-1}\mathbf{C}^T]^{-1}(\mathbf{C}\hat{\mathbf{B}}\mathbf{U} - \boldsymbol{\Theta}_0) \\
&= N(\mathbf{C}\hat{\mathbf{B}}\mathbf{U} - \boldsymbol{\Theta}_0)^T[\mathbf{C}(\ddot{\mathbf{X}}^T\mathbf{W}\ddot{\mathbf{X}})^{-1}\mathbf{C}^T]^{-1}(\mathbf{C}\hat{\mathbf{B}}\mathbf{U} - \boldsymbol{\Theta}_0) \qquad (8.47) \\
&= N\mathbf{H}^*,
\end{aligned}$$

the $df_U \times df_U$ sums of squares and cross-products matrix for the hypothesis. $\ddot{\mathbf{X}}$ and $\mathbf{W}$ were defined and discussed in Section 8.3. $\hat{\sigma}^2$ generalizes to $\mathbf{U}^T\boldsymbol{\Sigma}\mathbf{U} = \mathbf{E}/(N - r)$, where

$$\mathbf{E} = \mathbf{U}^T(\mathbf{Y} - \mathbf{X}\hat{\mathbf{B}})^T(\mathbf{Y} - \mathbf{X}\hat{\mathbf{B}})\mathbf{U}. \qquad (8.48)$$

The $s = \min(df_C, df_U)$ eigenvalues of $\mathbf{H}(\mathbf{H} + \mathbf{E})^{-1}$, $\hat{\rho}_k^2$, are the (generalized) squared canonical correlations. Many authors choose to work with the roots of $\mathbf{HE}^{-1}$, $\hat{\rho}_k^2/(1 - \hat{\rho}_k^2)$; mathematically, the choice is arbitrary. Many multivariate texts, including Seber's (1984), describe the Wilks, Hotelling-Lawley, Pillai-Bartlett, and Roy methods for forming test statistics from these eigenvalues and the transforms that convert the first three to approximate F statistics. Wilks' likelihood ratio statistic is the determinant of $\mathbf{E}(\mathbf{H} + \mathbf{E})^{-1}$, or equivalently,

$$W = \prod_{k=1}^{s} (1 - \hat{\rho}_k^2). \qquad (8.49)$$

Rao's transformation converts this to an $F(df_H, df_E)$ statistic,

$$F_W = \frac{(1 - W^{1/g})/df_H}{(W^{1/g})/df_E}, \qquad (8.50)$$

where

$$g = \begin{cases} 1, & df_H \leq 3, \\ [(df_H^2 - 4)/(df_C^2 + df_U^2 - 5)]^{1/2}, & df_H \geq 4, \end{cases} \qquad (8.51)$$

and

$$df_E = g[N - r - (df_U - df_C + 1)/2] - (df_H - 2)/2. \tag{8.52}$$

Muller and Peterson outlined a good approximation (exact if $df_U = 1$) for the noncentrality of F_W. Replacing $\hat{\mathbf{B}}$ with $\mathbf{B}$ in (8.47) gives us $\mathbf{H}_e = N\mathbf{H}_e^*$, the multivariate version of $SSH_e = N \cdot SSH_e^*$, the exemplary data form used in Section 8.3. Compute the eigenvalues of $\mathbf{H}_e^*\{\mathbf{H}_e^* + [(N - r)/N]\mathbf{U}^T\mathbf{\Sigma}\mathbf{U}\}^{-1}$, which are denoted ρ_{ek}^2. Use ρ_{ek}^2 in place of $\hat{\rho}_k^2$ in computing F_W to get F_{W_e}. Then F_W is distributed approximately as $F(df_H, df_E, \lambda)$, where

$$\lambda = df_H F_{W_e}. \tag{8.53}$$

Due to the $(N - r)/N$ term in the eigenproblem leading to λ, λ/N is not invariant to N, as it is in the univariate case. Thus the primary noncentrality cannot be defined. O'Brien and Shieh (under review) have proposed a modification to this method that does define a primary noncentrality and that may give more accurate power probabilities.

Muller and Peterson applied this idea to get power approximations for the F statistics associated with the Hotelling-Lawley and Pillai-Bartlett statistics as well. They summarized arguments supporting the method's numerical accuracy. Barton and Cramer (1989) reported Monte Carlo results that also supported the accuracy of the approximation. A general, practical power approximation for Roy's greatest root statistic does not exist, for here even its null distribution is difficult to characterize.

Example 5: Multivariate GLM for a Cross-Over Design
Dr. Katie Kohlimein is planning an experiment to assess how mental stress and an adrenaline-like hormone affect the metabolism of cholesterol in people age 30–45 with normal cholesterol levels. Gender differences are of key interest here, because of the much greater incidence of coronary heart disease in men (see Stoney, Matthews, McDonald, and Johnson, 1988). The study will compare the following conditions:

Control day (C):   Subject will have an 8-hour admission to the General Clinical Research Center (GCRC), having blood and urine sampled every hour.
Mental stress day (S):   Same as C, but in addition performs a series of stressful mental tasks for 2 hours.
Dosophrine day (D):   Same as C, but in addition is infused with 30 mg/kg/hour dosophrine continuously for 2 hours. Dosophrine is similar to adrenaline (epinephrine).

Subjects will go through each condition, one per month in a randomly determined order, thus forming a three-period crossover design. Each admission will be preceded by 2 weeks on a low-fat, low-cholesterol diet. At Hour 0 the subject will be given a "meal" very high in cholesterol. Baseline measurements will be taken at Hours 1 and 2 and averaged. Experimental manipulations will occur during Hours 3 and 4. Measurements at Hours 5 and 6 will be averaged to give short-term, posttreatment values. The outcome measure of interest for this power analysis will be $\Delta LDL56$, the change in low-density lipoprotein cholesterol (LDL-C) blood levels from Hours 1–2 to Hours 5–6. Equal numbers of men and women will be studied and equal numbers of subjects will get the six orders of treatment (CSD, CDS, . . . , DSC). Thus, the total sample size will be a factor of 12. Due to the expensive labwork, a tight budget calls for $N = 36$ (18 of each gender), but it may be possible to run $N = 48$. Crossover effects will be checked, but are considered unlikely. The main analysis will employ a MANOVA-based repeated measures structure, with one two-level between-subjects factor, Gender, and one three-level within-subjects (crossover) factor, Treatment.

Dr. Kohlimein makes conjectures for the mean $\Delta LDL56$ values, which can be arranged to form the matrix,

$$\mathbf{B} = \begin{array}{ccc} C & S & D \\ \end{array} \\ \begin{bmatrix} 3 & 12 & 8 \\ 1 & 5 & 7 \end{bmatrix} \begin{array}{l} \text{male} \\ \text{female} \end{array}.$$

She further specifies that the within-group variances are 25, 64, and 36 across the C, S, and D conditions. She also believes that the C/S and C/D correlations are .400 and the S/D correlation is .625 (5/8), thus forming the covariance matrix,

$$\mathbf{\Sigma} = \begin{array}{ccc} C & S & D \\ \end{array} \\ \begin{bmatrix} 25 & 16 & 12 \\ 16 & 64 & 30 \\ 12 & 30 & 36 \end{bmatrix} \begin{array}{l} C \\ S \\ D \end{array}.$$

Define $\mathbf{C}_A = [.5\ .5]$, $\mathbf{C}_G = [1\ -1]$,

$$\mathbf{U}_A = \begin{bmatrix} 1/3 \\ 1/3 \\ 1/3 \end{bmatrix} \quad \text{and} \quad \mathbf{U}_T = \begin{bmatrix} 1 & 0 \\ -1 & 1 \\ 0 & -1 \end{bmatrix},$$

where A = average, G = Gender, and T = Treatment. Then the null hypotheses that specify the Gender main, Treatment main, and Gender × Treatment interaction effects are H_G: $\mathbf{C}_G\mathbf{B}\mathbf{U}_A = \mathbf{0}$, H_T: $\mathbf{C}_A\mathbf{B}\mathbf{U}_T = \mathbf{0}$, and $H_{G \times T}$: $\mathbf{C}_G\mathbf{B}\mathbf{U}_T = \mathbf{0}$.

We compute the approximate λ and associated power using *MVpower*, a module of SAS IML statements (see Appendix). The input and selected

Listing 8.5 Computations for Dr. Kohlimein's cross-over study.

Listing 8.5.1.   SAS PROC IML statements to use MVpower module.

```
options nosource2;
title "Katie Kohlimein: Power for 2 (between) x 3 (within)";
proc iml;
%include MVpower;
Opt_off = {UNIGG};
*Define matrices *;
********************************
* BETA ROW1: Males         *
*      ROW2: Females       *
********************************;
* COLUMNS:  Control  Psych Stress  Dosophrine;
beta   = {    3          12            8    ,
              1           5            7    };

sigma = {     25         16           12    ,
              16         64           30    ,
              12         30           36    };

essenceX=I(2);      *creates 2 x 2 identity matrix;
repN={12 18 24};    *creates N sizes of 24, 36, and 48;
round=3;

Utreat =   { 1   0,
            -1   1,
             0  -1};

title2 "Gender main effect";
C = {1 -1};
U = {1,
     1,
     1}/3;
run power;

title2 "Treatment main effect";
U = Utreat;
C = {.5 .5};
run power;

title2 "Gender*Treatment interaction";
C = {1 -1};
U = Utreat;
run power;
```

Listing 8.5.2. Selected output

```
Katie Kohlimein: Power for 2 (between) x 3 (within)


                    Gender main effect

    CASE       ALPHA      TOTAL_N      WLK_PWR

     1         0.05         24          0.326
     2         0.05         36          0.467
     3         0.05         48          0.589



                 Treatment main effect

    CASE       ALPHA      TOTAL_N      WLK_PWR

     1         0.05         24          0.983
     2         0.05         36          0.999
     3         0.05         48          0.999



              Gender*Treatment interaction

    CASE       ALPHA      TOTAL_N      WLK_PWR

     1         0.05         24          0.461
     2         0.05         36          0.671
     3         0.05         48          0.814
```

output are given in Listing 8.5. In *MVpower*, $\mathbf{X}$ can be formed by repeating the "essenceX" matrix, $\dot{\mathbf{X}}$, "RepN" times. For example, if RepN is 3, then

$$\mathbf{X} = \begin{bmatrix} \dot{\mathbf{X}} \\ \dot{\mathbf{X}} \\ \dot{\mathbf{X}} \end{bmatrix}.$$

Because this is just a balanced two-group design (between-subjects), we set $\dot{\mathbf{X}}$ to be a 2×2 identity matrix. With RepN set at 12, 18, or 24, we create an $\mathbf{X}$ matrix for a cell-means model having N of 24, 36, or 48. The rest of the input should be easy to follow. *MVpower* has far more generality and complexity than we can deal with here.

With $\Omega > .99$, the test of the Treatment main effect has superb power at $N = 36$. But the Gender main and Gender*Treatment interaction effects are the key components of Dr. Kohlimein's research question, and thus she is troubled that $N = 36$ yields powers of only .47 and .67 under this scenario. Even with $N = 48$, the powers of the two tests are only .59 and .81. The latter sounds powerful, but represents a Type II error rate of .19. Of course, this is a rather limited power analysis at this point. Other reasonable scenarios for **B** and $\mathbf{\Sigma}$ should be studied as well. This might include the specification of tighter, planned contrasts having $df_U = 1$.

8.5 OTHER METHODS

This chapter focuses almost exclusively on performing sample-size analyses for situations in which the research questions call for studies designed to reject null hypotheses to be tested under fixed-effects linear models, both univariate and multivariate. What about analyzing sample size in other situations? In this section, we summarize some methods and cite key references that are helpful in other common situations. This includes methods appropriate for handling tests for simple random effects models and univariate repeated measures analysis. One large class of interest is the analysis of categorical variables, from comparing two independent proportions to more complex situations, such as comparing or modeling many proportions (logit analysis) or modeling contingency tables (log-linear models). Survival analyses are useful when the outcome measure is "time to event A" and the observations will be right censored because the subject is lost to follow-up or the study will end before the some subjects have experienced "event A." Finally, we note that sample-size considerations should sometimes focus on the widths of key confidence intervals, as when studies are being designed to show that treatments are indeed very similar rather than different.

8.5.1 Univariate Repeated Measures

Most texts on univariate experimental design discuss the so-called traditional univariate repeated measures analysis, also known as within-subjects designs and split-plot designs. Maxwell and Delaney (1990) were particularly thorough in comparing the univariate approach with the multivariate approach, which is outlined above in Example 5. The univariate method makes Subjects a random-effects blocking factor and thus becomes a mixed-model ANOVA. Its key assumption is that the vectors of repeated residuals from each subject are distributed according to a covariance matrix that has a structure called sphericity. Because sphericity is commonly violated, the Geisser-Greenhouse (1958) and Huynh-Feldt (1970) corrections to the mixed-model solution are often used. Muller and Barton (1989, 1991) and Muller et al. (1992) presented a scheme to compute approximate powers for these methods, but this work is beyond the scope of this chapter. The *MVpower* module contains special features to handle the univariate repeated measures problem.

8.5.2 Random-Effects ANOVAs

For ordinary random-effects designs, the nonnull distributions of the common F statistics are entirely different from their counterparts for fixed-effects designs—even when the null distributions are identical. Consider a one-way, random-effects design with J levels in a factor A, in which we test the variance component, H_0: $\sigma_A^2 = 0$ versus H_0: $\sigma_A^2 > 0$. Most texts note that the appropriate F-test (normal theory) is identical to that for the fixed-effects case, given here as (8.26). Scheffé (1959, p. 226) shows that if there are n cases in each level, F is distributed as $\{1 + n\sigma_A^2/\sigma^2\}F(J - 1, nJ - J, \lambda = 0)$. The term $\{1 + n\sigma_A^2/\sigma^2\}$ is the ratio of the numerator and denominator expected means squares (EMS). When H_0: $\sigma_A^2 = 0$ is true, $\{1 + n\sigma_A^2/\sigma^2\} = 1.0$, making the null case identical to that for fixed effects. When $\sigma_A^2 > 0$, the nonnull sampling distribution is just $\{1 + n\sigma_A^2/\sigma^2\}$ times a *central* F distribution. Thus, the power is

$$\Omega = \Pr[F(J - 1, nJ - J, \lambda = 0) \geq F_\alpha/\{1 + n\sigma_A^2/\sigma^2\}]. \tag{8.54}$$

The SAS code to do this is simply

```
EMSratio = 1 + n*VarCompA/(sigma**2);
F_alpha = FINV(1-alpha, J - 1, n*J - J, 0);
power = 1 - PROBF (F_alpha/EMSratio, J - 1, J*n - J, 0);
```

Prihoda (1983) generalized this idea somewhat and offered FORTRAN shareware.

8.5.3 Comparing Two Independent Proportions

Many studies require comparing two independent proportions, i.e., testing H_0: $\pi_1 = \pi_2$, using sample proportions, p_1 and p_2, estimated from $n_1 = w_1N$ and $n_2 = w_2N$ cases randomly drawn from (or assigned to) two groups. This is sometimes called the 2×2 comparative trial because it can be put into the context of a 2×2 contingency table in which one tests the independence of Group (control vs. experimental) and Outcome (success vs. failure). Just getting good p values is such a complex problem that over 25 different methods have been suggested for the generic 2×2 problem.

Some of these have several possible sample-size/power solutions. Here we give two power approximations based on the common t statistic and another for Fisher's exact test. The *OneWyPow* module performs the computations for all three.

Fisher's exact test is a *conditional* test in that it supposes that both margins of the 2×2 table are fixed. The methods based on the t statistic give *unconditional* tests. Debate on the merits of unconditional versus conditional tests has continued for decades with eminent statisticians taking both points of view. For the 2×2 comparative trial, however, we agree with Upton (1982) that the exact conditional test is "inappropriate," with its wide popularity coming largely from its designation of "exact." Suissa and Shuster (1985) stressed that unconditional tests are easier to interpret and explain to most nonstatisticians, who often do not understand the implications of making conditional inferences. Accordingly, Suissa and Shuster presented an *exact unconditional* test which, though computationally intensive, should one day become a common tool. For now, however, Fisher's exact conditional test is still very popular.

Unconditional Tests

The standard error of $(p_1 - p_2)$ is

$$\text{s.e.}[(p_1 - p_2)] = \left[\frac{w_2\pi_1(1 - \pi_1) + w_1\pi_2(1 - \pi_2)}{Nw_1w_2} \right]^{1/2}. \qquad (8.55)$$

This naturally leads to the *unpooled t* statistic,

$$t_u = N^{1/2} \left\{ \frac{[w_1w_2]^{1/2}(p_1 - p_2)}{\{N[w_2p_1(1 - p_1) + w_1p_2(1 - p_2)]/(N - 2)\}^{1/2}} \right\}. \qquad (8.56)$$

Suissa and Shuster presented exact critical values for $Z_u \equiv [N/(N - 2)]^{1/2}t_u$. When transformed to t_u, they agree well with those from the $t(N - 2)$ distribution. Accordingly, we take t_u be a $t(N - 2)$ variate with noncentrality parameter

$$\delta_u = N^{1/2} \left\{ \frac{[w_1w_2]^{1/2}(\pi_1 - \pi_2)}{[w_2\pi_1(1 - \pi_1) + w_1\pi_2(1 - \pi_2)]^{1/2}} \right\}. \qquad (8.57)$$

After studying this problem with respect to the robustness of the Type I error rates for balanced and unbalanced designs, D'Agostino, Chase, and Belanger (1988) recommended that the ordinary two-group t statistic be applied to data coded $Y = 0$ or 1, thus making $\hat{\mu}_j = p_j$. This is the *pooled t* statistic,

$$t_p = N^{1/2} \left\{ \frac{[w_1w_2]^{1/2}(p_1 - p_2)}{\{N[w_1p_1(1 - p_1) + w_2p_2(1 - p_2)]/(N - 2)\}^{1/2}} \right\}. \qquad (8.58)$$

As per Dozier and Muller (1993), t_p is taken to be a $t(N-2)$ random variable with noncentrality

$$\delta_p = N^{1/2} \left\{ \frac{[w_1 w_2]^{1/2}(\pi_1 - \pi_2)}{[w_1 \pi_1(1 - \pi_1) + w_2 \pi_2(1 - \pi_2)]^{1/2}} \right\}, \qquad (8.59)$$

which produced acceptable power estimates in their simulations.

Both noncentralities keep the form $\delta = N^{1/2}\delta^*$, where δ^*, the primary noncentrality, is not dependent on N. For balanced designs, $t_u = t_p$ and $\delta_u = \delta_p$. Letting

$$A = \frac{w_1 \pi_1(1 - \pi_1) + w_2 \pi_2(1 - \pi_2)}{w_2 \pi_1(1 - \pi_1) + w_1 \pi_2(1 - \pi_2)}, \qquad (8.60)$$

we see that $\delta_u = A^{1/2}\delta_p$. t_u will have greater (less) approximate noncentrality than t_p if and only if $A > 1$ ($A < 1$). By taking the first and second derivatives of δ_u and δ_p with respect to w_1, one can show that the optimal w_1 values for t_u and t_p are

$$\tilde{w}_{1u} = \left[1 + \left(\frac{\pi_2(1 - \pi_2)}{\pi_1(1 - \pi_1)} \right)^{1/2} \right]^{-1} \qquad \text{and}$$

$$\tilde{w}_{1p} = \left[1 + \left(\frac{\pi_1(1 - \pi_1)}{\pi_2(1 - \pi_2)} \right)^{1/2} \right]^{-1}, \qquad (8.61)$$

respectively. These results show that the unpooled test increases in power when the lower w_j weights correspond to the groups with π_j closer to 0 or 1, whereas the pooled test works in the opposite manner.

The Conditional Test

The Fisher (or Fisher-Irwin) exact test computes each p value by defining a specific hypergeometric distribution based on fixing both margins of the table. The power approximation presented here stems from the sequence of work by Casagrande, Pike, and Smith (1978), Fleiss, Tytun, and Ury (1980), and Diegert and Diegert (1981). It is based around Yates' (1934) statistic, which can be written

$$Z_y = N^{1/2} \left\{ \frac{[w_1 w_2]^{1/2}[(p_1 - p_2) - (2Nw_1 w_2)^{-1}]}{[\bar{p}(1 - \bar{p})]^{1/2}} \right\}, \qquad (8.62)$$

where $\bar{p} = w_1 p_1 + w_2 p_2$. Z_y is taken to be a standard normal variate, Z. For the directional test, which by our convention is $H_A: \pi_1 > \pi_2$, Z_y is significant at level α if $Z_y > Z_\alpha$. The nondirectional test simply uses $Z_{\alpha/2}$. Upton (1982) concluded that Z_y gives "extremely conservative" Type I error rates when considered as an unconditional test, but that it is "practically identical" to Fisher's exact conditional test. Suissa and Shuster (1985) gave exact results showing that Fisher's test requires larger sample sizes than does their exact version of t_u.

To compute approximate power for the directional test, let $r = w_2/w_1$; $\bar{\pi} = w_1\pi_1 + w_2\pi_2$; $h = (r + 1)/(r|\pi_1 - \pi_2|)$; $m = (r + 1)^2 h^2/N^2 - h + N/(r + 1)$. Compute

$$Z_\Omega = \frac{-(\pi_1 - \pi_2)(m\,r)^{1/2} + Z_\alpha[(r + 1)\bar{\pi}(1 - \bar{\pi})]^{1/2}}{[r\,\pi_1(1 - \pi_1) + \pi_2(1 - \pi_2)]^{1/2}}, \tag{8.63}$$

and then find the power, $\Omega_1 = \Pr[Z \geq Z_\Omega]$. For the nondirectional test, use $Z_{\alpha/2}$ instead of Z_α. This method undergirds the sample-size tables in Fleiss (1981), which are very handy but are limited to balanced designs. Thomas and Conlon (1992) presented an efficient algorithm to compute powers directly from the hypergeometric distribution. In particular, it should be used in cases involving a small expected frequency in one of the cells of the 2 × 2 table.

Example 6: Comparing Two Independent Proportions

One hundred and forty early-Alzheimer's patients are to be recruited for a randomized, placebo controlled, double-blinded clinical trial to test whether DS110891, a synthetic form of glutamate, improves their memory and learning, at least temporarily. All patients will continue to receive their standard care. Based on several cognitive measurements, each patient will be scored "improved" or "not improved" after 3 months on study. It is conjectured that 40% of the DS110891 patients versus 20% of the placebo patients will improve. To optimize the power of the t_u statistic, $w_1 = 55\%$ will be randomized to receive DS110891. Because $A = 1.04$, t_u will be slightly more powerful than t_p.

Listing 8.6 gives input and output related to the use of *OneWyPow*. With $N = 140$, the power for the .05-based, directional test is .842 for t_u and .829 for t_p. Z_y, which corresponds to Fisher's exact conditional test, has an approximate power of .780. For a balanced design, the power is .838 for both t_u and t_p and .786 for Z_y. Suissa and Shuster reported that with $N = 136$ and $\alpha = .048$, the power of their exact unconditional test is exactly .804 for this case. *OneWyPow* returns a value for t_u of .824 for these specifications.

Other Tests on Proportions: Log-Linear Models

Comparing two proportions is only one of the myriad of ways to analyze categorical data. The large class of methods known as log-linear models (which include logit models) provides the familiarity and flexibility of a linear models framework. These methods commonly employ log-likelihood-ratio χ^2 test statistics, and O'Brien (1986) discussed how the strategy of using

Listing 8.6 Computations for comparing two independent proportions.

SAS Input

```
options ls=72 nosource2;
%include OneWyPow;
title1 "DS110891 trial with Alzheimer's Patients";
cards;
pi .40 .20 .
weight .55 .45 .
alpha .01 .05 .
Ntotal 100 140 200 .
end
%include PowTab2;
```

SAS Output

```
---------------------------------------------------------------
|                         |              ALPHA                |
|                         |-----------------------------------|
|                         |    0.01      |     0.05           |
|                         |--------------+--------------------|
|                         |   Total N    |    Total N         |
|                         |--------------+--------------------|
|                         |100 |140 |200 |100 |140 |200 |
|                         |----+----+----+----+----+----|
|                         |Pow-|Pow-|Pow-|Pow-|Pow-|Pow-|
|                         | er | er | er | er | er | er |
|-------------------------+----+----+----+----+----+----|
|Unpooled     |2-tailed t |.357|.521|.718|.605|.752|.886|
|Approx.      |-----------+----+----+----+----+----+----|
|Uncond. Test |1-tailed t |.456|.620|.797|.721|.842|.936|
|-------------+-----------+----+----+----+----+----+----|
|Pooled       |2-tailed t |.341|.500|.696|.588|.735|.873|
|Approx.      |-----------+----+----+----+----+----+----|
|Uncond. Test |1-tailed t |.439|.600|.779|.706|.829|.928|
|-------------+-----------+----+----+----+----+----+----|
|Approx.      |2-tailed Z |.264|.420|.633|.504|.671|.837|
|Conditional  |-----------+----+----+----+----+----+----|
|Test         |1-tailed Z |.355|.524|.726|.633|.780|.905|
---------------------------------------------------------------
```

exemplary data extends well to these important methods. This strategy was summarized in Agresti's (1990) extensive text on categorical data analysis. The *PowSetUp* module is designed to create power tables using results obtained from an ordinary log-linear analysis of exemplary data.

8.5.4 Survival Analysis

There is a growing literature on methods for estimating power and determining sample-size for studies involving survival analysis, where the dependent measure is time until "death." Commonly, subjects are lost to follow-up after a known time on study or are still "living" at the time of data analysis. Such data are not missing, they are *censored*. We know that a given subject "lived" at least Y days, and we should use that information as fully as possible. Performing power analyses for such studies is a most complex problem. Analytical approximations have been developed for straightforward situations; see the summaries by Lachin (1981) and Donner (1984). Shuster (1990) developed an extensive set of tables useful for planning clinical trials. Goldman and Hillman (1992) presented a scheme that uses the "analysis" of exemplary data. Computer simulations are sometimes used to assess complex situations (Halpern and Brown, 1987). Simulation is an option that is always available when the situation is too complex or unique to warrant the effort required to develop "nice" analytical solutions.

8.5.5 "Accepting the Null" Using Appropriate Confidence Intervals

Finally, we mention that many studies are designed to show either that two or more treatments or groups are similar or that no relationship exists between or among variables. In medical research, bioequivalence studies are designed to test whether a new therapy has nearly the same average efficacy as existing standard therapies, albeit the new one offers fewer side effects, lower costs, or more reliability across patients. Thus, the researcher is expecting to "accept" null hypotheses about efficacy but reject null hypotheses about side effects, costs, and reliability. Because appropriate confidence intervals or regions often play the central role in such analyses, the sample-size analysis should ensure that those intervals or regions will be sufficiently small. Beal (1989) has developed a method and provided tables to handle situations involving the one-group and two-group t-test.

8.6 CONCLUSION

It is important and practical to perform sound, in-depth power analyses for proposed studies. Power analysis has many parallels with data analysis. Our common test statistics have nonnull distributions that can easily be characterized using terms and formulas that are familiar to the data analyst. Any general method for data analysis should have a parallel general method for power analysis. This chapter shows how this holds for tests falling under the univariate and multivariate fixed-effects linear models, with normal errors. We presented realistic examples for the two-group t-test, a matched-pairs t-test, the one-way ANOVA with contrasts, an analysis of covariance, and a repeated-measures analysis. The concepts developed here apply to all of the special cases of the univariate and multivariate general linear models with fixed effects, and to many other methods. In particular, we treated the comparison of two proportions in some depth. We briefly discussed how it is possible to use existing data to get more objective estimates for the effect sizes. Usually, however, the number of subjects in most pilot studies is too small to make this effort "safe" and worthwhile.

Computing systems need to be developed so that researchers can conduct full sensitivity analyses of the power over a range of reasonable conjectures for the critical population parameters. Whenever possible, power analysis and data analysis should share the same software systems. Wright and O'Brien (1988) discussed how options in SAS PROC GLM could be developed to make it perform both steps of the two-step process used in Example 4. Some of these notions have now been implemented in JMP®, a Macintosh application from the SAS Institute (1991).

ACKNOWLEDGMENT

Work by R. G. O'B. was supported in part by grants from the U.S. National Institutes of Health (GCRC: RR00082) and the UF Division of Sponsored Research, which funded Zhanying Bai and Yonghwan Um in their writing of one portion of the OneWyPow.sas shareware module. Dan Bowling provided help on numerous matters. Work by K. E. M. was supported in part by grants from the U.S. National Institutes of Health (NICHD: P30-HD03110-22, NCI: P01 CA47982-04, GCRC: RR00046).

APPENDIX: GETTING THE SOFTWARE MODULES

All computations in this chapter were performed within the SAS® System using modules (files containing SAS statements) that are processed via %INCLUDE statements in SAS input. At the time of publication, the developers are making these modules available on a "freeware" basis.

The *OneWyPow*, *PowSetUp*, and several *PowTab* modules handled all univariate methods (Examples 1–4 and 6). Their use and output is displayed in Listings 8.1–8.4 and 8.6. The modules run within the "base" SAS environment.

The *MVpower* module performed the computations for Example 5. See Listing 8.5. *MVpower* is a module of PROC IML statements. PROC IML is not part of base SAS.

Software for power and sample size analysis is available from numerous others. We have made no attempt to survey this growing field. Goldstein (1989) reviewed several programs for the PC-DOS environment; see also O'Brien (1988). JMP®, a Macintosh application from the SAS Institute (1991), has new power options.

Univariate Modules: OneWyPow, PowSetUp, FPowTab1, Etc.

Developer: Ralph G. O'Brien, Statistics/Biostatistics,
Box 100212, UF Health Sciences Center,
University of Florida, Gainesville, FL 32606-0212
904-392-8446, robrien@stat.ufl.edu

These univariate modules are freeware. They run within the base SAS environment. Versions of them have run on various computing platforms, including IBM VM/CMS, DEC Ultrix (UNIX), and PC-DOS. The files distributed contain all source statements and instructions on installation and use.

These modules have no warranty whatsoever. The developer would appreciate hearing about problems and suggestions for improvements regarding the software. He would also like to hear when they worked well for you. Please correspond via electronic mail (preferred) or telephone. Time rarely allows him to respond to "consulting" questions.

This freeware and their updates and successors may be obtained from the developer using three methods:

1. The best method is file transfer via the Internet network and "anonymous ftp," as described below. There is no fee whatsoever.
2. For a nominal service fee, the developer will transfer the files to your account over the Internet or BITNET networks. Please use email to initiate this.
3. For a nominal fee for service and materials, the developer will send a single 3.5-inch, 720K, DOS-formattted diskette. Please call to initiate this.

anonymous ftp

The best way to obtain this freeware is to use the Internet network to ftp-get the files on your own. "ftp" (lowercase letters may be required) stands for File Transfer Protocol and is now available on almost all networked computers. The details regarding where and how these files are stored will change over time. Accordingly, the instructions given here tell only how to get a short file that gives complete, up-to-date instruction on all this.

1. Get to the "front door" of the ftp serving workstation in the University of Florida's Department of Statistics by entering

    ftp ftp.stat.ufl.edu

    on your local Internet-connected computer.
2. Login using the name

    anonymous

For the password, enter your full system name, something of the form
    mmouse@wdw.orlando.fl
3. Get a copy of the short information memo by entering
    cd pub
    get ReadMe.power myfile.asc
This will put a file called ReadMe.power on your local computer using the file specification of your choice ("myfile.asc").
4. Enter
    close
    quit
to return gracefully to your local environment, where you can study the instructions in myfile.asc.
5. This method should work for a number of years, but computing systems have unpredictably short half-lives. If you encounter problems, call Ralph O'Brien or the Data Services Manager in the UF General Clinical Research Center (904 395-0111 Ext: 5-8565).

The Multivariate Module: MVpower

Codevelopers: Lynette L. Keyes, UNC, Chapel Hill

Keith E. Muller, Department of Biostatistics, CB#7400 University of North Carolina, Chapel Hill, NC 27599 919-966-7272, muller.bios@mhs.unc.edu

MVpower.sas is a module of SAS PROC IML statements that perform power analyses for the General Linear Multivariate Model and for the univariate approach to repeated measures, including the Geisser-Greenhouse and Huynh-Feldt solutions. The power approximation methods of Muller and Peterson (1984) are used for the general multivariate case. Exact results are provided for all univariate models and some multivariate models. The methods of Muller and Barton (1989, 1991) undergird the multivariate and univariate approaches to the power approximations for repeated measures. Key restrictions include the assumptions of Gaussian (normal) errors, fixed predictor values, a common design for all responses, and no missing data.

MVpower handles a very broad class of univariate and multivariate ANOVA and regression problems, because users specify directly the **X**, **B**, **Σ**, **C**, and **U** matrices that form the models and hypotheses outlined in Section 8.4. All matrices must be full rank. Little or no experience with PROC IML is required to use *MVpower*. PROC IML is a supplement to the "basic" distribution of SAS. If it is not yet available to you, you may order it from SAS Institute.

Information about obtaining the current version of *MVpower* is given in the ReadMe.power file obtainable via "anonymous ftp" as described above. Or you may contact the developers directly.

REFERENCES

Agresti, A. (1990). *Categorical Data Analysis*. Wiley, New York.

Barton, C. N. and Cramer, E. C. (1989). Hypothesis Testing in Multivariate Linear Models with Randomly Missing Data. *Comm. Stat.—Simul. Comp.*, *B18*: 875–895.

Beal, S. L. (1989). Sample Size Determination for Confidence Intervals on the Population Means and on the Difference Between Two Population Means. *Biometrics*, *45*: 969–977.

Blanchard, E. B., Appelbaum, K. A., Radnitz, C. L., Morrill, B., Michultka, D., Kirsch, C., Guarnieri, P., Hillhouse, J., Evans, D. D., Jaccard, J., and Barron, K. D. (1990). A Controlled Evaluation of Thermal Biofeedback and Thermal Biofeedback Combined with Cognitive Therapy in the Treatment of Vascular Headache. *J. Cons. Clin. Psych.*, *2*: 216–224.

Casagrande, J. T., Pike, M. C., and Smith, P. G. (1978). An Improved Approximate Formula for Calculating Sample Sizes for Comparing Two Binomial Distributions. *Biometrics*, *34*: 483–486.

Cohen, J. (1988). *Statistical Power Analysis for the Behavioral Sciences* (2nd ed.). Lawrence Erlbaum, Hillsdale, New Jersey.

Cohen, J. (1992). A Power Primer. *Psych. Bull.*, *112*: 155–159.

Cohen, S., Tyrrell, D. A. J., and Smith, A. P. (1991). Psychological Stress and Susceptibility to the Common Cold. *N. Engl. J. Med.*, *325*: 606–612.

D'Agostino, R. B., Chase, W., and Belanger, A. (1988). The Appropriateness of Some Common Procedures for Testing the Equality of Two Independent Binomial Populations. *Amer. Statist.*, *42*: 198–202.

Diegert, C. and Diegert, K. V. (1981). Note on Inversion of Casagrande-Pike-Smith Approximate Sample-Size Formula for Fisher-Irwin Test on 2×2 Tables. *Biometrics*, *37*: 595.

Donner, A. (1984). Approaches to Sample Size Estimation in the Design of Clinical Trials—A Review. *Statist. Med.*, *3*: 199–214.

Dozier, W. G. and Muller, K. E. (1993). Small-Sample Power of Uncorrected and Satterthwaite Corrected *t* Tests for Comparing Two Proportions. *Comm. Stat.—Simul. Comp.*, *B22*.

Fleiss, J. L. (1981). *Statistical Methods for Rates and Proportions* (2nd ed.). Wiley, New York.

Fleiss, J. L., Tytun, A., and Ury, H. K. (1980). A Simple Approximation for Calculating Sample Sizes for Comparing Independent Proportions. *Biometrics*, *36*: 343–346.

Fleming, T. R., Harrington, D. P., and O'Brien, P. C. (1984). Designs for Group Sequential Tests. *Controlled Clin. Trials*, *5*: 348–361.

Freiman, J. A., Chalmers, T. C., Smith, H., and Kuebler, R. R. (1979). The Importance of Beta, the Type II Error, and Sample Size in the Design and

Interpretation of the Randomized Clinical Trial. *N. Engl. J. Med.*, *299*: 690–694. [Updated and reprinted in Bailar, J. C. III and Mosteller, F. (eds.) (1992). *Medical Uses of Statistics* (2nd ed.). NEJM Books, Boston, pp. 357–373.]

Gatsonis, C. and Sampson, A. R. (1989). Multiple Correlation: Exact Power and Sample Size Calculations. *Psych. Bull.*, *106*: 516–524.

Geisser, S. and Greenhouse, S. W. (1958). An Extension of Box's Results on the Use of the *F* Distribution in Multivariate Analysis. *Ann. Math. Statist.*, *29*: 885–891.

Goldman, A. I. and Hillman, D. W. (1992). Exemplary Data: Sample Size and Power in the Design of Event-Time Clinical Trials. *Controlled Clin. Trials*, *13*: 256–271.

Goldstein, R. (1989). Power and Sample Size via MS/PC-DOS Computers. *Amer. Statist.*, *43*: 253–260.

Halpern, J. and Brown, B. W. (1987). Designing Clinical Trials with Arbitrary Specification of Survival Functions and for the Log Rank and Generalized Wilcoxon Text. *Controlled Clin. Trials*, *8*: 177–189.

Hedges, L. V. (1981). Distribution Theory for Glass's Estimator of Effect Size and Related Estimators. *J. Educ. Statist.*, *6*: 107–128.

Huynh, H. and Feldt, L. S. (1970). Conditions Under Which Mean Square Ratios in Repeated Measurements Have Exact *F*-Distribution. *J. Amer. Statist. Assoc.*, *65*: 1582–1589.

Jemmott III, J. B., Hellman, C., McClelland, D. C., Locke, S. E., Kraus, L., Williams, R. M., and Valeri, C. R. (1990). Motivational Syndromes Associated with Natural Killer Cell Activity. *J. Behav. Med.*, *13*: 53–73.

Kraemer, H. and Thiemann, S. (1987). *How Many Subjects?* Sage, Beverly Hills, California.

Lachin, J. M. (1981). Introduction to Sample Size Determination and Power Analysis for Clinical Trials. *Controlled Clin. Trials*, *2*: 93–113.

Maxwell, S. E. and Delaney, H. D. (1990). *Designing Experiments and Analyzing Data: A Model Comparison Perspective*. Wadsworth, Belmont, California.

Melton, K. (1986). A Procedure for Initiating Process Control. Unpublished Ph.D. dissertation, University of Tennessee, Knoxville.

Muller, K. E. and Barton, C. N. (1989). Approximate Power for Repeated Measures ANOVA Lacking Sphericity. *J. Amer. Statist. Assoc.*, *84*: 549–555.

Muller, K. E. and Barton, C. N. (1991). Correction to "Approximate Power for Repeated Measures ANOVA Lacking Sphericity." *J. Amer. Statist. Assoc.*, *86*: 255–256.

Muller, K. E. and Peterson, B. L. (1984). Practical Methods for Computing Power in Testing the Multivariate General Linear Hypothesis. *Comp. Statist. Data Anal.*, *2*: 143–158.

Muller, K. E., LaVange, L. M., Ramey, S. L., and Ramey, C. T. (1992). Power Calculations for General Linear Multivariate Models Including Repeated Measures Applications. *J. Amer. Statist. Assoc.*, *87*: 1209–1226.

O'Brien, R. G. (1979). A General ANOVA Method for Robust Tests of Additive Models for Variances. *J. Amer. Statist. Assoc.*, *74*: 877–881.

O'Brien, R. G. (1981). A Simple Test for Variance Effects in Experimental Designs. *Psych. Bull.*, *89*: 570–574.

O'Brien, R. G. (1986). Using the SAS System to Perform Power Analyses for Log-Linear Models. *Proceedings of the Eleventh SAS Users Group International Conference.* SAS Institute, Cary, North Carolina, pp. 778–784.

O'Brien, R. G. (1988). Review of PowerPack 2.22. *Amer. Statist.*, *42*: 266–270.

O'Brien, R. G. and Kaiser, M. K. (1985). MANOVA Method for Analyzing Repeated Measures Designs: An Extensive Primer. *Psych. Bull.*, *97*: 316–333.

O'Brien, R. G. and Shieh, G. (under review). Pragmatic, Unifying Algorithm Gives Power Probabilities for Common F Tests of the Multivariate General Linear Hypothesis.

Pocock, S. J. and Simon, R. (1975). Sequential Treatment Assignment with Balancing for Prognostic Factors in the Controlled Clinical Trial. *Biometrics*, *31*: 103–115.

Prihoda, T. J. (1983). Convenient Power Analyses for Complex Analysis of Variance Models. *Proceedings of the ASA Statistical Computing Section*, pp. 267–271.

Rosenthal, R. and Rosnow, R. L. (1985). *Contrast Analysis: Focused Comparisons in the Analysis of Variance.* Cambridge University Press, Cambridge, U.K.

SAS Institute. (1990). *SAS® Language Reference, V. 6.* SAS Institute, Inc., Cary, North Carolina.

SAS Institute. (1991). *JMP® Users Guide, V. 2.0.* SAS Institute, Inc., Cary, North Carolina.

Scheffé, H. (1959). *The Analysis of Variance.* Wiley, New York.

Seber, G. A. F. (1984). *Multivariate Observations.* Wiley, New York.

Sedlmeier, P. and Gigerenzer, G. (1989). Do Studies of Statistical Power Have an Effect on the Power of Studies? *Psych. Bull.*, *105*: 309–316.

Shuster, J. J. (1990). *Handbook of Sample Size Guidelines for Clinical Trials.* CRC Press, Boca Raton, Florida.

Stoney, C. M., Matthews, K. A., McDonald, R. H., and Johnson, C. A. (1988). Sex Differences in Lipid, Lipoprotein, Cardiovascular, and Neuroendocrine Responses to Acute Stress. *Psychophys*, *25*: 645–656.

Suissa, S. and Shuster, J. J. (1985). Exact Unconditional Sample Sizes for the 2 × 2 Comparative Trial. *J. Roy. Statist. Soc.*, *A148*: 317–327.

Thomas, R. G. and Conlon, M. (1992). Sample Size Determination Based on Fisher's Exact Test for Use in 2 × 2 Comparative Trials with Low Event Rates. *Controlled Clin. Trials.*, *13*: 134–147.

Upton, G. J. G. (1982). A Comparison of Alternative Tests for the 2 × 2 Comparative Trial. *J. Roy. Statist. Soc.*, *A145*: 86–105.

Venables, W. (1975). Calculation of Confidence Intervals for Noncentrality Parameters. *J. Roy. Statist. Soc.*, *B37*: 406–412.

Wright, S. P. and O'Brien, R. G. (1988). Power Analysis in an Enhanced GLM Procedure: What It Might Look Like. *Proceedings of the Thirteenth SAS Users Group International Conference*, SAS Institute, Cary, North Carolina, pp. 1097–1102.

Yates, F. (1934). Contingency Tables Involving Small Numbers and the χ^2 Test. *J. Roy. Stat. Soc. Suppl.*, *1*: 217–235.

9
Robustness in ANOVA

RAND WILCOX University of Southern California,
Los Angeles, California

9.1 INTRODUCTION

Two fundamental assumptions associated with the usual analysis of variance
model for independent groups are that the groups have equal variances
and that observations within each group have a normal distribution. Twenty
years ago there was little evidence that violating either of these assumptions
would cause a serious problem. In fact, popular textbooks reassured the
reader that there was nothing to worry about, and if there was a cause for
concern, using equal sample sizes would eliminate any problems. This is
not to say, however, that 20 years ago there was no cause for concern. In
fact, articles by Student (1927), Pearson (1931), Box (1953), and Tukey
(1960) hinted that nonnormality might be a serious problem, especially
in reducing power. Nevertheless, most researchers apparently felt that
there was no compelling reason to label standard ANOVA procedures as
unsatisfactory.

Today it is abundantly clear that the usual F test, and related multiple
comparison procedures, are not robust. In fact, for reasons described below, violating assumptions can have serious consequences in terms of both
Type I errors and power. This is true when violating the usual equal variance assumption *as well as the usual assumption of normality*. Section 9.2
describes a history of the problems. Included is a brief description of some
of the technical details needed to arrive at better solutions for dealing with
nonnormality. Section 9.3 illustrates some of the problems that can arise,
and Section 9.4 outlines recent developments for dealing with the problems
described in Sections 9.2 and 9.3. Problems associated with unequal var-

iances are reviewed in Section 9.2, but the emphasis in this chapter, particularly in Section 9.4, is on recently developed methods for dealing with nonnormality. The reason for this emphasis is that nonnormality can be an especially serious problem in terms of power, and the procedures described in Section 9.4 are relatively new and unfamiliar to most applied researchers. The newer procedures also have advantages in terms of Type I errors, but their efficiency and power provide a more compelling reason for choosing them over the traditional techniques.

It should be stressed that some of the more obvious approaches to nonnormality can be unsatisfactory. One of these is to use a conventional nonparametric technique; another is to test for outliers, remove them if they are found, and proceed using a conventional procedure for comparing means. For example, the Mann-Whitney (1947) method for comparing two independent groups is both biased and inconsistent (Kendall and Stuart, 1973). That is, there are situations in which the null hypothesis is false, yet the probability of rejecting the hypothesis is less than the nominal level; and there are also situations in which the null hypothesis is false, yet the power of the test does not approach one as the sample sizes go to infinity. Zaremba (1962) proposed a correction, but the resulting procedure can have low power relative to other procedures. It is not being suggested that nonparametric procedures are completely unsatisfactory. In fact, recently proposed methods where observations are converted to ranks appear to have some positive features, but there are technical issues that still need to be resolved. Details of these technical issues are described in Section 9.2.8. Comments on simply discarding observations and using a conventional procedure for comparing means can be found in Section 9.2.6. For the moment it is merely noted that this approach can be unsatisfactory and that better methods are available.

9.2 ISSUES RELATED TO UNEQUAL VARIANCES AND NONNORMALITY

For J independent groups, let μ_j and σ_j be the corresponding means and standard deviations, $j = 1, \ldots, J$. Also let X_{ij} be the ith observation randomly sampled from the jth group. The usual assumptions are that the X_{ij}'s have a normal distribution with a common standard deviation $\sigma = \sigma_1 \cdots = \sigma_J$. The equal variance assumption has received most of the attention in the behavioral science literature, and it has become increasingly clear that violating this assumption can cause serious problems (Keppel, 1991; Tomarken and Serlin, 1986; Wilcox, 1987a, 1987b; Wilcox, Charlin and Thompson, 1986). A brief history of the problem, plus a description of recent developments, is given in the next three subsections of this chapter. The remainder of this section deals with issues related to nonnormality. Recently developed methods for dealing with unequal variances and non-normality are described in Section 9.4.

9.2.1 The Effects of Unequal Variances

First consider J independent groups, and for the moment assume observations are sampled from a normal distribution. For many years it was thought that violating the equal variance assumption would have little or no effect on the probability of a Type I error. This belief stemmed from an overgeneralization of some results reported by Box (1954). Let $\sigma_{(1)} \le \cdots \le \sigma_{(J)}$ be the population standard deviations written in ascending order. In addition to various theoretical results, Box reported some numerical results on how the F test is affected when the variances are unequal. His numerical results were limited to situations with $\tau = \sigma_{(J)}/\sigma_{(1)} \le \sqrt{3}$. Box never claimed that the F test is robust to violations of the equal variance assumption, but his results indicated that it is robust if τ is less than or equal to $\sqrt{3}$, and some researchers apparently believed that this made the F test acceptable in practice.

For the special case of $J = 2$ groups, analytical results indicate that the F test is robust when equal sample sizes are used and when the sample sizes are not too small (Ramsey, 1980). More specifically, with sample sizes of 15 observations from both groups at $\alpha = .05$, the actual probability of a Type I error will not exceed .06 when sampling from normal distributions. However, with unequal sample sizes, problems can arise. For example, with sample sizes of 11 and 21, and $\tau = 4$, Wilcox et al. (1986) found that, when testing at the $\alpha = .05$ level, the actual probability of a Type I error is approximately .155 when there is a negative pairing of sample sizes and variances, that is, when the group with the smallest sample size has the largest variance. One reaction might be that this problem can be ignored when larger sample sizes are used, but when are the sample sizes large enough? When distributions are nonnormal, the situation is much worse for reasons given below.

About 20 years after Box's paper, evidence began to mount that the F test was indeed sensitive to unequal variances and that difficulties could arise in applied settings. The first step was to look at situations with larger values of τ than those used by Box. Brown and Forsythe (1974a,1974b) considered situations with $\tau = 3$, and they found situations where the F test was unsatisfactory in terms of Type I errors. They proposed a solution that appeared to correct the problem. (For an extension of their procedure to a two-way design, see Brown and Forsythe, 1974b.) They also found that Welch's (1951) method gave good results. Progress had been made, but what about situations where $\tau > 3$? Is $\tau > 3$ a realistic situation, and, if so, what happens to the Welch (1951) and Brown–Forsythe (1974a) solutions for handling unequal variances? To answer the first question, Wilcox (1987a) surveyed some published studies and found situations with $\tau > 4$. Thus, the inflated Type I errors found in the two-group case with $\tau = 4$ (Wilcox et al., 1986) correspond to what would seem to be a realistic situation. Of course, there is some possibility that the estimates of τ found by Wilcox (1987a) exceed the actual values of τ, in which case there is some hope that the results reported by Brown and Forsythe still apply. However, this is not very reassuring for the applied researcher who wants to control the probability of a Type I error. At a minimum, a method for comparing means should be robust when $\tau = 4$, and ideally it should perform well when τ is even larger.

Wilcox et al. (1986) expanded on the results reported by Brown and Forsythe (1974b) by considering larger τ values. Two unexpected results were obtained regarding the conventional F test. The first was that F became increasingly sensitive to unequal variances as J, the number of groups, increased. For example, with $J = 2$, the actual probability of a Type I error was found to be as high as .155 when testing at the .05 level, while for $J = 6$, it was as high as .309! The other unexpected result was that for $J > 2$, Type I errors were inflated even when using equal sample sizes. For example, with $J = 4$, and equal sample sizes of 50, the actual probability of a Type I error can be as high as .088 when testing at the .05 level.

Wilcox et al. found that for that $\tau > 3$, the Welch and Brown–Forsythe procedures also begin to break down, at least for $J > 2$. With $J = 4$ groups the probability of a Type I error for either procedure can exceed .08 when testing at the .05 level. Welch's procedure performed reasonably well with equal sample sizes; the actual probability of a Type I error never exceeded .065. However, a better solution is clearly needed. A positive feature is that Welch's procedure seems to perform very well for $J = 2$ groups, at least in terms of Type I errors, but for nonnormal distributions with unequal sample sizes, problems can arise. It should be stressed, however, that Welch's procedure is not being advocated even when equal sample sizes are used. The reason is that Welch's procedure can have relatively low power in commonly occurring situations in which distributions have heavy tails. Student's t-test has even more serious problems, which are described below.

James (1951) considered a general approach to the problem of unequal variances, and he suggested that two special cases be used in practice. His first-order method is known to be unsatisfactory, but his second-order method has been found to give very good control over the probability of a Type I error (Dijkstra and Werter, 1981; Wilcox, 1989). A possible objection to James's procedure is that the required computations are rather involved. A FORTRAN program can be written to perform James's method on a computer, but such a program is not readily available to applied researchers, nor has it been included in standard statistical packages. A simpler yet accurate alternative to James's solution would seem useful. Such a procedure, derived by Wilcox (1989), is easy to use, and it appears to be slightly more accurate than James's solution in terms of Type I errors. Both James's and Wilcox's procedures have about the same amount of power, but Wilcox's method has the advantage that it is easily extended to higher-way designs. Moreover, Wilcox's procedure performs well even with $\tau = 6$. A description of Wilcox's procedure is given in Section 9.4. For some recent alternative solutions, see Matuszewski and Sotres (1986) and Krutchkoff (1988).

9.2.2 Testing for Unequal Variances

A natural reaction to the problems associated with unequal variances is to try to salvage the F test by first testing for equal variances. However, all indications are that this approach is unsatisfactory (Markowski and Markowski, 1990; Moser, Stevens, and Watts, 1989; Wilcox et al., 1986). There are at least two problems. First, even when normality is assumed, tests for equal variances can have too little power to detect situations where the F test should be abandoned, even when testing at the .25 level. Second, most tests for equal variances are not robust to nonnormality. Conover, Johnson, and Johnson (1981) examined many tests for equal variances. Some of these appeared to be robust, but Wilcox (1990b) found situations where the methods recommended by Conover et al. are unsatisfactory. In fact, the Box–Scheffé method (Box, 1953; Scheffé, 1959) was the only procedure found to be satisfactory among the methods examined by Wilcox (1990b). However, even this procedure can be unsatisfactory when there are unequal sample sizes. Other problems are that the Box–Scheffé procedure is not invariant under permutations of the observations, and it has relatively low power.

9.2.3 Issues Related to Nonnormality and Type I Errors

This subsection provides a brief description of how nonnormality can affect Type I errors. First consider Welch's test for two independent groups. Let $\overline{X}_j$ be the sample mean for the jth group, let $s^2 = s_1^2 + s_2^2$, and let ϑ_{kj} be the kth cumulant corresponding to the jth group. Welch's test statistic is

$$W = \frac{\overline{X}_1 + \overline{X}_2}{\sqrt{s_1^2/n_1 + s_2^2/n_2}}.$$

From Tan (1982) or Kendall and Stuart (1973), the effect of nonnormality on W, in terms of Type I errors, is primarily due to the correlation between $\overline{X}_1 - \overline{X}_2$ and s^2. The covariance between these two terms is approximately

$$\text{COV}(\overline{X}_1 - \overline{X}_2, s^2) = \frac{(n_1 - 1)\vartheta_{31}}{n_1} - \frac{(n_2 - 1)\vartheta_{32}}{n_2}.$$

If the skewnesses corresponding to the two groups are identical, in which case $\vartheta_{31} = \vartheta_{32}$, and if the sample sizes are equal, then $\text{COV}(\overline{X}_1 - \overline{X}_2, s^2)$ = 0. This suggests that for this special case, Welch's test is robust in terms of Type I errors, and all indications are that this speculation is correct. However, for unequal variances and different skewnesses, problems can arise as shown in Section 9.3. In fact, Welch's test can be negatively biased. That is, it is possible that the null hypothesis is false, yet the probability of a Type I error is less than the nominal level.

It is noted that Student's t-test is also affected by unequal sample sizes and unequal skewnesses, but the situation is only worse. In particular, Cressie and Whitford (1986) show that, in general, Student's t-test is not even asymptotically correct!

Another approach to controlling the Type I error probability is to use a more robust estimate of the mean in the hope that the effect of unequal skewnesses would be reduced. Tiku (1982) describes a method for comparing groups where the mean is estimated based on trimming. However, Wilcox (1990a) found Tiku's procedure to be even more sensitive to unequal skewnesses than Welch's method. Both 10% and 20% trimming were considered.

9.2.4 Nonnormality and Power

Nonnormality can have a devastating effect on power. Two features make this a concern for the applied researcher. First, heavy-tailed distributions are pervasive in actual data. Investigators in the behavioral sciences found that distributions often have unexpectedly heavy tails (Micceri, 1989; Wilcox, 1990a). Interestingly, Tukey (1960) argued that in applied problems, heavy-tailed distributions are to be expected for various reasons. The second feature is that it is not always obvious, based on observed data, whether sampling is from a heavy-tailed distribution. In fact, even if a graph of the actual distribution under study were available, it still might not be evident that a heavy-tailed distribution is being investigated.

Tukey illustrates this point with a contaminated normal distribution given by

$$\Phi_{\varepsilon,v}(x) = (1 - \varepsilon)\Phi(x) + \varepsilon\Phi(x/v),$$

where $\Phi(x)$ is the standard normal distribution, ε is the proportion of contamination, and v is the scaling constant for the second normal distribution. Suppose $\varepsilon = .1$, and $v = 10$. At first glance it might appear that there is a large difference between the normal and contaminated normal distributions. However,

$$d(\Phi, \Phi_{\varepsilon,v}) = \sup|\Phi_{\varepsilon,v}(x) - \Phi(x)| \leq .04.$$

That is, the two distributions do not differ by more than .04 for any value of x. Tukey shows a graph of these two distributions illustrating that they look nearly identical.

Heavy-tailed distributions are of concern because they inflate the variance, which in turn reduces power. For example, the standard normal distribution has variance 1, while the contaminated normal distribution with $\varepsilon = .1$ and $v = 10$ has variance 10.9. The effect on power can be tremendous, as illustrated in Section 9.3. What is needed is a method for comparing groups that is not unduly affected by heavy tails. Suppose that sampling is from a normal distribution and that Welch's procedure has high power when, say, $\mu_1 - \mu_2 = 1$. The ideal procedure would have high power for this same situation and continue to have high power when sampling from a distribution only slightly different from a standard normal distribution with heavier tails. That is, a small shift away from the normal model should not have a big impact on power. Such efficient procedures are described in the next subsection as well as in Section 9.4.

9.2.5 Measuring Effect Size

It is noted that heavy-tailed distributions have important implications about how a researcher might measure effect size. To illustrate this point, consider two normal distributions with means μ_1 and μ_2 and a common variance σ^2. In this case the most common measure of effect size is $\delta = (\mu_1 - \mu_2)/\sigma$. Suppose $\mu_1 = 3$, $\mu_2 = 0$, and $\sigma = 1$. Then $\delta = 3$, and most researchers would probably judge this to be a large effect. Now consider contaminated normal distributions with the same means, but with $\varepsilon = .1$ and $v = 10$. As indicated in the previous section, the distributions have been changed by only a slight amount, in terms of the metric d, yet δ has been reduced to $3/\sqrt{10.9} = .91$. Suppose $\varepsilon = .1$ and $v = 20$. Then $d(\Phi, \Phi_{\varepsilon,v}) = .045$, but this distribution has variance 40.9. Thus, as before, a relatively small change in the tail of a distribution results in a large increase in the variance.

Consequently, the effect size reduces to $\delta = .47$. As previously indicated, heavy-tailed distributions are common in applied work, so any measure of effect size based in part on the variance must be interpreted with caution.

9.2.6 Yuen's Trimmed *t*-test

Yuen (1974) suggested a method for comparing groups that counters the effect of heavy tails. Her procedure is based on the *g*-times trimmed mean, which, for a random sample $X_1, \ldots, X_n$, is defined to be

$$\overline{X}_g = (X_{(g+1)} + \cdots + X_{(n-g)})/(n - 2g),$$

where $X_{(1)} \leq \cdots \leq X_{(n)}$ are the usual order statistics. Common choices for *g* are $[.1n]$ and $[.2n]$, where $[x]$ is the greatest integer less than or equal to x. Other choices for *g* might provide more power in various situations. Empirical methods for determining *g* are discussed by Hogg (1974). For some concerns about estimating *g*, written in a broader context, see Huber (1981, p. 7). The corresponding Winsorized mean is designated by $\overline{X}_w$.

Let $S_g^2 = SSD/(h - 1)$ where $h = n - 2g$ and

$$SSD = (g + 1)(X_{(g+1)} - \overline{X}_w)^2 + (X_{(g+2)} - \overline{X}_w)^2 + \\ \cdots + (X_{(n-g-1)} - \overline{X}_g)^2 + (g + 1)(X_{(n-g)} - \overline{X}_g)^2.$$

Yuen's test statistic is

$$Y = (\overline{X}_{g1} - \overline{X}_{g2}) \Big/ \sqrt{\frac{S_{g1}^2}{h_1} + \frac{S_{g2}^2}{h_2}},$$

where $\overline{X}_{gj}$ and S_{gj}^2 are the values of $\overline{X}_g$ and S_g^2 corresponding to the *j*th group. The null distribution of Y is approximated with Student's *t* distribution with estimated degrees of freedom f, where $1/f = c^2/(h_1 - 1) + (1 - c^2)(h_2 - 1)$, and $c = (S_{g1}^2/h_1)/(S_{g1}^2/h_1 + S_{g2}^2/h_2)$. It is noted that Yuen's test is designed to handle unequal variances by using estimated degrees of freedom similar to the estimate used by Welch (1951). Indirect support for Yuen's procedure is reported by Patel, Mudholkar, and Fernando (1988). Comparisons between Yuen's trimmed *t*-test and Welch's procedure are described in the next section.

Yuen's procedure illustrates an interesting technical point. If the *g* largest and *g* smallest observations are discarded, the temptation is to simply apply Welch's procedure to the remaining observations. If this were done, SSD would be replaced by

$$(X_{(g+2)} - \overline{X}_g)^2 + \cdots + (X_{(n-g-1)} - \overline{X}_g)^2.$$

However, theory associated with the trimmed mean indicates that SSD should be used instead (Staudte and Sheather, 1990), and simulations reported by Yuen indicate that SSD gives good results. If SSD were replaced by the expression just given, the estimated standard error of the trimmed mean would be too low, and Yuen's procedure would no longer provide good control over the probability of a Type I error without some other adjustment.

It is noted that Yuen's procedure is an example of what is called an L-statistic. L-statistics are just linear combinations of order statistics. A general issue is whether there are L-statistics that would provide any advantages over Yuen's trimmed t-test. In terms of power, the answer is yes. An example is Wilcox's (1991a) C procedure, described in Section 9.4, which is designed to compare to the medians of two or more groups.

9.2.7 *M*-Estimators of Location

A basic problem with procedures designed to compare means is that the sample mean has variance σ^2/n, and the value of σ^2 is sensitive to the tails of a distribution. As already indicated, a small shift from a normal distribution toward a contaminated normal can have a substantial effect on σ^2. A better method of comparing groups is first to search for a measure of location with a standard error that is relatively unaffected by small changes in the tails of a distribution. L-statistics represent one class of measures, but another class of statistics, called M-estimators, is beginning to take on increased importance when comparing two or more groups, M-estimators include the usual sample mean, as well as maximum likelihood estimators, as a special case. One very important feature is that M-estimators can be derived that have very nice theoretical properties. In particular, there are M-estimators with standard errors that are relatively unaffected by heavy tails. Moreover, "good" M-estimators can be derived under very general conditions, even when distributions are skewed.

The theory behind the derivation of robust M-estimators goes well beyond the scope of this chapter. Readers interested in technical details are referred to Huber (1981), Hampel, Ronchetti, Rousseeuw, and Stahel (1986), and Staudte and Sheather (1990). For a more elementary introduction to M-estimators, see Hoaglin, Mosteller, and Tukey (1983). Attention here will be focused on the one-step M-estimator because it is relatively easy to compute and it has the desired property of having a standard error that is not unduly affected by small changes in the tail of a distribution. Other M-estimators also have this property. Perhaps they have some advantage over the one-step M-estimator used here, but this remains to be seen.

The one-step M-estimator is based on Huber's (1964) Ψ function given by

$$\Psi(x) = \max[-k, \min(k, x)],$$

where k is a "tuning" constant. A common choice for k is 1.28, and this choice will be used here. Let M be the usual sample median corresponding to the observations $X_1, \ldots, X_n$. Compute the median absolute deviation statistic given by MAD equal to the median of the n values $|X_1 - M|$, $\ldots, |X_n - M|$. Set $\hat{\zeta} = MAD/.6745$. Letting $\Psi'(x)$ indicate the derivative of Ψ, and setting

$$U_i = (X_i - M)/\hat{\zeta},$$

$$A = \Sigma \, \Psi(U_i),$$

and

$$B = \Sigma \, \Psi'(U_i),$$

the one-step M-estimator is given by

$$\hat{\xi} = M + \frac{\hat{\zeta} A}{B}$$

$$= \frac{k\hat{\zeta}(i_2 - i_1) + \Sigma_{i=i_1+1}^{n-i_2} X_{(i)}}{(n - i_1 - i_2)},$$

where i_1 is the number of observations X_i satisfying $(X_i - M)/\hat{\zeta} < -k$, and i_2 is the number of observations satisfying $(X_i - M)/\hat{\zeta} > k$. When sampling from a symmetric distribution, an estimate of the variance of $\hat{\xi}$ is

$$\hat{\eta} = \frac{\Sigma \, \hat{\zeta} \Psi^2(U_i)}{[\Sigma \, \Psi'(U_i)]^2}.$$

For asymmetric distributions, this estimate seems satisfactory when comparing two groups provided that the bootstrap method described in Section 9.4 is used to control the probability of a Type I error. However, for more than two groups, an alternative estimate will be needed.

To put the one-step M-estimator on more familiar ground, it might help to note that it is a measure of location, just like the mean and median, and that typically the value of $\hat{\xi}$ is somewhere between $\overline{X}$ and M. The one-step M-estimator is based on a type of trimming, as is the trimmed mean, and they often have very similar values. The important point is that the one-step M-estimator can be much more efficient than the trimmed mean, and this can translate into more power when comparing groups. Another advantage of the one-step M-estimator is that it has a breakdown point of .5, where the breakdown point of an estimator refers to the proportion of bad outliers that it can cope with. In contrast, the trimmed mean, with 10% trimming, has a breakdown point of .1, and the sample mean has a breakdown point of 0. The breakdown point of the trimmed mean can be increased by increasing the amount of trimming, but this results in a loss of efficiency and power. Even with 10% trimming, there are situations where comparing groups based on $\hat{\xi}$, rather than Yuen's procedure, can yield substantially more power. Nevertheless, Yuen's procedure might be useful in practice. Both Yuen's procedure and methods for comparing one-step M-estimators appear to control the probability of a Type I error at $\alpha = .05$, provided the sample sizes are not too small. However, at $\alpha = .01$, Yuen's procedure continues to perform well, while methods for comparing one-step M-estimators can be a bit unsatisfactory. Possible methods for correcting this deficiency, as well as computational details of comparing one-step M-estimators, are described in Section 9.4.

9.2.8 Methods Based on Ranks

Still another approach to nonnormality is to compare groups based on ranks. For example, a simple strategy is to pool all the observations, assign ranks, and then apply Student's t-test to the ranks corresponding to each group. It turns out that this is tantamount to applying the Mann–Whitney test (Conover and Iman, 1981). One difficulty with this general approach is getting a consistent estimate of the standard error of the average ranks when the null hypothesis is false. In particular, the denominator of an appropriate test statistic might converge to the wrong value as the sample sizes increase. If distributions are symmetric, or if a shift model is assumed, solutions are available. If the distributions are identical, then it is fairly easy to control the probability of a Type I error. The problem is that if the distributions have different shapes, e.g., unequal skewnesses, unsatisfactory characteristics can result. One difficulty is that the power of the test might not converge to unity as the sample sizes increase. Another is that the test can be biased. Some recent results on this problem can be found in Fligner and Policello (1981), who proposed a method for comparing medians based on ranks. Their procedure is easy to use, is closely related to a procedure where Welch's test is applied to the ranks, and can have relatively high power when distributions have heavy tails. The Fligner–Policello procedure is unbiased and consistent if both distributions are symmetric. What is unclear is the extent to which the Fligner–Policello method continues to have nice properties when distributions have different shapes. Zaremba (1962) proposed a conservative test that deals with this issue, but, as indicated in Section 9.4, its power may be lower than that of some other techniques currently available. For more recent results on using ranks, see Thompson (1991) and Akritas (1991).

9.3 SOME ILLUSTRATIONS

This section uses both actual and artificial data to illustrate some of the problems that can arise from unequal variances and nonnormality. Brief consideration is given to normal distributions with unequal variances, but the emphasis is on the effects of nonnormality.

9.3.1 Normal Distributions with Unequal Variances

As already noted in Section 9.2, violating the equal variance assumption can have serious consequences in terms of Type I errors. This is true for both the F test and standard multiple comparison procedures. This subsection illustrates how unequal variances can affect power.

Suppose all pairwise comparisons are to be performed with $n = 11$ observations randomly sampled from each of $J = 3$ independent groups. Suppose the sample variances are $s_1^2 = s_2^2 = 1$ and $s_3^2 = 7$. Thus, the estimate of τ is $\sqrt{7} \approx 2.65$, and this is not an extreme case. The estimate of the usual mean square within groups is $MSWG = 3$ with $\nu = 30$ degrees of freedom. For $\alpha = .05$, the half-length of the resulting confidence interval is $3.49 \sqrt{3/11} \approx 1.823$, where 3.49 is the .95 quantile of the studentized range distribution. Suppose Dunnett's (1980) $T3$ procedure is used instead. Simply put, this means that Welch's test statistic is computed for each pair of groups, then the critical value is read from a table of the studentized maximum modulus distribution. Percentage points of this distribution are available from Bechhofer and Dunnett (1982) and Wilcox (1987b). For the first two groups in the illustration, the Welch estimated degrees of freedom are 20, the critical value is 2.59, and the half-length of the resulting confidence interval is $2.59 \sqrt{2/11} \approx 1.1$. Thus, for the first two groups, the length of the confidence interval using Dunnett's procedure is nearly half of what it is using Tukey's procedure, and this has obvious implications in terms of power. With $J > 3$ groups, an even more extreme power discrepancy between these two procedures can be observed.

A possible objection to this illustration is that there are situations in which Tukey's procedure has considerably more power than Dunnett's $T3$. It is certainly the case that Student's t-test can be much more significant than Welch's test as illustrated by Staudte and Sheather (1990). In particular, they describe a situation where Student's t-test is significant at the .0004 level, while Welch's test has significance level .012. The reason was that the larger sample was taken from the population with the smaller variance, and this tends to inflate the value of Student's t-test. Moreover, this is a situation where Student's t-test tends to be unsatisfactory in terms of Type I errors. For a detailed explanation, see Staudte and Sheather (1990, pp. 178–181). All indications are that Student's t-test has minor advantages in certain situations, but it has major disadvantages in other situations, so Welch's procedure seems to be the better of the two methods for comparing means.

9.3.2 Nonnormality and Type I Errors

The main threat to Welch's procedure, in terms of Type I errors, is a situation in which there are unequal sample sizes and different skewnesses. To illustrate this point, suppose one group has a normal distribution, while a second group has skewness equal to 2. From Micceri (1989) and Wilcox (1990a), this is not an extreme case. With sample sizes of $n_1 = 40$ and $n_2 = 12$, the probability of a Type I error is approximately .072 when testing at the .05 level (Wilcox, 1990a). With sample sizes of 80 and 20, the probability of a Type I error is approximately .073.

When applying the Box–Scheffé procedure, even more serious problems arise. Consider a random sample of size n, and for convenience assume n is even. Let $U_i = (X_{2i} - X_{2i-1})^2/2$, $i = 1, \ldots, n/2$. Then $E(U_i)$, the expected value of U, is equal to σ^2, the variance of X. Thus, inferences about the mean of U are tantamount to inferences about the variance of X. The variances of two groups can be compared by computing the U's for two groups and then comparing the means of the U's using Welch's technique. This is a special case of the Box–Scheffé method for comparing variances. The main point here is that there are situations where the probability of a Type I error exceeds .1 when testing at the .05 level (Wilcox, 1990a). Perhaps this problem becomes negligible when the smallest sample size is at least 30. However, this has not been established. It is noted that taking logarithms of the U's does not correct the problem.

One approach to improving on Welch's method is to include an adjustment based on estimates of skewness (Cressie and Whitford, 1986; Kleijnen, Kloppenburg, and Meeuwsen, 1986). Unfortunately, these methods have been found to be unsatisfactory (Wilcox, 1990a, in press a).

9.3.3 Heavy Tails and Power

To illustrate the effect of heavy tails, some actual (nonsimulated) data are used from a dissertation by Dana (1990). The study dealt with self-awareness. In one portion of the study, two groups were compared with the following observations:

Group 1: 77, 87, 88, 114, 151, 210, 219, 246, 253, 262, 296, 299, 306, 376, 428, 515, 666, 1310, 2266;

Group 2: 59, 106, 174, 207, 219, 237, 313, 365, 458, 497, 515, 529, 557, 615, 625, 645, 973, 1065, 3215.

The data from both groups indicate that the right portions of both distributions have heavy tails. Applying Welch's procedure with a two-sided alternative, the test statistic has an absolute value of .72 with a critical value of 2.03 based on 35 estimated degrees of freedom, and $\alpha = .05$. The sample means are 448 and 598, and the standard deviations are 595 and 688. The heavy tails are inflating the variance, which suggests a possible problem in terms of a Type II error. If, for example, it was desired to have power equal to .8 when the difference between the means is 200, methods in Wilcox (1987a) indicate that the first group would require 181 observations, while the second needs 242. Of course, a researcher could increase the sample sizes, but a more efficient approach is to use a method that has high power despite the heavy tails. Another concern is that the observed scores are highly skewed, suggesting that a measure of location, other than the mean, would be more appropriate. The hypothesis of equal medians is rejected using Wilcox's method C described in Section 9.4, again testing at the .05 level. Thus, there is a substantial difference between the conclusion reached by Welch's procedure and method C. Section 9.4 also describes what will be called method H for comparing one-step M-estimators. Applying method H to the data given above, the test statistic is $H = 1.98$, and the critical value is 2.22. Yuen's procedure is significant at the .14 level.

Table 9.1 illustrates the effect of heavy tails on five procedures for comparing two independent groups. Observations were generated from a standard normal distribution, which corresponds to distribution 1 in Table 9.1, and three contaminated normal distributions, $\Phi_{\varepsilon,v}(x)$, with $(\varepsilon,v) = (.1,3)$, $(.1,10)$, and $(.1,20)$, corresponding to distributions 2, 3, and 4, respectively. Sample sizes for both groups were 25. Table 9.1 shows the estimated power of each procedure when a constant, 1, is added to every observation in the first group. The columns headed C and H refer to results for methods C and H mentioned in the previous paragraph and described in Section 9.4.

First note that for normal distributions, there is little difference among all three procedures. However, even a slight departure from normality has a substantial effect on Welch's method. In particular, the power drops from .931 to .278 when $\varepsilon = .1$ and $v = 10$. As noted in the introduction, for this special case, $d(\Phi,\Phi_{\varepsilon,v}) \leq .04$, so according to the Kolmogorov metric, $d(\Phi,\Phi_{\varepsilon,v})$, the two distributions are nearly identical. Making the tails a little heavier by increasing ε to .2, the power drops to .162. Yuen's pro-

Table 9.1 Approximate Power for Contaminated Normal Distributions

Dist.	Welch	Yuen	Zaremba	C	H
1	.931	.914	.832	.865	.894
2	.744	.833	.713	.807	.835
3	.278	.705	.633	.778	.804
4	.162	.383	.434	.639	.614

cedure does much better, and for the first three entries in Table 9.1 it performs better than Zaremba's (1962) modification of the Mann–Whitney test. However, increasing ε to .2 causes a serious drop in power. Zaremba's procedure now has better power, but its power has been reduced by a substantial amount as well, and it is substantially worse than method C or H.

9.4 RECENT DEVELOPMENTS

This section describes recently developed methods for dealing with unequal variances and nonnormality. The first two subsections briefly consider methods that adjust for unequal variances. The remaining sections discuss methods for handling nonnormality.

9.4.1 Heteroscedastic ANOVA

Wilcox (1989) derived a new method for dealing with unequal variances when comparing the means of J independent groups. As shown in Section 9.2, all indications are that it controls the probability of a Type I error as well as or better than any other procedure, it is relatively easy to use, and most other procedures can be unsatisfactory, so when comparing means it is recommended. However, when dealing with nonnormality, it is often better to compare measures of location, other than means, using one of the methods described in Sections 9.4.3, 9.4.4, and 9.4.5. The main reason is that these alternative procedures can have substantially more power.

Wilcox's procedure is applied as follows. Let X_{ij}, $i = 1, \ldots, n_j$; $j = 1, \ldots, J$ be the ith observation randomly sampled from the jth group. Let s_j^2 be the usual sample variance, let

$$\tilde{Y}_j = \frac{X_{n_j}}{n_j} + \sum_{i=1}^{n_j-1} \left(1 - \frac{1}{n_j}\right) X_{ij}/(n_j + 1).$$

[It might appear that the term $(n_j + 1)$ in this last expression should be $(n_j - 1)$, but $(n_j + 1)$ is used because it provides slightly better control over the probability of a Type I error.] In words, for the jth group, add up the all the observations excluding the last observation, multiply this sum by $(1 - 1/n_j)$, and divide by $n_j + 1$. Next, divide the last observation by the sample size, add this to the sum just computed, and call the result $\tilde{Y}_j$. Finally, compute

$$D_j = n_j/s_j^2 ,$$
$$W = \Sigma\, D_j ,$$
$$\tilde{Y} = \Sigma\, D_j \tilde{Y}_j/W,$$

and

$$P = \Sigma \, D_j (\bar{Y}_j - \bar{Y})^2.$$

The null hypothesis is rejected if P exceeds the $1 - \alpha$ quantile of a chi-square distribution with $J - 1$ degrees of freedom.

9.4.2 Multiple Comparisons

As is the case with omnibus tests, unequal variances can create difficulties in terms of Type I errors and power when comparing means, and there is the additional problem of getting accurate probability coverage when computing confidence intervals. Summaries on multiple comparisons can be found in Hochberg and Tamhane (1987), Wilcox (1987), and Chapter 2 of this volume. Two of the best procedures for handling unequal variances are the $T3$ and C procedures proposed by Dunnett (1980). Here it is noted that the step-down procedure suggested by Wilcox (1991d) can have substantially more all-pairs power than Dunnett's procedures, where all-pairs power refers to the probability of detecting all true differences among all pairs of groups. A disadvantage of Wilcox's method is that confidence intervals cannot be computed. For an illustration of how to apply a step-down procedure, see Hochberg and Tamhane (1987).

9.4.3 New Results Related to Yuen's Procedure

Three aspects of Yuen's procedure were recently investigated by Wilcox (1991c). The first issue is whether Yuen's procedure continues to control the probability of a Type I error when distributions are skewed but otherwise identical. Yuen considered only symmetric distributions in her simulations, but Wilcox (1991c) found that in terms of probability coverage and Type I errors, Yuen's procedure continues to perform well when distributions are skewed. This is in contrast to Kafadar's (1982) method for comparing biweight measures of location (Wilcox, 1990c).

The second aspect is whether unequal skewnesses affect the probability of a Type I error. As noted in Section 9.2, Tiku's procedure, which uses trimming, is even more sensitive to unequal skewnesses than is Welch's method for comparing means. A positive feature of Yuen's procedure is that, based on simulations, it does not appear to break down in the same situations where Tiku's and Welch's procedures are unsatisfactory.

A third issue is how to extend Yuen's procedure to more than two groups. When performing all pairwise multiple comparisons, all indications are that very good results are obtained by simply performing Yuen's procedure on each pair of groups, and using a critical value read from the studentized maximum modulus distribution. That is, proceed as in Dunnett's *T*3 procedure, but replace the sample means with the trimmed means, replace the sample variances with the winzorized variances, and replace n_j with the "effective" sample sizes $h_j = n_j - 2g$. A similar strategy appears to perform very well for an omnibus test. In particular, use Welch's ANOVA procedure for $J \geq 2$ groups, and again replace the means, variances, and sample sizes with their trimmed counterparts. In fact, all indications are that, in terms of Type I errors, Yuen's procedure is less sensitive to nonnormality and unequal variances than any procedure currently available (Wilcox, 1991b).

9.4.4 Comparing Medians

Because measures often have highly skewed distributions, a natural reaction is that the median of the distribution might be a better measure of location. Various methods for comparing medians have been proposed (Hettmansperger, 1984; Lunneborg, 1986; Wilcox et al., 1986). These procedures are based on the usual sample median, which has the positive feature of a high breakdown point, but procedures based on the usual sample median have the disadvantage of having relatively low power. Accordingly, these procedures are not described here. A procedure with high power was recently proposed (Wilcox, 1991a), and a description of this procedure is given below.

Harrell and Davis (1982) proposed an estimate of the median that is more efficient than the usual estimate. Their basic idea was to use all of the order statistics rather than just the middle one or two. Let

$$W_i = I_{i/n}\left[\frac{(n+1)}{2}, \frac{(n+1)}{2}\right] - I_{(i-1)/n}\left[\frac{(n+1)}{2}, \frac{(n+1)}{2}\right],$$

where $I_x[p,q]$ is the incomplete beta function. The incomplete beta function can be evaluated with a subroutine BETDF in the International Mathematical Subroutine Libraries (IMSL, 1987). The Harrell–Davis estimator of the median is just

$$\hat{\theta} = \Sigma \, W_i X_{(i)}.$$

The motivation for $\hat{\theta}$ is that it estimates $EX_{((n+1)/2)}$, and $EX_{((n+1)/2)}$ converges to the population median. Hence, $\hat{\theta}$ is a consistent estimate of the population median. Yoshizawa, Sen, and Davis (1985) show that $\hat{\theta}$ is asymptotically normal. Harrell and Davis demonstrate that $\hat{\theta}$ can be more efficient than the usual sample median. Thus, a method for comparing groups based on $\hat{\theta}$ would be expected to have more power than procedures that rely on the usual sample median. One problem is finding an effective method of controlling the probability of a Type I error for such a test statistic. Once a method for controlling Type I errors has been found, the next problem is assessing the effect of heavy tails relative to other methods.

Before continuing, it should be mentioned that there are various alternatives to the Harrell–Davis estimator that might play an important role in the future. A brief discussion of these estimators is given in Wilcox (1991a). Parrish (1990) compared 10 quantile estimators in terms of bias and mean squared error and found that the Harrell–Davis estimator performed especially well. Recent work by Kaigh and Cheng (in press) might provide a good alternative, but this is not pursued here.

First, consider the problem of comparing the medians of $J = 2$ independent groups. Wilcox (in press) examined several methods for controlling the probability of a Type I error, but only the one successful method is described here. The procedure is based on a slight modification of a bootstrap-calibration procedure suggested by Loh (1987). For a summary of bootstrap techniques, see Efron (1982) and DiCiccio and Romano (1988). Let $\hat{\theta}_j$ be the Harrell–Davis estimate of θ_j, the population median corresponding to the jth group, and let V_j be some estimate, to be described momentarily, of the variance of $\hat{\theta}_j$. Then an appropriate test statistic is

$$Q = \frac{\hat{\theta}_1 - \hat{\theta}_2}{\sqrt{V_1 + V_2}}.$$

A simple strategy is to assume that Q has a standard normal distribution, but this approach is unsatisfactory, at least for sample sizes ≤ 25. A strategy that appears to give good results at $\alpha = .05$ and n not too small is to adjust the empirical distributions so that they have identical medians and then perform simulations to estimate the probability of a Type I error when observations are randomly sampled, with replacement, from the empirical distributions, assuming that Q has a standard normal distribution. Next, adjust the critical value so that the actual probability of a Type I error is closer to the nominal level. More specifically, if α is the nominal probability of a Type I error, and $\hat{\alpha}$ is the estimate of the actual probability of a Type I error when Q is assumed to have a standard normal distribution, then linear interpolation indicates that testing at the α_a level will give better results where

$$\alpha_a = \begin{cases} \alpha^2/\hat{\alpha} & \text{if } \hat{\alpha} \geq \alpha \\ \alpha + (1 - \alpha)(\alpha - \hat{\alpha})/(1 - \hat{\alpha}) & \text{if } \hat{\alpha} < \alpha \end{cases}$$

The motivation for using the bootstrap is that if $F_j(x)$ is the distribution corresponding to the jth group, and if the distributions were known, then the actual probability of a Type I error could be estimated, with simulations if necessary, when testing H_0 under the assumption that Q has a standard normal distribution. Of course, $F_j(x)$ is not known, but an estimate is available from the observed data, so the strategy is to use the estimate of the distributions to obtain an estimate of the actual probability of a Type I error.

More specifically, the value of $\hat{\alpha}$ is obtained as follows. For the first group, draw a random sample of size n_1 from the observations, where each observation has probability $1/n_1$ of being sampled and sampling is done with replacement. That is, generate n_1 observations from $X_{11}, \ldots, X_{n1}$ by randomly selecting observations with replacement. Let $X_{11}^*, \ldots,$ X_{n1}^* be the resulting values. Compute the Harrell–Davis estimator using this bootstrap sample yielding, say, $\hat{\theta}_1^*$. Repeat this process B times yielding $\hat{\theta}_{1b}^*$, $b = 1, \ldots, B$. Choosing B is discussed momentarily. An estimate of the variance of $\hat{\theta}_1$ is

$$V_1 = \Sigma(\hat{\theta}_{1b}^* - \bar{\theta}_1^*)^2/(B - 1),$$

where $\bar{\theta}_1^* = \Sigma \, \hat{\theta}_{1b}^*/B$. Perform the same operations for the second group yielding V_2. Then the test statistic, Q, is obtained.

A complication is that approximating the distribution of Q with a standard normal distribution is unsatisfactory when n is small. This problem is addressed using the calibration procedure just outlined. Let

$$Q_b = \frac{\hat{\theta}_{1b}^* - \hat{\theta}_{2b}^* - \hat{\theta}_1 + \hat{\theta}_2}{\sqrt{V_1 + V_2}}.$$

Note that this is tantamount to adjusting the empirical distributions of these two groups so that they have identical medians and then computing the test statistic, Q, for the bootstrap sample. Thus, $\hat{\alpha}$, an estimate of the actual probability of a Type I error when Q is assumed to have a standard normal distribution, is equal to $\#\{|Q_b| > z_{1-\alpha/2}\}/B$ where $\#\{|Q_b| > z_{1-\alpha/2}\}$ indicates the number of Q_b values such that $|Q_b| > z_{1-\alpha/2}$, where $z_{1-\alpha/2}$ is the $1 - \alpha/2$ quantile of the standard normal distribution. Having computed an estimate of α, α_a can be computed as described before, and the hypothesis of equal medians is rejected if $|Q| > z_{1-\alpha/2}$. This will be called method C, where C indicates calibration.

There remains the problem of choosing B, the number of resamplings to be carried out. Wilcox (in press) tried $B = 100$, 200, and 300 and found that using $B = 200$ provided substantially better results than $B = 100$, in terms of Type I errors, while increasing B to 300 yielded virtually no further improvement.

Method C appears to perform very well when $\alpha = .05$ and when sampling from normal distributions or distributions with heavier tails. When sampling from a uniform distribution with $n_1 = n_2 = 11$, the actual probability of a Type I error was estimated to be .071, based on a simulation with

1000 replications, while for $n_1 = n_2 = 25$ the estimate was .067. When the above procedure is extended to $J = 4$ groups, problems with light-tailed distributions increase. For example, with a uniform distribution and 25 observations sampled from all four groups, the probability of a Type I error was estimated to be .083. For a normal distribution with 11 observations sampled from each group the estimate is .078, while with 25 observations the estimate drops to .063. An important and encouraging result was that the procedure appears to perform very well for heavy-tailed distributions. Tukey (1960) suggests that light-tailed distributions are less likely to occur in practice than are heavy-tailed distributions, so method C may prove to be important.

In Section 9.3, actual data from a study of self-awareness are analyzed using Welch's procedure. As previously indicated, Welch's procedure was not even close to being significant, while method C is significant at the .05 level. The empirical distributions from both groups are highly skewed with heavy tails, suggesting that method C will provide good control over the probability of a Type I error, while Welch's procedure would be expected to have relatively low power. Of course, there is room for improvement. One problem is that it is unclear how heavy the tails of the empirical distributions have to be before method C can be expected to control the probability of a Type I error. For sample sizes greater than 25, better results can be expected, but how large does the sample size have to be when sampling from light-tailed distributions? One approach toward getting better results is to smooth the empirical distributions before drawing the bootstrap samples. Some results of this approach are reported in Wilcox (in press). Another possibility is to use Beran's (1987) prepivoting method. Hall, Martin, and Schucany (1989) got good results with this method when making inferences about the correlation coefficient. However, Beran's method requires a bootstrap nested within another bootstrap, making it expensive to study via simulations.

It is noted that method C can be modified to compare $J > 2$ groups. The details are nearly identical to the procedure for comparing M-estimators described in the next subsection; they are also described in Wilcox (in press), so to conserve space they are not given here.

9.4.5 Comparing Robust *M*-Estimators of Location

Two advantages of method C are that it compares well to other procedures in terms of power, particularly when distributions have heavy tails, and it attempts to compare groups in terms of a measure of location that might be preferred when distributions are skewed. Its negative feature is that it is based on a statistic with poor breakdown properties. An important issue is whether it is possible to compare groups in terms of a measure of location that has a high breakdown point and yet has relatively high power when distributions have heavy tails. One possibility is the biweight measure of location, which has high efficiency and a high breakdown point. A bootstrap method generally performs well, in terms of Type I errors, when comparing groups in terms of the biweight measure of location (Wilcox, 1990c). However, Wilcox (in press b) found that for heavy-tailed distributions, control over the probability of a Type I error can be unsatisfactory. The problem is related to the iterative method used to estimate the biweight measure of location. It might be possible to eliminate this problem by using one of the alternative estimation procedures suggested by Hampel et al. (1986), but this has not been pursued. Instead, attention has been placed on the one-step M-estimator, which also has a high breakdown point and high power, even when distributions have heavy tails.

Instead of using a bootstrap calibration method to control the probability of a Type I error, a bootstrap-percentile method is used. One reason for this is that the calibration method does not extend to $J > 2$ groups in a convenient fashion. The procedure is first described for two groups; then a slight modification for $J > 2$ groups is discussed.

Let $\hat{\xi}_j$ be the one-step M-estimate, and let $\hat{\eta}_j$ be the estimate of the variance of $\hat{\xi}_j$ corresponding to the jth group. The computational details are described in Section 9.2.7. As in method C, draw a bootstrap sample from the jth group and compute the one-step M-estimator, say $\hat{\xi}_j^*$, and the estimate of its variance, $\hat{\eta}_j^*$. Compute

$$H^* = \frac{|\hat{\xi}_1^* - \hat{\xi}_2^* - \hat{\xi}_1 + \hat{\xi}_2|}{\sqrt{\hat{\eta}_1 + \hat{\eta}_2}}.$$

Repeat this process B times, yielding H_b^*, $b = 1, \ldots, B$. Let $H_{(1)}^*, \ldots, H_{(B)}^*$ be the H_b^* values written in ascending order. The H_b^* values estimate the distribution of the test statistic

$$H = \frac{|\hat{\xi}_1 - \hat{\xi}_2|}{\sqrt{\hat{\eta}_1 + \hat{\eta}_2}},$$

when the null hypothesis is true. Thus, letting $m = [(1 - \alpha)B]$, where $[x]$ is the greatest integer less than or equal to x, an estimate of the $1 - \alpha$ quantile of the distribution of H is $\hat{h}_{1-\alpha} = H_{(m)}^*$. Hence, reject the null hypothesis if $H > \hat{h}_{1-\alpha}$.

Results in Hall (1986) indicate that when the bootstrap-percentile t-method is used, B should be chosen so that $(B + 1)^{-1}$ is a multiple of $1 - \alpha$. Wilcox (in press b) used $B = 399$ and got good results at $\alpha = .05$ for sample sizes of 21 or more in each group. This is in contrast to other situations, where the bootstrap-percentile method can require $B = 1000$ or larger (Efron, 1987).

Note that the only requirement for this procedure to be successful, in terms of Type I errors, is that $\hat{h}_{1-\alpha}$ be a good estimate of the $1 - \alpha$ quantile of the distribution of H. This does not require that $\hat{\eta}_j$ be a good estimate of the variance of $\hat{\xi}_j$. Thus, even for asymmetric distributions, this procedure might perform well, and all indications are that this is the case (Wilcox, in press b). In particular, Wilcox (in press b) examined situations with skewness as high as 2, distributions with heavy tails, distributions with unequal skewnesses, as well as distributions with unequal variances, and all indications are that the bootstrap-percentile procedure just described generally performs well at $\alpha = .05$ with $n \geq 21$. Smaller sample sizes result in "division by zero" errors in the bootstrap algorithm, and this is why the sample size is set to $n \geq 21$. The worst control over the probability of a Type I error occurs when both distributions have skewness equal to 2, one distribution is light-tailed, the other is heavy-tailed, the first has variance one, the second variance16, and the sample sizes are 21 and 41. For this case, the probability of a Type I error was estimated to be .069. There was only one other case where the estimate exceeded .06.

Next consider the more general case where there are $J > 2$ groups and the goal is to perform all pairwise comparisons. That is, the goal is to test $H_0 : \xi_j = \xi_k$ for all $j < k$, where ξ_j is the population value being estimated of $\hat{\xi}_j$. A simple extension to $J > 2$ groups is to generate a bootstrap sample from each of the J groups. Next, compute

$$H_{jk}^* = \frac{|\hat{\xi}_j^* - \hat{\xi}_k^* - \hat{\xi}_j + \hat{\xi}_k|}{\sqrt{\hat{\eta}_j^* + \hat{\eta}_k^*}},$$

and set $Q = \max\{H_{jk}^*\}$, the maximum being taken over all $j < k$. Repeating this process B times yields Q_b^*, $b = 1, \ldots, B$. Let $Q_{(1)}^* \leq \cdots \leq Q_{(B)}^*$ be the resulting order statistics, and let $\hat{q}_{1-\alpha} = Q_{(m)}^*$. Finally, compute

$$H_{jk} = \frac{|\hat{\xi}_j - \hat{\xi}_k|}{\sqrt{\hat{\eta}_j + \hat{\eta}_k}},$$

and reject $H_0 : \xi_j = \xi_k$ if $H > \hat{q}_{1-\alpha}$. However, this procedure tends to be too conservative. When testing at the .05 level with sample sizes of 40 from each of $J = 4$ groups, the actual probability of obtaining at least one Type I error can be as low as .025. Moreover, this procedure does not compare well to Dunnett's $T3$ method when distributions have heavy tails, even when the distributions are symmetric. Because one of the goals is to have relatively high power for heavy-tailed distributions, some other method is clearly needed.

The following modification for $J > 2$ groups gave much better results. For convenience, refer to this modification as method H_m. Proceed exactly as just indicated, but use a bootstrap estimate of the standard errors instead. That is, rather than compute $\hat{\eta}_j$ as was done when comparing $J = 2$ groups, use

$$\hat{\eta}_j = \sum_{b=1}^{B} \frac{(\hat{\xi}_{jb} - \bar{\xi}_j)^2}{(B - 1)}.$$

Table 9.2 shows the estimated probability of obtaining at least one Type I error when method H_m is used with normal distributions. For nonnormal distributions, the control over the probability of at least one Type I error was even better, the one exception being a uniform distribution. For $\alpha = .01$, the actual probability of getting at least one Type I error was estimated to be as high as .022. This was the highest value obtained among all the situations considered. For $\alpha = .05$, the highest value was .071, while for $\alpha = .10$ it was .117. For heavy-tailed distributions, the control over the probability of getting at least one Type I error seems to be very good— the estimates never exceeded .06 and never went below .04 when testing at the .05 level. For $\alpha = .01$ the estimates never exceeded .016. Additional details are available in Wilcox (1991c).

For normal distributions, all indications are that there is little difference between method H_m and Dunnett's $T3$. For example, if all $J = 4$ groups have normal distributions with variance equal to one, and if the first three have means equal to zero, while the fourth has a mean of one, the all-pairs power of Dunnett's $T3$ method is approximately .460. As indicated before, all-pairs power refers to the probability of detecting all true paired differences. Using method H_m, the all-pairs power is .443. However, for a contaminated normal distribution with $\varepsilon = .1$ and $v = 3$, the all-pairs power of Dunnett's $T3$ drops to .192, while for method H_m it drops to only .315. Increasing v to 10, Dunnett's $T3$ has all-pairs power of .018 while method H_m has .213.

Table 9.2 Probability of a Least One Type I Error When Performing All Pairwise Comparisons of M-Estimators, $J = 4$ Groups

σ's	n's	$\alpha = .01$	$\alpha = .05$	$\alpha = .10$
(1, 1, 1, 1)	(21, 21, 21, 21)	.022	.058	.104
(1, 1, 1, 1)	(41, 21, 21, 21)	.014	.059	.117
(4, 1, 1, 1)	(21, 21, 21, 21)	.021	.061	.118
(4, 1, 1, 1)	(41, 21, 21, 21)	.016	.056	.121
(4, 1, 1, 1)	(21, 21, 21, 41)	.018	.071	.117

Now consider the problem of testing $H_0 : \xi_1 = \cdots = \xi_J$. A simple statistic for testing this omnibus hypothesis is

$$Z = \sum \frac{n_j(\hat{\xi}_j - \bar{\xi})^2}{N},$$

where

$$\bar{\xi} = \sum \frac{\hat{\xi}_j}{J}$$

and

$$N = \sum n_j.$$

This test statistic was mentioned by Schrader and Hettmanpserger (1980), and it was studied by He, Simpson, and Portnoy (1990). Again a fundamental problem is determining an appropriate critical value. Wilcox (1991c) approached this problem by adopting a bootstrap-percentile method. The first step is to shift all J empirical distributions so that $\hat{\xi}_j = 0$. That is, shift the distributions so that H_0 is true. Next, generate bootstrap samples and compute Z yielding $Z_1^*, \ldots, Z_B^*$. The Z_b^* values estimate the null distribution of Z. Thus, reject H_0 if $Z > Z_{(m)}^*$. Table 9.3 shows the estimated probability of a Type I error when sampling from normal distributions. For heavier-tailed distributions, the control of Type I errors is even better. For example, when testing at the .05 level, the simulation estimates of the probability of a Type I error never exceeded .06.

The nonnormal distributions considered in Wilcox (1991c) included distributions with skewness and kurtosis values equal to $(\alpha_3, \alpha_4) = (0, 1.8)$, $(0, 9)$, $(2, 9)$, and $(2, 15.6)$. For equal sample sizes of 21 observations per group, the omnibus test for M-estimators had a probability of a Type I error ranging from .004 to .013 for $\alpha = .01$, .026 to .048 for $\alpha = .05$, and .067 to .087 for $\alpha = .10$. For pairwise comparisons, procedure H_m can still

Table 9.3 Probability of a Type I Error When Using the Omnibus Test for M-Estimators of Location

σ's	n's	$\alpha = .01$	$\alpha = .05$	$\alpha = .10$
(1, 1, 1, 1)	(21, 21, 21, 21)	.009	.048	.086
(1, 1, 1, 1)	(41, 21, 21, 21)	.008	.057	.093
(4, 1, 1, 1)	(21, 21, 21, 21)	.013	.054	.106
(4, 1, 1, 1)	(41, 21, 21, 21)	.013	.062	.109
(4, 1, 1, 1)	(21, 21, 21, 41)	.024	.061	.110

be a bit liberal. For $\alpha = .01$ the actual probability of a Type I error can be as high as .022, while for $\alpha = .05$ it can be as high as .061. Thus the omnibus test appears to be satisfactory when equal sample sizes are used. The pairwise comparison procedure might not be satisfactory for $\alpha = .01$, but for $\alpha = .05$ it appears to perform fairly well. Slight increases in the sample sizes may improve matters considerably, but this has not been determined. One hindrance is the tremendous cost of the simulations necessary to determine the small-sample properties of the procedure under consideration.

9.4.6 Comments on Other Experimental Designs

The primary focus in this chapter is on experimental designs involving independent groups. It is noted that heavy tails also cause problems in other experimental designs, such as comparing dependent groups and the analysis of covariance. Efforts are being made to deal with heavy-tailed distributions in various experimental designs.

Analysis of covariance provides one of the more difficult problems when trying to deal with heavy tails or violations of standard assumptions. Consider a single experimental group, and suppose pairs of observations, $(X_1, Y_1), \ldots , (X_n, Y_n)$ are obtained. The usual regression model assumes that the conditional expectation of Y takes the form $E(Y \mid X) = \beta_1 X + \beta_0$, and analysis of covariance attempts to compare the conditional mean of Y, given X, for two or more groups. A variety of problems can arise. For example, this model explicitly assumes that means are an appropriate measure of location, but some other measure might be better when distributions are skewed. A better approach may be to compare $M(Y \mid X)$, the median of Y given X. Brown and Mood (1950) describe a method for estimating $M(Y \mid X)$, but how can the probability of a Type I error be controlled when comparing two or more groups? Another problem is that if the Brown–Mood method were used, the resulting ANCOVA procedure would be expected to have low power. This is because the Brown–Mood method is based in part on the usual sample median. Wilcox (1991e) examined a modification of the Brown–Mood procedure that is based on the Harrell–Davis estimator. Included are some results on comparing two independent groups in terms of $M(Y \mid X)$ where it is assumed that $M(Y \mid X) = \beta_1 X + \beta_0$. When comparing the intercepts corresponding to two independent groups, the probability of a Type I error seems to be fairly well controlled with sample sizes of 40 or more. However, when comparing the slopes, the procedure appears to be slightly conservative. For example, when testing at the .05 level, the actual probability of a Type I error was estimated to be as low as .022 based on simulations with 1000 replications.

Currently, an extension of Welsch's (1980) robust regression technique to the analysis of covariance is being pursued. A detailed description of Welsch's method, including an illustration, can be found in Staudte and Sheather (1990). The method under investigation is based on a percentile-t bootstrap procedure incorporating the influence function estimate of the standard errors. Preliminary results indicate that the method controls the probability of a Type I error very well with only 20 observations per group, even when testing at the .01 level. There are many other robust regression techniques. Perhaps they have some advantage over the method currently pursued in dealing with analysis of covariance, but this has not been determined.

9.5 CONCLUDING REMARKS

The main points of this chapter can be summarized as follows:

• Heavy-tailed distributions are common in the behavioral sciences.
• Relative to other procedures, such as Yuen's trimmed t-test, methods for comparing means are seriously affected by heavy tails in terms of power.
• Currently, methods C and H are the most effective methods for maximizing power. They are both reasonably effective in terms of controlling the probability of a Type I error when $\alpha = .05$ and the sample sizes are not too small, but for $\alpha = .01$ they might be unsatisfactory. Methods based on ranks, such as the Fligner–Policello procedure, can also have high power when distributions have heavy tails, but it is unclear whether these procedures continue to have good properties when distributions have different shapes. As noted by Fligner and Policello, their test is consistent if distributions are symmetric. In this case, the Fligner–Policello procedure seems to be a good alternative to method C or H, but otherwise it is far from clear whether their procedure can be recommended.
• The most effective method for controlling the probability of a Type I error appears to be Yuen's procedure and its extension to more than two groups. Yuen's method has the added advantage of being relatively easy to compute, and it does fairly well in terms of power provided that the tails of the distributions are not too heavy. A negative feature is that if the tails of the distributions are too heavy, then its power can be low compared to method C or H.
• A tempting approach to testing hypotheses is to choose a procedure based on the characteristics of the sampling distributions. For example, a researcher might check for outliers, or look for any evidence that sampling is from a heavy-tailed distribution. If no evidence is found, use one of the standard methods for comparing means. At the moment, this approach cannot be recommended, at least when the goal is to use a procedure with

relatively high power in the event that sampling is from a heavy-tailed distribution. Of course, it might be that a researcher is specifically interested in comparing means, in which case a procedure like Yuen's is unsatisfactory. As for unequal variances, again the choice of a procedure should not be based on the observed data—use a procedure that is insensitive to unequal variances.

• Much remains to be done, and many questions need to be answered. For example, how large a sample is needed before methods C and H can be used with $\alpha = .01$? Is there a way to improve the control over the probability of a Type I error when using these two techniques? How should the methods described in this chapter be extended to other experimental designs?

REFERENCES

Akritas, M. G. (1991). Limitations of the Rank Transform Procedure: A Study of Repeated Measures Designs, Part I. *J. Amer. Statist. Assoc.*, *86*: 457–460.

Bechhofer, R. and Dunnett, C. W. (1982). Multiple Comparisons for Orthogonal Contrasts: Examples and Tables. *Technometrics*, *24*: 213–222.

Beran, R. (1987). Prepivoting to Reduce Level Error in Confidence Sets. *Biometrika*, *74*, 457–468.

Box, G. E. P. (1953). Non-Normality and Tests on Variances. *Biometrika*, *40*: 318–335.

Box, G. E. P. (1954). Some Theorems on Quadratic Forms Applied in the Study of Analysis of Variance Problems. I. Effects of Inequality of Variance in the One-Way Model. *Ann. Statist.*, *25*: 290–302.

Brown, G. W. and Mood, A. M. (1950). On Median Tests for Linear Hypotheses. *Proceedings of the 2nd Berkeley symposium*. Neyman, J. (ed.). University of California Press, Berkeley, pp. 159–166.

Brown, M. B. and Forsythe, A. B. (1974a). The Small Sample Behavior of Some Statistics Which Test the Equality of Several Means. *Technometrics*, *16*: 129–132.

Brown, M. B. and Forsythe, A. B. (1974b). The ANOVA and Multiple Comparisons for Data with Heterogeneous Variances. *Biometrics*, *30*: 719–724.

Conover, W. J. and Iman, R. L. (1981). Rank Transformations as a Bridge between Parametric and Nonparametric Studies. *Amer. Statist.*, *35*: 124–129.

Conover, W. J., Johnson, M. E., and Johnson, M. M. (1981). A Comparative Study of Tests for Homogeneity of Variances with Applications to the Outer Continental Shelf Bidding Data. *Technometrics*, *23*: 351–361.

Cressie, N. A. C. and Whitford, H. J. (1986). How to Use the Two-sample *t*-Test. *Biometrical. J.*, *28*: 131–148.

Dana, E. (1990). Salience of the Self and the Salience of a Standard: Attempts to Match Self to a Standard. Unpublished Ph.D. dissertation. Department of Psychology, University of Southern California.

DiCiccio, T. J. and Romano, J. P. (1988). A review of bootstrap confidence intervals. *Roy. Statist. Soc.*, *50*: 338–354.

Dijkstra, J. B. and Werter, P. S. P. J. (1981). Testing the Equality of Several Means When the Population Variances are Unequal. *Comm. Statist.–Simul. Comput.*, *B10*: 557–569.

Dunnett, C. W. (1980). Pairwise Multiple Comparisons in the Unequal Variance Case. *J. Amer. Statist. Assoc.*, *75*: 796–800.

Efron, B. (1987). Better Bootstrap Confidence Intervals. *J. Amer. Statist. Assoc.*, *82*: 171–185.

Fligner, M. A. and Policello, G. E. II (1981). Robust Rank Procedures for the Behrens–Fisher Problem. *J. Amer. Statist. Assoc.*, *76*: 162–168.

Hall, P. (1986). On the Number of Bootstrap Simulations Required to Construct a Confidence Interval. *Ann. Statist.*, *14*: 1453–1462.

Hall, P., Martin, M. A., and Schucany, W. R. (1989). Better Nonparametric Bootstrap Confidence Intervals for the Correlation Coefficient. *J. Statist. Comput. Simul.*, *B33*: 161–172.

Hampel, F. R., Ronchetti, E. M., Rousseeuw, P. J., and Stahel, W. A. (1986). *Robust Statistics*. Wiley, New York.

Harrell, F. E. and Davis, D. E. (1982). A New Distribution-free Quantile Estimator. *Biometrika*, *69*: 635–640.

He, X., Simpson, D. G., and Portnoy, S. L. (1990). Breakdown Robustness of Tests. *J. Amer. Statist. Assoc.*, *85*: 446–452.

Hettmansperger, T. P. (1984). Two-Sample Inference Based on One-Sample Sign Statistics. *Appl. Statist.*, *33*: 45–51.

Hoaglin, D. C., Mosteller, F., and Tukey, J. W. (1983). *Understanding Robust and Exploratory Data Analysis*. Wiley, New York.

Hogg, R. V. (1974). Adaptive Robust Procedures. A Partial Review and Some Suggestions for Future Applications and Theory. *J. Amer. Statist. Assoc.*, *69*: 909–923.

Hochberg, Y. and Tamhane, A. C. (1987). *Multiple Comparison Procedures*. Wiley, New York.

Huber, P. (1964). Robust Estimation of a Location Parameter. *Ann. Math. Statist.*, *35*: 73–101.

Huber, P. (1981). *Robust Statistics*. Wiley, New York.

IMSL. (1987). *IMSL User's Manual* (10th ed.). Houston, Texas.

James, G. S. (1951). The Comparison of Several Groups of Observations When the Ratios of the Population Variances Are Unknown. *Biometrika*, *38*: 324–329.

Kaigh, W. D. and Lachenbruch, P. A. (1982). A Generalized Quantile Estimator. *Comm. Statist.–Theor. Meth.*, *A11*: 2217–2238.

Kaigh, W. D. and Cheng, C. (in press). Subsampling Quantile Estimators and Uniformity Criteria. *Comm. Statist.–Theor. Meth.*, *A20*, to appear.

Kafadar, K. (1982). Using Biweight M-estimates in the Two Sample Problem. Part 1: Symmetric Populations. *Comm. Statist.–Theor. Meth.*, *A11*: 1883–1901.

Kendall, M. G. and Stuart, A. (1973). *The Advanced Theory of Statistics*, Vol. 2. Hafner, New York.

Keppel, G. (1991). *Design and Analysis*. Prentice Hall, Englewood Cliffs, New Jersey.

Kleijnen, J. P. C., Kloppenburg, G. L. J., and Meeuwsen, F. L. (1986). *Comm. Statist.–Simul. Comput.*, *B15*: 715–732.

Krutchkoff, R. G. (1988). One-way Fixed Effects Analysis of Variance When the Error Variances May Be Unequal. *J. Statist. Comput. Simul.*, *B30*: 259–271.

Loh,W. (1987). Calibrating Confidence Coefficients. *J. Amer. Statist. Assoc.*, *82*: 155–162.

Lunneborg, C. E. (1986). Confidence Intervals for a Quantile Contrast: Application of the Bootstrap. *J. Appl. Psych.*, *71*: 451–456.

Mann, H. B. and Whitney, D. R. (1947). On a Test of Whether One of Two Random Variables Is Stochastically Larger than the Other. *Ann. Math. Statist.*, *18*: 50–60.

Markowski, C. A. and Markowski, E. P. (1990). Conditions for the Effectiveness of a Preliminary Test of Variance. *Amer. Statist.*, *44*: 322–326.

Matuszewski, A. and Sotres, D. (1986). A Simple Test for the Behrens–Fisher Problem. *Comput. Statist. Data Anal.*, *3*: 241–249.

Micceri, T. (1989). The Unicorn, the Normal Curve, and Other Improbable Creatures. *Psych. Bull.*, *105*: 156–166.

Moser, B. K., Steven, G. R., and Watts, C. L. (1989). The two-Sample *t*-Test Versus Satterthwaite's Approximate *F* Test. *Comm. Statist.–Theor. Meth.*, *A18*: 3963–3975.

Parrish, R. S. (1990). Comparison of Quantile Estimators in Normal Sampling. *Biometrics*, *46*: 247–257.

Patel, K. P., Mudholkar, G. S., and Fernando, J. L. (1988). Student's *t* Approximation for Three Simple Robust Estimators. *J. Amer. Statist. Assoc.*, *83*: 1203–1210.

Pearson, E. S. (1931). The Analysis of Variance in Cases of Non-Normal Variation. *Biometrika*, *23*: 114–133.

Ramsey, P. H. (1980). Exact Type I Error Rates for Robustness of Student's *t* Test with Unequal Variance. *J. Educ. Statist.*, *5*: 337–350.

Scheffé, H. (1959). *The Analysis of Variance*. Wiley, New York.

Schrader, R. M. and Hettmansperger, T. P. (1980). Robust Analysis of Variance. *Biometrika*, *67*: 93–101.

Staudte, R. G. and Sheather, S. J. (1990). *Robust Estimation and Testing*. Wiley, New York.

Student (1927). Errors of Routine Analysis. *Biometrika*, *19*: 151–164.

Tan, W. Y. (1982). Sampling Distributions and Robustness of *t*, *F* and Variance-Ratio in Two Samples and ANOVA Models with Respect to Departures from Normality. *Comm. Statist.–Theor. Meth.*, *A11*: 2485–2511.

Thompson, G. L. (1991). A Unified Approach to Rank Tests for Multivariate and Repeated Measures Designs. *J. Amer. Statist. Assoc.*, *86*: 410–419.

Tiku, M. L. (1982). Robust Statistics for Testing Equality of Means or Variances. *Comm. Statist.–Theor. Meth.*, *A11*: 2543–2558.

Tomarken, A. and Serlin, R. (1986). Comparison of ANOVA Alternatives Under Variance Heterogeneity and Specific Noncentrality Structures. *Psych. Bull.*, *99*: 90–99.

Tukey, J. W. (1960). A Survey of Sampling from Contaminated Distributions. *Contributions to Probability and Statistics*, Olkin, I., Hoeffding, W., Ghurye, S., Madow, W., and Mann, H. (eds.). Stanford University Press, Stanford, California, pp. 448–485.

Welch, B. F. (1951). On the Comparison of Several Mean Values: An Alternative Approach. *Biometrika*, *38*: 330–336.

Welsch, R. E. (1980). Regression Sensitivity Analysis and Bounded Influence Estimation. *Evaluation of Econometric Models*. Kmenta, J. and Ramsey, J. B. (eds.). Academic Press, New York, pp. 153–167.

Wilcox, R. R., Charlin, V., and Thompson, K. T. (1986). New Monte Carlo Results on the ANOVA F, W, and F^* Statistics. *Comm. Statist.–Simul. Comput.*, *B15*: 933–944.

Wilcox, R. R. (1987a). New Designs in Analysis of Variance. *Ann. Rev. Psych.*, *38*: 29–60.

Wilcox, R. R. (1987b). *New Statistical Procedures for the Social Sciences*. Erlbaum, Hillsdale, New Jersey.

Wilcox, R. R. (1989). Adjusting for Unequal Variances When Comparing Means in One-Way and Two-Way Fixed Effects ANOVA Models. *J. Educ. Statist.*, *14*: 269–278.

Wilcox, R. R. (1990a). Comparing the Means of Two Independent Groups. *Biometrical J.*, *32*: 771–780.

Wilcox, R. R. (1990b). Comparing Variances and Means When Distributions Have Non-Identical Shapes. *Comm. Statist.–Simul. Comput.*, *B19*: 155–173.

Wilcox, R. R. (1900c). Comparing Biweight Measures of Location in the Two-Sample Problem. *Comm. Statist.–Simul. Comput.*, *B19*: 1231–1246.

Wilcox, R. R. (in press). Testing Whether Independent Groups Have Equal Medians. *Psychometrika*, *56*: to appear.

Wilcox, R. R. (1991b). Yuen's Trimmed t-Test: Some Results and Extensions to More Than Two Groups. Unpublished technical report.

Wilcox, R. R. (1991c). Comparing One-step M-Estimators of Location When There Are More Than Two Groups. Unpublished technical report.

Wilcox, R. R. (1991d). A Step-Down Heteroscadastic Multiple Comparison Procedure. *Comm. Statist.–Theor. Meth.*, *A20*: 1087–1098.

Wilcox, R. R. (1991e). Nonparametric Analysis of Covariance Based on Predicted Medians. *Brit. J. Math. Statist. Psych.*, *44*: 221–230.

Wilcox, R. R. (in press a). An Improved Method for Comparing Variances When Distributions Have Non-Identical Shapes. *Comput. Statist. Data Anal.*, to appear.

Wilcox, R. R. (in press b). Comparing One-Step M-Estimators of Location Corresponding to Two Independent Groups. *Psychometrika*, to appear.

Yoshizawa, C. N., Sen, P. K., and Davis, C. E. (1985). Asymptotic Equivalence of the Harrell–Davis Median Estimator and the Sample Median. *Comm. Statist.–Theor. Meth.*, *A14*: 2129–2136.

Yuen, K. K. (1974). The Two-Sample Trimmed t for Unequal Population Variances. *Biometrika*, *61*: 165–170.

Zaremba, S. K. (1962). A Generalization of Wilcoxon's Test. *Monatshefte Math.*, *66*: 359–370.

10
Unbalanced Data and Cell Means Models

SHAYLE R. SEARLE College of Agriculture and Life Sciences, Cornell University, Ithaca, New York

10.1 ABSTRACT

Data classifiable into two or more levels of one or more factors and having unequal numbers of observations in the subclasses are called unbalanced data. The complications of their analysis by linear model and analysis of variance techniques (relative to analyzing equal-subclass-numbers data, i.e., balanced data) are described in terms of overparameterization, hypothesis testing, estimable functions, fitting submodels, partitioning sums of squares, all-cells-filled data, weighted squares of means, some-cells-empty data, with-interaction and no-interaction models, connected data, and subset analysis.

10.2 THE 1-WAY CLASSIFICATION

10.2.1 A Traditional Model

10.2.1.1 An Example

Suppose an attitudinal test on 7, 8, and 10 children chosen randomly from three different grades in a high school produces average scores of 76, 71, and 80, respectively. If y_{ij} is the score of the jth child in grade i, then a familiar model equation appropriate to the analysis of variance of such data would be

$$y_{ij} = \mu + \alpha_i + e_{ij} . \tag{10.1}$$

Paper number BU-618 in the Biometrics Unit, Cornell University.

In this equation μ could be described as an overall mean, α_i as the effect on a child's score due to that child being in grade i, and e_{ij} as a random error term. This situation is called the 1-way classification because the data can be classified only one way, according to classroom grade.

10.2.1.2 Least Squares Equations

Consider the problem of estimating μ and the α_i's. Using the well-accepted method of estimation by least squares, we get the following equations for the preceding data.

$$
\begin{aligned}
27\mu^0 + 7\alpha_1^0 + 8\alpha_2^0 + 10\alpha_3^0 &= 1800 \\
7\mu^0 + 7\alpha_1^0 &= 532 \\
8\mu^0 + 8\alpha_2^0 &= 568 \\
10\mu^0 + 10\alpha_3^0 &= 700
\end{aligned}
\tag{10.2}
$$

A noteworthy feature of (10.2) is that the last three equations add to the first. Therefore there are effectively only three equations (not four) in four unknowns; and therefore there is an infinite number of solutions to (10.2). To emphasize this the unknowns in (10.2) are symbolized with superscript zeros rather than using the hat notation traditionally used for estimation, e.g., μ^0 rather than $\hat{\mu}$. This highlights μ^0 as being part of just a solution to (10.2) and distinguishes it from the traditional meaning of $\hat{\mu}$ as being a least squares estimator of μ.

Four solutions to (10.2) are shown in Table 10.1. It is clear from Table 10.1 that no one of the solutions therein is of any use as a vehicle for estimating any individual parameter because for each of them there is a wide range of values that constitute a solution to the least squares equations (10.2), indeed a range of infinite breadth.

10.2.1.3 Overparameterization

The situation of having many solutions to least squares equations such as (10.2) arises from the fact that in the model equation (10.1) there are (for

Table 10.1 Four Solutions to Eq. (10.2)

Solution element	Solution			
	1	2	3	4
μ^0	0	$72\frac{1}{3}$	72	3000
α_1^0	76	$3\frac{2}{3}$	4	-2924
α_2^0	71	$-1\frac{1}{3}$	-1	-2929
α_3^0	70	$-2\frac{1}{3}$	-2	-2930

the data envisaged) four parameters, μ, α_1, α_2, and α_3, whereas there are only three observed means from which to try estimating them. This is called overparameterization. And it is the *bête-noir* of traditional analysis of variance models, of which (10.1) is the simplest example.

Within the framework of such models these are two techniques for circumventing the difficulty of having multiple solutions to the least squares equations. They are (1) imposing restrictions on the parameters of the model and (2) confining attention solely to estimating estimable functions of the parameters.

10.2.1.4   A Restriction on Parameters

In texts on the design and analysis of experiments, one often sees in the case of model (10.1), for instance, the restriction $\alpha_1 + \alpha_2 + \alpha_3 = 0$ imposed on the parameters. The purpose of such a restriction (nowadays often called a Σ-restriction) is to redefine the parameters so that there are effectively only as many of them as there are observed cell means from which to estimate them; in our case, three. And then, in concert with that restriction, there is only one solution to the least squares equations. In our example it is solution number 2 of Table 10.1.

Another restriction sometimes used with (10.1) and (10.2) is $7\alpha_1 + 8\alpha_2 + 10\alpha_3 = 0$, which leads to solution 3 of Table 10.1; and the restriction $\mu = 0$ yields solution 1.

None of these restrictions is especially satisfying, especially when one tries generalizing to data classified in two or more ways and with interactions being included. Moreover, if two people with the same data choose to settle for different restrictions they will usually have different solutions to the least squares equations. And if they then try to use those solutions as parameter estimates they will have to face one another with different values as estimates of the same parameter. This can be avoided, and without the need for any restrictions, by confining attention to estimable functions of the parameters.

10.2.1.5   Estimable Functions of Parameters

Observe that for each of the solutions in Table 10.1, $\mu^0 + \alpha_1^0 = 76$ and $\alpha_1^0 - \alpha_2^0 = 5$. This is because the functions $\mu + \alpha_1$ and $\alpha_1 - \alpha_2$ are each what is called an estimable function; and because of this $\mu^0 + \alpha_1^0 = 76$, not just for the solutions in Table 10.1, but for every one of the infinite number of solutions to equations (10.2).

This is no place for detailing the general theory of estimable functions [see, e.g., Searle (1971, Section 5.4) and Searle (1987, Section 8.7)]. It suffices to say that the expected value of each observation is an estimable function and so is any linear combination of those estimable functions. Thus with model equation (10.1) we usually start with $E(y_{ij}) = \mu + \alpha_i$

and define e_{ij} as $e_{ij} = y_{ij} - E(y_{ij})$, where $E(\cdot)$ represents expectation over repeated sampling. Therefore $\mu + \alpha_i$ is estimable for each i of the data and so is every linear combination of the $(\mu + \alpha_i)$ expressions; for example, $\alpha_1 - \alpha_2$ is an estimable function because it equals $(\mu + \alpha_1) - (\mu + \alpha_2)$.

The all-important features of an estimable function are two: that its least squares estimator is the same function of the elements of *any* solution of the least squares equations as the estimable function is of the parameters, and that estimate is the same for all solutions. For example, the least squares estimator of $\mu + \alpha_i$ is $\mu^0 + \alpha_i^0$ for μ^0 and α_i^0 being the elements of any solution of (10.2). Indeed, this is not only the least squares estimator, it is more than that: it is the *best linear unbiased estimator* (BLUE), an estimator that has the attractive properties of being a linear combination of the observations, of being unbiased, and, among all such unbiased linear combinations, of being the unique one that has minimum variance. For this example, because

$$\mu + \alpha_i \text{ is estimable,} \qquad \text{BLUE}(\mu + \alpha_i) = \mu^0 + \alpha_i^0 \,,$$

and

$$\alpha_1 - \alpha_2 \text{ is estimable,} \qquad \text{BLUE}(\alpha_1 - \alpha_2) = \alpha_1^0 - \alpha_2^0 \,.$$

10.2.2   A Cell Means Model

The preceding discussion all points in one direction: since $\mu + \alpha_i$ is the basic feature of the model equation (10.1) that can be estimated, why not give it a symbol, μ_i say, and so without any reference to μ or α_i have the model based on $E(y_{ij}) = \mu_i$; and with $e_{ij} = y_{ij} - E(y_{ij}) = y_{ij} - \mu_i$ have the model equation

$$y_{ij} = \mu_i + e_{ij} \,. \tag{10.3}$$

Then μ_i is thought of as the population mean of the population of possible y_{ij} values of which the observed y_{ij}'s are considered a random sample. In this way we call (10.3) the model equation for a cell means model. It is the simplest example of a cell means model.

10.2.2.1   Estimation

Least squares estimation applied to (10.3) yields

$$\text{BLUE}(\mu_i) = \bar{y}_{i.} = \frac{\sum_{j=1}^{n_i} y_{ij}}{n_i} \,, \tag{10.4}$$

where n_i is the number of observations in cell (or class, or group) i. And on attributing, as is so often done, a variance σ^2 to each e_{ij}, with zero covariance between different e_{ij}'s, the variance of the estimator is

$$\text{var}[\text{BLUE}(\mu_i)] = \sigma^2/n_i . \qquad (10.5)$$

Then, for any combination of μ_i's, denoted by $\Sigma \lambda_i \mu_i$

$$\text{BLUE}(\Sigma \lambda_i \mu_i) = \Sigma \lambda_i \bar{y}_{i.}, \text{with var}[\text{BLUE}(\Sigma \lambda_i \mu_i)] = \sigma^2 \Sigma \left(\frac{\lambda_i}{n_i} \right) . \qquad (10.6)$$

Notice how easily understood everything is when expressed in terms of the μ_i's, arising from the fact that μ_i is simply the (population) mean of class i—something which is conceptually very straightforward. Moreover, the complexities of estimable functions do not arise at all, and every linear combination of μ_i's that interests us can be estimated, and all with, as yet, no introduction of normality. This we now do by introducing some vectors and matrices to condense notation.

10.2.2.2 Normality Assumption

Define $\mathbf{e}$ as the vector of error terms in the whatever model we deal with, be it (10.1) or (10.3) for the 1-way classification or any of the several different models used for the 2-way classification in Section 10.3 of this chapter. Then, whenever normality or normality assumptions are referred to we mean that the elements of $\mathbf{e}$, the error terms in the model, are normally distributed, each with zero mean and variance σ^2 (or σ_e^2 in Section 10.13), and that the covariance between every pair of different error terms is zero; i.e., because of normality, the error terms are independent. Thus we can write that error terms are i.i.d. $\mathcal{N}(0, \sigma^2)$, meaning that they are normally, independently, and identically distributed with zero mean and variance σ^2. An equivalent statement is $\mathbf{e} \sim \mathcal{N}(\mathbf{0}, \sigma^2 \mathbf{I}_N)$, meaning that $\mathbf{e}$ has a multivariate normal distribution with mean $\mathbf{0}$ and dispersion matrix $\sigma^2 \mathbf{I}_N$ where $\mathbf{I}_N$ is an identity matrix of order N, where $N = \Sigma_{i=1}^a n_i$. Then $\mathbf{y} \sim [E(\mathbf{y}), \sigma^2 \mathbf{I}_N]$.

10.2.2.3 Hypothesis Testing

Under the preceding normality assumption, which shall be referred to frequently, we can test any linear hypothesis $H : \mathbf{K}' \boldsymbol{\mu} = \mathbf{m}$ for $\boldsymbol{\mu}$ being the $a \times 1$ vector of μ_i's of (10.3). This can be done providing $\mathbf{K}'$ has full row rank, no greater than a, and where $\mathbf{m}$ is a vector of whatever constants are appropriate to the hypothesis of interest. Then, for

$$\hat{\sigma}^2 = \frac{\Sigma_i \Sigma_j (y_{ij} - \bar{y}_{i.})^2}{N - a} \qquad (10.7)$$

the F-statistic for testing

$H: \mathbf{K}'\boldsymbol{\mu} = \mathbf{m}$

is, for $\bar{\mathbf{y}}$ being the vector of the $\bar{y}_i$-means,

$$F = \frac{(\mathbf{K}'\bar{\mathbf{y}} - \mathbf{m})'(\mathbf{K}'\mathbf{D}\{1/n_i\}\mathbf{K})^{-1}(\mathbf{K}'\bar{\mathbf{y}} - \mathbf{m})}{\hat{\sigma}^2 r_{\mathbf{K}}} \tag{10.8}$$

where $r_{\mathbf{K}}$ is the rank of $\mathbf{K}'$, in this case the number of rows in the full row rank matrix $\mathbf{K}'$, and $\mathbf{D}\{1/n_i\}$ is a diagonal matrix of diagonal elements $1/n_i$.

10.2.2.4 Analysis of Variance

The well-known analysis of variance table for the 1-way classification is as shown in Table 10.2.

The means in the calculations of Table 10.2 are

$$\bar{y}_{i.} = \frac{\sum_{j=1}^{n_i} y_{ij}}{n_i} \quad \text{and} \quad \bar{y}_{..} = \frac{\sum_{i=1}^{a} \sum_{j=1}^{n_i} y_{ij}}{N}. \tag{10.9}$$

And the F-statistic

$$F = \frac{\text{MSA}}{\text{MSE}} \quad \text{tests the hypothesis} \quad H: \mu_1 = \mu_2 = \dots = \mu_a .$$

Notice again how easy it is to understand this hypothesis: that the population cell means are all equal. It is, of course, equivalent to the hypothesis $H: \alpha_1 = \alpha_2 = \dots = \alpha_a$ in the overparameterized model; this is so even though it may seem somewhat self-contradictory because in that model the α_i's themselves are not estimable functions, but the $(\mu + \alpha_i)$ terms are.

Table 10.2 Analysis of Variance for the 1-Way Classification of a Classes with n_i Observations in Class i

Term	Degrees of freedom	Sum of squares	Mean square	F statistic
Between classes	$a - 1$	$\text{SSA} = \sum_{i=1}^{a} n_i(\bar{y}_{i.} - \bar{y}_{..})^2$	$\text{MSA} = \dfrac{\text{SSA}}{a - 1}$	$F = \dfrac{\text{MSA}}{\text{MSE}}$
Within classes	$N - a$	$\text{SSE} = \sum_{i=1}^{a} \sum_{j=1}^{n_i} (y_{ij} - \bar{y}_{i.})^2$	$\text{MSE} = \dfrac{\text{SSE}}{N - a}$	
Total corrected for the mean	$N - 1$	$\text{SST}_{\text{m}} = \sum_{i=1}^{a} \sum_{j=1}^{n_i} (y_{ij} - \bar{y}_{..})^2$		

And the hypothesis $H: \mu_1 = \mu_2 = \ldots = \mu_a$ for $\mu_i = \mu + \alpha_i$ is $H:\mu + \alpha_1 = \ldots = \mu + \alpha_a$, which reduces to $H:\alpha_1 = \alpha_2 = \ldots = \alpha_a$.

Notice that in all of this nothing is lost in using the cell means model of Eq. (10.2), even though at first sight that equation seems so elementary that one might be tempted into thinking it could yield very little. In fact, not only is nothing lost, but something important is gained: considerable clarity. It is so easy to think in terms of just cell means, the μ_i's; and in doing so one is totally relieved of worries about numerous solutions to least squares equations, restrictions on parameters, and estimable functions, all of which are headaches of the overparameterized models.

10.3 THE 2-WAY CROSSED CLASSIFICATION

10.3.1 Introduction

When data can be classified in two different ways they are described as coming from a 2-way classification, and when every class in one classification can occur in combination with every class of the other classification the situation is called a 2-way crossed classification. For example, suppose our attitudinal survey in three different high school grades is extended from one school to five, with a sample of 10 children picked at random from each grade in each school—one survey score on each of 150 children. Let y_{ijk} be the score of child k in grade j in school i, where $i = 1, 2, \ldots, 5$, $j = 1, 2, \ldots, 10$. These data are now classified in two ways: by grade and by school, with each grade occurring in each school, and so this situation is called a 2-way crossed classification.

Suppose the data are arrayed in a rows-by-columns table with rows representing schools and columns grades. We henceforth adopt this generic terminology of rows and columns with, in general, a denoting the number of rows and b the number of columns, and for the sake of generality n_{ij} will denote the number of observations at the intersection of row i and column j—cell (i, j).

10.3.1.1 A Model Equation

The model equation for this situation is developed by starting with the expected value of y_{ijk} defined as

$$E(y_{ijk}) = \mu + \alpha_i + \beta_j + \gamma_{ij}, \qquad (10.10)$$

where μ is a general mean, and α_i, β_j, and γ_{ij} are the effects on score of a child's being, respectively, in school i, in grade j, and in the ij school-grade combination. We then define a random error term corresponding to y_{ijk} as

$$e_{ijk} = y_{ijk} - E(y_{ijk}), \qquad (10.11)$$

which yields the model equation

$$y_{ijk} = \mu + \alpha_i + \beta_j + \gamma_{ij} + e_{ijk} , \tag{10.12}$$

for $i = 1, \ldots , a; j = 1, \ldots , b;$ and $k = 1, \ldots , n_{ij} .$

10.3.1.2 Overparameterization

It is clear from (10.12) that the number of parameters there is 1 for μ, a for the α_i's, b for the β_j's, and can be ab for the interaction parameters γ_{ij}; a total of $1 + a + b + ab$. But at most there are only ab cells that contain data; and so there are no more than ab observed cell means of the form

$$\bar{y}_{ij.} = \frac{1}{n_{ij}} \sum_{k=1}^{n_{ij}} y_{ijk} . \tag{10.13}$$

And the row, column, and grand means are

$$\bar{y}_{i..} = \frac{\sum_{j=1}^{b} \sum_{k=1}^{n_{ij}} y_{ijk}}{n_{i.}}, \qquad \bar{y}_{.j.} = \frac{\sum_{i=1}^{a} \sum_{k=1}^{n_{ij}} y_{ijk}}{n_{.j}}, \qquad \text{and}$$

$$\bar{y}_{...} = \frac{\sum_{i=1}^{a} \sum_{j=1}^{b} \sum_{k=1}^{n_{ij}} y_{ijk}}{N} \tag{10.14}$$

where

$$n_{i.} = \sum_{j=1}^{b} n_{ij} , \qquad n_{.j} = \sum_{i=1}^{a} n_{ij} , \qquad \text{and} \quad N = \sum_{i=1}^{a} \sum_{j=1}^{b} n_{ij} .$$

Notation: For the sake of brevity, the summations $\Sigma_{i=1}^{a}$, $\Sigma_{j=1}^{b}$, and $\Sigma_{k=1}^{n_{ij}}$ will henceforth be written as Σ_i, Σ_j, and Σ_k, respectively, except where clarity demands otherwise.

When every ij-cell contains obsevations there are $1 + a + b + ab$ parameters in the model equations for the data; and this exceeds ab, the number of observed cell means $\bar{y}_{ij.}$, which are the fundamental data for estimating the parameters. So here again we have an overparameterized model, and overparameterized in a manner considerably more complicated than is that of (10.1).

10.3.1.3 Basic Estimation

If we go through the details of applying least squares to (10.12) for estimating μ and the α_i's, the β_j's, and the γ_{ij}'s, we arrive at the result that

$$\text{BLUE}(\mu + \alpha_i + \beta_j + \gamma_{ij}) = \bar{y}_{ij.} \tag{10.15}$$

with its variance being σ^2/n_{ij}. This is in concert with $\mu + \alpha_i + \beta_j + \gamma_{ij}$ being the basic estimable function, in accord with (10.10).

10.3.2 Balanced Data

When every (i, j) cell has the same number of observations, n say, i.e., $n_{ij} = n$ for all i and j ($\forall\ i$ and j), the data are described as balanced data. For such data the means in (10.13) and (10.14) simplify a little through every n_{ij} being n. In particular the following relationships hold:

$$\bar{y}_{i..} = \frac{1}{b} \Sigma_j\, \bar{y}_{ij.}\,, \qquad \bar{y}_{.j.} = \frac{1}{a} \Sigma_i\, \bar{y}_{ij.}$$

and   (10.16)

$$\bar{y}_{...} = \frac{1}{ab} \Sigma_i \Sigma_j\, \bar{y}_{ij.} = \frac{1}{a} \Sigma_i\, \bar{y}_{i..} = \frac{1}{b} \Sigma_j\, \bar{y}_{.j.}\,;$$

and they do not hold when the n_{ij} values are not all equal.

10.3.2.1 The Σ-Restricted Model

Define parameter averages as follows:

$$\bar{\alpha}_{.} = \Sigma_i\, \alpha_i/a, \qquad \bar{\beta}_{.} = \Sigma_j\, \beta_j/b, \qquad \bar{\gamma}_{.j} = \Sigma_i\, \gamma_{ij}/a, \quad \text{and} \quad \bar{\gamma}_{i.} = \Sigma_j\, \gamma_{ij}/b.$$

Now define new parameters

$$\mu' = \mu + \bar{\alpha}_{.} + \bar{\beta}_{.} + \bar{\gamma}_{..}$$

$$\alpha'_i = \alpha_i - \bar{\alpha}_{.} + \bar{\gamma}_{i.} - \bar{\gamma}_{..}\,, \qquad \beta'_j = \beta_j - \bar{\beta}_{.} + \bar{\gamma}_{.j} - \bar{\gamma}_{..} \qquad (10.17)$$

and

$$\gamma'_{ij} = \gamma_{ij} - \bar{\gamma}_{i.} - \bar{\gamma}_{.j} + \bar{\gamma}_{..}\,.$$

Then it is easily seen that

$$\Sigma_i\, \alpha'_i = 0, \qquad \Sigma_j\, \beta'_j = 0,$$

$$\Sigma_i\, \gamma'_{ij} = 0\ \ \forall\ j, \quad \text{and} \quad \Sigma_j\, \gamma'_{ij} = 0\ \ \forall\ i, \qquad\qquad (10.18)$$

and that

$$\mu + \alpha_i + \beta_j + \gamma_{ij} = \mu' + \alpha'_i + \beta'_j + \gamma'_{ij}\,.$$

Therefore the model equation $E(y_{ijk}) = \mu + \alpha_i + \beta_j + \gamma_{ij}$ of (10.10) is equivalent to

$$E(y_{ijk}) = \mu' + \alpha'_i + \beta'_j + \gamma'_{ij}\,, \qquad\qquad (10.19)$$

with the primed parameters satisfying the Σ-restrictions (10.18) which are a natural extension of Section 10.2.1.1. Thus (10.18) and (10.19) represent what we call a Σ-restricted model.

10.3.2.2 Estimation

Now consider the basic estimation result in (10.15). In light of (10.19) it can be rewritten as

$$\text{BLUE}(\mu' + \alpha_i' + \beta_j' + \gamma_{ij}') = \bar{y}_{ij.} \ . \tag{10.20}$$

Averaging both sides of this equation over $j = 1, \ldots, b$, and using (10.18) in doing so, gives

$$\text{BLUE}(\mu + \alpha_i') = \bar{y}_{i..} \ . \tag{10.21}$$

And averaging this over $i = 1, \ldots, a$ and using $\Sigma_i \ \alpha_i' = 0$ gives

$$\text{BLUE}(\mu) = \bar{y}_{...}$$

and hence

$$\text{BLUE}(\alpha_i') = \bar{y}_{i..} - \bar{y}_{...} \ . \tag{10.22}$$

Similarly

$$\text{BLUE}(\beta_j') = \bar{y}_{.j.} - \bar{y}_{...}$$

and

$$\text{BLUE}(\gamma_{ij}') = \bar{y}_{ij.} - \bar{y}_{i..} - \bar{y}_{.j.} + \bar{y}_{...} \ .$$

These are the standard estimation results for balanced data using the Σ-restrictions. This is a somewhat nonstandard mode of derivation, but it has merit: it demonstrates relationships (10.17) between an overparameterized model and the Σ-restricted model—relationships, be it noted, that are unrelated to data. It also shows how simple averaging of the basic estimation result (10.20) leads to (10.22). This does depend on data because, for example, in going from (10.20) to (10.21) the key feature is that with balanced data all cells contain data and so the averaging of the left-hand side of (10.20) includes $\Sigma_j \ \beta_j'/b$, which by (10.18) is zero. And averaging the right-hand side of (10.20) leads to $\bar{y}_{i..}$ of (10.21) because of (10.16). Were there to be even one empty cell in row i, in column t, say, that averaging would include $(\Sigma_j \ \beta_j' - \beta_t')/b = -\beta_t'/b$, which is not zero. Hence the averaging would not lead to (10.21). Further comments along these lines are made in Section 10.3.4.3.

10.3.2.3 Analysis of Variance

The development of analysis of variance by R. A. Fisher was for balanced data and it apparently began (e.g., Urquhart et al., 1970) not with model equations such as (10.1) or (10.12) but with the obvious identity:

$$y_{ijk} - \bar{y}_{...} \equiv (\bar{y}_{i..} - \bar{y}_{...}) + (\bar{y}_{.j.} - \bar{y}_{...})$$
$$+ (\bar{y}_{ij.} - \bar{y}_{i..} - \bar{y}_{.j.} + \bar{y}_{...}) + (y_{ijk} - \bar{y}_{ij.}). \tag{10.23}$$

Then, with an interest in sums of squares of the general form $\Sigma_{t=1}^{r}(x_t - \bar{x})^2$ for $\bar{x} = \Sigma_{t=1}^{r} x_t/r$, Fisher showed that when $n_{ij} \equiv n$,

$$
\begin{aligned}
\Sigma_i \Sigma_j \Sigma_k (y_{ijk} - \bar{y}_{...})^2 \equiv{} & \Sigma_i \Sigma_j \Sigma_k (\bar{y}_{i..} - \bar{y}_{...})^2 + \Sigma_i \Sigma_j \Sigma_k (\bar{y}_{.j.} - \bar{y}_{...})^2 \\
& + \Sigma_i \Sigma_j \Sigma_k (\bar{y}_{ij.} - \bar{y}_{i..} - \bar{y}_{.j.} + \bar{y}_{...})^2 \\
& + \Sigma_i \Sigma_j \Sigma_k (y_{ijk} - \bar{y}_{ij.})^2. \quad\quad (10.24)
\end{aligned}
$$

This became the basis of the well-known analysis of variance table for balanced data from a 2-way crossed classification model with interaction.

The ultimate development of that table from (10.24) starts with simply summarizing that identity. Then, on assuming normality (Section 10.2.2.2), each sum of squares on the right-hand side of (10.24) is distributed proportional to a χ^2-distribution, independently of the others; and from this come the familiar F-statistics and the hypotheses they test. These and (10.24) are then summarized in tabular form as the familiar analysis of variance table for balanced data as in Table 10.3.

Fisher has an interesting comment on such a table. In a letter dated January 6, 1934 (on display at the 50th Anniversary Conference of the Statistics Department at Iowa State University, June 1983), Fisher writes to Snedecor that

> . . . the analysis of variance is (not a mathematical theorem but) a simple method of arranging arithmetical facts so as to isolate and display the essential features of a body of data with the utmost simplicity.

That the analysis of variance table is indeed, as Fisher says, no more than "a simple method of arranging arithmetical facts" is worth emphasizing in these days of computer-generated tables, which too many computer package users are inclined to treat uncritically as sacrosanct. What is important about this, though, is that while computers are efficient at doing arithmetic, the intervention of human thinking is always required for making valid interpretation.

The column of mean squares in Table 10.3 yields the F-statistics $F_A = \text{MSA}/\hat{\sigma}^2$, $F_B = \text{MSB}/\hat{\sigma}^2$, and $F_{AB} = \text{MSAB}/\hat{\sigma}^2$, and in the presence of Σ-restrictions they test, respectively, hypotheses

$$
H\!: \alpha_i' \, all \, equal, \quad\quad H\!: \beta_j' \, all \, equal, \quad\quad \text{and} \quad H\!: \gamma_{ij}' all \, equal. \quad\quad (10.25)
$$

All of this is fine; for balanced data the analysis of variance is well known, well understood, and discussed in detail in numerous texts (e.g., Winer, 1971).

Table 10.3 Analysis of Variance Table for a 2-Way Crossed Classification of a Rows, b Columns, and n Observations per Cell

Term	Degrees of freedom	Sum of squares	Mean square
Rows	$a - 1$	$\text{SSA} = \sum_i \sum_j \sum_k (\bar{y}_{i..} - \bar{y}_{...})^2$	$\text{MSA} = \dfrac{\text{SSA}}{a-1}$
Column	$b - 1$	$\text{SSB} = \sum_i \sum_j \sum_k (\bar{y}_{.j.} - \bar{y}_{...})^2$	$\text{MSB} = \dfrac{\text{SSB}}{b-1}$
Interaction	$(a-1)(b-1)$	$\text{SSAB} = \sum_i \sum_j \sum_k (\bar{y}_{ij.} - \bar{y}_{i..} - \bar{y}_{.j.} - \bar{y}_{...})^2$	$\text{MSAB} = \dfrac{\text{SSAB}}{(a-1)(b-1)}$
Residual	$ab(n-1)$	$\text{SSE} = \sum_i \sum_j \sum_k (y_{ijk} - \bar{y}_{ij.})^2$	$\hat{\sigma}^2 = \dfrac{\text{SSE}}{ab(n-1)}$
Total corrected for the mean	$abn - 1$	$\text{SST}_m = \sum_i \sum_j \sum_k (y_{ijk} - \bar{y}_{...})^2$	

10.3.3   Unbalanced Data: All Cells Filled

Now we consider unbalanced data, in which the number of observations is not the same in every (i, j) cell. Thus the ab values of n_{ij} are not all the same and, indeed, some of them may be zero; i.e., some cells may contain no data at all. Analysis of variance methodology is much more complicated and difficult to understand for unbalanced data than for balanced data. This is so because analysis of variance methodology for unbalanced data is not just a short and simple adaptation of the analysis of variance for balanced data. We discuss this more difficult situation, of unbalanced data, in terms of the 2-way crossed classification because that is the simplest case for illustrating most of the difficulties arising with unbalanced data.

A dichotomy of unbalanced data that is very important and yet frequently underemphasized, is that of whether all cells have data in them or not. When they all do contain data, i.e., $n_{ij} > 0$ ∀ i and j, we have what is called all-cells-filled data; when some cells have no data, i.e., $n_{ij} \geq 0$ ∀ i and j but with at least one $n_{ij} = 0$, we have some-cells-empty data. This distinction is very important in terms of what is useful in analysis of variance methodology. This section is confined to all-cells-filled data.

10.3.3.1   Estimation

Again the basic estimation result (10.15) still holds, and because (10.17) and (10.18) have nothing to do with data, (10.20) still holds:

$$\text{BLUE}(\mu' + \alpha_i' + \beta_j' + \gamma_{ij}') = \bar{y}_{ij.} \,. \tag{10.26}$$

Averaging this in the manner in which (10.22) was derived now gives

$$\text{BLUE}(\mu' + \alpha_i') = \frac{1}{b} \Sigma_j \, \bar{y}_{ij.} \,.$$

Because the n_{ij}'s are not all equal, the right-hand side here does not equal $\bar{y}_{i..}$; it stays as is. And in place of (10.22) this leads to

$$\text{BLUE}(\mu') = \frac{1}{ab} \Sigma_i \Sigma_j \, \bar{y}_{ij.} \,,$$

$$\text{BLUE}(\alpha_i') = \frac{1}{b} \Sigma_j \, \bar{y}_{ij.} - \frac{1}{ab} \Sigma_i \Sigma_j \, \bar{y}_{ij.} \,, \tag{10.27}$$

$$\text{BLUE}(\beta_j') = \frac{1}{a} \Sigma_i \, \bar{y}_{ij.} - \frac{1}{ab} \Sigma_i \Sigma_j \, \bar{y}_{ij.}$$

and

$$\text{BLUE}(\gamma_{ij}') = \bar{y}_{ij.} - \frac{1}{a} \Sigma_i \, \bar{y}_{ij.} - \frac{1}{b} \Sigma_j \, \bar{y}_{ij.} + \frac{1}{ab} \Sigma_i \Sigma_j \, \bar{y}_{ij.} \,.$$

As in (10.22) these results are for the Σ-restricted model (10.18) and (10.19). And all linear combinations of these equations hold, too. Also, (10.27) reduces to (10.22) for balanced data, based on (10.16).

10.3.3.2 An Analysis of Variance Analog

Fisher implicitly used (10.15) for balanced data, and it also holds for unbalanced data. But, and it is a big "but," the sum of squares identity in (10.24) does *not* hold for unbalanced data. The trifling example of Table 10.4 illustrates this. The reader can readily verify that the left-hand side of (10.24) is 1016 but the right-hand side is $270 + 120 + 38 + 648 = 1076 \neq 1016$. Therefore (10.24) cannot be used as a basis for partitioning the left-hand side of (10.24) as is done for balanced data in Table 10.3.

Hence an alternative to (10.24) must be put up if we want to stay within the traditional analysis of variance framework of partitioning SST_m (the sum of squares corrected for the mean) into a sum of sums of squares. For unbalanced data the first two terms in Table 10.3 are

$$SSA = \Sigma_i n_{i.}(\bar{y}_{i..} - \bar{y}_{...})^2 \quad \text{and} \quad SSB = \Sigma_j n_{.j}(\bar{y}_{.j.} - \bar{y}_{...})^2 \qquad (10.28)$$

and the residual term is as there,

$$SSE = \Sigma_i\Sigma_j\Sigma_k(y_{ijk} - \bar{y}_{ij.})^2. \qquad (10.29)$$

One derivation of a term which, when added to the three terms in (10.28) and (10.29) would give SST_m, would be to subtract those three from SST_m, so defining

$$SSAB^* = SST_m - SSA - SSB - SSE.$$

This, after a little simplification, reduces to

$$SSAB^* = \Sigma_i\Sigma_j n_{ij}\bar{y}_{ij.}^2 - \Sigma_i n_{i.}\bar{y}_{i..}^2 - \Sigma_j n_{.j}\bar{y}_{.j.}^2 + n_{..}\bar{y}_{...}^2, \qquad (10.30)$$

and then, of course,

$$SST_m = SSA + SSB + SSAB^* + SSE. \qquad (10.31)$$

Table 10.4 Five Observations in a 2-Way Classification of 2 Rows and 2 Columns

6	4	$y_{1..} = 10$	$(n_{1.} = 2)$	$\bar{y}_{1..} = 5$
6, 42	12	$y_{2..} = 60$	$(n_{2.} = 3)$	$\bar{y}_{2..} = 20$
$y_{.1.} = 54$	$y_{.2.} = 16$	$\bar{y}_{...} = 70$	$(n_{..} = 5)$	$\bar{y}_{...} = 14$
$(n_{.1} = 3)$	$(n_{.2} = 2)$			
$\bar{y}_{.1.} = 18$	$\bar{y}_{.2.} = 8$			

In comparing SSAB* with the expanded form of SSAB in Table 10.3 for balanced data, namely

$$\text{SSAB} = \Sigma_i \Sigma_j \Sigma_k (\bar{y}_{ij.} - \bar{y}_{i..} - \bar{y}_{.j.} + \bar{y}_{...})^2, \qquad (10.32)$$
$$= \Sigma_i \Sigma_j n \bar{y}_{ij.}^2 - \Sigma_i bn \bar{y}_{i..}^2 - \Sigma_i an \bar{y}_{.j.}^2 + abn \bar{y}_{...}^2,$$

we see that SSAB* is an unbalanced data form of SSAB in the sense that the squared means in SSAB* of (10.30) occur there exactly as they do in SSAB of (10.32), and their coefficients in (10.30) are unbalanced data forms of those in (10.32). This seems to suggest that the partitioning of SST_m in (10.31), which is (10.24) with SSAB* replacing SSAB, would be suitable for unbalanced data. But it is not, because SSAB* is not a sum of squares. This is so because SSAB* can be negative, as evidenced by the fact that for the data of Table 10.4:

$$\text{SSAB}^* = 6^2 + 4^2 + 2(24)^2 + 12^2 - 2(5^2)$$
$$- 3(20^2) - 3(18^2) - 2(8^2) + 5(14^2) = -22.$$

Hence (10.31) is not a partitioning of SST_m into a sum of squares. We therefore look elsewhere for an alternative to (10.16) and turn to fitting submodels of the overparameterized model of equation (10.12).

10.3.3.3 Fitting Sub-Models

Since, for example, μ and $\mu + \alpha_i$ are each part of the model $E(y_{ijk}) = \mu + \alpha_i + \beta_j + \gamma_{ij}$, we think of $E(y_{ijk}) = \mu$ and $E(y_{ijk}) = \mu + \alpha_i$ as sub-models. Then, denoting the reductions in sum of squares due to fitting these sub-models as $R(\mu)$ and $R(\mu, \alpha)$, respectively, the difference between these two reductions is represented as

$$R(\alpha \mid \mu) = R(\mu, \alpha) - R(\mu).$$

This idea is extended [e.g., Searle (1971; Sections 6.3, 7.1 and 7.2) and Searle (1987, Section 2.7)] to also have $R(\beta \mid \mu, \alpha)$ and $R(\gamma \mid \mu, \alpha, \beta)$. They then provide a partitioning of SST_m in the form

$$\text{SST}_m = R(\alpha \mid \mu) + R(\beta \mid \mu, \alpha) + R(\gamma \mid \mu, \alpha, \beta) + \text{SSE}. \qquad (10.33)$$

And a similar partitioning is

$$\text{SST}_m = R(\beta \mid \mu) + R(\alpha \mid \mu, \beta) + R(\gamma \mid \mu, \alpha, \beta) + \text{SSE}. \qquad (10.34)$$

Note that in (10.34) only the first two terms differ from those in (10.33). And each of the terms in (10.33) and (10.34) can be shown to be a sum of squares.

10.3.3.4 Computing Formulas

Computing formulas for most of the terms in (10.33) and (10.34) are easy. The one term that is troublesome is $R(\beta \mid \mu, \alpha)$. In matrix and vector notation it is [see, e.g., Searle (1971), Chapter 7, equations (63)–(65), and, equivalently, Searle (1987), Chapter 9, equations (67), (74), and (100)]

$$R(\beta \mid \mu, \alpha) = \mathbf{r}'\mathbf{C}^{-1}\mathbf{r}$$

where

$$\mathbf{r} = \{r_j\} \quad \text{for } j = 1, \ldots, b - 1 \quad \text{with} \quad r_j = y_{.j.} - \sum_{i=1}^{a} n_{ij}\bar{y}_{i..}$$

and

$$\mathbf{C} = \{c_{jj'}\} \quad \text{for} \quad j, j' = 1, \ldots, b - 1$$

with

$$c_{jj} = n_{.j} - \sum_{i=1}^{a} \frac{n_{ij}^2}{n_{i.}} \quad \text{and} \quad c_{jj'} = - \sum_{i=1}^{a} \frac{n_{ij}n_{ij'}}{n_{i.}} \quad \text{for } j \neq j'.$$

Then the two partitionings of SST_m in (10.33) and (10.34) can be computed as follows (Searle, 1987, Chapter 9).

$$\left. \begin{array}{l} R(\alpha \mid \mu) = \text{SSA} \\ R(\beta \mid \mu, \alpha) = \mathbf{r}'\mathbf{C}^{-1}\mathbf{r} \end{array} \right\} \text{or} \left\{ \begin{array}{l} R(\beta \mid \mu) = \text{SSB} \\ R(\alpha \mid \mu, \beta) = \text{SSA} + \mathbf{r}'\mathbf{C}^{-1}\mathbf{r} - \text{SSB} \end{array} \right. \quad (10.35)$$

$$R(\gamma \mid \mu, \alpha, \beta) = \sum_{i=1}^{a} \sum_{j=1}^{b} n_{ij}\bar{y}_{ij.}^2 - \sum_{i=1}^{a} n_{i.}\,\bar{y}_{i..}^2 - \mathbf{r}'\mathbf{C}^{-1}\mathbf{r}$$

$$\text{SSE} = \Sigma_i \, \Sigma_j \, \Sigma_k \, (y_{ijk} - \bar{y}_{ij.})^2.$$

10.3.3.5 Normality, F-Statistics, and Hypotheses

On assuming normality (Section 10.2.2.2), the distribution of SSE is proportional to a χ^2-distribution, and each of the other sums of squares is independent of SSE. Therefore, for example, with N observations y_{ijk} and all ab cells containing data, and estimating σ^2 as

$$\hat{\sigma}^2 = \frac{\text{SSE}}{N - ab}, \quad (10.36)$$

each of

$$F_{\alpha \mid \mu} = \frac{R(\alpha \mid \mu)}{(a - 1)\hat{\sigma}^2} = \frac{\text{SSA}}{(a - 1)\hat{\sigma}^2} \quad \text{and} \quad F_{\beta \mid \mu, \alpha} = \frac{R(\beta \mid \mu, \alpha)}{(b - 1)\hat{\sigma}^2} \quad (10.37)$$

has an F-distribution that can be used to test hypotheses. The catch is, however, that in general these hypotheses are worthless. They are as follows;

$$F_{\alpha|\mu} \quad \text{tests} \quad H: \alpha_i + \sum_{j=1}^{b} n_{ij}(\beta_j + \gamma_{ij})/n_{i.} \quad \text{equal} \quad \forall\, i = 1, \dots, a$$

$$F_{\beta|\mu,\alpha} \quad \text{tests} \quad H: \sum_{i=1}^{a} n_{ij}(\beta_j + \gamma_{ij}) = \sum_{i=1}^{a} \sum_{j'=1}^{b} \frac{n_{ij} n_{ij'}}{n_{i.}}(\beta_{j'} + \gamma_{ij'}) \tag{10.38}$$

$$\text{for} \quad j = 1, \dots, b - 1.$$

Several facets of these hypotheses are worth noting. First, neither of them is clean, simple, and useful as is $H: \alpha_i'$ *all equal* of (10.25) for balanced data in the presence of Σ-restrictions. Neither do Σ-restrictions applied to (10.38) induce useful hypotheses. Second, if data are balanced, i.e., all n_{ij} equaling n, (10.38) does reduce to (10.25) in the presence of Σ-restrictions. Third, for unbalanced data, with (10.38) being as it stands, the relatively complicated nature of the hypotheses therein is all too apparent. Fourth, particularly odious is the occurrence in (10.38) of the n_{ij}-values as part of each hypothesis. This means the hypotheses are determined by the data, not by the observations (the y_{ijk}) themselves, but by the numbers of them in the cells. This is not appropriate. The scientific method makes progress via hypotheses that are not set up by data, not even by the numbers of them.

The one hypothesis coming out of (10.35) and (10.36) that does not involve the n_{ij}-values is that tested by

$$F_{\gamma|\mu,\alpha,\beta} = \frac{R(\gamma \mid \mu, \alpha, \beta)}{(a-1)(b-1)}, \tag{10.39}$$

namely,

$$H: \gamma_{ij} - \gamma_{ij'} - \gamma_{i'j} + \gamma_{i'j'} = 0 \quad \forall\, i \neq i' \quad \text{and} \quad j \neq j'. \tag{10.40}$$

This is, it is to be emphasized, for all-cells-filled data.

10.3.4 The Cell Means Model: All-Cells-Filled Data

The unwieldiness of the algebra in the statements of the hypotheses in (10.38), together with the estimability result in (10.15), prompts consideration of the cell means model. This, similar to (10.3) and (10.4) of the 1-way classification, starts with $E(y_{ijk}) = \mu_{ij}$ and on defining $e_{ijk} = y_{ijk} - E(y_{ijk})$ gives

$$y_{ijk} = \mu_{ij} + e_{ijk} \tag{10.41}$$

where μ_{ij} is the population mean of the (i, j) cell. It is from this cell that we think of the data y_{ijk} for $k = 1, \dots, n_{ij}$ as being a random sample.

10.3.4.1 Estimation

Estimation of μ_{ij} in (10.41) is just like in (10.4) and (10.5):

$$\text{BLUE}(\mu_{ij}) = \bar{y}_{ij.} \quad \text{with} \quad \text{var}[\text{BLUE}(\mu_{ij})] = \sigma^2/n_{ij}. \tag{10.42}$$

And as with (10.16) for the 1-way classification, every linear combination of the μ_{ij}-terms is estimable:

$$\text{BLUE}(\Sigma_{i,j}\,\lambda_{ij}\mu_{ij}) = \sum_{i,j} \lambda_{ij}\bar{y}_{ij.}$$

with

$$\text{var}[\text{BLUE}\,(\Sigma_{i,j}\,\lambda_{ij}\mu_{ij})] = \sigma^2 \sum_{i,j} \frac{\lambda_{ij}^2}{n_{ij}}. \tag{10.43}$$

It is to be noted that with the cell means model of (10.41) there are no parameters such as α_i and β_j, which in the overparameterized model are often referred to as main effects. In their place, however, we now have means of cell means, e.g.,

$$\bar{\mu}_{i.} = \frac{1}{b}\Sigma_j\,\mu_{ij} \quad \text{and} \quad \bar{\mu}_{.j} = \frac{1}{a}\Sigma_i\,\mu_{ij}. \tag{10.44}$$

At first thought these may appear somewhat unsatisfactory: no main effect α_i clearly defined and no interaction parameter γ_{ij} accounted for. On the other hand, $\bar{\mu}_{i.}$ in (10.44) is dead easy to understand: it is the mean of all the cell means in row i. As such it is a form of row mean. And it is in no way clouded by n_{ij}-values.

Moreover, because of (10.43), the $\bar{\mu}_{i.}$ and $\bar{\mu}_{.j}$ of (10.44) are estimable and easily estimated:

$$\text{BLUE}(\bar{\mu}_i) = \frac{1}{b}\Sigma_j\,\bar{y}_{ij.} \quad \text{and} \quad \text{BLUE}(\bar{\mu}_{.j}) = \frac{1}{a}\Sigma_i\,\bar{y}_{ij.}. \tag{10.45}$$

10.3.4.2 Recommendation I
Use (10.42)–(10.45) for estimation.

10.3.4.3 Hypothesis Testing: Akin to Analysis of Variance

In contrast to the messy hypotheses of (10.38), the cell means model and the easily understood averages of means in (10.44) permit us to test clean, simple, and useful hypotheses about those averages in (10.44)—and they are hypotheses that are not plagued by having n_{ij}-values in their specification as are the hypotheses of the overparameterized model in (10.38). The procedure is as follows.

On assuming normality, the hypothesis

$$H: \bar{\mu}_{1.} = \bar{\mu}_{2.} = \ldots = \bar{\mu}_{i.} = \ldots = \bar{\mu}_{a.} \tag{10.46}$$

can be tested by calculating

$$\text{SSA}_W = \Sigma_i \, w_i (\bar{y}_{i..} - \Sigma_i \, w_i \bar{y}_{i..}/\Sigma_i \, w_i)^2 \tag{10.47}$$

for

$$\bar{y}_{i..} = \frac{1}{b} \Sigma_j \, \bar{y}_{ij.} \quad \text{and} \quad \frac{1}{w_i} = \frac{1}{b^2} \Sigma_j \frac{1}{n_{ij}}.$$

Thus

$$F_A = \frac{\text{SSA}_W}{(a-1)\hat{\sigma}^2} \quad \text{tests} \quad H: \bar{\mu}_{i.} \text{ all equal.} \tag{10.48}$$

Similarly, with

$$\text{SSB}_W = \Sigma_j \, v_j (\bar{y}_{.j.} - \Sigma_j \, v_j \bar{y}_{.j.}/\Sigma_j \, v_j)^2 \tag{10.49}$$

for

$$\bar{y}_{.j.} = \frac{1}{a} \Sigma_i \, \bar{y}_{ij.} \quad \text{and} \quad \frac{1}{v_j} = \frac{1}{a^2} \Sigma_i \frac{1}{n_{ij}}$$

then

$$F_B = \frac{\text{SSB}_W}{(b-1)\hat{\sigma}^2} \quad \text{tests} \quad H: \bar{\mu}_{.j} \text{ all equal.} \tag{10.50}$$

As usual, $\hat{\sigma}^2$ is the pooled within-cell estimated variance, defined in (10.36) based on (10.35); and $\bar{\mu}_{i.}$ and $\bar{\mu}_{.j}$ are defined in (10.44).

The hypotheses in (10.48) and (10.50) are far more easily understood than those in (10.38). Moreover, their formulation does not involve the n_{ij}-values. For example, the hypothesis in (10.48) is that the row averages $\Sigma_j \, \mu_{ij}/b$ of the cell means are all equal. Although this is not the same as $H: \alpha_i$ *all equal* of the overparameterized model, it is a broadly useful concept: that the average μ_{ij} in a row is the same for all rows. Nevertheless, it may not seem very satisfactory when viewed from a perspective that is steeped in main effects and interactions. An obvious deficiency seems to be the absence of interaction effects. To ascertain their occurrence in (10.48) write

$$\mu_{ij} = \mu + \alpha_i + \beta_j + \gamma_{ij}.$$

Then note that one of the equality statements of the hypothesis (10.48), say

$$\bar{\mu}_{1.} = \bar{\mu}_{2.},$$

becomes

$$\mu + \alpha_1 + \bar{\beta}_. + \bar{\gamma}_{1.} = \mu + \alpha_2 + \bar{\beta}_. + \bar{\gamma}_{2.},$$

i.e.,

$$\alpha_1 + \bar{\gamma}_{1.} = \alpha_2 + \bar{\gamma}_{2..}$$

Thus (10.46) becomes

$$H: \alpha_i + \bar{\gamma}_{i.} \text{ all equal.} \tag{10.51}$$

Hence (10.48) is testing, in terms of the overparameterized model, that the α_i's plus their corresponding average interaction effects, the $\bar{\gamma}_{i.}$'s, are all equal. Because $\bar{\gamma}_{i.} = (1/b) \sum_j \gamma_{ij}$ one can, if one so desires, will away these average interaction effects from (10.51) by applying the Σ-restrictions to the γ_{ij}'s, namely, $\sum_j \gamma_{ij} = 0 \ \forall \ i$. I prefer *not* to do that, because whether or not one uses the elementary algebra $\sum_j \gamma_{ij} = 0$, the mean $\bar{\mu}_{i.}$ does include interactions for row i averaged over all the columns. It therefore seems appropriate to think of $H: \bar{\mu}_{i.}$ *all equal* as being the hypothesis of equality over rows, of row effects plus averaged interactions. More succinctly, it is simply equality of the row averages of the cell means.

We note in passing that the sums of squares SSA_W and SSB_W of (10.47) and (10.49) are those of the weighted squares of means analysis suggested by Yates (1934).

In the overparameterized model where the messy hypotheses of (10.38) correspond to sums of squares in the partitionings (10.33) and (10.34) of SST_m, the hypothesis (10.40) tested by $R(\gamma \mid \mu, \alpha, \beta)$ is the one hypothesis stemming from those partitionings which can be specified without involving n_{ij}-values. Yet in terms of γ_{ij}'s, which have been loosely defined as interaction effects, that hypothesis is not altogether easily comprehended. But in terms of the cell means model it is

$$H: \mu_{ij} - \mu_{ij'} - \mu_{i'j} + \mu_{i'j'} = 0 \quad \forall \ i \neq i \text{ and } j \neq j'.$$

And this is easily understood, because

$$\theta_{ij,i'j'} \equiv \mu_{ij} - \mu_{i'j} - \mu_{ij'} + \mu_{i'j'} \tag{10.52}$$

for $i \neq i'$ and $j \neq j'$ is precisely what we usually think of as the interaction between rows i and i' and columns j and j'. [Detailed discussion is available in Searle (1987, Section 4.8)]. And, although there are $\frac{1}{4}ab(a-1)(b-1)$ different θ's when there are a rows and b columns of data, then only $(a-1)(b-1)$ of them are linearly independent (*loc cit.*).

10.3.4.4   No Partitioning of the Total Sum of Squares

Notice that in the preceding subsection we first set out a useful, easily understood hypothesis, e.g., H: $\bar{\mu}_{i.}$ *all equal*, and then for testing it displayed an F-statistic based on SSA_W. This is a correct usage of hypothesis-testing concepts. In contrast, in (10.37) and (10.38), we developed certain sums of squares and then asked what hypotheses they tested. In that case the sums of squares, e.g., $R(\alpha \mid \mu)$, $R(\beta \mid \mu, \alpha)$, $R(\gamma \mid \mu, \alpha, \beta)$, and SSE were a partitioning of SST_m. But such a partitioning does not exist for SSW_W, SSB_W, $R(\gamma \mid \mu, \alpha, \beta)$, and SSE. They do not sum to SST_m. And there is nothing wrong in this. Using sums of squares in F-statistics is not dependent on their adding to SST_m. The utility of sums of squares in this context depends on the hypotheses that the corresponding F-statistics test; and clearly (10.48) and (10.50) based on SSA_W and SSB_W, are far more useful than (10.38) based on $R(\alpha \mid \mu)$ and $R(\beta \mid \mu, \alpha)$.

10.3.4.5   Recommendation II

For the 2-way crossed classification, with-interaction models, with all-cells-filled data use, from (10.48), (10.50) and (10.40), use

$$\frac{SSA_W}{(a-1)\hat{\sigma}^2} \sim F_{a-1, N-ab} \qquad \text{to test} \qquad H: \bar{\mu}_{i.} \text{ all equal},$$

$$\frac{SSB_W}{(b-1)\hat{\sigma}^2} \sim F_{b-1, N-ab} \qquad \text{to test} \qquad H: \bar{\mu}_{.j} \text{ all equal},$$

$$\frac{R(\gamma \mid \mu, \alpha, \beta)}{(a-1)(b-1)\hat{\sigma}^2} \sim F_{(a-1)(b-1), N-ab} \qquad \text{to test}$$

$$H: \mu_{ij} - \mu_{ij'} - \mu_{i'j} + \mu_{i'j'} = 0, \quad \forall\, i \neq i' \text{ and } j \neq j'.$$

10.3.4.6   Hypothesis Testing: The General Case

We write the estimation result (10.42) in vector terms:

$$\text{BLUE}(\boldsymbol{\mu}) = \bar{\mathbf{y}} \qquad \text{and} \qquad \text{var}[\text{BLUE}(\boldsymbol{\mu})] = \sigma^2 \mathbf{D}\{1/n_{ij}\} \qquad (10.53)$$

where $\boldsymbol{\mu}$ is the vector of cell means μ_{ij}, $\bar{\mathbf{y}}$ is the vector of observed cell means $\bar{y}_{ij.}$, and $\mathbf{D}\{1/n_{ij}\}$ is the diagonal matrix of elements $1/n_{ij}$. Then exactly as in (10.8), the general linear hypothesis

$$H: \mathbf{K}'\boldsymbol{\mu} = \mathbf{m} \qquad (10.54)$$

can be tested using

$$F = (\mathbf{K}'\bar{\mathbf{y}} - \mathbf{m})'[\mathbf{K}'\mathbf{D}\{1/n_{ij}\}\mathbf{K}]^{-1}(\mathbf{K}'\bar{\mathbf{y}} - \mathbf{m})/r_K \hat{\sigma}^2. \qquad (10.55)$$

This is to be compared with the tabulated F-distribution on r_K and $N - ab$ degrees of freedom.

10.3.4.7 Recommendation III
Use (10.55) to test (10.54).

10.3.4.8 What If Interactions Are Not Significant?
An oft-asked question is "If interactions are not significant, should the interaction sum of squares be pooled with the error sum of squares?" Equivalently, if $F_{\gamma|\mu,\alpha,\beta}$ of (10.39) is deemed nonsignificant, should σ^2 be estimated as

$$\bar{\sigma}^2 = \frac{SSE + R(\beta \mid \mu, \alpha, \beta)}{(N - ab) + (a - 1(b - 1)} \tag{10.56}$$

rather than, as in (10.36), as

$$\hat{\sigma}^2 = \frac{SSE}{N - ab}? \tag{10.57}$$

My opinion is that there is no unequivocal answer to this question. If one believes that there are truly no interactions, then (10.56) is appropriate. But then the question can be asked: "Is this a situation of truly *knowing* that there are no interactions?" If the answer is "Yes," then (10.56) is certainly appropriate. On the other hand, it seems to me that this second question will usually be answered negatively, in which case, even though our test based on $F_{\gamma|\mu,\alpha,\beta}$ has failed to spotlight interaction, the conservative approach is to use $\hat{\sigma}^2$ of (10.57). Not that (10.57) is necessarily larger (or smaller) than (10.56), but it is an estimator that takes into account the possibility of interactions being present.

10.3.4.9 The No-Interaction Model
If one accepts the hypothesis of no interaction or if one simply starts out with a no-interaction model

$$y_{ijk} = \mu + \alpha_i + \beta_j + e_{ijk}, \tag{10.58}$$

then both $F_{\alpha|\mu,\beta}$ and F_A of (10.39) and (10.48), respectively, are available for testing $H: \alpha_i$ *equal* $\forall_i$, using $\hat{\sigma}^2$ of (10.56) in place of $\hat{\sigma}^2$ in the F-values. The decision of which of these two F's to use could rest upon $F_{\alpha|\mu,\beta}$ being the likelihood ratio test or upon the magnitude of their noncentrality parameters, which are proportional to

$$\Sigma_i \, \Sigma_j \, n_{ij}\left(\alpha_i - \frac{\Sigma_{i'} \, n_{i'j}\alpha_{i'}}{n_{.j}}\right)^2 \quad \text{for} \quad F_{\alpha|\mu,\beta}$$

and

$$\Sigma_i \, w_i\left(\alpha_i - \frac{\Sigma_{i'} \, w_{i'}\alpha_{i'}}{\Sigma_{i'} \, w_{i'}}\right)^2 \quad \text{for} \quad F_A$$

with w_i being as used in F_A of (10.48). Clearly, these expressions depend on the magnitudes of both the n_{ij}-values and the α_i's themselves. Since both these F's have the same degrees of freedom, the probability of a Type II error (of not rejecting the hypothesis when it is not true) is smaller for a larger noncentrality parameter. If one knew the values of the α_i's, these parameters could be calculated.

Similar comments apply, of course, to $R(\beta \mid \mu, \alpha)$ and SSB_W.

An important feature of no-interaction models is that differences between levels of each main effect factor can be estimated. Thus both $\alpha_i - \alpha_{i'}$ for $i \neq i'$ and $\beta_j - \beta_{j'}$ for $j \neq j'$ can be estimated for every i, i' and j, j' pair. Thus

$$\text{BLUE}(\alpha_i - \alpha_{i'}) = \alpha_i^0 - \alpha_{i'}^0$$

and   (10.59)

$$\text{BLUE}(\beta_j - \beta_{j'}) = \beta_j^0 - \beta_{j'}^0,$$

where, on reverting to **C** and **r** defined prior to (10.35), one way of calculating α_i^0 and β_j^0 is

$$\begin{bmatrix} \beta_1^0 \\ \vdots \\ \beta_{b-1}^0 \end{bmatrix} = \mathbf{C}^{-1}\mathbf{r} \quad \text{and} \quad \beta_b^0 = 0;$$ (10.60)

and

$$\alpha_i^0 = \bar{y}_{i..} - \Sigma_j \, n_{ij}\beta_j^0/n_{i.}.$$ (10.61)

[Details of deriving these results are in Searle (1971, Chapter 7), equations (14) and (16) and, equivalently, Searle, (1987, Chapter 9), equations (75) and (77).]

10.3.4.10 Recommendation IV

In the no-interaction model use

$$F_{\alpha\mid\mu,\beta} = \frac{R(\alpha \mid \mu, \beta)}{(a - 1)\bar{\sigma}^2} \quad \text{to test } H: \alpha_i \text{ all equal}$$

and

$$F_{\beta\mid\mu,\alpha} = \frac{R(\beta \mid \mu, \alpha)}{(b - 1)\bar{\sigma}^2} \quad \text{to test } H: \beta_j \text{ all equal}.$$

Use $\hat{\sigma}^2$ in place of $\bar{\sigma}^2$ if desiring to be more conservative.

Use (10.60) and (10.61) for estimating (10.59) and linear functions thereof.

10.3.5 The Cell-Means Model: Some-Cells-Empty Data

The hypotheses in (10.38) for the overparameterized model are messy enough even when all cells are filled. For some-cells-empty data they are even messier in the sense that with some n_{ij}-values being zero, parameters corresponding thereto simply do not occur in (10.38). This drives us to the cell means model even more so than with all-cells-filled data because with

$$E(y_{ijk}) = \mu_{ij} \tag{10.62}$$

it is clear that we can consider only those μ_{ij}-terms on which we have data; i.e., (10.62) applies only for ij-values for which $n_{ij} > 0$, namely for only the cells that contain data. This may seem obvious (especially once said) but it has important consequences, not in complicating algebra but in appreciating what analysis is useful and what is not.

10.3.5.1 Estimation

Thus for each datum we still have

$$y_{ijk} = \mu_{ij} + e_{ijk}$$

of (10.41), but (10.42) and (10.43) are amended by adding the condition "for $n_{ij} > 0$":

$$\text{for } n_{ij} > 0, \quad \text{BLUE}(\mu_{ij}) = \bar{y}_{ij.}, \text{ with var}\{\text{BLUE}(\mu_{ij})\} = \sigma^2/n_{ij}; \tag{10.63}$$

and for λ_{ij} corresponding only to $n_{ij} > 0$

$$\text{BLUE}\left(\Sigma_{i,j}\lambda_{ij}\mu_{ij}\right) = \Sigma_{i,j}\lambda_{ij}\bar{y}_{ij.} \text{ with var}[\text{BLUE}(\Sigma_{i,j}\lambda_{ij}\mu_{ij})] = \sigma^2\Sigma_{i,j}\lambda_{ij}^2/n_{ij} . \tag{10.64}$$

All this means that with some-cells-empty data we can estimate the μ_{ij}'s of only the filled cells and linear combinations of those filled-cell μ_{ij}'s.

A first important consequence of the $n_{ij} > 0$ condition is that although all $\bar{\mu}_{i.}$ and $\bar{\mu}_{.j}$ of (10.44) still exist, those of them that involve any empty cell cannot be estimated. Thus if cell r, s is empty, then $\bar{\mu}_{r.}$ and $\bar{\mu}_{.s}$ cannot be estimated.

Example

Suppose in the case of 2 rows and 3 columns the filled cells in the 2×3 grid representation thereof are as indicated by check marks in Grid 1.

Then $\bar{\mu}_{1.} = (\mu_{11} + \mu_{12} + \mu_{13})/3$ and $\bar{\mu}_{2.} = (\mu_{21} + \mu_{22} + \mu_{23})/3$ both exist; but because cell $(2, 3)$ has no data, μ_{23} cannot be estimated. So while

$$\text{BLUE}(\bar{\mu}_{1.}) = (\bar{y}_{11.} + \bar{y}_{12.} + \bar{y}_{13.})/3,$$

$\text{BLUE}(\bar{\mu}_{2.})$ does not exist for the data of Grid 1.

Grid 1

	j = 1	j = 2	j = 3
i = 1	✓	✓	✓
i = 2	✓	✓	

So how, we will ask, can one compare rows in some manner? This cannot be done using $\bar{\mu}_{1.}$ and $\bar{\mu}_{2.}$ since $\bar{\mu}_{2.}$ cannot be estimated. Faced with this, we go back to the underlying model $E(y_{ijk}) = \mu_{ij}$ and ask "What can be estimated?" The immediate answer is "μ_{ij} for every filled cell, and every linear combination of μ_{ij}'s of the filled cells." This being so, the next question is "Which linear combinations of μ_{ij}'s of filled cells might be of interest for comparing rows?" Scanning Grid 1, it is clear that

$$\bar{\mu}_{1.}^* = \tfrac{1}{2}(\mu_{11} + \mu_{12}) \qquad \text{and} \qquad \bar{\mu}_{2.}^* = \tfrac{1}{2}(\mu_{21} + \mu_{22}) \qquad (10.65)$$

are both estimable and that the BLUE of $\bar{\mu}_{1.}^* - \bar{\mu}_{2.}^*$ will provide information about the difference between rows 1 and 2, albeit over only columns 1 and 2. Thus, although $\bar{\mu}_{1.}$ might be used to define an overall row mean, in (10.65) a different definition, $\bar{\mu}_{1.}^*$, is used for comparing rows 1 and 2. The need for this different definition is forced on us by the paucity of the data, because it is obvious that rows 1 and 2 cannot be compared in column 3 using data from Grid 1. We therefore compare rows 1 and 2 using the BLUEs of (10.65) in the form $\tfrac{1}{2}(\bar{y}_{11.} + \bar{y}_{12.}) - \tfrac{1}{2}(\bar{y}_{21.} + \bar{y}_{22.})$; and this uses only part of the total available data. This using of subsets of data is discussed further in Section 10.3.6

10.3.5.2 Recommendation V
Use (10.63) for estimating each μ_{ij} corresponding to a filled cell and (10.64) for linear combinations thereof.

10.3.5.3 Hypothesis Testing: Akin to Analysis of Variance
Having seen in (10.38) for all-cells-filled data that hypotheses tested by all analysis-of-variance style sums of squares [except $R(\gamma \mid \mu, \alpha, \beta)$] are of little use in with-interaction models, these hypotheses can be even more emphatically dismissed for some-cells-empty data. Indeed, SSA_w and SSB_w, the weighted squares of mean sums of squares, can also be dismissed. This is because the expressions (10.47) and (10.49) cannot be calculated when even one n_{ij} is zero. [The corresponding $1/w_i = \Sigma_j(1/n_{ij})/b^2$ will be infinite.] Thus with some-cells-empty data, for with-interaction models, the hypotheses $H:\bar{\mu}_{i.}$ *all equal* and $H:\bar{\mu}_{.j}$ *all equal* cannot be tested.

10.3.5.4 Estimable Interactions

With a rows and b columns there are $\frac{1}{4} ab(a - 1)(b - 1)$ interactions of which a maximum of only $(a - 1)(b - 1)$ are linearly independent (see Section 10.3.4.3). This is true for both all-cells-filled data and some-cells-empty data. But whereas with all-cells-filled data, any set of $(a - 1)(b - 1)$ linearly independent interactions $\theta_{ij,i'j'}$ can be estimated (from which all interactions can then be estimated), with some-cells-empty data fewer than $(a - 1)(b - 1)$ linearly independent interactions can be estimated. For example, in Grid 1 only one interaction is estimable, namely

$$\theta_{11,12} = \mu_{11} - \mu_{12} - \mu_{21} + \mu_{22} \text{ with } \hat{\theta}_{11,12} = \bar{y}_{11.} - \bar{y}_{12.} - \bar{y}_{21.} + \bar{y}_{22.} \,. \qquad (10.66)$$

In contrast, although the other two interactions of the grid,

$$\theta_{11,23} = \mu_{11} - \mu_{13} - \mu_{21} + \mu_{23} \text{ and } \theta_{12,23} = \mu_{12} - \mu_{13} - \mu_{22} + \mu_{23}, \qquad (10.67)$$

can be defined, neither of them is estimable because no BLUE of μ_{23} is available from data of Grid 1. In this simple 2×3 case there are only $\frac{1}{4}2(3)1(2) = 3$ interactions and no more than $1(2) = 2$ can constitute a linearly independent set. Hence the three interactions in (10.66) and (10.67) comprise all possible interactions for Grid 1, and any two are linearly independent. Therefore there is a relationship between the three, for example,

$$\theta_{11,22} = \theta_{11,23} - \theta_{12,23} \,. \qquad (10.68)$$

Hence, since $\theta_{11,22}$ is estimable (the only estimable interaction for Grid 1), its BLUE can also be considered as the BLUE of the difference between the other two interactions as in (10.68); but neither of those two is estimable.

This example illustrates that, for some-cells-empty data, not all of the $\theta_{ij,i'j'}$'s are estimable. Therefore, with such data, the hypothesis $H: all$ $\theta_{ij,i'j'}$ *zero* cannot be tested. This is in sharp contrast to the all-cells-filled data case, where $R(\gamma \mid \mu, \alpha, \beta)$ does provide a test of that hypothesis.

10.3.5.5 Connected Data

Before describing what is tested by $R(\gamma \mid \mu, \alpha, \beta)$ with some-cells-empty data, brief reference must be made to the property of such data called connectedness. For the two-way cross classification, data are said to be connected when all the filled cells of their grid can be jointed by a continuous line, consisting solely of horizontal and vertical segments, that has changes of direction only in filled cells. Grids 2 and 3 illustrate situations when data are, respectively, connected and disconnected (not connected). Consequences of connectedness, or lack thereof, play an important role in what follows.

Grid 2
Connected Data

Row	Column			
	1	2	3	4
1	√	√	—	√
2	√	√	√	
3			√	√

Grid 3
Disconnected Data

Row	Column				
	1	2	3	4	5
1	√	—	√		
2	√			√	√
3			√		
4	√				√

10.3.5.6 The Sum of Squares for Interaction

The sum of squares for interaction is $R(\gamma \mid \mu, \alpha, \beta)$. With all-cells-filled data, its degrees of freedom are $(a - 1)(b - 1) = ab - a - b + 1$, whereas for some-cells-empty data they are fewer, namely, $s - a - b + 1$ with $s < ab$ being the number of filled cells. Provided the some-cells-empty data are connected (all-cells-filled data always are), $R(\gamma \mid \mu, \alpha, \beta)$ tests a hypothesis about $\theta_{ij,i'j'}$'s, but it is not the hypothesis that they are all zero. For example, with Grid 1, since as in (10.66) $\theta_{11,22}$ is the only estimable θ, the hypothesis tested is $H: \theta_{11,22} = 0$. And this corresponds to $s - a - b + 1$ being unity since it is $5 - 2 - 3 + 1 = 1$.

Grid 2 extends this situation. There we have $s - a - b + 1 = 8 - 3 - 4 + 1 = 2$. Scrutiny of Grid 2 quickly reveals an estimable θ:

$$\theta_{11,22} = \mu_{11} - \mu_{12} - \mu_{21} + \mu_{22} . \tag{10.69}$$

But no other θ in Grid 2 is estimable. Yet there are to be two degrees of freedom for $R(\gamma \mid \mu, \alpha, \beta)$. Clearly $H: \theta_{11,22} = 0$ can account for one of them. And, although neither $\theta_{11,23}$ nor $\theta_{13,34}$ is estimable, because each involves μ_{13}, which corresponds to an empty cell and so is not estimable, the sum of these two θ's is estimable because it does not involve μ_{13} since that occurs negatively in one θ and positively in the other, i.e.,

$$\theta_{11,23} + \theta_{13,34} = (\mu_{11} - \mu_{13} - \mu_{21} + \mu_{23}) + (\mu_{13} - \mu_{14} - \mu_{33} + \mu_{34})$$
$$= \mu_{11} - \mu_{21} + \mu_{23} - \mu_{14} - \mu_{33} + \mu_{34} . \tag{10.70}$$

Thus for Grid 2 the hypothesis tested ty $R(\gamma \mid \mu, \alpha, \beta)$ is

$$H: \begin{cases} \theta_{11,22} = 0 \\ \theta_{11,23} + \theta_{13,34} = 0 \end{cases} ; \tag{10.71}$$

notice, it is not testing H: *all interactions zero*.

We can also notice that this is not the only specification of the hypothesis. The following sum of two nonestimable θ's is also estimable:

$$\theta_{12,24} + \theta_{23,34} = (\mu_{12} - \mu_{14} - \mu_{22} + \mu_{24}) + (\mu_{23} - \mu_{24} - \mu_{33} + \mu_{34})$$

$$= \mu_{12} - \mu_{14} - \mu_{22} + \mu_{23} - \mu_{33} + \mu_{34} . \qquad (10.72)$$

Hence the hypothesis (10.71) can also be stated as

$$H: \begin{cases} \theta_{11,22} = 0 \\ \theta_{12,24} + \theta_{23,34} = 0 \end{cases} . \qquad (10.73)$$

Although at first sight this may not seem to be the same as (10.71), it is, for the following reason. (10.71) involves (10.69) and (10.70), whereas (10.73) involves (10.69) and (10.72). Each pair of the functions (10.69), (10.70), and (10.72) is a linearly independent pair of functions suitable for the hypothesis (10.71), but not all three of them are linearly independent: the sum of (10.69) and (10.72) is (10.70). Furthermore, no such linearly independent pair of functions is sufficient for deriving all interactions available from a 3×4 grid, of which Grid 2 is just one pattern of possible data being considered. Deriving all available interactions requires having not two but $(a - 1)(b - 1) = 2(3) = 6$ linearly independent θ's. Therefore, even though (10.71) and (10.73) are equivalent, they are not testing $H: all$ $\theta_{ij,i'j'}$ zero.

The generalization of all this for some-cells-empty data is this: provided those data are connected, the hypothesis tested by $R(\gamma \mid \mu, \alpha, \beta)$ is

$$H: \left\{ \begin{array}{l} \textit{Any set of } s - a - b + 1 \textit{ linearly independent} \\ \textit{functions of } \theta_{ij,i'j'}\textit{s where such functions are either} \\ \textit{estimable } \theta s \textit{ or estimable sums or differences of } \theta s \end{array} \right\} \textit{ are all zero.}$$

$$(10.74)$$

Confirmation of this statement can be found, for example, in Searle (1971, p. 311).

If the connectedness condition does not exist among the data, the hypothesis (10.74) will not apply. This is because disconnected data consist of two or more sets of connected data which need to be analyzed on a within-set basis. Grid 3 is an example. Degrees of freedom for wrongly analyzing such data while ignoring disconnectedness and for analyzing them as two disconnected sets are shown in Table 10.5. Scrutiny of Grid 3 reveals that there is clearly one estimable interaction, that for rows 2 and 4 and columns 2 and 4. In Table 10.5 this is not provided for when the data are wrongly analyzed, but it is in the analysis labeled Set II.

Table 10.5 Degrees of Freedom in the Analysis of Variance for Grid 3

| | Degrees of freedom | | |
| | Analyzed wrongly, as one set of data, ignoring | Analyzed as disconnected data | |
Sum of squares	disconnectedness	Set I (3 cells)	Set II (5 cells)
$R(\mu)$	1	1	1
$R(\alpha \mid \mu)$	3	1	1
$R(\beta \mid \mu, \alpha)$	4	1	2
$R(\gamma \mid \mu, \alpha, \beta)$	0	0	1

Any time that $s - a - b + 1$ is zero (as for Grid 3) or negative (as for Grid 4), one should be suspicious that connectedness is lacking. And even $s - a - b + 1$ being positive is not necessarily assurance that data are connected, as the example of Grid 5 will illustrate.

Grid 4

✓	✓	
✓		
		✓

Grid 5

✓	✓		
✓	✓		
		✓	✓
		✓	✓

10.3.5.7 Recommendation VI

Acknowledge that with some-cells-empty data, the interaction sum of squares $R(\gamma \mid \mu, \alpha, \beta)$ does not test $H: all\ \theta_{ij,i'j'}$ zero. It tests the hypothesis specified in (10.74).

10.3.6 Subset Analyses for Some-Cells-Empty Data and With-Interaction Models

10.3.6.1 Difficulties

A first difficulty with some-cells-empty data is that there is no test of the hypothesis $H: all\ interactions\ zero$. A second difficulty is that for the with-interaction model, the row and column averages of cell means such as $\overline{\mu}_{i.} = (\mu_{i1} + \mu_{i2} + \cdots + \mu_{ib})/b$ cannot be estimated for every row, as illustrated in Section 10.2.2.1. Therefore, as illustrated in (10.65), we are driven by the fact of having empty cells to sometimes defining a row mean in different ways for different purposes.

10.3.6.2 What Is of Interest?

The salient, general question, exemplified just prior to (10.65) for comparing rows in Grid 1, is "What linear functions of μ_{ij}'s of filled cells are of interest for whatever purpose we have in mind?"

This is the all-important question for analyzing some-cells-empty data using a with-interaction model. It can be answered in any manner we wish so far as linear functions of μ_{ij}'s of filled cells are concerned. All such functions are estimable. Exactly which ones are to be used depends on what is of interest to the person whose data are being analyzed. We call that person the investigator.

10.3.6.3 The Investigator's Role

It is the investigator's knowledge of the study for which the data were collected, as well as of the data themselves and their source, that produces what is of interest insofar as linear functions of filled-cell μ_{ij}'s are concerned. Thus it is, with some-cells-empty data and with-interaction cell mean models, that the question of what to estimate (and test hypotheses about) rests squarely on the investigator. Once decisions are made about which linear functions of cell means (of filled cells) are of interest, the mechanics of estimation and hypothesis testing are absolutely straightforward. And making such decisions is not the task of just the statistician— far from it. Investigators familiar with the data, their source, and the data-gathering process have to be involved. Not only do these aspects of data have to be taken into account when deciding on what linear functions of cell means (of filled cells) are of interest, but the purpose for which the linear function is intended must also be considered. This means that sometimes a feature of interest has to be defined differently for different purposes. Just as in (10.65) for Grid 1, to estimate a row mean for row 1 we might use $\overline{\mu}_{1.} = \frac{1}{3}(\mu_{11} + \mu_{12} + \mu_{13})$, so in comparing rows 1 and 2 we would probably use $\overline{\mu}_{1.}^* - \overline{\mu}_{2.}^*$ of (10.65), which can be estimated. We could, of course, also use the estimable

$$\overline{\mu}_{1.} - \overline{\mu}_{2.}^* = \frac{1}{3}(\mu_{11} + \mu_{12} + \mu_{13}) - \frac{1}{2}(\mu_{21} + \mu_{22}), \qquad (10.75)$$

realizing in doing so that it is a comparison that, for each column, involves different weights for the two rows. Generally speaking, (10.75) is not very appealing, compared to $\overline{\mu}_{1.}^* - \overline{\mu}_{2.}^*$. Nevertheless, if an investigator finds (10.75) to be of interest then, since it is a linear combination of μs of filled cells, it is estimable (it has a BLUE) and hypotheses about it can be tested.

10.3.6.4 A Procedure

The first step is for the investigator to decide what kinds of comparisons among (linear functions of) cell means are of interest. To the exent that these include cell means of empty cells, the linear functions have to be adapted to include cell means of only the filled cells. What has to be done is to look at (the pattern, so to speak, of) which cells are filled and which are not, and from that pattern decide what comparisons among the cell means of filled cells might be of interest in terms of providing at least some information about comparisons of interest to the investigator. For example, if in Grid 2 we want to compare rows 2 and 3, the only data that provide a comparison in the same column are those in column 3; i.e., $\mu_{23} - \mu_{33}$ is an estimable function. True, that function provides information that is not very substantive about rows 2 and 3, but the paucity of filled cells in Grid 2 would usually be apparent to any investigator having such data, who would therefore realize the impracticability of comparing rows 2 and 3 over all four columns.

At the heart of this process is the investigator's knowledge of the data; in the presence of empty cells, the investigator must contribute knowledge to deciding what combinations of means (of filled cells) are of interest. No longer can the automatic hypotheses like "equality of rows" be tested. Considerable thought must be given, under the spotlight of having some cells empty, to what combinations of the filled cells are interesting. Knowledge of the data and the pattern of filled cells both have to be utilized. What one would *like* to consider, if all cells were filled, has to be tempered by what *can* be considered in the light of some cells being empty. The consulting statistician and the investigator, working together, must formulate what they think are interesting linear combinations of cell means of cells that contain data. These combinations can then be estimated, the sampling variance of each estimate can be estimated, and tests of hypotheses can be made about them.

10.3.6.5 Subset Analyses

It is clear that estimating neither $\frac{1}{2}(\mu_{11} + \mu_{12} - \mu_{21} - \mu_{22})$ similar to (10.65), nor $\frac{1}{2}(\mu_{11} + \mu_{12}) - \frac{1}{3}(\mu_{21} + \mu_{22} + \mu_{23})$ similar to (10.75), would use all the data available in rows 1 and 2 of Grid 2. Each uses just a subset of the data. We therefore use the name *subset analyses*.

Subset analyses of some-cells-empty data are certainly not as informative as either the analysis of variance of balanced data or the weighted squares of means analysis of unbalanced but all-cells-filled data (see Section 10.6.5). But for unbalanced and some-cells-empty data they are much more useful than the analyses of variance of unbalanced data that are an extension of those of balanced data (see Section 10.7.5). Furthermore, subset analyses for some-cells-empty data are vastly easier to understand, to interpret, and to explain to decision makers than are the analyses that are extensions from balanced data. Not only are the latter difficult to interpret, but in most cases of some-cells-empty data they are of no real interest.

10.3.6.6 Interactions

The preceding discussion has been largely from the viewpoint of being interested in row (and, of course, column) effects. But it applies even more forcefully to interactions. We have already seen for some-cells-empty data that not all interactions are estimable and the traditional sum of squares "for interaction" is not testing the hypothesis of all interactions zero. Despite this, an investigator's desire to study the occurrence of interactions is often very strong; and whereas the F-statistic based on $R(\gamma \mid \mu, \alpha, \beta)$ is of little help in this connection when dealing with some-cells-empty data, the use of subset analyses can be very helpful indeed. This is achieved by looking at the data grid to see which sets of filled cells suggest themselves as possibilities for subset analyses that might yield information about interactions. The information so obtained may not be as far-reaching as when all cells are filled, but it will be better than nothing (which is only what a no-interaction model can yield, insofar as interactions are concerned), and it will nearly always be better than using $R(\gamma \mid \mu, \alpha, \beta)$, which can, in the face of empty cells, be very difficult to interpret.

The crux of the procedure, when wishing to consider interactions with some-cells-empty data, is therefore to view the data from the perspective of a cell means model and to seek subsets of data that can yield information about interactions. Then, as consultant to a client who insists (as do some clients) on considering interactions when data have empty cells, the statistician can offer clarification in the form of helping the client decide which subsets of data might provide analyses of interest.

10.3.6.7 An Example

Suppose data occurred as indicated in Grid 6, of six rows and eight columns, with only 19 of the 48 cells containing data.

Grid 6

	1	2	3	4	5	6	7	8
I	✓			✓		✓	✓	
II				✓	✓			✓
III				✓				✓
IV	✓	✓		✓		✓	✓	✓
V								✓
VI			✓	✓				✓

We use this example to illustrate how we can find our way toward analyses that may provide more useful interpretation than does analyzing the full data set "warts and all," in this case the warts being the large number of empty cells: 60% are empty. In this regard the very grid itself is useful, because it provides opportunity to scrutinize just which cells have data and whether or not any of them form subsets of the data that may be open to straightforward analysis. Such scrutiny reveals that columns 2, 3, and 5 and row V each have but a single filled cell. Setting these data aside leaves Grid 6a:

Grid 6a

	1	4	6	7	8
I	✓	✓	✓	✓	
II		✓			✓
III		✓			✓
IV	✓	✓	✓	✓	✓
VI		✓			✓

This is easily seen to divide into two subsets of data as in grids 6aa and 6ab:

Grid 6aa

	1	4	6	7
I	✓	✓	✓	✓
IV	✓	✓	✓	✓

Grid 6ab

	4	8
II	✓	✓
III	✓	✓
IV	✓	✓
V	✓	✓

These subsets have but one cell in common and account for all six of the degrees of freedom for interaction available in the analysis of the full data set. But now, in directing attention to these two subsets, we see quite clearly what interactions are being considered.

10.3.6.8 Difficulties with Subsets

Two difficulties with subset analyses are readily evident. One is that a data set may not always yield subsets that are useful in the way that those of Grid 6 appear to be. For example, consider Grid 7:

Grid 7

	1	2	3	4
1	✓			✓
2		✓		✓
3	✓	✓	✓	

$R(\gamma \mid \mu, \alpha, \beta)$ will have 1 degree of freedom, and the corresponding F-statistic will test a hypothesis stated, as follows, in three equivalent ways:

$$H: \mu_{11} - \mu_{14} - \mu_{21} + \mu_{24} + (\mu_{21} - \mu_{22} - \mu_{31} + \mu_{32}) = 0,$$

$$H: \mu_{11} - \mu_{14} - \mu_{31} + \mu_{34} - (\mu_{22} - \mu_{24} - \mu_{32} + \mu_{34}) = 0,$$

$$H: \mu_{11} + \mu_{24} + \mu_{32} - (\mu_{14} + \mu_{22} + \mu_{31}) \qquad = 0.$$

The first of these involves the sum of two interactions, whereas the second involves the difference between two interactions. Whatever the utility of these may be (if any), scrutiny of Grid 7 reveals that no subsets of the data manifest themselves as being candidates for informative analyses. When this kind of situation occurs the statistician can do little more than persuade the investigator that this is so, and fall back on the no-interaction model. Of course, with data sets larger than that of Grid 7, coming to the conclusion of no useful subsets may not be as easy as it is there, and much resourcefulness might be needed before such a conclusion can be firmly established.

A second and obvious difficulty inherent in subset analyses is that a data set might well be divisible into two or more different subset analyses. For example, Grid 8 can easily be divided into two subsets in two different ways: one way consists of rows I and II, and row III; and the other is columns 1 and 4, and columns 2 and 3.

Grid 8

	1	2	3	4
I	√	√	√	√
II	√	√	√	√
III	√			√

This situation emphasizes what is so important about analyzing unbalanced data, especially some-cells-empty data; there is seldom just a single, correct way of doing a statistical analysis. Therefore the first responsibility of a consulting statistician to those who have garnered such data is to impress upon them that analyzing those data has no single, easy umbrella of interpretation. Within that umbra, the statistician can certainly provide advice as to which analyses might be helpful; and two different statisticians may well have two different lines of advice for analyzing the same data set. As with lawyers, the advice of statisticians is not necessarily uniform, let alone uniformly right or uniformly wrong.

10.3.6.9 The Investigator

Naturally, the investigator must contribute to deciding on possible divisibility of a total data set into subset analyses. In the preceding examples the pattern of empty cells has been the sole criterion for suggesting subsets. That is always a useful criterion because it can lead to analyses that are interpretable. But it must not be the sole criterion. Investigators must be urged, from their prior knowledge of similar data and of the context from which the present data have come, to decide what specific levels of the factors (or pooled combinations thereof) are of prime interest, especially in the context of interactions. Indeed, the statistician's advice should be in terms of helping investigators, nay even cajoling them, perhaps, into deciding what specific subset of filled cells might be of real interest. One would hope that when faced with empty cells, the combined efforts of statistician and investigator would usually reveal some data subsets that provide both easy analysis and straightforward interpretation through the use of the cell means model. Examples of doing this in animal science and in agronomy research are to be found, respectively, in Urquhart and Weeks (1978) and Meredith and Cady (1988).

10.3.6.10 Recommendation VII

For some-cells-empty data consider interaction by using subset analyses as described in this section and illustrated by the example in subsection g.

10.3.7 No-Interaction Models for Some-Cells-Empty Data

10.3.7.1 Using the No-Interaction Model

Since, as has been described, the F-statistic based on $R(\gamma \mid \mu, \alpha, \beta)$ is, for some-cells-empty data, not testing that all interactions are zero, the procedure of assuming a model that truly has no interactions is even more tenuous here than it is with all-cells-filled data, as discussed in Section 6h. Nevertheless, using the no-interaction model with some-cells-empty data can provide information that is not otherwise available from such data, namely, some overall indication of the importance of differences among row effects (and among column effects) in the *assumed absence* of interactions—and provided, of course, that the some-cells-empty data are connected. It is all too true that if indeed there are interactions between the underlying factors and we use a no-interaction model, then any information about rows and columns based on that no-interaction model may be misleading. On the other hand, as seen in the discussion of Grid 1 at (10.65) and (10.75), using the with-interaction model with some-cells-empty data can provide very little information about differences among rows (or columns) generally. Therefore it is my feeling that with some-cells-empty data one should not only carry out subset analyses but also consider using the no-interaction model

$$E(y_{ijk}) = \mu + \alpha_i + \beta_j .$$

Provided some-cells-empty data are connected (Section 10.3.5.5), the fitting of this is exactly the same as described in Section 10.3.4.8 for all-cells-filled data. Equations (10.60) and (10.61) provide solutions to the least squares equation and they can be used to estimate estimable functions as in (10.59). The only effect on (10.60) and (10.61) is that the n_{ij}-value for each empty cell will be zero.

10.3.7.2 Estimable Functions

One of the prime advantages of the no-interaction model for some-cells-empty data is that, provided the data are connected, differences between row effects and between column effects can always be estimated, i.e., (10.59) holds true. Although this result is for a model that ignores inter-actions it is a clean result and, within that setting, a very useful one. Therefore, in my opinion, it always merits consideration.

10.3.7.3 Hypothesis Testing

The preceding comment is equally applicable to the F-statistics of Rec-ommendation IV (Section 10.3.4.10), namely, that they provide tests of the hypotheses $H{:}\alpha_i$ *all equal* and $H{:}\beta_j$ *all equal*. After all, if $F_{\alpha|\mu,\beta}$ was significant at the .0001 level, might we not be encouraged to look in some detail at the row data to see, even in the absence of information about interactions, if there is a possible explanation for an otherwise unexpected result? This is what data analysis is all about.

10.3.7.4 Estimating Residual Variance

Even when using a no-interaction model it is always possible, for connected data, to calculate $R(\gamma \mid \mu, \alpha, \beta)$. Even though this sum of squares does not lead to testing the hypothesis $H{:}interactions$ *all zero*, its availability raises the question of whether to use SSE $+ R(\gamma \mid \mu, \alpha, \beta)$ or just SSE as the sum of squares for estimating σ^2. This is the same question as whether to use $\tilde{\sigma}^2$ or $\hat{\sigma}^2$ of (10.56) and (10.57), respectively. The remarks following those equations apply equally as well here as there.

10.4 SOME NOTES ON OTHER TOPICS

From the host of topics allied to what has already been dealt with, only four are considered here and then but briefly.

10.4.1 More Than Two Factors

The 2-way crossed classification has been used as the basis of the seven preceding sections because it is the simplest situation that permits satis-factory description of the difficulties associated with unbalanced data, empty cells, interactions, and cell means models. Those difficulties simply get worse when there are more than two factors. But notice that it is interpretation that gets worse, not computing. With today's computing power there is almost no limit to being able to compute reductions in sums of squares of the form $R(\cdot \mid \cdot)$. For large amounts of unbalanced data, with many factors and interactions, computers readily provide us with as many $R(\cdot \mid \cdot)$ terms as we might think we want. The trouble is in interpreting them, in having interpretations that are useful. For example, extensions of (10.38) to three and more factors are clearly of not much use. So we mention but four aspects of multifactor data that are deemed useful.

10.4.1.1 Weighted Squares of Means Analyses

In the 2-way classification a cell is the "intersection" of one row and one column. With more than two factors a cell is similarly the intersection of one level of each factor. When all such cells contain data, the SSA_w and SSB_w sums of squares of the weighted squares of means analysis of Section 10.3.4.3 extend quite naturally to having more than two factors. For example, with three factors A, B, and C, for $E(y_{ijkt}) = \mu_{ijk}$, there is a sum of squares SSA_w similar to (10.47) that tests $H:\overline{\mu}_{i..}$ *all equal*, similar to (10.48). And there is also a sum of squares which can be labeled $SSAB_w$ that tests $H:\overline{\mu}_{ij.} - \overline{\mu}_{ij'.} - \overline{\mu}_{i'j.} + \overline{\mu}_{i'j'.}$ *all zero*, analogous to the interaction hypothesis tested by $R(\gamma \mid \mu, \alpha, \beta)$ in the 2-way classification (see Recommendation II). Details and a numerical example of a 3-way classification are given in Searle (1987, Section 10.2).

10.4.1.2 Subset Analysis

Subset analysis for some-cells-empty data as suggested in Section 10.3.6 is easy for two factors because the occurrence of data and of empty cells is readily visible in a rows-by-columns grid from which subsets of data (sets of contiguous filled cells) are easily identified. A similar procedure for multifactor data is easily envisaged, but carrying it out will take a little more effort. With just three or four factors, though, especially if they have only a few levels, identifying suitable subsets, if they exist, should not be too difficult. And from such subsets one may be able to obtain at least a little information about interactions.

10.4.1.3 Connectedness

Connectedness of data is defined in Section 10.3.5.5 for a 2-way classification. Unfortunately, that straightfoward geometric definition cannot be extended when there are more than two factors. One then has to use an algebraic definition given by Weeks and Williams (1964), which they show in their 1965 errata provides only a sufficient and not a necessary condition for data to be connected (in the sense of intrafactor differences between main effects being estimable). Further discussion is available in Searle (1971, Section 8.1e) and Searle (1987, Section 5.3e).

10.4.1.4 Main-Effects-Only Models

The point is made in Section 10.3.7.2 for the 2-way classification that provided some-cells-empty are connected, then in the no-interaction model one can estimate differences between row effects ($\alpha_i - \alpha_{i'}$) and between column effects ($\beta_j - \beta_{j'}$) as in (10.59). One can also test equality of the α's and of the β's, i.e., equality of row effects and of column effects (see Section 10.3.4.10).

These results extend directly to multifactor data. For the k-way crossed classification, with the data being connected, consider a model with no interactions of any sort. We call this a main-effects-only model. For this situation one can estimate differences between levels of a factor and also test equality of the effects of a factor. And this can be done for every factor.

Let the factors be represented by $\alpha, \beta, \gamma, \eta, \tau, \ldots$. Then if, for example, α_i^0 and β_j^0 are elements in the solutions to the least squares equations for the complete main-effects-only model, we have

$$\text{BLUE}(\alpha_i - \alpha_{i'}) = \alpha_i^0 - \alpha_{i'}^0 \quad \text{and} \quad \text{BLUE}(\beta_j - \beta_{j'}) = \beta_j^0 - \beta_{j'}^0$$

just as in (10.59)—but now, of course, in having more than two factors, the solution elements α_i^0 and β_j^0 are not derived from (10.60) and (10.61). They will usually be obtained from a suitable computing package (see Section 10.4.3).

In this same situation testing the hypothesis $H:\alpha_i$ *all equal* will be based on $R(\alpha \mid \mu, \beta, \gamma, \eta, \tau, \ldots)$ and likewise $H:\beta_j$ *all equal* will be based on $R(\beta \mid \mu, \alpha, \gamma, \eta, \tau, \ldots)$. Each of these reductions in sums of squares can well be described as a reduction due to one factor adjusted for all others. This concept also occurs in describing certain computer output (see Section 10.4.3). Searle (1971, Section 8.1c) and Searle (1987, Section 10.3) have more details on this topic.

It is my opinion that this main-effects-only analysis is always worth doing for data with many classifications, a large number of cells, and a high proportion of them empty [e.g., 9 factors, 56 main effects, 5,474,304 cells, and less than 0.16% of them containing data (8577 observations); see Searle (1971, Section 8.1)] this analysis does, of course, have deficiencies. These include both ignoring the possibility of interactions and knowing that the sums of squares $R(\cdot \mid \cdot)$ that will be used will be correlated, to some extent, although little or nothing is known about the magnitude of that correlation.

10.4.2 Analysis of Covariance

This is a huge topic, and not necessarily an easy one. Until our present computer age it was severely limited by its presentation being confined to easily computed procedures, often based on tinkering with well-known analysis of variance tables. An example is given in Section 11.1 of Searle (1987).

The simplest example of analysis of covariance is the 1-way classification of Sections 10.2.1 and 10.2.2 extended to

$$E(y_{ij}) = \mu_i + bz_{ij} \tag{10.76}$$

where z_{ij} is the observed covariate corresponding to the response datum y_{ij}. And the usual intraclass regression extension of this is

$$E(y_{ij}) = \mu_i + b_i z_{ij}, \tag{10.77}$$

having a slope parameter, b_i, for each class. Details of these models for unbalanced data are in Searle (1971, Section 8.2.b) and even more extensively in Searle (1987, Chapter 6).

With today's computing power, covariance analysis can easily be calculated for models much more complicated than just (10.76) and (10.77); e.g.,

$$E(y_{ijk}) = \mu + \alpha_i + \beta_j + b_i z_{ijk} + b_j^* w_{ijk}$$

where z_{ijk} and w_{ijk} are two covariables, with a slope parameter b_i for z_{ijk} for each i-class and one b_j^* for w_{ijk} for each j-class. And the arithmetic for estimation and analysis of variance calculations, even for unbalanced data, is readily obtainable from a number of computing packages. The basis of the estimation calculations is as follows. Suppose $E(y) = \mathbf{X}\boldsymbol{\beta}$ is for a no-covariable model, and $E(y) = \mathbf{X}\boldsymbol{\beta} + \mathbf{Zb}$ is a corresponding with-covariables model, where columns of $\mathbf{Z}$ are the observed covariables. Then solutions to the least squares equations are

$$\boldsymbol{\beta}^0 = (\mathbf{X}'\mathbf{X})^-\mathbf{X}'(\mathbf{y} - \mathbf{Z}\hat{\mathbf{b}})$$

and

$$\hat{\mathbf{b}} = (\mathbf{R}'\mathbf{R})^{-1}\mathbf{R}'\mathbf{y} \qquad \text{for } \mathbf{R} = [\mathbf{I} - \mathbf{X}(\mathbf{X}'\mathbf{X})^-\mathbf{X}']\mathbf{Z},$$

where the tth column of $\mathbf{R}$ is the residual after fitting, for computing purposes only, $E(\mathbf{z}_t) = \mathbf{X}\boldsymbol{\beta}$ with $\mathbf{z}_t$ being the tth column of $\mathbf{Z}$. Full details of this approach, including analysis of variance tables, are laid out in Searle (1971, Section 8.2) and Searle (1987, Section 11.2) along with application to the 1-way classification of (10.76) and (10.77).

10.4.3 Computing with SAS GLM

A computing package which includes among its output many of the calculated values that linear model analysis requires is SAS GLM. Descriptions of that output are available in many places: Freund, Littell, and

Spector (1986) and other SAS Institute publications; Searle and Yerex (1987); and, more briefly, Searle (1987, Sections 12.2, 3 and 4c). Therefore just a few short notes are given here, in terms of the 2-way classification overparameterized model $E(y_{ijk}) = \mu + \alpha_i + \beta_j + \gamma_{ij}$ of (10.10), with the factor labels A, B, and A $*$ B being presented to the computing package in that sequence.

10.4.3.1 Sums of Squares

SAS GLM has four different sets of sums of squares labeled Types I, II, III, and IV. A summary of what these are is shown in Table 10.6. A few brief comments are in order.

Type I. The parenthetical title "sequential" is self-evident: the sums of squares come from fitting μ, then $\mu + \alpha_i$, then $\mu + \alpha_i + \beta_j$, and finally $\mu + \alpha_i + \beta_j + \gamma_{ij}$.

Type II. These sums of squares are for fitting each factor adjusted for all appropriate others. The meaning of appropriate here is that adjustment is made *neither* for interactions that involve the factor of interest [e.g., it is $R(\alpha \mid \mu, \beta)$ not $R(\alpha \mid \mu, \beta, \gamma)$] *nor* for factors nested within that factor.

Type III. The symbols under the heading "some cells empty" represent sums of squares for a model somewhat akin to the Σ-restricted model of Section 10.4.1 for balanced data, adapted to unbalanced data. For all-cells-filled data that adaptation yields [see Searle, Speed, and Henderson (1981, Appendix B)] SSA_w and SSB_w, as in Table 10.6, and as such is very useful (see Recommendation II, Section 10.3.4.5). But for some-cells-empty data the adaptation is of little or no use (see Searle and Henderson, 1983).

Type IV. Sums of squares Types I, II, and III are derived directly from fitting different models and submodels. Type IV operates differently. From the pattern of filled cells, the computing routine sets up hypotheses and calculates the numerator sums of squares for testing them. As an example, consider Grids 9 and 9a, where the occurrence of a μ instead of a check mark indicates the presence of data.

Grid 9

μ_{11}		μ_{13}	μ_{14}
μ_{21}	μ_{22}		
	μ_{32}	μ_{33}	μ_{34}

Grid 9a

μ_{11}		μ_{13}	μ_{14}
	μ_{32}	μ_{33}	μ_{34}
μ_{21}	μ_{22}		

Table 10.6 Sums of Squares from SAS GLM for a 2-Way Crossed Classification $E(y_{ijk}) = \mu + \alpha_i + \beta_j + \gamma_{ij}$

Label	Type I[a] (Sequential)	Type II (Adjusted)	Type III — All cells filled	Type III — Some cells empty	Type IV
A	$R(\alpha \mid \mu)$	$R(\alpha \mid \mu, \beta)$	SSA_w	$R(\dot\alpha \mid \dot\mu, \dot\beta, \dot\gamma)$	[c]
B	$R(\beta \mid \mu, \alpha)$	$R(\beta \mid \mu, \alpha)$	SSB_w	$R(\dot\beta \mid \dot\mu, \dot\alpha, \dot\gamma)$	
A * B[b]	$R(\gamma \mid \mu, \alpha, \beta)$	$R(\gamma \mid \mu, \alpha, \beta)$	$R(\gamma \mid \mu, \alpha, \beta)$	$R(\gamma \mid \mu, \alpha, \beta)$	$R(\gamma \mid \mu, \alpha, \beta)$

[a]Types I–IV are the same for balanced data.
[b]Types II–IV are the same when there is no interaction, whereupon there is no A * B line.
[c]Types III and IV are the same for all-cells-filled data; for some-cells-empty data, SAS GLM selects hypotheses.

The hypothesis from which the Type IV row sum of squares is calculated for Grid 9 is

$$H:\begin{cases} \mu_{33} + \mu_{34} - (\mu_{13} + \mu_{14}) = 0 \\ \mu_{32} - \mu_{22} = 0 \end{cases}. \tag{10.78}$$

Grid 9a is for exactly the same data as would be in Grid 9, except that the second and third rows have been interchanged. This should not affect any good system of calculating sums of squares that could be used with the intention (or hope) of providing inference-making procedures. Yet, with the Type IV algorithm, it does make a difference. The hypothesis from which the Type IV row sums of squares is calculated for Grid 9a is

$$H:\begin{cases} \mu_{22} - \mu_{32} = 0 \\ \mu_{21} - \mu_{11} = 0 \end{cases}. \tag{10.79}$$

This is entirely different from (10.78).

Note that this algorithmic approach is not necessarily any part of a traditional analysis of variance partitioning of the total sum of squares. Nor does it necessarily involve all the data. For (10.78) the cells in column 1 will not be used; and for (10.79) those in columns 3 and 4 will not be used.

All in all, the Type IV sums of squares for some-cells-empty data seem to be totally worthless. For other data they are identical to one or more of Types I, II, and III (see footnotes to Table 10.6), whereupon they may or may not be useful. More detail is available in Searle (1987, Chapter 13).

10.4.3.2   A General Estimable Function

The SAS GLM routine provides information from which the user can generate estimable functions. For example, for the 1-way classification with three classes it will provide, under the heading "general estimable function," two columns. One column has the labels Intercept (meaning μ) and three others, labeled (in default) A1, A2, and A3, meaning α_1, α_2, and α_3. The other column will have entries L1, L2, L3, and L1 $-$ L2 $-$ L3. These two columns are put together by multiplying corresponding entries. This gives, on using l_1 in place of L1 and so on,

$$l_1\mu + l_2\alpha_1 + l_3\alpha_2 + (l_1 - l_2 - l_3)\alpha_3.$$

This can be rewritten as

$$(l_1 - l_2 - l_3)(\mu + \alpha_3) + l_2(\mu + \alpha_1) + l_3(\mu + \alpha_3),$$

which we see is a linear function of the estimable functions $\mu + \alpha_i$ and so is itself estimable—for *any* values that one might care to give to l_1, l_2, and l_3. This is why it is called a general estimable function.

10.4.3.3 Estimable Functions for Sums of Squares

For each sum of squares [but not $R(\mu)$ or SSE] in each of its four types of sums of squares, SAS GLM outputs an estimable function in the same two-column format as described in the preceding section. Its use is that it provides a means of setting up the hypothesis that will be tested by using that sum of squares as the numerator of an *F*-statistic. For example, the estimable function for a 1-way classification having three classes will, for $R(\alpha \mid \mu)$, which is SSA $= \Sigma_{i=1}^3 n_i(\bar{y}_{i.} - \bar{y}_{..})^2$ of Table 10.2, be

$$f = l_2\alpha_1 + l_3\alpha_2 + (-l_2 - l_3)\alpha_3.$$

It is used as follows. f has two different l's, corresponding to there being two degrees of freedom for SSA. Any two linearly independent sets of numbers can be given to those two l's. Use them in f to create two forms of f, e.g.,

$$l_2 = 1 \text{ and } l_3 = 0 \quad \text{give} \quad f_1 = \alpha_1 - \alpha_3,$$

and

$$l_2 = 0 \text{ and } l_3 = 1 \quad \text{give} \quad f_2 = \alpha_2 - \alpha_3.$$

Then the hypothesis tested by $R(\alpha \mid \mu)$ is

$$H: \begin{cases} f_1 = 0 \\ f_2 = 0 \end{cases}, \quad \text{which is} \quad H: \begin{cases} \alpha_1 - \alpha_3 = 0 \\ \alpha_2 - \alpha_3 = 0 \end{cases},$$

i.e., $H: \alpha_i$ *all equal*. The generalization of this is that in each case there will be r different l's in the output, and using r linearly independent sets of r values for those l's to create $f_1, \ldots, f_r$, the corresponding hypothesis is $H: f_1 = 0 = f_2 = \cdots = f_r$.

10.4.3.4 Solutions to Least Squares Equations

For the parameters in whatever overparameterized model (for that is what SAS GLM is geared to) is being used, the output includes a list of values labeled ESTIMATE alongside a PARAMETER list, the latter using labels as indicated in Section 10.4.3.2. This labeling would seem to suggest in the 1-way classification, for example, that the ESTIMATE value alongside the parameter α_1 would be the least squares estimate of α_1. Unfortunately, that is *not so*. It is the BLUE of something else. In a 1-way classification of three classes the ESTIMATE values corresponding to parameters μ, α_1, α_2, and α_3 are BLUE($\mu + \alpha_3$), BLUE($\alpha_1 + \alpha_3$), BLUE($\alpha_2 + \alpha_3$), and BLUE($\alpha_3 - \alpha_3$) = 0.

The universal form of this is that for the general linear model expressed as $E(\mathbf{y}) = \mathbf{X\beta}$, the column of SAS GLM output values that is labeled ESTIMATE is not the BLUE of $\mathbf{\beta}$: it is BLUE$[(\mathbf{X'X})^{-}\mathbf{X'X\beta}]$. This means, of course, that the corresponding F-statistics output is not for testing that elements of $\mathbf{\beta}$ are zero but that elements of $(\mathbf{X'X})^{-}\mathbf{X'X\beta}$ are zero. Details can be found in Searle (1987, Section 12.4c).

10.4.4 Variance Components Models

All of the preceding presentation is concerned with what are called fixed-effects models. This means, for example, in the 2-way crossed classification no-interaction model based on the equation

$$y_{ij} = \mu + \alpha_i + \beta_j + e_{ij}, \tag{10.80}$$

that μ, α_i, and β_j are deemed to be fixed constants. They are unknown, and usually unknowable, and we seek estimates of them.

There are also situations where, for example, the β_j's are considered not as fixed constants but as unknowable realized values of random variables. Thus suppose y_{ij} of (10.80) is the test score on three mathematics tests ($i = 1, 2, 3$) for 40 randomly chosen freshmen entering a high school ($j = 1, 2, \ldots, 40$). Then β_j would be considered random, and the interest would be in estimating its variance, σ_β^2, say. And, because the variance of y_{ij} would be $\sigma_\beta^2 + \sigma_e^2$, each variance in that sum is called a variance component.

In this example of test scores on three tests, the α_i in (10.80) are effects due to the three different tests, and so are fixed effects. Thus with the β_j's being random effects we have a mixture of fixed and random effects and so such a model is called a mixed model. μ is always a fixed effect, and models in which it is the only fixed effect are called random-effects models, or random models. Likewise, error terms such as e_{ij} are always random, and models in which they are the only random terms are called fixed-effects models, or fixed models.

In contrast to fixed-effects models where there is only one variance (that of the error terms) and it is easily estimated, mixed and random models have two or more variance components to be estimated. Whereas this estimation is reasonably straightforward for balanced data, it is full of complications for unbalanced data. For example, there are many different methods of estimation. Searle (1971, Chapter 10) has a now somewhat outdated discussion of these complications. Searle (1987, Chapter 13) has a review chapter, and Searle, Casella, and McCulloch (1992) devote their entire book to the subject.

REFERENCES

Freund, R. J., Littell, R. D., and Spector, P. (1986). *SAS System for Linear Models*. SAS Institute, Cary, North Carolina.

Meredith, M. P. and Cady, F. B. (1988). Pragmatic Analysis of Factorial Experiments with Unequal Replication. Unpublished Technical Report BU-969-M, Biometrics Unit, Cornell University, Ithaca, New York.

Searle, S. R. (1971). *Linear Models*. Wiley, New York.

Searle, S. R. (1987). *Linear Models for Unbalanced Data*. Wiley, New York.

Searle, S. R. and Henderson, H. V. (1983). Faults in a Computing Algorithm for Reparameterizing Linear Models. *Comm. Statist.–Simul. Comput.*, *B12*: 67–76.

Searle, S. R. and Yerex, R. P. (1987). ACO_2: SAS GLM: Annotated Computer Output for Analysis of Variance of Unbalanced Data. Unpublished Technical Report BU-949-M, Biometrics Unit, Cornell University, Ithaca, New York.

Searle, S. R., Casella, G., and McCulloch, C. E. (1992). *Variance Components*. Wiley, New York.

Searle, S. R., Speed, F. M., and Henderson, H. V. (1981). Some Computational and Model Equivalences in Analysis of Variance of Unequal-Subclass-Numbers Data. *The Amer. Statist.*, *35*: 11–33.

Urquhart, N. S. and Weeks, D. L. (1978). Linear Models in Messy Data: Some Problems and Alternatives. *Biometrics*, *34*: 696–705.

Urquhart, N. S., Weeks, D. L., and Henderson, C. R. (1970). Estimation Associated with Linear Models: A Revisitation. Unpublished Technical Report BU-195, Biometrics Unit, Cornell University, Ithaca, New York.

Weeks, D. L. and Williams, D. R. (1964). A Note on the Determination of Connectedness in an *N*-Way Cross Classification: I. *Technometrics*, *6*: 319–324.

Weeks, D. L. and Williams, D. R. (1965). A Note on the Determination of Connectedness in an *N*-Way Cross Classification: II. *Technometrics*, *7*: 281.

Winer, B. J. (1971). *Statistical Principles in Experimental Design*. McGraw-Hill, New York.

Yates, F. (1934). The Analyses of Multiple Classifications with Unequal Number of Observations in the Different Classes. *J. Amer. Statist. Assoc.*, *29*: 51–66.

11
Tukey's Nonadditivity Test

ANANT M. KSHIRSAGAR   University of Michigan, Ann Arbor, Michigan

11.1   INTRODUCTION

With one observation per cell in a two-way classification, the presence of interaction between the two underlying factors cannot be tested. Tukey (1949) proposed a test for this which later came to be associated with the multiplicative interaction term. This test was later generalized in various directions in the general linear model: to models with more than one term representing interaction effects, to multivariate analysis of variance and covariance, and to growth curve models. A review of this development is given here.

Consider pq observations y_{ij} ($i = 1, 2, \ldots, p; j = 1, 2, \ldots, q$) arranged in p rows corresponding to the levels of a factor A and q columns corresponding to the levels of another factor B. The usual model for such a two-way classification analysis of variance is

$$y_{ij} = \mu + \alpha_i + \beta_j + \varepsilon_{ij} \tag{11.1}$$

where μ is the general mean, α_i is the effect of the ith level of A, β_j is the effect of the jth level of B, and ε_{ij}'s are random errors assumed to be independently and normally distributed with zero means and a common but unknown variance σ^2. The estimable parameters [see, for example, Searle (1971) or Kshirsagar (1983)] in this additive model are contrasts of α_i's, contrasts of β_j's, and the function $\mu + \bar{\alpha} + \bar{\beta}$, where $\bar{\alpha}$ and $\bar{\beta}$ are the means of the α_i's and β_j's, respectively. The degrees of freedom ($d.f.$) of the error sum of squares (SSE) are then

$$pq - (p - 1) - (q - 1) - 1 = (p - 1)(q - 1)$$

421

as there are $(p - 1)$ linearly independent contrasts of the α_i's and $(q - 1)$ linearly independent contrasts of the β_j's. However, if the effects of the two factors A and B are not additive and there is an interaction between them, the model is

$$y_{ij} = \mu + \alpha_i + \beta_j + \gamma_{ij} + \varepsilon_{ij} \qquad (11.2)$$

and then the number of estimable parameters is increased by $(p - 1)$ $(q - 1)$ as linear functions of the type $\Sigma_i \Sigma_j l_{ij}\gamma_{ij}$ where $\Sigma_i l_{ij} = \Sigma_j l_{ij} = 0$ are also estimable. We are, then, left with no degree of freedom for the SSE. In the model (11.2) where we have only one observation per cell, the parameters are estimable but we cannot test any hypothesis about the estimable parameters. In the additive model (11.1), we can estimate and test hypotheses about the effects of A and B but cannot include the interaction term. We need more than one observation in cells to be able to do both. The introduction of the general interaction term γ_{ij} overparameterizes the model.

Tukey (1949) suggested an ingenious method to test for nonadditivity without giving any specific model. However, his test has been now associated with the model that assumes the interaction between factors A and B to be of the form

$$\lambda(\alpha_i - \overline{\alpha})(\beta_j - \overline{\beta}). \qquad (11.3)$$

The interaction is multiplicative and the model is

$$y_{ij} = \mu + \alpha_i + \beta_j + \lambda(\alpha_i - \overline{\alpha})(\beta_j - \overline{\beta}). \qquad (11.4)$$

(Some people impose the restrictions $\Sigma_i \alpha_i = 0$ and $\Sigma_j \beta_j = 0$, and then α_i and β_j themselves are estimable and one can take the interaction to be $\lambda\alpha_i\beta_j$. I prefer not to impose any restrictions on the parameters but to consider estimable functions of parameters.) Note that (11.4) is *not a linear model*. But Tukey (1949) [see also Scheffé (1959)] provided a test for the hypothesis

$$H_0: \lambda = 0 \qquad (11.5)$$

in the model (11.4) by using the independence of the best linear unbiased estimates (BLUEs) of $\alpha_i - \overline{\alpha}$, $\beta_j - \overline{\beta}$, and the residuals

$$e_{ij} = y_{ij} - y_{i.} - y_{.j} + y_{..} \qquad (11.6)$$

in the additive model (11.1), where

$$y_{i.} = \frac{1}{q}\sum_{j=1}^{p} y_{ij}, \qquad y_{.j} = \frac{1}{p}\sum_{i=1}^{p} y_{ij}, \qquad y_{..} = \frac{1}{pq}\sum_{i=1}^{p}\sum_{j=1}^{p} y_{ij}.$$

It is an F test with one degree of freedom for the numerator χ^2 and $(p - 1)(q - 1) - 1$ degrees of freedom for the denominator χ^2. This Tukey's test with one degree of freedom is a test for the additive model (11.1) against an alternative hypothesis which assumes the nonadditivity of a particular type (11.3). The alternative model assumes the interaction to be a function of the main effects α_i and β_j.

Actually, we need not restrict the interaction to being multiplicative. Tukey's original test has been generalzied to situations in which the non-additivity term could be λ times any "known" function f, say

$$\lambda f(\alpha_i - \overline{\alpha}, \beta_j - \overline{\beta}). \tag{11.7}$$

For example, it could be

$$\lambda(\alpha_i - \overline{\alpha})/(\beta_j - \overline{\beta}) \qquad \text{or even} \qquad \lambda \sin^{-1}(\alpha_i - \overline{\alpha})/\cosh(\beta_j - \overline{\beta}).$$

However, in practice, f is rarely known and this generalization is often of mathematical interest only. If the function f is unknown, it can be expanded by Taylor series, and the simplest term in the expansion representing the interaction is $\lambda(\alpha_i - \overline{\alpha})(\beta_j - \overline{\beta})$. So, in a sense, Eq. (11.4) is an approximation to Eq. (11.7).

Again, there is only one term (11.3) in the model, but it is possible to extend this idea (Milliken and Graybill, 1970; Johnson and Graybill, 1972; Rao, 1965) and introduce more terms of the Taylor series expansion. For example, we can take the nonadditive model to be

$$\begin{aligned}
y_{ij} = {}& \mu + \alpha_i + \beta_j + \lambda_1(\alpha_i - \overline{\alpha})(\beta_j - \overline{\beta}) \\
& + \lambda_2(\alpha_i - \overline{\alpha})^2(\beta_j - \overline{\beta})^2 + \lambda_3(\alpha_i - \overline{\alpha})(\beta_j - \overline{\beta})^2 + \varepsilon_{ij} \tag{11.8}
\end{aligned}$$

and test the hypothesis $\lambda_1 = \lambda_2 = \lambda_3 = 0$. The F test will then have 3 for the numerator degrees of freedom and $(p - 1)(q - 1) - 3$ for the denominator degrees of freedom. The SSE, $\Sigma_i \Sigma_j e_{ij}^2$, for the model (11.1) is split into two parts, with one part corresponding to the nonadditivity terms involving $\lambda_1, \lambda_2, \lambda_3, \ldots$, etc. and the other part corresponding to the remainder. The more λ's one introduces in the model, the less will be the d.f. for the remainder SSE and the usefulness of the procedure will become questionable. The extreme case is the one where we introduce γ_{ij} and no d.f. is left for the SSE. Usually, the simple model (11.4) is adopted and the hypothesis $\lambda = 0$ is tested. We give the test in the next section.

11.2 TUKEY'S NONADDITIVITY TEST

For the model (11.1), the usual analysis of variance table entries are provided by Table 11.1. In the table, R_i stands for the total of the ith row,

Table 11.1 Analysis of Variance

Source	d.f.	Sum of squares
Row: A	$p - 1$	$SSR = \Sigma_i R_i^2/q - g^2/pq$
Columns: B	$q - 1$	$SSC = \Sigma_j C_j^2/p - g^2/pq$
Error	$(p - 1)(q - 1)$	$SSE = SST - SSR - SSC = \Sigma_i \Sigma_j e_{ij}^2$
Total (corrected)	$pq - 1$	$SST = \Sigma_i \Sigma_j y_{ij}^2 - g^2/pq$

$\Sigma_j y_{ij}$; C_j stands for the total of the jth column, $\Sigma_i y_{ij}$; and g stands for the grand total. In addition, SSR, SSC, SSE, and SST stand for the row sum of squares, column sum of squares, error sum of squares, and total sum of squares, respectively. Tukey's F statistic based on one d.f. for testing the hypothesis $\lambda = 0$ is

$$F = \frac{(p - 1)(q - 1) - 1}{1} \frac{Z^2}{(SSE - Z^2)} \tag{11.9}$$

where

$$Z^2 = pq \left[\sum_{i=1}^{p} \sum_{j=1}^{q} y_{ij}(y_{i.} - y_{..})(y_{.j} - y_{..}) \right]^2 \bigg/ (SSR)(SSC). \tag{11.10}$$

The hypothesis is rejected at a level of significance α, if the observed F exceeds the $100(1 - \alpha)\%$ point of an F with 1 and $(p - 1)(q - 1) - 1$ degrees of freedom.

The justification of this test is based on the fact that when $\lambda = 0$, e_{ij}, $\hat{\alpha}_i = y_{i.} - y_{..}$, and $\hat{\beta}_j$ and $y_{.j} - y_{..}$ are independent. Recall, $\hat{\alpha}_i$ and $\hat{\beta}_j$ are the BLUEs of $\alpha_i - \bar{\alpha}$ and $\beta_j - \bar{\beta}$, respectively. If we keep $\hat{\alpha}_i$ and $\hat{\beta}_j$ ($i = 1, \ldots, p; j = 1, \ldots, q$) fixed, Z^2 and $SSE - Z^2$ are independent quadratic forms in the normal variables e_{ij}. This can be proved by using a result of Shanbhag (1966) for quadratic forms of normal variables with a singular covariance matrix. Since the conditional F distribution when $\hat{\alpha}_i$ and $\hat{\beta}_j$ are fixed does not involve these conditioning variables, the absolute distribution is also an F distribution (Kshirsagar, 1983).

A numerical example is given below for illustration of this test. The 16 observations in Table 11.2 correspond to four ethnic groups and four socioeconomic-status (SES) classifications:

1. High SES males (HM)
2. High SES females (HF)
3. Low SES males (LM)
4. Low SES females (LF)

Table 11.2 Scores in General Knowledge for 16 Individuals

Socio-economic status	Ethnic group				Raw totals
	1	2	3	4	
HM	70	75	80	85	$R_1 = 310$
HF	75	80	85	90	$R_2 = 330$
LM	81	82	90	97	$R_3 = 350$
LF	85	90	93	101	$R_4 = 369$
Column total	$C_1 = 311$	$C_2 = 327$	$C_3 = 348$	$C_4 = 373$	$g = 1359$

The observations are scores in General Knowledge. The analysis of variance with the additive model for these data is presented in Table 11.3. The values of $\hat{\alpha}_i (i = 1, 2, 3, 4)$ are

$$\hat{\alpha}_1 = -7.4375, \qquad \hat{\alpha}_2 = -2.4375, \qquad \hat{\alpha}_3 = 2.5625, \qquad \hat{\alpha}_4 = 7.3125.$$

The values of $\hat{\beta}_j$ $(j = 1, 2, 3, 4)$ are

$$\hat{\beta}_1 = -7.1875, \qquad \hat{\beta}_2 = -3.1875, \qquad \hat{\beta}_3 = 2.0625, \qquad \hat{\beta}_4 = 8.3125.$$

The residuals e_{ij} are given in Table 11.4, and the values of $\hat{\alpha}_1 \hat{\beta}_j$ are given in Table 11.5. From this,

$$\sum_{i=1}^{4} \sum_{j=1}^{4} e_{ij} \hat{\alpha}_i \hat{\beta}_j = \sum_{i=1}^{4} \sum_{j=1}^{4} y_{ij} \hat{\alpha}_i \hat{\beta}_j = 79.31$$

and

$$Z^2 = \frac{pq(\Sigma_i \Sigma_j y_{ij} \hat{\alpha}_i \hat{\beta}_j)^2}{(SSR)(SSC)} = 0.3836$$

and the F statistic for nonadditivity is

$$F = \frac{8 Z^2}{SSE - Z^2} = 0.2421,$$

Table 11.3 Analysis of Variance

Source	d.f.	Sum of squares
Rows	$p - 1 = 3$	SSR = 485.1875
Columns	$q - 1 = 3$	SSC = 540.6875
Error	9	SSE = 13.0625
Total	$pq - 1 = 15$	SST = 1038.9375

426 Kshirsagar

Table 11.4 Residuals (e_{ij})

$i \backslash j$	1	2	3	4
1	-0.3125	0.6875	0.4375	-0.8125
2	-0.3125	0.6875	0.4375	-0.8125
3	0.6875	-2.3125	0.4375	1.1875
4	-0.0625	0.9375	-1.3125	0.4374

which is less than 5.32, the 5% value of F with 1 and 8 degrees of freedom. The hypothesis $\lambda = 0$ is not rejected.

11.3 TUKEY'S TEST FOR A GENERAL LINEAR MODEL

We considered Tukey's test for a two-way classification. However, it can be used for a general linear model and hence for any general analysis of variance or experimental designs model. Milliken and Graybill (1970) and Rao (1965) have given this generalization. Consider the general linear model

$$\mathbf{y} = \mathbf{X}\boldsymbol{\beta} + \boldsymbol{\varepsilon} \tag{11.11}$$

where $\mathbf{y}$ is $n \times 1$, $\mathbf{X}$ is $n \times p$ of rank r, $\boldsymbol{\beta}$ is $p \times 1$, and $\boldsymbol{\varepsilon}$ is $n \times 1$. $\mathbf{y}$ is the vector of observations, $\mathbf{X}$ is the known design matrix, $\boldsymbol{\beta}$ are unknown parameters, and the elements of $\boldsymbol{\varepsilon}$ are random errors, assumed to have independent normal distributions with zero means and a common but unknown variance σ^2. In this setup, $\mathbf{X}\boldsymbol{\beta}$ and its linear combinations are estimable. The BLUE of $\mathbf{X}\boldsymbol{\beta}$ is $\mathbf{X}\hat{\boldsymbol{\beta}}$, where

$$\hat{\boldsymbol{\beta}} = (\mathbf{X}'\mathbf{X})^{-} \mathbf{X}'\mathbf{y} \tag{11.12}$$

Table 11.5 Values of $\hat{\alpha}_i\hat{\beta}_j$

$i \backslash j$	1	2	3	4
1	53.4750	23.7070	-15.3398	-61.8242
2	17,5195	7.7695	-5.0273	-20.2617
3	-18.4180	-8.1680	5.2852	21.3008
4	-52.5586	-23.3086	15.0820	60.7852

and $(\mathbf{X'X})^-$ is a generalized inverse of $\mathbf{X'X}$. [It satisfies

$$(\mathbf{X'X})(\mathbf{X'X})^-(\mathbf{X'X}) = (\mathbf{X'X}).]$$

The error sum of squares is

$$\text{SSE} = \mathbf{y'y} - \hat{\boldsymbol{\beta}}'\mathbf{X'y} \qquad (11.13)$$

with $(n - r)$ degrees of freedom. Now consider the model

$$\mathbf{y} = \mathbf{X\boldsymbol{\beta}} + \mathbf{F}(\mathbf{X\boldsymbol{\beta}})\boldsymbol{\lambda} + \boldsymbol{\varepsilon} \qquad (11.14)$$

where $\mathbf{F}(\mathbf{X\boldsymbol{\beta}})\boldsymbol{\lambda}$ is the "nonadditivity" contribution, $\mathbf{F}$ being an $n \times k$ matrix such that every element $F_{ij}(\mathbf{X\boldsymbol{\beta}})$ of $\mathbf{F}$ is a function of the estimable parametric functions $\mathbf{X\boldsymbol{\beta}}$ and the functional form F_{ij} is known. $\boldsymbol{\lambda}$ is $k \times 1$. The nonadditivity test examines the hypothesis $\boldsymbol{\lambda} = \mathbf{0}$ and is based on the statistic

$$F = \frac{(n - r) - b}{b} \frac{U}{(\text{SSE} - U)} \qquad (11.15)$$

where

$$\mathbf{d} = (\mathbf{I} - \mathbf{X}(\mathbf{X'X})^- \mathbf{X'})\mathbf{y} \qquad (11.16)$$
$$= \mathbf{y} - \mathbf{X}\hat{\boldsymbol{\beta}},$$

$$\hat{\mathbf{F}} = \mathbf{F}(\mathbf{X}\hat{\boldsymbol{\beta}}), \qquad (11.17)$$

$$\mathbf{M} = (\mathbf{I} - \mathbf{X}(\mathbf{X'X})^- \mathbf{X'})\hat{\mathbf{F}}, \qquad (11.18)$$

$$U = \mathbf{d'M}(\mathbf{M'M})^- \mathbf{M'd}, \qquad (11.19)$$

and

$$b = \text{rank } M. \qquad (11.20)$$

The hypothesis $\boldsymbol{\lambda} = \mathbf{0}$ is rejected if the value of F in (11.15) exceeds the $100(1 - \alpha)\%$ point of the F distribution with b and $(n - r) - b$ degrees of freedom. Usually b is the same as k.

The proof of this result is a straightforward generalization of the earlier result for a two-way classification with only one term $\lambda(\alpha_i - \bar{\alpha})(\beta_j - \bar{\beta})$ as nonadditivity. U and $(\text{SSE} - U)$ are independent quadratic forms. This is proved by keeping $\mathbf{X}\hat{\boldsymbol{\beta}}$ fixed first and using the independence of $\mathbf{X}\hat{\boldsymbol{\beta}}$ and $\mathbf{y} - \mathbf{X}\hat{\boldsymbol{\beta}}$.

In practice, it is easier to get the F statistic (11.15) from the following steps:

1. Minimize $(\mathbf{y} - \mathbf{X}\hat{\boldsymbol{\beta}})' (\mathbf{y} - \mathbf{X}\hat{\boldsymbol{\beta}})$ with respect to $\hat{\boldsymbol{\beta}}$ and obtain any solution $\hat{\boldsymbol{\beta}}$ of the normal questions.

2. obtain the error sum of squares by using Eq. (11.13).
3. Obtain the residuals $\mathbf{y} - \mathbf{X}\hat{\boldsymbol{\beta}} = \mathbf{d}$.
4. Obtain $\mathbf{F}(\mathbf{X}\hat{\boldsymbol{\beta}})$ by substituting $\hat{\boldsymbol{\beta}}$ for $\boldsymbol{\beta}$ in every element of $\mathbf{F}$. Let $\hat{f}_{il}$ denote the element in the ith row and lth column of $\hat{\mathbf{F}}$ and let $\hat{\mathbf{f}}_l$ denote the lth column of $\hat{\mathbf{F}}$.
5. To obtain $\mathbf{M}$, observe that the relation between $\mathbf{y}$ and $\mathbf{d}$ is the same as that between any column of $\hat{\mathbf{F}}$ and the corresponding column of $\hat{\mathbf{M}}$. So, if $\mathbf{y}$ is replaced by the corresponding elements of $\hat{\mathbf{f}}_l$ in this relation, the result is $\mathbf{m}_l$, the lth column of $\mathbf{M}$. Specifically, if, in step 3 above,

$$d_i = \psi_i(y_1, \ldots, y_n)$$

gives the ith element of $\mathbf{d}$, then

$$m_{il} = \psi_i(\hat{f}_{1l}, \hat{f}_{2l}, \ldots, \hat{f}_{nl}) \qquad (l = 1, 2, \ldots, k)$$

gives the element in the ith row and lth column of $\mathbf{M}$ and hence $\mathbf{M}$.
6. To compute U as required in (11.19), pretend that you are considering the linear model

$$\mathbf{d} = \mathbf{M}\boldsymbol{\lambda} + \text{error} \tag{11.21}$$

and obtain the regression sum of squares for this model, namely,

$$\hat{\boldsymbol{\lambda}}'\mathbf{M}'\mathbf{d},$$

where $\hat{\boldsymbol{\lambda}}$ is any solution of the equations we solve to minimize

$$(\mathbf{d} - \mathbf{M}\hat{\boldsymbol{\lambda}})' (\mathbf{d} - \mathbf{M}\hat{\boldsymbol{\lambda}}),$$

i.e., a solution to the equations $\mathbf{M}'\mathbf{d} = (\mathbf{M}'\mathbf{M})\hat{\boldsymbol{\lambda}}$.
7. Now find $\text{SSE} - U$ and use the F test of (11.15). If one uses a computer program such as SAS (SAS Institute, 1989), the above steps can be carried out by first using the additive model (11.11) and obtaining $\hat{\boldsymbol{\beta}}$, SSE, and storing the residuals $\mathbf{d}$. Then substitute $\hat{\boldsymbol{\beta}}$ for $\boldsymbol{\beta}$ in every element of $\mathbf{F}$ to obtain $\hat{\mathbf{F}}$. Now replace the observations $\mathbf{y}$ by every column of $\hat{\mathbf{F}}$ turn by turn in the model (11.11) and obtain the residuals. These columns will now form $\mathbf{M}$. Now use the model (11.21) and obtain the regression sum of squares. This is U, from which the nonadditivity test can be carried out.

As an illustration, consider the two-way classification model (11.1) and the same data as in Section 11.2; but let the nonadditivity model now be

$$y_{ij} = \mu + \alpha_i + \beta_j + \lambda_1(\alpha_i - \overline{\alpha})(\beta_j - \overline{\beta})$$

$$+ \lambda_2(\alpha_i - \overline{\alpha})^2(\beta_j - \overline{\beta}) + \lambda_3(\alpha_i - \overline{\alpha})(\beta_j - \overline{\beta})^2 + \varepsilon_{ij}. \tag{11.22}$$

We know, from standard analysis of variance theory, that for the model (11.1),

$$\hat{\alpha}_i = y_{i.} - y_{..}, \qquad \hat{\beta}_j = y_{.j} - y_{..}, \tag{11.23}$$

and the residuals are

$$e_{ij} = y_{ij} - y_{i.} - y_{.j} + y_{..}. \tag{11.24}$$

If we write (11.22) as (11.14), $\mathbf{y}$ is the vector of the pq observations y_{ij}, and the first column of $\hat{\mathbf{F}}$ is the vector of values of

$$Z_{1ij} = (\hat{\alpha}_i - \hat{\bar{\alpha}})(\hat{\beta}_j - \hat{\bar{\beta}}) = \hat{\alpha}_i\hat{\beta}_j, \tag{11.25}$$

the second column of $\hat{\mathbf{F}}$ is the vector of values of

$$Z_{2ij} = (\hat{\alpha}_i - \hat{\bar{\alpha}})^2(\hat{\beta}_j - \hat{\bar{\beta}}) = \hat{\alpha}_i^2\hat{\beta}_j, \tag{11.26}$$

and the third column in the vector of values of

$$Z_{3ij} = (\hat{\alpha}_i - \hat{\bar{\alpha}})(\hat{\beta}_j - \hat{\bar{\beta}}) = \hat{\alpha}_i\hat{\beta}_j^2. \tag{11.27}$$

To obtain $\mathbf{M}$, we note the relation (11.24) between the residuals e_{ij} and the observations y_{ij}. Hence the lth column ($l = 1, 2, 3$) of $\mathbf{M}$ is obtained from the values of

$$Z_{lij} - Z_{li.} - Z_{l.j} + Z_{l..} \tag{11.28}$$

which, however, reduces to Z_{lij} itself because $Z_{li.}$, $Z_{l.j}$, and $Z_{l..}$ are all zeros due to the restraints $\Sigma_i\, \hat{\alpha}_i = \Sigma_j\, \hat{\beta}_j = 0$.

So finally we consider the model (11.21), which, in this case, is simply

$$e_{ij} = \lambda_1\hat{\alpha}_i\hat{\beta}_j + \lambda_2\hat{\alpha}_i^2\hat{\beta}_j + \lambda_3\hat{\alpha}_i\hat{\beta}_j^2 \tag{11.29}$$

$$(i = 1, 2, \ldots, p; j = 1, 2, \ldots, q).$$

In other words, we carry out a multiple regression of the pq values of e_{ij} on three variables Z_l ($l = 1, 2, 3$) whose values are Z_{lij} as given by (11.25) to (11.27), *without intercept*. The regression sum of squares is U. The actual calculations are shown below.

The values of e_{ij} and $\hat{\alpha}_i\hat{\beta}_j$ are already given in Tables 11.4 and 11.5. The values of $\hat{\alpha}_i^2\hat{\beta}_j$ and $\hat{\alpha}_i\hat{\beta}_j^2$ are given in Tables 11.6 and 11.7.

Running a multiple regression of the 16 values of e_{ij} on the corresponding values of $Z_1 = \hat{\alpha}_i\hat{\beta}_j$, $Z_2 = \hat{\alpha}_i^2\hat{\beta}_j$, and $Z_3 = \hat{\alpha}_i\hat{\beta}_j^2$, as given in Tables 11.5 and 11.6 without intercept, we find the regression sum of squares to be

$$U = 1.6387.$$

The F statistic for testing the hypothesis

$$H_0: \lambda_1 = \lambda_2 = \lambda_3 = 0$$

Table 11.6 Values of $\hat{\alpha}_i^2\hat{\beta}_j$

$i \backslash j$	1	2	3	4
1	-397.587	-176.321	114.090	459.818
2	-42.704	-18.938	12.254	49.388
3	-47.196	-20.930	13.543	54.583
4	-384.335	-170.444	110.287	444.491

is, therefore,

$$F = \frac{(p-1)(q-1)-3}{3}\frac{U}{SSE-U}$$
$$= \frac{6}{3}\frac{1.6387}{(13.0625-1.6387)}$$
$$= 0.26186$$

which is less than 4.76, the 5% value of F with 3 and 6 degrees of freedom.

11.4 EXTENSION TO MULTIVARIATE ANALYSIS OF VARIANCE (MANOVA) AND COVARIANCE

Tukey's test has been extended to the multivariate linear model by McDonald (1972), Khuri (1985), and Kshirsagar (1988). Consider the multivariate model

$$E(Y) = X\beta + Z\gamma + F(X\beta)\Lambda \qquad (11.30)$$

where Y is $n \times p$, X is $n \times m$, β is $m \times p$, Z is $n \times q$, γ is $q \times p$, F is $n \times k$, and Λ is $k \times p$. Y represents the n observations on p variables, X is the design matrix, Z is the matrix of observations on covariates, and

Table 11.7 Values of $\hat{\alpha}_i\hat{\beta}_j^2$

$i \backslash j$	1	2	3	4
1	-384.222	-75.566	-31.638	-513.914
2	-125.922	-24.765	-10.369	-168.426
3	132.379	26.035	10.901	177.063
4	377.765	74.296	31.107	505.277

$\mathbf{F}(\mathbf{X}\boldsymbol{\beta})\boldsymbol{\Lambda}$ represents the nonadditivity portion. The elements of $\mathbf{F}$ are, as before, functions of the estimable parameters $\mathbf{X}\boldsymbol{\beta}$. The hypothesis of additivity implies $\boldsymbol{\Lambda} = \mathbf{0}$ and can be tested by a generalization of Tukey's test to Wilks's Λ test (or Hotelling's generalized T^2 or Pillai's V). First, the least squares estimates $\hat{\boldsymbol{\beta}}$ and $\hat{\boldsymbol{\gamma}}$, namely

$$\hat{\boldsymbol{\gamma}} = (\mathbf{Z'PZ})^{-1}(\mathbf{Z'PY}), \tag{11.31}$$

$$\hat{\boldsymbol{\beta}} = (\mathbf{X'X})^{-}\mathbf{X'Y} - (\mathbf{X'X})^{-}\mathbf{X'Z}\hat{\boldsymbol{\gamma}}, \tag{11.32}$$

where

$$\mathbf{P} = \mathbf{I} - \mathbf{X}(\mathbf{X'X})^{-}\mathbf{X'} \tag{11.33}$$

for the model without the term $\mathbf{F}(\mathbf{X}\boldsymbol{\beta})\boldsymbol{\Lambda}$ are calculated. Then the matrix of residuals

$$\mathbf{D} = \mathbf{Y} - \mathbf{X}\hat{\boldsymbol{\beta}} - \mathbf{Z}\hat{\boldsymbol{\gamma}} \tag{11.34}$$

is obtained. The error matrix for the model without the term $\mathbf{F}\boldsymbol{\Lambda}$ is $\mathbf{D'D}$. It is then split into two parts $\mathbf{B}$ and $\mathbf{A}$. $\mathbf{B}$ is the hypothesis matrix for the hypothesis $\boldsymbol{\Lambda} = \mathbf{0}$, and $\mathbf{A}$ is the remainder of the error matrix. The following are the actual expressions for $\mathbf{B}$ and $\mathbf{A}$. Let

$$\hat{\mathbf{F}} = \mathbf{F}(\mathbf{X}\hat{\boldsymbol{\beta}}) \tag{11.35}$$

$$\mathbf{M} = \mathbf{P}(\mathbf{I} - \mathbf{Z}(\mathbf{Z'PZ})^{-1}\mathbf{Z'})\hat{\mathbf{F}}. \tag{11.36}$$

Then

$$\mathbf{B} = \mathbf{D'M}(\mathbf{M'M})^{-}\mathbf{M'D} \tag{11.37}$$

and

$$\mathbf{A} = \mathbf{D'D} - \mathbf{B}. \tag{11.38}$$

Analogous to (11.21), $\mathbf{B}$ is the regression matrix in the multivariate regression of $\mathbf{D}$ on $\mathbf{M}$, and $\mathbf{A}$ is the error matrix. Wilks's Λ for testing $\boldsymbol{\Lambda} = \mathbf{0}$ is then

$$\Lambda = |\,\mathbf{A}\,|/|\,\mathbf{A} + \mathbf{B}\,| \tag{11.39}$$

and its distribution, in the null case, depends on the following parameters (Kshirsagar, 1972):

$$d_m = p, \tag{11.40}$$

$$d_H = \text{rank } M \tag{11.41}$$

and

$$d_E = n - \text{rank } [\mathbf{X} \mid \mathbf{Z}]. \tag{11.42}$$

The exact percentage points of

$$-\{(d_H + d_E) - \tfrac{1}{2}(d_m + d_H + 1)\} \log_e \lambda \qquad (11.43)$$

are available in Kres (1983). An exact F test, when either d_H or d_m is 1 or 2, is also available (see Kshirsagar, 1972).

Factorization of this Wilks's Λ, to find out whether the nonadditivity is due to a single given linear combination of the responses, is given by Kshirsagar (1988).

Extension of the Tukey's test to growth curve models of Potthoff and Roy (1964), with nonadditivity, namely,

$$E(\mathbf{Y}) = \mathbf{A} \, \xi \, \mathbf{B} + \mathbf{F} \, \Lambda \, \mathbf{B} \qquad (11.44)$$

has also been given by Kshirsagar (in press).

11.5 MISCELLANEOUS OTHER RESULTS AND COMMENTS

St. Laurent (1990a) has shown the equivalence of the Milliken–Graybill procedure in Section 11.3 with the score test for testing $\Lambda = \mathbf{0}$ (Cox and Hinkley, 1974). The score test is also known as the Lagrange multiplier test and is a widely applicable method of test construction providing an alternative to the likelihood ratio test. Sapra (1990) notes that the Milliken–Graybill procedure is asymptotically equivalent to a conditional moment test introduced in the econometrics literature by Newey (1985) and Tauchen (1985). However, the score test, as noted by St. Laurent (1990b), has the added advantage that its exact distribution is known. St. Laurent shows that the score test statistic S can be expressed as nR^2, where R^2 is the squared multiple correlation coefficient in the regression of the residuals $\mathbf{d}$ on the columns of the matrix $\mathbf{M}$ and is, therefore, a computationally simple method. The Tukey test and its Milliken–Graybill generalizations thus have the support of the reputed likelihood methodology of tests.

Mandel's (1961) bundle-of-lines model, Freeman's (1973) model of interaction between genotypic and environmental factors in genetics, McAleer's (1983) test of a linear model specification against several nonnested, nonlinear alternatives, Andrews' (1971) test of the consistency of the data with a proposed transformation, and St. Laurent's (1987) class of tests for nonlinearity in the responses in regression are all either particular cases of application or the Milliken–Graybill generalized test.

Milliken and Graybill (1971) give detailed instruction with an example for Tukey's test in a two-way classification model when some cells are missing and with, at most, one observation per cell.

Pettitt (1989) gives a short and interesting review of how the nonadditive model (11.4) has been used in statistical modeling and data analysis. If the hypothesis $\lambda = 0$ is rejected, then λ in the model (11.4) is estimated by

$$\hat{\lambda} = \frac{\sum_{i=1}^{p} \sum_{j=1}^{q} \hat{\lambda}_i \hat{\beta}_j e_{ij}}{(\sum_{i=1}^{p} \sum_{j=1}^{q} \hat{\alpha}_i \hat{\beta}_j)^2}. \tag{11.45}$$

Pettit suggests the power transformation y^θ of y, with the use of $1 - \hat{\lambda}\bar{y}$ as an estimate of θ. This transformation might produce an analysis where the additive model fits reasonably well. He presents various interpretations of the model (11.4), including the suggestion of a new link function.

The estimate $\hat{\lambda}$ above is obtained by treating μ, α_i, and β_j known in (11.4) and estimating λ by the usual least squares procedure. The unknown parameters are then replaced by their estimates in this estimate to obtain (11.44). This is an ad hoc method of estimating λ. Sinha, Saharay, and Mukhopadhyay (1990) consider the estimate of λ in (11.4), in particular, and λ in (11.14), in general, in greater detail. The estimation part of λ was more or less ignored by Tukey as well as by Milliken and Graybill and others; and attention was focused on hypothesis testing only. Sinha et al. (1990) point out that λ is not necessarily estimable for any design matrix X. They investigate the estimability of λ and develop reasonable optimality criteria and compare different designs for efficient estimation of λ.

The power function of Tukey's test has been studied by Ghosh and Sharma (1963).

11.6 CONCLUDING REMARKS

Model specification and empirical model building are important aspects of regression analysis, analysis of variance, and design of experiments. If one wishes to find out whether a linear model is augmented by one or more terms consisting of nonlinear functions of the unknown parameters which represent interaction-type effects, there are hardly any exact tests or methods available. Tukey (1949) and, later, Milliken and Graybill (1970) provided a procedure for this purpose which is now being used widely. This procedure is found useful in a variety of situations. However, not all interactions are functions of other parameters such as main effects of the factors. Tukey's test may not be suitable in such models, and alternative procedures might be needed. The merit of Tukey's test is that it is an exact test for at least one type of nonadditivity. Due to its relationship with the score test (St. Laurent, 1990a), it can be viewed as an example of a generally accepted, likelihood-based method of test construction.

REFERENCES

Andrews, D. F. (1971). Significance Tests Based on Residuals. *Biometrika*, *58*: 139–142.

Cox, D. R. and Hinkley, D. V. (1974). *Theoretical Statistics*, Chapman & Hall, London.

Freeman, G. H. (1973). Statistical Methods for the Analysis of Genotype-Environment Interactions. *Heredity*, *31*: 339–349.

Ghosh, M. N. and Sharma, D. (1963). Power of Tukey's Test for Non-Additivity. *J. Roy. Statist. Soc.*, *B25*: 213–219.

Johnson, D. E. and Graybill, F. A. (1972). An Analysis of a Two-Way Model with Interaction and No Replication. *J. Amer. Statist. Assoc.*, *67*: 862–868.

Khuri, A. I. (1985). A Test for Lack of Fits of a Linear Multiresponse Model. *Technometrics*, *27*: 213–214.

Kres, H. (1983). *Statistical Tables for Multivariate Analysis*. Springer-Verlag, New York.

Kshirsagar, A. M. (1972). *Multivariate Analysis*. Marcel Dekker, New York.

Kshirsagar, A. M. (1983). *A Course in Linear Models*. Marcel Dekker, New York.

Kshirsagar, A. M. (1988). A Note on Multivariate Linear Models with Non-Additivity. *Austral. J. Statist.*, *30*: 1–7.

Kshirsagar, A. M. (in press). A Test for Non-Additivity in Growth Curve Models. *J. Gujarath Statist. Assoc. India* (Khatri Memorial Volume).

Mandel, J. (1961). Non-Additivity in Two-Way Analysis of Variances. *J. Amer. Statist. Assoc.*, *56*: 878–888.

McAleer, M. (1983). Exact Tests of a Model Against Nonnested Alternatives. *Biometrika*, *70*: 285–290.

McDonald, R. (1972). A Multivariate Extensions of Tukey's One Degree of Freedom for Non-Additivity. *J. Amer. Statist. Assoc.*, *67*: 674–675.

Milliken, G. A. and Graybill, F. A. (1970). Extensions of the General Linear Hypothesis Model. *J. Amer. Statist. Assoc.*, *65*: 797–807.

Milliken, G. A. and Graybill, F. A. (1971). Tests for Interaction in the Two-Way Model with Missing Data. *Biometrics*, *27*: 1079–1083.

Newey, W. K. (1985). Maximum Likelihood Specification Testing and Conditional Moment Tests. *Econometrica*, *53*: 1047–1070.

Pettit, A. N. (1989). One Degree of Freedom for Non-Additivity: Applications with Generalized Linear Models and Link Function. *Biometrics*, *45*: 1153–1162.

Potthoff, R. F. and Roy, S. N. (1964). A Generalized Multivariate Analysis of Variance Models Useful Especially for Growth Curve Problems. *Biometrika*, *51*: 313–326.

Rao, C. R. (1965). *Linear Statistical Inference and Its Applciations*. Wiley, New York.

St. Laurent, R. T. (1987). Detecting Curvature in the Response in Regression. Unpublished Ph.D. dissertation, School of Statistics, University of Minnesota.

St. Laurent, R. T. (1990a). The Equivalence of the Milliken–Graybill Procedure and the Score Test. *Amer. Statist.*, *44*: 36–57.

St. Laurent, R. T. (1990b). Response to S. K. Saproe. (Letter to the editor.) *Amer. Statist.*, *44*: 328.

Sapra, S. K. (1990). Interpreting the Milliken–Graybill Test as a Conditional Moment Test. *Amer. Statist.*, *44*: 328.

SAS Institute (1989). *SAS/STAT User's Guide*. (Version 6.06), SAS Institute, Cary, North Carolina.

Scheffé, H. (1959). *The Analysis of Variance*. Wiley, New York.

Searle, S. R. (1971). *Linear Models*. Wiley, New York.

Shanbhag, D. N. (1966). On the Independence of Quadratic Forms. *J. Roy. Statist. Soc.*, *B28*: 582.

Sinha, B. K., Saharay, R., and Mukhopadhyay, A. C. (1990). Non-Additive Linear Models: Estimability and Efficient Estimations of Interaction. *Comm. Statist.*, *A19*: 739–764.

Tauchen, G. (1985). Diagnostic Testing and Evaluation of Maximum Likelihood Models. *J. of Econometrics*, *30*: 415–443.

Tukey, J. W. (1949). One Degree of Freedom for Non-Additivity. *Biometrics*, *5*: 232–242.

12
Analysis of Time-Dependent Observations

LYNNE K. EDWARDS   University of Minnesota, Minneapolis, Minnesota

12.1   INTRODUCTION

One of the most frequently used research methods in the biological, behavioral, and social sciences makes repeated observations of the same individuals. Several different labels are used for these studies, at times reflecting the disciplines with which the studies are identified: repeated measures designs, changeover designs, crossover trials, response or growth curve studies, time series designs, longitudinal studies, and studies with serial measurements. Studies in which the repeated factors are experimentally induced are referred to as *repeated measures designs*, and studies which are nonexperimental and in which time is the repeated factor are referred to as *longitudinal studies*. But such distinctions are far from absolute.

In this chapter, we focus on methods of analysis for data arising from serial measurements taken from multiple subjects with relatively few time points. Because time-dependent observations are a subset of general repeated observations and because many of the statistical models and analyses developed for repeated observations are also applicable to time-dependent observations, we first present some examples in which various repeated observations are made. We then examine some unique features

Portions of this chapter were reproduced from "Fitting a Serial Correlation Pattern to Repeated Observations" by Lynne K. Edwards (1991), *J. Educ. Statist.*, *16*, 53–76, with permission from the American Educational Research Association.

of time-dependent observations as opposed to repeated observations with randomized treatments. In the following section, we categorize models for time-dependent observations: multivariate, univariate mixed effects, and time series, followed by a numerical illustration comparing a *univariate mixed-effects* model to a *time series* model. Finally, we summarize selected recent developments and their references.

12.2 REPEATED OBSERVATIONS

12.2.1 Examples

Repeated observations may be sought from each entity (subject) in several different contexts. A sample of students may be asked to take two versions of the same test in a counterbalanced sequence. Such a study may be referred to as a 2×2 crossover (changeover) study, a 2×2 Latin square with repeated measurements, or an AB-BA study. The purpose of the study is to produce A and B equivalent forms of a test.

Another variation of an AB study is a pre-post (or before-after) study in which the pretreatment and posttreatment measures are compared for possible treatment effects. We can generalize a before-after study even further either to a multiple follow-up study or to a multiple-treatment study. A multiple follow-up study may take a form in which a group of subjects is measured at intake, given an intervention, and then monitored repeatedly at some intervals. A multiple-treatment study may involve patients undergoing several different drug treatments with the treatments being randomly sequenced for each patient. In the latter case (known as a clinical trial), the experimenter often provides a long enough interval between treatments to prevent carryover or residual effects from confounding the true treatment effects. But when the order of treatments cannot be randomized or if the sequence or Time effects themselves are the foci of the study (as in a multiple follow-up case), we take serial measurements across time by definition.

Yet another context which is specific to developmental studies is a multiple follow-up study in which individual growth is monitored at some intervals over a period of time. Such a study may specifically be called a *growth curve* study. In a study similar except for being done in an industrial setting, the performance of an instrument may be examined as a function of time; and in pharmacological trials, effects of a drug may be monitored across a time axis. A study of this type is referred to as a *response curve* or *response surface* study. For additional examples involving serial measurements in various disciplines, readers are referred to Koch, Amara, Stokes, and Gillings (1980). Discussions specific to crossover designs can be found in Chapter 5 of this book, in Matthews (1988), and in Jones and Kenward (1989). Many new and traditional analyses and models for developmental studies are found in Johnston, Rocher, and Susanne (1979), Nesselroade and Baltes (1979), and von Eye (1990). General discussion of and references for repeated measures designs and analyses can be found in Crowder and Hand (1990), Cullis and McGilchrist (1990), Diggle (1988), Diggle and Donnelly (1989), Geary (1989), Geisser (1980), Koch, Elashoff, and Amara (1988), Louis (1988), Timm (1980), and Verbyla and Cullis (1990).

12.2.2 Repeated Observations Versus Time-Dependent Observations

Measuring the same entity repeatedly under all treatment conditions is an efficient method for assessing treatment effects because individual differences are thus partitioned out of the experimental error. In contexts in which either testing Time effects of modeling intraindividual change or growth is the primary focus, serial measurements are taken by default. In either case, the response measures from the same individuals are correlated across time (or across treatment levels), and they are assumed to be uncorrelated across different subjects. This dependence in intraindividual measures causes the ordinary univariate F test to be no longer exact unless some covariance structure can be assumed. However, the standard MANOVA (without imposing any covariance structure) can analyze both randomized repeated treatment data and time-dependent data. Yet another approach for time-dependent observations (but not for randomized treatment data) is to adopt one of the time series models.

The focus of this chapter is on data arising from serial measurements of multiple subjects with relatively few time points. Such studies are called either repeated measures studies or longitudinal studies. When the repeated factor is time (and thus no randomization is possible), the plausible covariance matrices of the observations are slightly different from those applicable to repeated measures experiments. For example, if treatment levels are randomly sequenced for each subject, it may be plausible to postulate a compound symmetric covariance matrix (i.e., equal correlations and equal variances). However, if all subjects are measured under t time points, it becomes plausible that the observations with closer temporal proximity are more highly correlated than those farther apart.

Yet another category of analysis exists in which a single subject is observed repeatedly while some intervention is continuously applied to the subject. In this chapter, however, we will exclude such single-subject or case studies because in these studies parameterization of the individual's behavior and/or the specification of the covariance matrix is not typically sought.

12.3 MODELS FOR TIME-DEPENDENT OBSERVATIONS

Restricting ourselves to time-dependent observations for which we would apply analysis of variance procedures (i.e., tests on the overall Time effects, specific contrasts on Time effects, and fitting polynomials), there are a few plausible models for such a data set. When repeated observations are made on one dependent variable, Y, at fixed time intervals, the data are traditionally analyzed using one of three models: a general multivariate model (Anderson, 1984; Crowder and Hand, 1990; Games, 1990; Geisser, 1980; Muirhead, 1982); a univariate mixed-effects model which does not impose time-dependent structure to data (Barcikowski and Robey, 1984; Collier, Baker, Mandeville, and Hayes, 1967; Davidson, 1972; Greenhouse and Geisser, 1959); or a time series model (Anderson, 1971, 1978; Anderson, Jensen, and Schou, 1981; Azzalini, 1984, 1989; Frederiksen and Rotondo, 1979; Hearne, Clark, and Hatch, 1983; Pantula and Pollock, 1985; Schmitz, 1990).

We now examine three types of models for time-dependent observations: a multivariate model, a traditional univariate mixed-effects model, and a time series model, an AR(1), in particular. As a numerical illustration, an AR(1) model and a univariate mixed-effects model are compared using the same data. (Although the models without measurement errors are presented, random measurement errors may be added to these models.)

12.3.1 A Multivariate Model

In a multivariate model, each vector of responses y_i obtained from the ith subject, where $i = 1, \ldots, N$, is multivariate normal with mean μ_i ($t_i \times 1$) and an error covariance matrix Σ_i ($t_i \times t_i$), where t_i measures are taken from the ith individual and no specific structure is imposed on Σ_i, aside from a typical positive-definite structure. The model is greatly simplified when $t_i = t$ and $\Sigma_i = \Sigma$.

With a reasonable sample size, a multivariate method is a viable option. Multivariate procedures and statistics for testing overall effects as well as for making specific contrasts on repeated factor(s) are discussed fully in Chapters 4 and 7 of this book.

12.3.2   A Traditional Univariate Mixed-Effects Model

A univariate mixed-effects model is applicable to data with a time-dependent structure as well. One of the differences between a univariate mixed-effects model and a time series model is in their assumptions about the correlations between the observations taken at two time points. A univariate mixed-effects model essentially assumes an arbitrary covariance matrix for the data. However, a certain covariance matrix pattern is required for the F test to be valid in examining the overall Time effects or contrasts using the conventional pooled MSerror.

Assuming that serial measurements from N subjects are made at t equally spaced time points, the Y_{ij}, the response of subject i at time j can be expressed as

$$Y_{ij} = \mu + \alpha_i + \beta_j + \varepsilon_{ij}, \qquad (12.1)$$
$$i = 1, \ldots, N; j = 1, \ldots, t,$$

where μ is the overall mean, α_i is the ith subject effect, β_j is the jth occasion effect, and ε_{ij} is the error. We assume α_i is random and distributed as $N(0, \sigma_\alpha^2)$, β_j is fixed with a usual constraint, $\Sigma_{j=1}^{t} \beta_j = 0$, and the error is distributed randomly as $N(0, \sigma^2)$. The observations taken from the same individual are correlated. Therefore the covariances are nonzero across j, but the observations taken from different individuals are expected to be uncorrelated across i. In matrix notation, this covariance structure, Σ, has to satisfy the error sphericity condition—a combined requirement that orthonormalized contrasts on Time effects must be orthogonal and homogeneous in their variances.

This sphericity condition has been well documented (Huynh and Feldt, 1970; Rouanet and Lepine, 1970). (For further detail, readers are referred to Chapter 4 of this book.) If the condition is not met, the univariate statistic for testing Time effects is approximately distributed as the F with εv_1 and εv_2 degrees of freedom (d.f.), where ε is the d.f. correction factor (Box, 1954). The correction factor is bounded between $1/(t - 1)$ and 1, where t is the number of time points. Greenhouse and Geisser (1959) suggest estimating ε by a sample covariance matrix, $\hat{\Sigma}$, and adjusting the d.f. of the F tests by $\hat{\varepsilon}$.

The model (12.1) can be made more general by incorporating group effects (as in a split-plot design), unequal time spacing, differential Time effects, a random time factor, and/or a measurement error. [The subject $\times$ time interaction is not represented in this model, but the denominator MS for testing Time effects is $MS_{S \times T}$. For detail of the $E(MS)$ for a univariate mixed-effects model, see Chapter 1.]

12.3.3 A Time Series Model

A time series structure may be postulated for any data set with time-dependent observations. For securing stable estimates of parameters, i.e., the variance and the correlation, in a time series model, many time points are needed. Although there are several different classes of time series models, we focus here on the first-order autoregressive model, AR(1), and defer comment on the general autoregressive moving average (ARMA) models and state space methodology to Section 12.4.

A time series analysis typically hypothesizes the correlation between the observations from two time points as having a systematic pattern. In their simplest form, the observations taken from two immediately adjacent time points (assuming they are positively correlated) are most highly correlated; and as the amount of time between observations increases, the correlation decreases proportionally. The assumption of error sphericity is applicable in examining the overall Time effects or contrasts using the pooled MSerror.

Specifying the covariance matrix pattern is advantageous in detecting Time effects because one can, by fitting the patterned covariance structure to the data, adopt a potentially more powerful method. For examples of different patterned covariance structures, see Jennrich and Schluchter (1986).

12.3.3.1 AR(1) Model

A simplest form of a time series structure for equally spaced time data is the first-order autoregressive model, AR(1). In a stationary AR(1), sometimes described as having a simplex pattern, the variance is constant across t time points and the correlations gradually decrease over time. For example, the highest correlation .80 is observed between two adjacent time points such as time 1 and time 2, or time 2 and time 3, and the next highest correlation, $.80^2 = .64$ is observed between time 1 and time 3 or between time 2 and time 4.

Let us assume Y_j to have a simplex pattern for each i. Then the simplest form of serial dependence can be expressed as

$$Y_j - \mu = \rho(Y_{j-1} - \mu) + \varepsilon_j, \qquad j = 1, \ldots, t, \tag{12.2}$$

where ρ is the correlation between the two observations with one time unit apart and the ε_j are independent and $N(0, \sigma^2)$. The covariance matrix, $\mathbf{\Sigma}$, of size $t \times t$ with a patterned correlation pattern, $\mathbf{R}$, can be expressed as

$$\mathbf{\Sigma} = \sigma^2[1/(1 - \rho^2)]\mathbf{V},$$

where $\sigma^2[1/(1 - \rho^2)]$ is the common variance and $\mathbf{V} = \mathbf{D}^{1/2} \mathbf{R} \mathbf{D}^{1/2}$. Here, t responses from each subject form a geometric series. With $|\rho| < 1$, this series converges uniformly, and by assuming $\rho > 0$, $\mathbf{R}$ represents an exponential decay. $\mathbf{R}$ and $\mathbf{D}$ are expressed as

$$
\mathbf{R} = \begin{bmatrix}
1 & \rho & \rho^2 & \cdots & \rho^{t-1} \\
\rho & 1 & \rho & \cdots & \rho^{t-2} \\
\cdot & & \cdot & \cdot & \cdot \\
\cdot & & \cdot & \cdot & \cdot \\
\cdot & & \cdot & \cdot & \cdot \\
\rho^{t-1} & \rho^{t-2} & \rho^{t-3} & \cdots & 1
\end{bmatrix} \quad \text{and}
$$

$$
\mathbf{D} = \begin{bmatrix}
\phi_1^2 & 0 & 0 & \cdots & 0 \\
0 & \phi_2^2 & & \cdots & 0 \\
\cdot & \cdot & \phi_3^2 & \cdot & \cdot \\
\cdot & \cdot & \cdot & \cdot & \cdot \\
\cdot & \cdot & \cdot & \cdot & \cdot \\
\cdot & \cdot & \cdot & \cdot & \cdot \\
0 & 0 & \cdots & \cdots & \phi_t^2
\end{bmatrix}.
$$

In a stationary AR(1) model, $\phi_j^2 = 1$ for all j, and $\mathbf{D} = \mathbf{I}$. Thus $\mathbf{V}$ is identical to the patterned correlation matrix, $\mathbf{R}$. Such a model is sometimes called a stationary Markov simplex (Frederiksen and Rotondo, 1979) or a serial correlation pattern (Wallenstein and Fleiss, 1979).

Wallenstein and Fleiss derived the lower-bound expression for the correction factor ε as a function of time points t and the correlation ρ. They have shown that when the covariance matrix has a simplex pattern, the lower bound of the d.f. correction factor equals $\varepsilon_\rho = 5(t + 1)/(2t^2 + 7)$, while the lower bound of the correction factor from an arbitrary covariance matrix equals $\varepsilon = 1/(t - 1)$. Because the min. ε_ρ is larger than the min. ε, the F test adjusted by the former provides a more power test of Time effects than the corresponding test adjusted by the latter.

Their finding, however, is most useful if one always adopts the lower-bound correction factor for evaluating Time effects. Overly conservative results come from simulation studies when the lower bound of the correction factor, the min. ε, is uniformly used (Collier et al., 1967; Rogan, Keselman, and Mendoza, 1979). Similarly, uniformly adopting the lower bound ε_ρ is expected to result in lack of power, although the lower bound of ε_ρ is larger than the lower bound of ε. Therefore, if the time-dependent structure is to be useful, a sample estimate, $\hat{\varepsilon}_\rho$, instead of the min. ε_ρ, has to be computed for each data set.

There are additional stages involved in fitting a serial correlation pattern to repeated observations before $\hat{\varepsilon}_\rho$ can be computed.

Step 1.   The sample variance and the correlation coefficient are estimated based on the assumption of a serial correlation pattern. Using these two sample estimates, a patterned covariance matrix is derived.

Step 2. The fit between the original covariance matrix and the derived patterned matrix is tested by the likelihood ratio test. If the likelihood ratio test (asymptotically distributed as chi-square) is nonsignificant, the null hypothesis of simplex pattern is subsequently retained, and a sample correction factor $\hat{\epsilon}_\rho$ is computed from the derived patterned covariance matrix. If a simplex pattern is not assumed and, therefore, no specific pattern is imposed on the covariance matrix, a sample correlation factor $\hat{\epsilon}$ is computed from the original covariance matrix.

Step 3. Using either correction factor thus derived, the probability level of the overall F test for Time effects is adjusted.

12.3.3.2 Likelihood Ratio Test for a Serial Correlation Pattern

Based on the work of Koopmans (1942), Hearne et al. (1983, pp. 240–241) present a likelihood ratio (LR) test statistic for examining whether a given covariance matrix has a serial correlation pattern. A single group of subjects is assumed here for simplicity, but the generalization to multigroup cases is given by Hearne et al.

Let the LR test be $l = L(\hat{\omega})/L(\hat{\Omega})$, where the numerator likelihood estimates are constrained in the null hypothesis space, namely by $\hat{\sigma}^2$ and $\hat{\rho}$, and the denominator estimates are in the general unrestricted space. By expressing y_{ij} as a response in the form of a deviation from the sample mean, we eliminate the unknown nuisance parameter, μ. First, we need to solve for $\hat{\sigma}^2$ and $\hat{\rho}$ by setting two partial differential equations to zero:

$$(\partial/\partial\rho) \ln L(\omega) = 0 \quad \text{and} \quad (\partial/\partial\sigma^2) \ln L(\omega) = 0.$$

From $(\partial/\partial\rho) \ln L(\omega) = 0$, we have

$$-N\rho\sigma^2 + (1 - \rho^2)(A_1 - \rho A_2) = 0, \tag{12.3}$$

where

$$A_1 = \frac{N}{N-1} \sum_{i=1}^{N} \sum_{j=2}^{t} y_{ij}y_{i(j-1)}, \tag{12.4}$$

and

$$A_2 = \frac{N}{N-1} \sum_{i=1}^{N} \sum_{j=2}^{t-1} y_{ij}^2. \tag{12.5}$$

From $(\partial/\partial\sigma^2) \ln L(\omega) = 0$, we then have

$$-Nt\sigma^2 + A_2 + A_3 - 2A_1\rho + A_2\rho^2 = 0, \tag{12.6}$$

where

$$A_3 = \frac{N}{N-1} \sum_{i=1}^{N} (y_{i1}^2 + y_{it}^2). \tag{12.7}$$

By solving (12.6) for σ^2 and substituting the result into (12.3), we obtain

$$\left(1 - \frac{1}{t}\right)A_2\rho^3 - \left(1 - \frac{2}{t}\right)A_1\rho^2$$

$$- \left\{\left(1 + \frac{1}{t}\right)A_2 + \frac{A_3}{t}\right\}\rho + A_1 = 0, \tag{12.8}$$

which is a polynomial function of ρ. By solving (12.8), ρ is obtained and this solution in turn is used to solve for σ^2. The likelihood ratio test statistic, $-2 \ln l$, is approximately distributed as χ^2 with $\frac{1}{2}t(t+1) - 2$ d.f. Hearne et al.'s A_1, A_2, and A_3 correspond to Koopman's Nm, Nn, and Nl, respectively.

12.3.3.3 A Numerical Example

The same data are used to compare the traditional univariate approach and the AR(1)-based approach. The modified likelihood ratio test, LR*, is also compared against the LR test. The data are from a larger study described by Egeland, Pianta, and O'Brien (in press) in which the total scores from the Peabody Individual Achievement Test (PIAT) (covering math, reading recognition, reading comprehension, spelling, and general information) were obtained for the same children at Grade 1, Grade 2, and Grade 3 levels. A composite total score obtained for each individual at each of the three grade levels was age adjusted using a national standardization sample.

The data in this example are a subset from the aforementioned study in which 22 girls ($N = 22$) were measured at three equally spaced time points ($t = 3$). The means of the total PIAT scores were 105.09, 103.09, and 103.27 and the corresponding standard deviations were 12.29, 10.17, and 11.44 for Grade 1, Grade 2, and Grade 3, respectively. The ANOVA summary table for this data set is presented in Table 12.1.

The sample covariance and correlation matrices are

$$\hat{\Sigma} = \begin{bmatrix} 150.94 & & \\ 107.52 & 103.52 & \\ 124.55 & 107.59 & 130.87 \end{bmatrix}$$

Table 12.1   Analysis of Variance Summary Table[a]

Sources	SS	df	MS	F	p
Subjects	7452.48	21	354.88		
Time	53.82	2	26.91	1.77	.1833
S × T	639.52	42	15.23		
Total	8145.82	65			

[a]The probability level reported assumes perfect sphericity.
Source: L. K. Edwards (1991) with permission from the American Educational Research Association.

and

$$\hat{\mathbf{R}} = \begin{bmatrix} 1.00 & & \\ .86 & 1.00 & \\ .89 & .92 & 1.00 \end{bmatrix}.$$

The three steps described in Section 12.3.3.1 are as follows:

Step 1.   Estimation of $\hat{\sigma}^2$ and $\hat{\rho}$:
For the time series approach, the correlation and variance are estimated following Eqs. (12.3) through (12.8). The following values are obtained: $\hat{\sigma}^2 = 27.93$ and $\hat{\rho} = .89$.
The estimated covariance and correlation matrices, assuming the simplex pattern, are

$$\hat{\mathbf{\Sigma}}_\rho = \begin{bmatrix} 137.46 & & \\ 122.71 & 137.46 & \\ 109.53 & 122.71 & 137.46 \end{bmatrix}$$

and

$$\hat{\mathbf{R}}_\rho = \begin{bmatrix} 1.00 & & \\ .89 & 1.00 & \\ .80 & .89 & 1.00 \end{bmatrix}.$$

Step 2.   LR test of fit:
The LR test is then performed to examine the fit of the assumed simplex pattern to the data. The likelihood ratio is defined as a powered ratio of the two determinants: $l = |\hat{\mathbf{\Sigma}}|^{1/2N}/|\hat{\mathbf{\Sigma}}_\rho|^{1/2N}$, where $|\hat{\mathbf{\Sigma}}| = 60471.40$ and $|\hat{\mathbf{\Sigma}}_\rho| = 107212.53$. Subsequently, $-2 \ln l = 12.60$, which asymptotically distributes as χ^2 with $\frac{1}{2} t(t + 1) - 2 = 4$ degrees of freedom. The probability associated with the ob-

served χ^2 value is 0.013. Therefore, if $\alpha = 0.05$ is used, the hypothesis of simplex pattern is rejected. Consequently, a simplex pattern will not be fitted to the data.

Step 2'. LR* test of fit:
Alternatively, the modified likelihood ratio (LR*) test can be used to examine the simplex hypothesis. This modified test may be preferable to the LR test because the LR test is known to be biased. In Muirhead (1982, pp. 353–359), the modified likelihood ratio, LR* is expressed as

$$l^* = \left(\frac{\exp}{n}\right)^{tn/2} \text{etr}\left(-\frac{1}{2} SS_0^{-1}\right)(|SS_0^{-1}|)^{n/2},$$

where exp is an exponential function, etr (**A**) is an exponential function of the trace of **A** matrix, **S** is a sample covariance matrix, S_0^{-1} is an inverse of an estimated patterned covariance matrix, and $n = N - 1$ are the degrees of freedom associated with **S**. In this example, the trace of $SS_0^{-1} = 3.0$, $|SS_0^{-1}| = 0.56$, and $n = 21$. Subsequently, $-2 M \ln l^* = 11.41$, where $M = n - [2t^2 + 1 - 1/(t + 1)]/6 = 19.92$. $-2 M \ln l^*$ distributes asymptotically as χ^2 with 4 d.f. and $p = 0.022$ is obtained for the simplex hypothesis. Again, if $\alpha = 0.05$ is used, the hypothesis of simplex pattern is rejected.

Step 3. Adjusted F test:
Because both the LR and LR* tests have rejected a simplex hypothesis, we then compute the sample d.f. correction factor, $\hat{\varepsilon}$, from the unconstrained sample covariance matrix, yielding .8676. The adjusted F test for Time effects will then have 1.74 and 36.44 as its d.f. and $p = .1885$, resulting in a conclusion of nonsignificance.

Although in this example there is no need to compute the correction factor under the assumption of the simplex pattern, for the purpose of illustration $\hat{\varepsilon}_\rho = .8262$ is computed from the estimated covariance matrix. The adjusted F test for Time effects using $\hat{\varepsilon}_\rho$ would have been 1.65 and 34.70 as its d.f. and $p = .1900$, resulting again in a conclusion of nonsignificance. Since in this example the $\hat{\varepsilon}$ and $\hat{\varepsilon}_\rho$ are almost identical in size (.8676 and .8262, respectively), even if the LR or LR* test of fit is ignored, the conclusions for Time effects using the $\hat{\varepsilon}$-adjusted and $\hat{\varepsilon}_\rho$-adjusted F tests would have been essentially the same.

12.3.3.4 A Simulation Study

Although the numerical example showed a negligible difference between a univariate mixed-effects model approach and an AR(1) approach, such is not always the case. A simulation study reported in Edwards (1991) is summarized here to show the general differences between those two approaches.

A nonstationary model, rather than a stationary model, may be more realistic in educational and psychological studies. For this reason, data with a patterned correlation matrix (**R**) but with nonstationary variances (**D** ≠ **I**) were also considered. For example, in studies where learning or maturation effects are involved, the variance may increase if the initial individual differences increase with time, i.e., $\sigma_j^2 < \sigma_{j'}^2$ for $j < j'$. Postulating that variances may increase (or decrease) monotonically across time, a nonstationary AR(1) model (a nonstationary Markov simplex) will have the form, $\Sigma = \sigma^2[1/(1 - \rho^2)]$ **V** and **V** = $D^{1/2} R D^{1/2}$, where

$$
V = \begin{bmatrix}
\phi_1^2 & \phi_1\phi_2\rho & \phi_1\phi_3\rho^2 & \cdots & \phi_1\phi_t\rho^{t-1} \\
\phi_1\phi_2\rho & \phi_2^2 & \phi_2\phi_3\rho & \cdots & \phi_2\phi_t\rho^{t-2} \\
\cdot & \cdot & \phi_3^2 & & \cdot \\
\cdot & & \cdot & \cdot & \cdot \\
\cdot & & \cdot & & \cdot \\
\cdot & & \cdot & & \cdot \\
\phi_1\phi_t\rho^{t-1} & \phi_2\phi_t\rho^{t-2} & \cdots & \cdots & \phi_t^2
\end{bmatrix}.
$$

Three types of data structures were examined via a Monte Carlo simulation: a simplex pattern, a moderately nonstationary simplex pattern, and a severely nonstationary simplex pattern. When the variance was moderately nonstationary ($\phi_j^2 = 1, 1.2$, and 1.5 for $j = 1, 2$, and 3, respectively), $\hat{\varepsilon}_\rho$ was almost always larger than $\hat{\varepsilon}$. This tendency accelerated as t increased (Table 12.2). However, these were occasions on which the simplex pattern did not hold, and, consequently, adopting $\hat{\varepsilon}_\rho$ uniformly would have led to erroneous conclusions.

The results indicated that postulating a serial correlation pattern would not always be advantageous over not assuming such a pattern. First of all, the LR or LR* test shows many false-negative results. For the chi-squared test based on the LR*, a sample size of about 20 or larger seems to ensure its proper performance while the counterpart for the LR requires 100. Second, when the serial correlation is high ($\rho > .50$) and the number of time points is small, up to 44% of the covariance matrices give advantage to $\hat{\varepsilon}$ over $\hat{\varepsilon}_\rho$. Third, when the serial correlation pattern is not warranted, $\hat{\varepsilon}_\rho$ consistently shows a positive bias when compared to the population correction factor ε. In other words, specifying a serial correlation pattern

Table 12.2 Moderately Nonstationary Population Covariance Matrices with $t = 3$[a]

	0.000	0.250	0.500	0.750	0.900	
			ε			
	0.986	0.962	0.908	0.842	0.792	Lowerbound
$t = 3$ at $N = 6$						
$\hat{\varepsilon}$	0.773	0.764	0.750	0.721	0.717	0.500
$\hat{\varepsilon}_\rho$	0.960	0.958	0.926	0.875	0.834	0.800
$\hat{\varepsilon} \geq \hat{\varepsilon}_\rho$	0.074	0.068	0.115	0.158	0.221	
LR$_{.05}$ sig or $\hat{\varepsilon} \geq \hat{\varepsilon}_\rho$	0.337	0.333	0.375	0.416	0.486	
(LR$^*_{.05}$ sig or $\hat{\varepsilon} \geq \hat{\varepsilon}_\rho$)	(0.161)	(0.152)	(0.198)	(0.252)	(0.326)	
$t = 3$ at $N = 12$						
$\hat{\varepsilon}$	0.863	0.851	0.829	0.786	0.761	0.500
$\hat{\varepsilon}_\rho$	0.981	0.968	0.928	0.868	0.830	0.800
$\hat{\varepsilon} \geq \hat{\varepsilon}_\rho$	0.075	0.105	0.202	0.262	0.280	
LR$_{.05}$ sig or $\hat{\varepsilon} \geq \hat{\varepsilon}_\rho$	0.219	0.224	0.338	0.425	0.466	
(LR$^*_{.05}$ sig or $\hat{\varepsilon} \geq \hat{\varepsilon}_\rho$)	(0.159)	(0.171)	(0.287)	(0.352)	(0.400)	
$t = 3$ at $N = 24$						
$\hat{\varepsilon}$	0.920	0.903	0.869	0.820	0.775	0.500
$\hat{\varepsilon}_\rho$	0.991	0.973	0.928	0.866	0.827	0.800
$\hat{\varepsilon} \geq \hat{\varepsilon}_\rho$	0.064	0.150	0.276	0.317	0.277	
LR$_{.05}$ sig or $\hat{\varepsilon} \geq \hat{\varepsilon}_\rho$	0.190	0.282	0.399	0.460	0.505	
(LR$^*_{.05}$ sig or $\hat{\varepsilon} \geq \hat{\varepsilon}_\rho$)	(0.160)	(0.253)	(0.375)	(0.434)	(0.464)	

[a]The results are based on 2000 replications. There were three repeated observations per subject and the population covariance matrix deviated from a simplex pattern due to moderately heterogeneous variances.
Source: Adapted from L. K. Edwards (1991) with permission from the American Educational Research Association.

incorrectly without examining the fit by the LR or LR* test may produce inflated Type I error rates in the F test for Time effects.

A preliminary check on the homogeneous variances and the size of the correlation is in order (Abraham and Vijayan, 1988). Care must be also taken to estimate ρ with a smaller bias (Azzalini and Frigo, 1991). When a simplex pattern is satisfied and the correlation is low, the AR(1) model has an advantage over the univariate mixed-effects model. Even a moderate amount of heterogeneity in variances can result in positively biased $\hat{\varepsilon}_\rho$, and because this bias does not dissipate with large sample sizes, extreme caution is warranted in adopting a stationary AR(1) model.

12.4 OTHER ISSUES AND DEVELOPMENTS

We have examined equal time intervals with no missing values. However, in a most general case, potentially every subject is observed at a different time and at unequal intervals. The interval of observations is not necessarily constant over time in many studies. The unequal spacing may be by design or by purposes of economy which allow measurements of subjects only at critical time points and, thus, eliminate the measurements during the periods when little change is expected. The fact that not every subject has a complete set of measurements results in an incomplete data structure. We will now briefly summarize selected new developments and some of their references for dealing with these irregularities. We recommend Diggle and Donnelly (1989) and Koch et al. (1980) for more comprehensive coverage of references.

12.4.1 The General Two-Stage Model

In many studies, the interval of observations is not necessarily constant over time, nor does every subject have complete measurements. Some models are more flexible than others in dealing with such irregularities.

When individual and population characteristics can be explicitly identified, the two-stage model, originally presented in Laird and Ware (1982) as one of the multistage random-effects models, may be a reasonable choice. In this model, the covariance structure is specified based on the distribution of subject profiles. The two-stage general mixed-effects model represents a family of models for serial measurements that include as special cases both growth curve models and repeated measures designs (see, e.g., Laird, Lange, and Stram, 1987).

The general two-stage model is expressed as

$$\mathbf{y}_i = \mathbf{X}_i\boldsymbol{\alpha} + \mathbf{Z}_i\mathbf{b}_i + \mathbf{e}_i, \tag{12.9}$$

where $\mathbf{e}_i$ are distributed as $N(\mathbf{0}, \mathbf{R}_i)$, $\mathbf{R}_i$ is a $t_i \times t_i$ positive-definite covariance matrix, and t_i observations are taken from the ith subject. $\boldsymbol{\alpha}$ is a $p \times 1$ vector of unknown fixed parameters, $\mathbf{X}_i$ and $\mathbf{Z}_i$ are a known design matrices of size $t_i \times p$ and $t_i \times q$, respectively. Subsequently, the $\mathbf{y}_i$ are independent normals with mean $\mathbf{X}_i\boldsymbol{\alpha}$ and covariance matrix, $\boldsymbol{\Sigma}_i = \mathbf{R}_i + \mathbf{Z}_i\mathbf{D}\,\mathbf{Z}_i^T$. When $\mathbf{R}_i = \sigma^2\mathbf{I}$, where $\mathbf{I}$ is the identity matrix, this model simplifies even further.

The solution is sought in two stages. At Stage 1, $\boldsymbol{\alpha}$ and $\mathbf{b}_i$ are unknown but considered fixed and the $\mathbf{e}_i$ are assumed independent. At Stage 2, the

b_i are considered random and distributed independently as $N(0, D)$, with D being a $q \times q$ positive-semidefinite covariance matrix.

As a simpler alternative to the above model, a model applicable to both equal and unequal time intervals and with missing observations was proposed by Rosner, Munoz, Tager, Speizer, and Weiss (1985) and Rosner and Munoz (1988). However, Stanek, Shetterley, Allen, Pelto, and Chavez (1989) criticized this model on the basis that it produces estimators that are dependent on the autocorrelation and, in turn, reduces their interpretability. Further readings on the two-stage model can be found in Laird and Ware (1982), Laird, Lange, and Stram (1987), Louis (1988), and Stanek (1990).

12.4.2 ARMA Models and State Space Methodology

We have discussed AR models and, in particular, AR(1), but we have not touched on more general autoregressive moving average (ARMA) models. Further readings on ARMA models are available in Jones (1980), Rochon (1992), and Rochon and Helms (1989). In particular, noting the restrictive nature of an ARMA covariance structure with constant variances and autocovariances, Rochon (1992) extended the ARMA model which incorporates heteroscedasticity. The state space methodology can be found in Aoki (1990) and Findley (1981), and its use in fitting an ARMA model to behavioral-epidemiological data can be found in Jones (1981, 1983, 1985) and in Jones and Ackerson (1990).

12.4.3 Missing Values, Random-Coefficients Models, and Unequal Spacing

From the general mixed-effects model approach, the EM algorithms of Dempster, Laird, and Rubin (1977) have been used in obtaining the maximum likelihood (ML) estimators (Laird et al., 1987; Lindstrom and Bates, 1990). The EM algorithm has been noted as one of the useful techniques for handling incomplete data or data with missing observations. Yet another approach to missing values using the ante-dependence structure has been proposed by Kenward (1987) and further pursued by Patel (1991).

For modeling irregularly timed observations, see Azencott and Dacunha-Castelle (1986), Brillinger (1983), Jones and Boadi-Boateng (1991), Jones (1983, 1985), Jones and Ackerson (1990), and Robinson (1977, 1983). For random-coefficient models, see e.g., Chi and Reinsel (1989) and Rosenberg (1973). The problem of separating the serial correlation from the random subject component is discussed in Jones (1990). For handling missing values using ARMA models, see the references cited in the previous segment.

12.4.4 Trend Analysis and Curve Fitting

Aside from testing the omnibus Time effects, a trend analysis or estimating the profile may be of interest. Discussion of both linear and nonlinear curve fitting to time-dependent observations can be found in Hui (1984) and Thissen and Bock (1990).

12.4.5 Multilevel Analysis

From the multilevel analysis perspective, Goldstein and McDonald (1988) discuss a general model for the analysis of multivariate, multilevel data in which both repeated measures designs and multiple time series are treated as special cases. The relationship of selected traditional analysis of variance designs and the multilevel analysis based on the hierarchical linear models are discussed in Chapter 13 of this book.

12.4.6 Categorical Measures and Nonparametric Methods

Developments in nonparametric techniques, including various ranking methods and their pros and cons, can be found in Agresti and Pendergast (1986), Akritas (1991), Clogg, Eliason, and Grego (1990), Kepner and Robinson (1988), Tandon and Moeschberger (1989), and Thompson (1991). For a summary of current directions and references for categorical repeated data, readers are referred to Diggle and Donnelly (1989) and Koch et al. (1980).

12.5 CONCLUDING REMARKS

We have examined several models for time-dependent observations. Some models are not specific to time-dependent observations but are applicable to general repeated measures. Others incorporate a specific time-dependent structure in the data. We have summarized the current efforts into a multivariate model with no specific covariance structure, a univariate mixed-effects model and its extensions, and a time series model with a particular covariance structure.

As a numerical example, we examined a real data set with three time points using both a univariate mixed-effects model and an AR(1) model. The results suggested that postulating a stationary AR(1) model (or a stationary simplex model) is not always advantageous over the univariate mixed-effects model with an arbitrary covariance structure.

We summarized selected recent developments and their references. Among them, the two-stage general linear model proposed by Laird and Ware (1982) requires explicitly identifying the individual and population characteristics, whereas a traditional mixed-effects model specifies only the covariance matrix for the group as a whole. However, the difficulty in specifying the two-stage model may be compensated by the interpretability of the results and its flexibility in handling irregular data. In time series models, we see some new developments in dealing with missing values and unequal time points (e.g., Jones and Boadi-Boteng, 1991).

Analyzing time-dependent observations is truly a multidisciplinary endeavor in which biostatisticians, econometricians, and psychometricians are currently devising and evaluating models that are flexible and general enough to handle the irregularities in real data (e.g., unequal time spacing and incomplete data structure). We have examined only a small portion of this extremely diversely approached topic area which in the near future will profoundly affect the way we study time-dependent observations.

REFERENCES

Abraham, B. and Vijayan, K. (1988). A Statistic to Check Model Accuracy in Time Series. *Comm. Statist. – Theor. Meth.*, *A17*: 4271–4278.

Agresti, A. and Pendergast, J. (1986). Comparing Mean Ranks for Repeated Measures Data. *Comm. Statist. – Theor. Meth.*, *A15*: 1417–1433.

Akritas, M. G. (1991). Limitations of the Rank Transform Procedures: A Study of Repeated Measures Designs, Part I. *J. Amer. Statist. Assoc.*, *86*: 457–460.

Anderson, A. H., Jensen, E. B., and Schou, G. (1981). Two-Way Analysis of Variance with Correlated Errors. *Internat. Statist. Rev.*, *49*: 153–167.

Anderson, T. W. (1971). *The Statistical Analysis of Time Series*. Wiley, New York.

Anderson, T. W. (1978). Repeated Measurements on Autoregressive Processes. *J. Amer. Statist. Assoc.*, *73*: 371–378.

Anderson, T. W. (1984). *An Introduction to Multivariate Statistical Analysis* (2nd ed.). Wiley, New York.

Aoki, M. (1990). *State Space Modeling of Time Series* (2nd ed.). Springer-Verlag, New York.

Azencott, R. and Dacunha-Castelle, D. (1986). *Series of Irregular Observations: Forecasting and Model Building*. Springer-Verlag, New York.

Azzalini, A. (1984). Estimation and Hypothesis Testing for Collections of Autoregressive Time Series. *Biometrika*, *71*: 85–90.

Azzalini, A. (1989). An Analysis of Variance Table for Repeated Measurements with Unknown Autoregressive Parameter. *Appl. Statist.*, *38*: 402–411.

Azzalini, A. and Frigo, A. C. (1991). An Explicit Nearly Unbiased Estimate of the AR(1) Parameter for Repeated Measurements. *J. Time Series Anal.*, *12*: 273–281.

Barcikowski, R. S. and Robey, R. R. (1984). Decisions in Single Group Repeated Measure Analysis: Statistical Tests and Three Computer Packages. *Amer. Statist.*, *38*: 148–150.

Box, G. E. P. (1954). Some Theorems on Quadratic Forms Applied in the Study of Analysis of Variance Problems. II. Effects of Inequality of Variance and

Correlation Between Errors in the Two-Way Classification. *Ann. Math. Statist.*, *25*: 484–498.

Brillinger, D. R. (1983). Statistical Inference for Irregularly Observed Processes. *Time Series Analysis of Irregularly Observed Data.* Parzen, E. (ed.). Springer-Verlag, New York, pp. 38–57.

Chi, E. M. and Reinsel, G. C. (1989). Models for Longitudinal Data with Random Effects and AR(1) Errors. *J. Amer. Statist. Assoc.*, *84*: 452–459.

Clogg, C. C., Eliason, S. R., and Grego, J. M. (1990). Models for the Analysis of Change in Discrete Variables. *Statistical Methods in Longitudinal Research*, Vol. II. von Eye, A. (ed.). Academic Press, New York, pp. 409–442.

Collier, R. O., Baker, F. B., Mandeville, G. K., and Hayes, T. F. (1967). Estimates of Test Size for Several Test Procedures Based on Conventional Variance Ratios in the Repeated Measures Design. *Psychometrika, 32*: 339–353.

Crowder, M. J. and Hand, D. J. (1990). *Analysis of Repeated Measures.* Chapman & Hall, London.

Cullis, B. R. and McGilchrist, C. (1990). A Model for the Analysis of Growth Data from Designed Experiments. *Biometrics, 46*: 131–142.

Davidson, M. L. (1972). Univariate Versus Multivariate Tests in Repeated Measures Experiments. *Psych. Bull.*, *77*: 446–452.

Dempster, A. P., Laird, N. M., and Rubin, D. B. (1977). Maximum Likelihood from Incomplete Data via the EM Algorithm. *J. Roy. Statist. Soc.*, *B39*: 1–38.

Diggle, P. J. (1988). An Approach to the Analysis of Repeated Measurements. *Biometrics, 44*: 959–971.

Diggle, P. J. and Donnelly, J. B. (1989). A Selected Bibliography on the Analysis of Repeated Measurements and Related Areas. *Austral. J. Statist.*, *31*: 183–193.

Edwards, L. K. (1991). Fitting a Serial Correlation Pattern to Repeated Observations. *J. Educ. Statist.*, *16*: 53–76.

Egeland, B., Pianta, R., and O'Brien, M. (in press). Maternal Intrusiveness in Infancy and Child Maladaptation in Early School Years. *Dev. Psychopath.*

Findley, D. F. (ed.) (1981). *Applied Time Series Analysis II.* Academic Press, New York.

Frederiksen, C. H. and Rotondo, J. A. (1979). Time-Series Models and the Study of Longitudinal Change. *Longitudinal Research in the Study of Behavior and Development.* Nesselroade, J. R. and Baltes, P. B. (eds.). Academic Press, New York, pp. 111–154.

Games, P. A. (1990). Alternative Analyses of Repeated-Measure Designs by ANOVA and MANOVA. *Statistical Methods in Longitudinal Research*, Vol. I. von Eye, A. (ed.). Academic Press, New York, pp. 81–122.

Geary, D. N. (1989). Modeling the Covariance Structure of Repeated Measurements. *Biometrics, 45*: 1183–1195.

Geisser, S. (1980). Growth Curve Analysis. *Handbook of Statistics*, Vol. 1: *Analysis of Variance.* Krishnaiah, P. R. (ed.). North-Holland, New York, pp. 89–116.

Goldstein, H. and McDonald, R. P. (1988). A General Model for the Analysis of Multilevel Data. *Psychometrika, 53*: 455–467.

Greenhouse, S. W. and Geisser, S. (1959). On Methods in the Analysis of Profile Data. *Psychometrika*, *24*: 95–112.

Hearne, E. M., Clark, G. M., and Hatch, J. P. (1983). A Test for Serial Correlation in Univariate Repeated-Measures. *Biometrics*, *39*: 237–243.

Hui, S. L. (1984). Curve Fitting for Repeated Measurements Made at Irregular Time-Points. *Biometrics*, *40*: 691–697.

Huynh, H. and Feldt, L. S. (1970). Conditions Under Which Mean Square Ratios in Repeated Measures Designs Have Exact *F*-Distributions. *J. Amer. Statist. Assoc.*, *65*: 1582–1589.

Jennrich, R. J. and Schluchter, M. D. (1986). Unbalanced Repeated-Measures Models with Structured Covariance Matrices. *Biometrics*, *42*: 805–820.

Johnston, F., Rocher, A. F., and Susanne, C. (eds.). (1979). *Methodologies for the Analysis of Human Growth and Development*. Plenum Publishing, New York.

Jones, B. and Kenward, M. G. (1989). *Design and Analysis of Crossover Trials*. Chapman & Hall, London.

Jones, R. H. (1980). Maximum Likelihood Fitting of ARMA Models to Time Series with Missing Observations. *Technometrics*, *22*: 389–395.

Jones, R. H. (1981). Fitting Continuous-Time Autoregressions to Discrete Data. *Applied Time Series Analysis II*. Findley, D. F. (ed.). Academic Press, New York, pp. 651–682.

Jones, R. H. (1983). Fitting Multivariate Models to Unequally Spaced Data. *Time Series Analysis of Irregularly Observed Data*. Parzen, E. (ed.). Springer-Verlag, New York, pp. 158–188.

Jones, R. H. (1985). Time Series Analysis with Unequally Spaced Data. *Handbook of Statistics*, Vol. 5, *Time Series in the Time Domain*. Hannan, E. J., Krishnaiah, P. R., and Rao, M. M. (eds.). North-Holland, New York, pp. 157–177.

Jones, R. H. (1990). Serial Correlation or Random Subject Effects? *Comm. Statist.-Simul. Comput.*, *B19*: 1105–1123.

Jones, R. H. and Ackerson, L. M. (1990). Serial Correlation in Unequally Spaced Longitudinal Data. *Biometrika*, *77*: 721–731.

Jones, R. H. and Boadi-Boateng, F. (1991). Unequally Spaced Longitudinal Data with AR(1) Serial Correlation. *Biometrics*, *47*: 161–175.

Kenward, M. G. (1987). A Method for Comparing Profiles of Repeated Measurements. *Appl. Statist.*, *36*: 296–308.

Kepner, J. L. and Robinson, D. H. (1988). Nonparametric Methods for Detecting Treatment Effects in Repeated-Measures Designs. *J. Amer. Statist. Assoc.*, *83*: 456–461.

Koch, G. G., Amara, I. A., Stokes, M. E., and Gillings, D. B. (1980). Some Views on Parametric and Non-Parametric Analysis for Repeated Measurements and Selected Bibliography. *Internat. Statist. Rev.*, *48*: 249–265.

Koch, G. G., Elashoff, J., and Amara, I. A. (1988). Repeated Measurements: Design and Analysis. *Encycl. Statist. Sci.*, *8*: 46–73.

Koopmans, T. (1942). Serial Correlation and Quadratic Forms in Normal Variables. *Ann. Math. Statist.*, *13*: 14–33.

Laird, N., Lange, N., and Stram, D. (1987). Maximum Likelihood Computations with Repeated Measures: Application of the EM Algorithm. *J. Amer. Statist. Assoc.*, *82*: 97–105.

Laird, N. M. and Ware, J. H. (1982). Random-Effects Models for Longitudinal Data. *Biometrics*, *38*: 963–974.

Lindstrom, M. J. and Bates, D. M. (1990). Nonlinear Mixed Effects Models for Repeated Measures Data. *Biometrics*, *46*: 673–687.

Louis, T. A. (1988). General Methods for Analyzing Repeated Measures. *Statist. Med.*, *7*: 29–45.

Matthews, J. N. S. (1988). Recent Developments in Crossover Designs. *Internat. Statist. Rev.*, *56*: 117–127.

Muirhead, R. J. (1982). *Aspects of Multivariate Statistical Theory*. Wiley, New York.

Nesselroade, J. R. and Baltes, P. B. (eds.) (1979). *Longitudinal Research in the Study of Behavior and Development*. Academic Press, New York.

Pantula, S. G. and Pollock, K. H. (1985). Nested Analysis of Variance with Autocorrelated Errors. *Biometrics*, *41*: 909–920.

Patel, H. I. (1991). Analysis of Incomplete Data from a Clinical Trial with Repeated Measurements. *Biometrika*, *78*: 609–619.

Robinson, P. M. (1977). Estimation of a Time Series Model from Unequally Spaced Data. *Stochast. Proc. Appl.*, *6*: 9–24.

Robinson, P. M. (1983). Multiple Time Series Analysis of Irregularly Spaced Data. *Time Series Analysis of Irregularly Observed Data*. Parzen, E. (ed.). Springer-Verlag, New York, pp. 276–289.

Rochon, J. (1992). ARMA Covariance Structures with Time Heteroscedasticity for Repeated Measures Experiments. *J. Amer. Statist. Assoc.*, *87*: 777–784.

Rochon, J. and Helms, R. W. (1989). Maximum Likelihood Estimation for Incomplete Repeated Measures Experiments Under an ARMA Covariance Structure. *Biometrics*, *45*: 207–218.

Rogan, J. C., Keselman, H. J., and Mendoza, J. L. (1979). Analysis of Repeated Measurements. *Brit. J. Math. Statist. Psych.*, *32*: 269–286.

Rosenberg, B. (1973). Linear Regression with Randomly Dispersed Parameters. *Biometrika*, *60*: 65–72.

Rosner, B. and Munoz, A. (1988). Autoregressive Modelling for the Analysis of Longitudinal Data with Unequally Spaced Examinations. *Statist. Med.*, *7*: 59–71.

Rosner, B., Munoz, A., Tager, I., Speizer, F., and Weiss, S. (1985). The Use of an Autoregressive Model for the Analysis of Longitudinal Data in Epidemiologic Studies. *Statist. Med.*, *4*: 457–467.

Rouanet, H. and Lepine, D. (1970). Comparison Between Treatments in a Repeated-Measurement Design: ANOVA and Multivariate Methods. *Brit. J. Math. Statist. Psychol.*, *23*: 147–163.

Schmitz, B. (1990). Univariate and Multivariate Time-Series Models: The Analysis of Intraindividual Variability and Intraindividual Relationships. *Statistical Methods in Longitudinal Research*, Vol. II. von Eye, A. (ed.). Academic Press, New York, pp. 351–386.

Stanek, E. J. (1990). A Two-Step Method for Understanding and Fitting Growth Curve Models. *Statist. Meth.*, *9*: 841–851.

Stanek, E. J., Shetterley, S. S., Allen, L. H., Pelto, G. H., and Chavez, A. (1989). A Cautionary Note on the Use of Autoregressive Models in Analysis of Longitudinal Data. *Statist. Med.*, *8*: 1523–1528.

Tandon, P. K. and Moeschberger, M. L. (1989). Comparison of Nonparametric and Parametric Methods in Repeated Measures Designs: A Simulation Study. *Comm. Statist.–Simul. Comput.*, *B18*: 777–792.

Thissen, D. and Bock, R. D. (1990). Linear and Nonlinear Curve Fitting. *Statistical Methods in Longitudinal Research*, Vol. II. von Eye, A. (ed.). Academic Press, New York, pp. 289–318.

Thompson, G. L. (1991). A Unified Approach to Rank Tests for Multivariate and Repeated Measures Designs. *J. Amer. Statist. Assoc.*, *86*: 410–419.

Timm, N. (1980). Multivariate Analysis of Variance of Repeated Measurements. *Handbook of Statistics*, Vol. 1, *Analysis of Variance*. Krishnaiah, P. R. (ed.). North-Holland, New York, pp. 41–88.

Verbyla, A. P. and Cullis, B. R. (1990). Modelling in Repeated Measures Experiments. *Appl. Statist.*, *39*: 341–356.

von Eye, A. (ed.). (1990). *Statistical Methods in Longitudinal Research*, Vols. I, II. Academic Press, New York.

Wallenstein, S. and Fleiss, J. L. (1979). Repeated Measurements Analysis of Variance When the Correlations Have a Certain Pattern. *Psychometrika*, *44*: 229–233.

13
Hierarchical Linear Models and Experimental Design

STEPHEN W. RAUDENBUSH Michigan State University, East Lansing, Michigan

13.1 INTRODUCTION

13.1.1 Purpose of the Chapter

This chapter demonstrates how data analysis based on recently developed hierarchical linear models (1) duplicates the results of standard ANOVA models for an important class of experimental designs and (2) extends the study of fixed and random effects to include unbalanced data, predictors that are either continuous or discrete, and random effects that covary. We illustrate these principles by reanalyzing data from Kirk's (1982) widely used text and data from 103 primary schools in Thailand. We consider one-way designs, two-factor nested and crossed designs, and randomized block (repeated measures) designs. The hierarchical linear model, like the ANOVA and standard regression models, is itself a special case of a more general mixed linear model. In this chapter, we briefly consider how the general mixed linear model can be tailored to analyze data for which hierarchical linear models do not apply.

13.1.2 Mixed Model Analysis of Variance

Of all the statistical approaches taught in graduate schools of education and psychology, probably none receives more attention than does the analysis of variance (ANOVA). The ANOVA is elegant in its computational simplicity, but, more important, it facilitates estimation in models having fixed effects, random effects, or both. This flexibility makes it appropriate for a tremendous variety of designs, including nested designs and repeated measures designs.

Nested designs are common in experimental behavioral and social research. In education, a common design involves students nested within classrooms and classrooms nested within methods of instruction. In counseling psychology, a standard design assigns subjects to therapists and therapists to treatments. However, nested designs also arise routinely in nonexperimental research such as multistage social surveys using census tracts, blocks, or schools as clusters. Similarly, repeated measures designs are common in experiments such as clinical trials where patients are administered different drugs on different occasions; but such designs also arise frequently in nonexperimental settings including longitudinal surveys and program evaluations using follow-up interviews.

Each of the above settings usually calls for a "mixed" statistical model, that is, a model including both fixed and random effects. In a study of students nested within classrooms and classrooms nested within methods of instruction, classroom effects will logically be specified as random and method effects as fixed. Repeated measures designs call for estimation of random effects of persons and fixed effects of treatments or programs. Multisite program evaluations also yield data appropriate for mixed models. For example, Raffe's (1991) study of a new vocational education initiative in Britain involved a treatment-control comparison at each of 19 sites. This is really a randomized block design in which the sites are the blocks, viewed as random, and the two programs are fixed. The students nested within site-by-program cells are "replications" (see Kirk, 1982, p. 293).

13.1.3 Limitations of ANOVA

Although the ANOVA methods commonly taught in experimental design courses work well for balanced designs having discrete independent variables, they are not widely applicable when the data are unbalanced and some predictors are continuous. Balanced designs with discrete independent variables arise primarily in carefully designed, small-scale experiments. However, in field experiments, quasi-experiments, and surveys, unbalanced data and a mix of discrete and continuous predictors will be the rule rather than the exception. Because of this, researchers often turn away from the mixed-model ANOVA to embrace multiple regression as an alternative.

13.1.4 Applicability of Multiple Regression

For fixed-effects designs, multiple regression is a more general analytic technique than ANOVA. Any fixed-effects analysis of variance can be reformulated as a regression problem, and the regression analysis will exactly reproduce the ANOVA results. In addition, regression allows a mix of discrete and continuous predictors. In contrast, the ANOVA cannot generally reproduce the regression results because the ANOVA cannot incorporate continuously measured predictor variables.

Of course, both ANOVA and regression are special cases of a "general linear model" (see e.g., Kirk, 1982, Chapter 5). This general linear model offers several approaches to estimation of deficient rank ANOVA models including specification of constraints on the parameters, use of the generalized inverse, and full-rank reparameterization via regression. For many researchers, the regression approach has become the method of choice. Not only does regression allow a single, simple framework for handling both continuous and discrete predictors, it also provides a natural and sensible way to handle unequal cell sizes.

Unfortunately, the benefits of standard regression analysis are available only in fixed-effects models. Standard computing packages for regression, based on ordinary least squares estimation, are inappropriate when some factors are viewed as random. Maximum likelihood estimation via SAS and BMDP allows estimation only for some special cases of a mixed model (Kang, 1991, Chapter IV).

The inappropriate use of multiple regression for data yielded by nested designs has a long and disreputable history in education, sociology, and psychology (cf. Robinson, 1950; Cronbach and Webb, 1975; Burstein, 1980). Yet this use has continued. Torn between the limitations of the mixed-model ANOVA and the fixed-effects regression, researchers have often clung to regression, especially in nonexperimental studies where it is necessary to control for multiple continuously measured covariates in order to avoid misspecifying estimates of theoretically interesting predictors.

In repeated measures analysis, multivariate methods have offered a reasonable alternative to the univariate mixed-model ANOVA. Within-subjects contrasts are represented as multiple dependent variables, allowing the investigator to make less stringent assumptions about the covariance structure of the repeated measures than are possible using the univariate mixed-model ANOVA (Bock, 1975). However, the standard multivariate approach does not accommodate missing data, uneven spacing of time series observations, or time-varying covariates (Ware, 1985).

The difficulty in extending general linear model analysis to designs having fixed and random effects has been technical. When designs are completely balanced, it is possible to estimate efficiently the fixed effects (for example, regression coefficients) without information about the variance components. Once the fixed effects are estimated, it is not difficult to estimate the variance components. However, in unbalanced designs, efficient estimates of the fixed effects and the variance components are mutually dependent, and some sort of iterative procedure is needed to arrive at a unique and efficient set of estimates.

The more general analytic methods described in this chapter are based on such iterative methods, which have become available as a result of advances in statistical theory and computation. Statisticians are now making rapid progress in algorithmic development, essentially enabling researchers to "catch up" with advances in computing. The result is an array of new analytic approaches rapidly becoming accessible to researchers.

13.1.5 Hierarchical Linear Models

This chapter considers hierarchical linear models, a class of models that combine the advantages of the mixed-model ANOVA, with its flexible modeling of fixed and random effects, and regression, with its advantages in dealing with unbalanced data and predictors that are discrete or continuous. Results based on hierarchical linear models duplicate the results of many classical ANOVA models (Kirk, 1982; Winer, 1971) and expand possibilities for data analysis.

Hierarchical linear models have been referred to as random coefficient models (Rosenberg, 1973), multilevel linear models (Mason, Wong, and Entwisle, 1983; Goldstein, 1987), covariance components models (Dempster, Rubin, and Tsutakawa, 1981), and unbalanced models with nested random effects (Longford, 1987). However, the term hierarchical linear models captures two defining features of the models. First, the data appropriate for such models are hierarchically structured, with first-level units nested within second-level units, second-level units nested within third-level units, and so on. Second, the parameters of such models may be viewed as having a hierarchical linear structure. The investigator may specify a level-one model, the parameters of which characterize linear relationships occurring between level-one units. These parameters are then viewed as varying across level-two units as a function of level-two characteristics. Higher levels may be added, in principle, without limit, although to date no published applications involve more than three levels.

The connection between hierarchical linear models and classical experimental design models has not been well understood in the past. Some assume that hierarchical models apply only to designs having nested factors and not to designs having crossed factors, but this intuition is false. What distinguishes hierarchical linear models is that *the random factors are nested and never crossed*. However, fixed factors can be crossed with random factors (or with each other) and random factors may be nested within fixed factors. The data can be unbalanced at any level, and continuous predictors can be defined at any level. Both discrete and continuous predictors can be specified as having random effects, and these random effects are allowed to covary. We shall return later to a broader class of models under the rubric of the general mixed linear model, which subsumes hierarchical linear models but also includes models with crossed random factors.

13.1.6 Organization of the Chapter

The next section formalizes the two-level hierarchical linear model. Subsequent sections apply this model to one-way designs, two-factor nested designs, two-factor crossed designs, and randomized blocks (repeated measures) designs. In the case of balanced data, hierarchical model results are shown to duplicate classical results, as demonstrated by reanalysis of data from Kirk (1982). In each case, the capacity of the hierarchical analysis to generalize application of the classical model is demonstrated on large-scale field data that would pose serious problems for a classical approach. In the final section, we consider more complex designs and indicate how a general mixed model fills in gaps left even by the hierarchical model.

13.2 THE TWO-LEVEL HIERARCHICAL LINEAR MODEL

In this section we formulate a general two-level hierarchical linear model. For concreteness, let us assume that the design involves students nested within schools. The level-one model specifies how student-level predictors relate to the student-level outcome. At level two, each of the regression coefficients defined in the level-one model, including the intercept, may be predicted by school-level predictors, and each may, in addition, have a random component of variation. The combined level-one and level-two models constitute a mixed linear model with fixed and random regression coefficients. A number of classical ANOVA and regression models may be shown to represent simplifications of this mixed model, as illustrated in Sections 13.3–13.6.

13.2.1 The Level-One Model

At level one, the student level, the outcome y_{ij} for student i in school j ($i = 1, \ldots, n_j; j = 1, \ldots, J$), varies as a function of student characteristics, $X_{qij}, q = 1, \ldots, Q$, and a random error r_{ij} according to the linear regression model

$$y_{ij} = \beta_{0j} + \Sigma \, \beta_{qj} X_{qij} + r_{ij} , \qquad r_{ij} \sim N(0, \sigma^2), \qquad (13.1)$$

where β_{0j} is the intercept and each $\beta_{qj}, q = 1, \ldots, Q$, is a regression coefficient indicating the strength of association between each X_{qij} and the

outcome within school j. Note that the intercept and the regression slopes
are each subscripted by j, allowing them to vary from school to school.
The error of prediction of y_{ij} by the X's is r_{ij}, which is assumed normally
distributed and, for simplicity, homoscedastic.

13.2.2 The Level-Two Model

At level two (the school level), each regression coefficient, β_{qj}, $q = 0, 1,$
$\ldots, Q$, defined by the level-one model, becomes an outcome variable to
be predicted by school-level characteristics, W_{sj}, $s = 1, \ldots, S$, according
to the regression model

$$\beta_{qj} = \Theta_{q0} + \Sigma \, \Theta_{qs} W_{sj} + u_{qj} , \qquad (13.2)$$

where Θ_{q0} is an intercept; each Θ_{qs}, $s = 1, \ldots, S$, is a regression slope
specifying the strength of association between each W_{sj} and the outcome
β_{qj}; and the random effects are assumed sampled from a $(Q + 1)$-variate
normal distribution, where each u_{qj}, $q = 1, 2, \ldots, Q$, has a mean of 0
and a variance τ_{qq}, and the covariance between u_{qj} and $u_{q'j}$ is $\tau_{qq'}$.

The user is faced with a considerable number of options in modeling
each β_{qj}, $q = 0, 1, \ldots, Q$, in Eq. (13.2). If every W_{sj} is assumed to have
no effect, the regression coefficients Θ_{qs}, $s = 1, \ldots, S$, are set to zero.
If the random effect u_{qj} is also constrained to zero, then $\beta_{qj} = \Theta_{q0}$, i.e.,
β_{qj}, is fixed across all schools. This option is useful when the user wishes
to constrain a regression to be homogeneous. If the user believes that one
or more of the W's does predict β_{qj}, but wishes to constrain u_{qj} to zero,
then β_{qj} does indeed vary, but strictly as a function of the W's: its variation
is nonrandom. Of course, β_{qj} could also could have a random component
of variation (u_{qj} not constrained to zero) but might not be predicted by
the W's. Its variation is strictly random, with no predictable component.
Clearly, Eq. (13.2) can be simplified in a large number of ways, which
enables the user to tailor the model to conform to a variety of conventional
linear models, as we shall demonstrate in the next four sections.

13.2.3 Research Applications

Applications of the two-level model in cross-sectional school effects research are presented by Aitkin and Longford (1986), deLeeuw and Kreft
(1986), Raudenbush and Bryk (1986), and Goldstein (1987). We consider
applications of this type in Sections 13.3 and 13.4.

In research on growth, Eq. (13.1) may be viewed as a within-subject
model relating the outcome Y_{ij} of person j at time i to age or time where
the X's might be polynomial trend coefficients. In that case the β's are the
growth parameters of subject j, to be predicted in Eq. (13.2) by subject

background characteristics or between-subjects treatments (the W's). Applications in growth research (Laird and Ware, 1982; Bryk and Raudenbush, 1987) will be considered in Section 13.6.

In research synthesis, Eq. (13.1) is a within-study model, perhaps relating treatment contrasts (the X's) to subject outcomes (Y). The treatment effects defined in that equation (the β's) then become the outcomes in the between-study model [Eq. (13.2)] where the W's are study characteristics that predict different effects. Raudenbush and Bryk (1985) and Becker (1988) used two-level models in this way.

A book-length treatment of the use of these models in studying organizational effects, growth, and research synthesis is given in Bryk and Raudenbush (1992). It also provides a detailed account of the statistical theory underlying estimation. The present chapter is novel in clarifying linkages between hierarchical linear models and classical models for experimental design.

13.3 THE ONE-WAY ANALYSIS OF VARIANCE

13.3.1 Classical Approach

13.3.1.1 The One-Way ANOVA Model

The usual statistical model for the one-way classification may be written as

$$y_{ij} = \mu + \alpha_j + r_{ij}, \qquad r_{ij} \sim N(0, \sigma^2), \tag{13.3}$$

where y_{ij} is the observation for subject i assigned to level j of the independent variable ($i = 1, \ldots, n_j; j = 1, \ldots, J$); μ is the grand mean; α_j is the effect associated with level j; and r_{ij} is assumed normally distributed with a mean of 0 and homogeneous variance, σ^2. In the fixed-effects case, the effect α_j is a fixed constant and the constraint $\Sigma\, n_j \alpha_j = 0$ is introduced. In the random-effects approach, each α_j is typically assumed independently, normally distributed, i.e., $\alpha_j \sim N(0, \tau^2)$.

13.3.1.2 Data and Results

The data in Table 13.1 were collected by Nelson (1976) and reprinted by Kirk (1982, p. 168). Fifty children were randomly assigned to five experimental methods of training in discrimination among blocks; the outcome was the number of blocks correctly identified. Thus, $n_j = n = 10$ for all j, $J = 5$, and the total sample size is $nJ = 50$.

We consider the results under the random-effects assumption simply because it offers an interesting comparison to the hierarchical linear model

Table 13.1 One-Way ANOVA Data and Source Table

<table>
<tr><th colspan="5">DATA</th></tr>
<tr><th>Method 1</th><th>Method 2</th><th>Method 3</th><th>Method 4</th><th>Method 5</th></tr>
<tr><td>0</td><td>2</td><td>2</td><td>2</td><td>1</td></tr>
<tr><td>1</td><td>3</td><td>3</td><td>4</td><td>0</td></tr>
<tr><td>3</td><td>4</td><td>4</td><td>5</td><td>2</td></tr>
<tr><td>1</td><td>2</td><td>4</td><td>3</td><td>1</td></tr>
<tr><td>1</td><td>1</td><td>2</td><td>2</td><td>1</td></tr>
<tr><td>2</td><td>1</td><td>1</td><td>1</td><td>2</td></tr>
<tr><td>2</td><td>2</td><td>2</td><td>3</td><td>1</td></tr>
<tr><td>1</td><td>2</td><td>3</td><td>3</td><td>0</td></tr>
<tr><td>1</td><td>3</td><td>2</td><td>2</td><td>1</td></tr>
<tr><td>2</td><td>4</td><td>2</td><td>4</td><td>3</td></tr>
</table>

<table>
<tr><th colspan="6">ANOVA Source Table</th></tr>
<tr><th>Source</th><th>df</th><th>Sum of squares</th><th>Mean square</th><th>E(mean square)</th><th>F</th></tr>
<tr><td>Between groups</td><td>4</td><td>21.88</td><td>5.47</td><td>$n\tau^2 + \sigma^2$</td><td>5.37</td></tr>
<tr><td>Within groups</td><td>45</td><td>45.80</td><td>1.02</td><td>σ^2</td><td></td></tr>
<tr><td>Total</td><td>49</td><td>67.68</td><td></td><td></td><td></td></tr>
</table>

results. The grand mean estimate and its estimated standard error are

$$\hat{\mu} = \Sigma\,\Sigma\,y_{ij}/50 = 2.08,$$
$$s(\hat{\mu}) = \sqrt{(MS_b/50)} = .331.$$

The ANOVA source table shows evidence of significant variation among groups, $F(4, 45) = 5.37$, $p < .001$. Under the random-effects model, this suggests rejection of the null hypothesis H_0: $\tau^2 = 0$. To estimate the variance components, the observed mean squares are set equal to their expected values given in Table 13.1:

$$MS_w = \hat{\sigma}^2,$$
$$MS_b = n\hat{\tau}^2 + \hat{\sigma}^2,$$

leading to the solutions

$$\hat{\sigma}^2 = MS_w = 1.02,$$
$$\hat{\tau}^2 = (MS_b - MS_w)/n = .445,$$

where MS_b and MS_w refer to mean squares between and within, respectively. This result implies that the proportion of variation between groups is about

.445/(.445 + 1.02) = .30 or 30%. This proportion is also the estimated intraclass correlation coefficient, that is, the estimated correlation between pairs of observations sharing membership in the same group, j.

13.3.2 Analysis via the Hierarchical Linear Model

13.3.2.1 The Model

Let us first set every regression coefficient in Eq. (13.1), save the intercept, to zero, so that the level-one model becomes

$$y_{ij} = \beta_{0j} + r_{ij}, \qquad r_{ij} \sim N(0, \sigma^2). \tag{13.4}$$

According to this model, the outcome for subject i in group j is predicted only by the intercept, β_{0j}, which is the group mean, so that the variance σ^2 is the within-group variance. Only one parameter, β_{0j}, is to be predicted at level two [Eq. (13.2)], and the model for that parameter is similarly simplified so that all regression coefficients except the intercept are set to zero:

$$\beta_{0j} = \Theta_{00} + u_{0j}, \qquad u_{0j} \sim N(0, \tau^2). \tag{13.5}$$

Here Θ_{00} is the grand mean and u_{0j} is the effect associated with level j. In the random-effects model, that effect is typically assumed normally distributed with a mean of zero. In terms of Eq. (13.2), this variance might be denoted by τ_{00}, but we use the notation τ^2 for conformity to standard ANOVA usage. In the fixed-effects models, each u_{0j} is a fixed constant. Combining Eqs. (13.4) and (13.5) yields the single model

$$y_{ij} = \Theta_{00} + u_{0j} + r_{ij}, \tag{13.6}$$

with $r_{ij} \sim N(0, \sigma^2)$ and $u_{0j} \sim N(0, \tau^2)$. This is clearly the one-way random effects ANOVA, identical to Eq. (13.3) with $\Theta_{00} = \mu$, and $\alpha_j = u_{0j}$.

13.3.2.2 Estimation

For all hierarchical analyses in this chapter we employ the computer program of Bryk, Raudenbush, Seltzer, and Congdon (1988), which computes restricted maximum likelihood (ML) estimates by means of an iterative approach known as the EM algorithm (Dempster, Laird, and Rubin, 1977). (We discuss the logic of this approach in some detail here; readers uninterested in the logic of estimation and hypothesis testing may skip to the results in Section 13.3.2.4.)

The restricted likelihood is the full likelihood with the fixed effects integrated out, leaving the likelihood of the data strictly as a function of the variance-covariance components (see Dempster et al., 1981). Estimates of variance-covariance components based on the restricted likelihood have the distinct advantage of taking into account uncertainty about the fixed effects. In balanced designs, this means that variance estimates are corrected for loss of degrees of freedom due to estimation of fixed effects (e.g., to estimate a residual variance in regression, the residual sum of squares is divided by $n - p$ rather than n, where n is the sample size and p is the number of regression coefficients including the intercept). The reader is warned that computing programs based on the full rather than the restricted likelihood will not duplicate exactly the standard ANOVA results for balanced designs.

In the case of the one-way analysis of variance with random effects, the ML grand mean estimate is a precision-weighted average of the group means. Let $\bar{y}_{.j} = \Sigma\, y_{ij}/n_j$ denote the sample mean for group j. Then the variance of $\bar{y}_{.j}$ is

$$V_j = \text{Var}(\bar{y}_{.j}) = \tau^2 + \sigma^2/n_j. \tag{13.7}$$

The precision of $\bar{y}_{.j}$ is then defined as the inverse of its variance

$$P_j = V_j^{-1},$$

and, with σ^2 and τ^2 known, the ML estimate and also the unique, minimum variance unbiased estimate of the grand mean is the precision-weighted average

$$\hat{\theta}_{00} = \Sigma\, P_j \bar{y}_{.j}/\Sigma\, P_j. \tag{13.8}$$

Note that the best estimate is *not* the grand mean in the sample. Nor is the best estimate the arithmetic mean of the sample means. Raudenbush (1984) showed that Eq. (13.8) is a compromise between the grand mean $\Sigma\,\Sigma\, y_{ij}/\Sigma\, n_j$ and the mean of the means $\Sigma\, \bar{y}_{.j}/J$. Equation (13.8) approaches the grand mean as τ^2 approaches zero (holding n_j constant) and approaches the mean of the means as τ^2 increases. Also, as n_j increases (with τ^2 held constant), Eq. (13.8) approaches the mean of the means. When the data are balanced, the two estimates are identical.

When σ^2 and τ^2 are not known, the ML estimate of the grand mean is Eq. (13.8) with ML estimates substituted for σ^2 and τ^2. However, when the data are balanced ($n_j = n$ for all j), P_j is constant for every group; that is, every group mean has the same precision and Eq. (13.8) becomes the grand mean

$$\hat{\theta}_{00} = \bar{y}_{..} = \Sigma\, \bar{y}_{.j}/J. \tag{13.9}$$

As Eq. (13.9) shows, in the case of balanced data, knowledge about the variance components σ^2 and τ^2 is not needed to obtain the ML estimate of the grand mean. This principle is important and generalizes to more complex designs. ML estimates of fixed effects do not depend on variance components when the data are balanced, so that, in the balanced case, the standard ANOVA estimators and the ML estimators of fixed effects converge.

With σ^2 and τ^2 known, the precision of the ML estimate is the sum of the respective precisions of the group means

$$\text{Precision}(\hat{\Theta}_{00}) = \Sigma\, P_j,$$

so that the sampling variance of $\hat{\Theta}_{00}$ is the inverse of its precision

$$\text{Var}(\hat{\Theta}_{00}) = 1/\Sigma\, P_j.$$

However, when the data are balanced, the precisions are equal and

$$\text{Var}(\hat{\Theta}_{00}) = \frac{1}{JP} = \frac{\tau^2 + \sigma^2/n}{J} = \frac{n\tau^2 + \sigma^2}{Jn}. \tag{13.10}$$

The reader will notice that the numerator in Eq. (13.10) is equivalent to the expected MS_b (see Table 13.1) so that the restricted ML estimate is

$$\text{Estimated Var}(\hat{\Theta}_{00}) = MS_b/(Jn), \tag{13.11}$$

and the estimated standard error of $\hat{\Theta}_{00}$ is the square root of Eq. (13.11). We shall illustrate the principle emerging from this simple example in subsequent sections. In balanced designs, standard errors for fixed-effects estimates will be simple functions of sufficient statistics such as mean squares.

Standard texts assert without qualification that the best estimate of a group mean in the one-way ANOVA is simply the group's sample mean, $\bar{y}_{.j}$. However, as Lindley and Smith (1972) pointed out, this point estimate can be improved upon whenever the design includes more than two groups. An alternative point estimator of the group mean, β_{0j}, is the weighted average

$$\beta_{0j}^* = \pi_j\bar{y}_{.j} + (1 - \pi_j)\hat{\Theta}_{00}, \tag{13.12}$$

where $\pi_j = \tau^2/(\tau^2 + \sigma^2/n_j)$. Note that β_{0j}^* is a weighted average of group j's sample mean, $\bar{y}_{.j}$, and the grand mean, $\hat{\Theta}_{00}$. The weight accorded $\bar{y}_{.j}$ is π_j, which, in fact, is the reliability of $\bar{y}_{.j}$ as an estimate of β_{0j}, where reliability is measured by the ratio of the "true-score" variance, $\text{Var}(\beta_{0j}) = \tau^2$, to the "observed-score" variance, $\text{Var}(\bar{y}_{.j}) = \tau^2 + \sigma^2/n_j$. Note that the observed-score variance includes both the true-score variance, τ^2, and the error variance, σ^2/n_j, of $\bar{y}_{.j}$ as an estimate of β_{0j}. Whenever the error variance is large relative to τ^2, π_j will be small and β_{0j}^* will rely heavily on the grand mean. However, when the error variance is small relative to τ^2, π_j will be large and β_{0j}^* will rely heavily on the group's sample mean. The weighted average, β_{0j}^*, has a smaller expected mean squared error of estimation than does the sample mean, $\bar{y}_{.j}$ (Efron and Morris, 1975).

Note that Eq. (13.12) requires knowledge of the variances σ^2 and τ^2. However, when these are unknown, results of the hierarchical linear model analysis yield maximum likelihood estimates of these variances, even when the data are unbalanced. These estimates can be substituted into Eq. (13.12) to obtain what are sometimes called "empirical Bayes" estimates (Morris, 1983).

We caution that these empirical Bayes estimates are justifiable only if the investigator has no a priori hypotheses regarding the expected magnitude of the group means. For example, if the treatments represent duration of exposure so that larger durations are expected to lead to greater effect sizes, these empirical Bayes estimates should not be used. In this case, "conditional" empirical Bayes estimators are available that represent a compromise between the sample mean and its predicted value based on a particular estimated *a priori* contrast (see Raudenbush, 1988).

13.3.2.3 Hypothesis Testing

A simple test of the null hypothesis of no group effects, i.e.,

$$H_0: \tau^2 = 0$$

is given by the statistic

$$H = \Sigma \, \hat{P}_j (\bar{y}_{.j} - \bar{y}_{..})^2, \tag{13.13}$$

where $\hat{P}_j = n_j/\hat{\sigma}^2$. The statistic H has a large sample chi-square distribution with $J - 1$ degrees of freedom under the null hypothesis. In the case of balanced data, this sum of precision-weighted squared differences reduces to

$$H = \hat{P} \, \Sigma (\bar{y}_{.j} - \bar{y}_{..})^2 = (J - 1)MS_b/MS_w, \tag{13.14}$$

revealing clearly that $H/(J - 1)$ is the usual F statistic for testing group differences in ANOVA.

13.3.2.4 Results

The results of the hierarchical analysis (Table 13.2) are mathematically identical to the results based on the usual ANOVA, although they are presented in a different form. The point estimates for all parameters are the same as those in Table 13.1. Note that the chi-square test for group differences is $H = 21.50$ and that $H/(J - 1) = 21.50/4 = 5.37$, the F statistic in the ANOVA table (Table 13.1).

13.3.2.5 Estimation of Group Effects

Table 13.3 lists the standard group effect estimators, $\bar{y}_{.j}$, and the empirical Bayes estimators, β_{0j}^*, for the data in Table 13.1. Note that the empirical

Table 13.2 One-Way ANOVA Results Based on the Hierarchical Linear Model

Fixed effects		
Parameter	Coefficient	Standard error
Grand mean, Θ_{00}	2.08	.331

Random effects			
Parameter	Estimate	Chi-square	DF
τ^2	.445	21.50	4
σ^2	1.02		

Bayes estimators are closer to the grand mean of 2.08 than are the five sample means.

In this case, the differences between the standard and empirical Bayes estimators are small because the reliability of the sample means is reasonably high at .81. Hence, the group mean $\bar{y}_{.j}$ is weighted heavily relative to the grand mean $\bar{y}_{..}$ in composing the empirical Bayes estimator, β_j^* [Eq. (13.12)].

13.3.3 Generalization

To illustrate application of the one-way random effects ANOVA in large-scale survey data, we consider the results of Raudenbush, Eamsukkawat, Di-Ibor, Kamali, and Taoklam (1991), who analyzed data from 103 small primary schools in rural Thailand. Each school had just one sixth-grade classroom, and the outcome variable was total academic achievement as measured by a test covering the five major areas of the Thai curriculum.

Table 13.3 Standard and Empirical Bayes Estimates of Group Effects in the One-Way ANOVA

Group	$\bar{y}_{.j}$	β_{0j}^*
1	1.50	1.55
2	2.40	2.34
3	2.50	2.42
4	2.90	2.75
5	1.20	1.37

Table 13.4 One-Way ANOVA Results of
Thailand Classroom Data Based on the Hierarchical
Linear Model

Fixed effects		
Parameter	Coefficient	Standard error
Grand mean, Θ_{00}	-24.45	6.25

Random effects			
Parameter	Estimate	Chi-square	DF
τ^2	3785	1947.8	102
σ^2	4099		

The outcome variable was standardized nationally to have a mean of 0 and a standard deviation of 100. Classrooms averaged 19.2 students with a minimum of 9 and a maximum of 35. The one-way ANOVA provided a number of useful preliminary results in their analysis (see Table 13.4).

The results show, first, that average achievement in these small rural schools is significantly less than the national average of 0 ($\hat{\Theta}_{00} = -24.45$, se = 6.25). Second, the between-classroom variance component is large. The proportion of variance between classrooms (equivalent to the intra-classroom correlation) is estimated to be $3785/(3785 + 4099) = .48$ or 48%. This proportion of variation between classrooms is much larger than that typically reported in U.S. primary schools (cf. Bryk and Raudenbush, 1988) and, not surprisingly, is highly statistically significant, with an approximate chi-square of 1947.8, in this case equal to $F(102, 1875) = 1947.8/102 = 19.1$, $p < .001$. Note that the denominator degrees of freedom are equal to the total sample size of 1978 minus 103, the number of classrooms.

Because of the large variation among classrooms (i.e., τ^2 is large), a typical classroom's sample mean has a high reliability of about .94 as an estimate of its unknown true mean. Hence, the empirical Bayes estimators of the classroom means will be very similar to the classroom sample means as shown in Figure 13.1. In contrast, in studies where τ^2 is small, little of the variation among the classroom sample means reflects variation in the true means. In such studies, the empirical Bayes distribution looks much less dispersed than does the distribution of these classroom means (see Willms and Raudenbush, 1989, for an example).

The example illustrates a generalization of the standard one-way random effects ANOVA in that maximum likelihood estimates of the grand mean

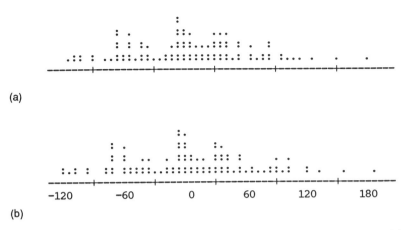

Figure 13.1 Empirical Bayes estimators and classroom sample means. (a) Empirical Bayes estimates; (b) class sample means.

and the variance components are available even though the data are unbalanced. These estimated variance components, besides enabling an estimate of the reliability of the group means, facilitate computation and interpretation of the empirical Bayes estimators.

13.4 THE TWO-FACTOR NESTED DESIGN

As mentioned in the introduction, two-factor nested designs are common in education (e.g., students within classrooms, classrooms within instructional methods), counseling psychology (e.g., patients within counselors, counselors within therapies), and sociology, for example, cross-national studies of demography in which women are nested within countries and countries within types of family-planning efforts (Mason et al., 1983). In these designs, classrooms, counselors, or countries are typically viewed as random factors because only a sample of the interesting levels is available and one wishes to make generalizations of such levels to the universe of levels. Noniterative maximum likelihood estimates are available in the balanced case treated in standard experimental design texts. Hierarchical linear models generalize maximum likelihood estimation to the case of unbalanced data, and covariates measured at each level can be discrete or continuous and can have either fixed or random effects. In this section, we (1) consider the classical analytic approach, (2) show how the hierarchical linear model can be formulated to duplicate the classical results in the balanced case, and (3) illustrate how the hierarchical analysis generalizes to cases involving unbalanced data and incorporation of discrete and continuous covariates.

13.4.1 Classical Approach

13.4.1.1 The Model

The standard model for the two-factor nested design may be written

$$y_{ijk} = \mu + \alpha_k + \pi_{j(k)} + r_{ijk}, \qquad \pi_{j(k)} \sim N(0, \tau^2), \qquad (13.15)$$
$$r_{ijk} \sim N(0, \sigma^2),$$

where y_{ijk} is the outcome for subject i nested within level j of the random factor, which is, in turn, nested within level k of the fixed factor ($i = 1$, ..., n_{jk}; $j = 1, \ldots, J_k$; $k = 1, \ldots, K$); μ is the grand mean; α_k is the effect associated with the kth level of the fixed factor; $\pi_{j(k)}$ is the effect associated with the jth level of the random factor within the kth level of the fixed factor; and r_{ijk} is the random (within-cell) error. In the case of balanced data ($n_{jk} = n$ for every level of the random factor), the standard analysis of variance method and the method of restricted maximum likelihood coincide.

13.4.1.2 Example and Results

Table 13.5 lists artificial data provided in Kirk (p. 460). For ease of understanding we shall refer to the design as involving classes nested within methods of instruction with the outcome being the number of correct responses on a posttreatment cognitive test. There are four classes within each of two methods.

The estimated grand mean and its estimated standard error are given by

$$\hat{\mu} = \bar{y}_{...} = 5.375,$$
$$se(\hat{\mu}) = \sqrt{(MS_{classes}/32)} = .738.$$

The estimated contrast between methods 1 and 2 is

$$\hat{\alpha}_1 - \hat{\alpha}_2 = 3.75,$$

and its significance is tested by the ratio of mean squares (methods to classes) yielding $F(1, 6) = 6.46, p < .05$. Variance estimates are given by equating the expected mean squares to their observed values and solving as in the one-way ANOVA, yielding

$$\hat{\sigma}^2 = MS_{wcell} = 0.77,$$
$$\hat{\tau}^2 = (MS_{classes} - MS_{wcell})/n = 4.16.$$

Table 13.5 Nested Two-Factor Data and Source Table

				Data			
	Method 1				Method 2		
Class 1	Class 2	Class 3	Class 4	Class 5	Class 6	Class 7	Class 8
3	1	5	2	7	4	7	10
6	2	6	3	8	5	8	10
3	2	5	4	7	4	9	9
3	2	6	3	6	3	8	11

			ANOVA source table		
Source	df	Sum of Squares	Mean square	E(mean square)	F
Methods	1	112.50	112.50	$\sigma^2 + n\tau^2 + nJ \Sigma \alpha_k^2/(K-1)$	6.46
Classes	6	104.50	17.42	$\sigma^2 + n\tau^2$	22.59
Within cell	24	18.50	0.77	σ^2	
Total	31	235.50			

As in the one-way ANOVA case, negative values of $\hat{\tau}^2$ are set to 0. (An advantage of estimation for the hierarchical model based on the EM algorithm is that estimates will be nonnegative.) Note that τ^2 is the residual between-classes variance after removing the effects of the method factor, so it will differ from the estimate resulting from a one-way ANOVA. A test of the null hypothesis of no class-level variance, i.e., H_0: $\tau^2 = 0$, is the F statistic computed as the ratio of mean squares (classes to within-cell), in this case yielding $F(6, 24) = 22.59$, $p < .01$.

13.4.2 Analysis by Means of a Hierarchical Linear Model

13.4.2.1 The Model

As in the case of the one-way ANOVA, we first set every regression coefficient in Eq. (13.1), save the intercept to zero, so that the level-one model is

$$y_{ij} = \beta_{0j} + r_{ij}, \qquad r_{ij} \sim N(0, \sigma^2), \tag{13.16}$$

where y_{ij} is the outcome for subject i within class j ($i = 1, \ldots, n_j; j =$

$1, \ldots, J$), β_{0j} is the mean of class j, and r_{ij} is the residual assumed normally distributed with a mean of zero and within-class variance σ^2.

The level-two (between-class) model is a regression model in which the class mean β_{0j} is the outcome and the predictor is a contrast between the two treatments (methods of instruction). This level-two model,

$$\beta_{0j} = \Theta_{00} + \Theta_{01}W_j + u_{0j}, \qquad u_{0j} \sim N(0, \tau^2), \tag{13.17}$$

has $W_j = 1/2$ for classes experiencing instructional method 1 and $W_j = -1/2$ for classes experiencing instructional method 2. Hence, the correspondences between the hierarchical model and the ANOVA model are

$$\Theta_{00} = \mu, \qquad \Theta_{01} = \alpha_1 - \alpha_1, \qquad u_{0j} = \pi_{j(k)}.$$

Combining Eqs. (13.16) and (13.17) yields the single model

$$y_{ij} = \Theta_{00} + \Theta_{01}W_j + u_{0j} + r_{ij}. \tag{13.18}$$

Note that $K - 1$ contrasts must be included to represent the variation among the K methods in order to duplicate the ANOVA results.

13.4.2.2 Results

The results of the analysis via the hierarchical model using Table 13.6 are again mathematically identical to the results based on the usual ANOVA, although they are presented in a different form. The reader will again notice that squaring the t-statistic associated with the method contrast will yield the corresponding F of the ANOVA in Table 13.5, and dividing the chi-square for the class variance by its degrees of freedom will yield the corresponding F in the ANOVA table.

Table 13.6 Nested Two-Factor Results Based on the Hierarchical Linear Model

	Fixed effects		
Parameter	Coefficient	Standard error	t
Grand mean, Θ_{00}	5.375	.738	—
Method contrast, Θ_{01}	3.750	1.48	2.54
	Variance components		
Parameter	Estimate	Chi-square	df
τ^2	4.16	135.6	6
σ^2	0.77		

13.4.3 Generalization

We again turn to the analysis of the primary school data from Thailand. Our goal is to examine the difference in achievement between classrooms taught by teachers with a bachelor's degree and classes taught by teachers with less than a bachelor's degree. Complicating the analysis are the following facts: (1) the data are unbalanced; and (2) because classes were not assigned at random to levels of teacher education, it will be important to control for potentially confounding variables. These covariates may exist *at both the student and the class levels* and they may be either continuously or discretely measured. Hence, although the design is indeed a nested two-factor design, the balanced two-factor nested ANOVA will be inappropriate.

Raudenbush, Kidchanapanish, and Kang (1991) addressed these issues with the following level-one model:

$$y_{ij} = \beta_{0j} + \beta_{1j}(\text{GPA})_{ij} + \beta_{2j}(\text{Repetition})_{ij} + \beta_{3j}(\text{SES})_{ij}$$
$$+ \beta_{4j}(\text{Time})_{ij} + \beta_{5j}(\text{Dialect})_{ij} + \beta_{6j}(\text{Breakfast})_{ij} + r_{ij}, \quad (13.19)$$

where GPA is fifth-grade point average, Repetition is an indicator for having ever repeated a grade, SES is a measure of socioeconomic status, Time is the time needed to travel to school, Sex is an indicator for males, Dialect is an indicator for speaking Central Thai dialect, and Breakfast is an indicator for eating breakfast daily.

At level two, only the intercept was allowed to be random. Other predictors could have been specified to have random coefficients, but they were not. In addition, only the intercept was specified as being predicted by school characteristics, so that the level-two model was

$$\beta_{0j} = \Theta_{00} + \Theta_{01}(\text{Infrastructure})_j + \Theta_{02}(\text{Remoteness})_j$$
$$+ \Theta_{03}(\text{Supervision})_j + \Theta_{04}(\text{Bachelor's})_j$$
$$+ u_{0j}, \quad \beta_{qj} = \Theta_{q0} \text{ for } q > 0, \quad (13.20)$$

where Infrastructure is a scale measuring the modernity of the community in which the school is located; Remoteness is a composite measure of the school's distance from the district capitol, the highway, and the market; Supervision is the intensity of the supervision provided the teacher; and Bachelor's is an indicator (1 = bachelor's degree; 0 = less than bachelor's).

The results (Table 13.7) indicate, after controlling for covariates, a mean difference of 32.0 points in favor of classes of teachers with a bachelor's degree. The among-classroom standard deviation (Table 13.4) is $\sqrt{(3785)} = 61.5$, indicating that the standardized effect size associated with a bachelor's degree is about $32/61.5 = .52$ in units of the classroom

Table 13.7 Two-Factor Nested Results for Thailand Classroom Data Based on the Hierarchical Linear Model

	Fixed effects					
	One-way ANOVA			Full model		
Predictors	Coeff	se	t	Coeff	se	t
School/classroom level						
Intercept	−24.45	6.25	−3.92	−50.29	11.55	−4.53
Infrastructure				37.25	15.80	2.36
Remoteness				6.53	2.76	2.37
Bachelor's				31.99	12.27	2.61
Internal supervision				10.25	4.71	2.18
Student level						
GPA				25.67	1.15	22.29
Repetition				−22.86	3.77	−6.06
SES				4.86	3.22	1.51
Time to school				−4.60	2.32	−1.98
Sex				−2.09	2.51	−0.83
Dialect				19.01	8.33	2.82
Breakfast				12.37	4.49	2.75

	Variance components	
Parameter	One-way ANOVA	Full model
Between class, $\hat{\tau}^2$	3785	2693
% Explained	0.0	28.8
Within class, $\hat{\sigma}^2$	4099	3072
% Explained	0.0	25.0

standard deviation. The effect size is about .36 in units of the overall standard deviation (note the overall standard deviation of the outcome shown in Table 13.4 is just less than 90). Notice also, that the estimated residual variances ($\hat{\sigma}^2$ at the student level and $\hat{\tau}^2$ at the classroom level) are reduced by 28.8% and 25.0%, respectively, revealing how the explanatory power of the model may be monitored in this type of analysis. The analysis has allowed for control of continuous and discrete covariates at two levels of aggregation and produces ML variance components estimates even though the data are unbalanced. These variance components estimates not only are interesting in themselves but also are required to obtain estimates of the regression coefficients, estimated via a generalized least squares approach that weights the data from each school in proportion to its precision (Aitkin and Longford, 1986; Raudenbush, 1988).

13.5 THE TWO-FACTOR CROSSED DESIGN (WITH REPLICATIONS WITHIN CELLS)

Experimental design texts distinguish among three types of models for two-factor designs: Model I (both factors fixed), Model II (both factors random), and Model III (the "mixed" case with one factor fixed and the other random). In Section 13.4 on nested factors, we considered only the mixed case, the case that arises most naturally in field research. In the present section on crossed two-factor designs, we again focus on the mixed case (Model III).

It is well known that the fixed-effects case (Model I) can readily be represented within the framework of standard multiple regression, an especially useful approach when the data are unbalanced. Multiple regression offers a flexible approach for incorporating effects of covariates measured either continuously or discretely. Available texts (for example, Cohen and Cohen, 1975) consider such applications in detail. Here we will simply indicate that standard multiple regression represents a special case of the hierarchical linear model [Eqs. (13.1) and (13.2)] in which all variance components except σ^2 have been set to zero.

On the other hand, the case in which both factors are random (Model III) cannot be represented within the hierarchical model; and generalizing Model III to include unbalanced data and covariates requires a more general mixed model (see Dempster, Rubin, and Tsutakawa, 1981). Goldstein (1987) has considered estimation for such models, and Kang (1991) has developed a workable computational approach based on the EM algorithm that allows fixed and random effects of rows, columns, and interactions where predictors may be either continuous or discrete and the data may be unbalanced.

Crossed two-factor designs with one factor fixed and one factor random arise commonly in social and behavioral research. For example, Raffe (1991) considers the case of a treatment-control contrast estimated within each of 19 sites in the context of evaluating Britain's innovative technical education initiative. In this context, sites and treatments are crossed with site being the random factor, treatment the fixed factor, and the site-by-treatment interaction also a random factor. The advantages of such a design over the two-factor nested design (sites nested within treatments) are well known. In the crossed design, site-by-treatment interactions are estimable, and, indeed, in Raffe's analysis, the varying effect of treatment across sites constituted the key focus of the inquiry. The main effect of treatment is estimated against within-site variation, making the analysis more powerful than that of a nested design when among-site variation is large. We note that meta-analyses of treatment-control studies are crossed designs of this type. A treatment-control contrast is estimated in each study, and the variation in these "effect sizes" across studies is the focus of inquiry.

As in the two-factor nested design, the two-factor crossed analysis of variance represents a workable approach when the data are balanced and no covariates are employed. We first illustrate this approach on artificial data. However, even in this simple case, the hierarchical analysis offers extra information such as the correlation between the random effect of site and the site-by-treatment interaction. After replicating the results of the ANOVA via the hierarchical approach, we consider a generalization to include unbalanced data and multiple within-site covariates, both continuous and discrete.

13.5.1 Classical Approach

13.5.1.1 The Model

The mixed model for the two-factor crossed design may be written as

$$y_{ijk} = \mu + \alpha_k + \pi_j + (\alpha\pi)_{jk} + r_{ijk}, \qquad \pi_j \sim N(0, \tau^2), \qquad (13.21)$$
$$(\alpha\pi)_{jk} \sim N(0, \delta^2), \qquad r_{ijk} \sim N(0, \sigma^2),$$

where y_{ijk} is the outcome for subject i nested within cell jk; μ is the grand mean; α_k is the effect associated with level k of the fixed effect; π_j is the effect associated with level j of the random effect; $(\alpha\pi)_{jk}$ is the interaction effect associated with cell jk; and r_{ijk} is the within-cell error ($i = 1, \ldots, n_{jk}$; $j = 1, \ldots, J$; $k = 1, \ldots, K$). The random terms, π_j, $(\alpha\pi)_{jk}$, and r_{ijk}, are assumed mutually independent, each with a mean of zero and variances τ^2, δ^2, and σ^2, respectively. As before, in the case of balanced data ($n_{jk} = n$ for every level of the random factor), the standard analysis of variance method and restricted maximum likelihood coincide except when the ML estimate of either τ^2 or δ^2 is at or very near zero (Hartley and Rao 1967).

13.5.1.2 Example and Results

Table 13.8 lists artificial data created by the author. Let us assume that these data resulted from an experiment in which 40 subjects were assigned at random to one of 10 tutors and that, within tutors, subjects were then assigned at random to instruction without or with practice on a cognitive task. The outcome is the score on a test measuring proficiency on the task. The result is a 2-by-10 cross classification with 2 levels of the fixed effect of practice ($K = 2$) and 10 levels of tutor ($J = 10$). The interesting question is not so much whether there is a main effect of practice (the effect is huge) but whether the magnitude of that effect varies across tutors. Table 13.8 provides the source table for the ANOVA.

Table 13.8 Crossed Two-Factor Data and Source Table

Data

					Tutor					
	1	2	3	4	5	6	7	8	9	10
Practice										
1 = No	65	70	62	56	62	45	56	82	53	82
	70	78	66	64	70	48	69	86	54	88
2 = Yes	140	159	163	139	127	141	130	139	128	156
	155	163	181	142	138	146	138	144	130	165

ANOVA source table

Source	df	Sum of squares	Mean square	E(mean square)	F
Practice	1	63840.1	63840.1	$\sigma^2 + n\delta^2 + nJ \Sigma \alpha_k^2/(K - 1)$	298.74
Tutors	9	4325.0	480.6	$\sigma^2 + nK\tau^2$	14.14
Practice × tutors	9	1923.4	213.7	$\sigma^2 + n\delta^2$	6.28
Within cell	20	679.0	34.0	σ^2	
Total	39	70767.5			

The estimated contrast between methods 1 and 2 is

$$\hat{\alpha}_1 - \hat{\alpha}_2 = 53.12,$$

and its significance is tested by the ratio of mean squares (practice to practice-by-tutors interaction), yielding $F(1, 9) = 298.74, p < .05$. Variance estimates are given by equating the expected mean squares to their observed values as in the one-way ANOVA, yielding

$$\hat{\sigma}^2 = MS(\text{within-cell}) = 34.0,$$
$$\hat{\delta}^2 = [MS(\text{practice} \times \text{tutors}) - MS(\text{within-cell})]/n = 89.6,$$
$$\hat{\tau}^2 = [MS(\text{tutors}) - MS(\text{within-cell})]/(nK) = 111.65.$$

Negative estimates of δ^2 and $\hat{\tau}^2$ would be set to 0. A test of the null hypothesis of no practice-by-tutors interaction, i.e., H_0: $\delta^2 = 0$, is the F statistic computed as the ratio of mean squares (practice-by-tutors interaction to within-cell), yielding, in this case, $F(9, 20) = 6.28$, $p < .01$. A test of the null hypothesis of no tutor effect, i.e., H_0: $\tau^2 = 0$, is given by the ratio of mean squares (tutors to within-cell), yielding $F(9, 20) = 14.14$, $p < .01$.

13.5.2 Analysis by Means of a Hierarchical Linear Model

13.5.2.1 The Model

Recall that, in the case of the two-factor (mixed) nested model, the fixed-effects contrasts were specified in the level-two model. In the two-factor (mixed) crossed model, the fixed factor is specified in the level-one (within-tutor) model, which, in general, is Eq. (13.1) with $(K - 1)$ X's. In our case, with $K = 2$, only one X is specified, so the level-one model becomes

$$y_{ij} = \beta_{0j} + \beta_{1j}X_{1ij} + r_{ij}, \qquad r_{ij} \sim N(0, \sigma^2), \tag{13.22}$$

where y_{ij} is the outcome for subject i having tutor j, β_{0j} is the mean for the jth tutor, β_{1j} is the contrast between the practice and no-practice conditions within tutor j, $X_{1ij} = 1$ for subjects of tutor j having practice and -1 for those having no practice, and r_{ij} is the within-cell error.

Notice that the level-one model defines two parameters that are allowed to vary across tutors: β_{0j}, the tutor mean, and β_{1j}, the treatment contrast. To replicate the results of the ANOVA, the level-two model is formulated to allow these to vary randomly across tutors:

$$\beta_{0j} = \Theta_{00} + u_{0j} \quad \text{and} \quad \beta_{1j} = \Theta_{10} + u_{1j}, \tag{13.23}$$

where Θ_{00} is the grand mean, u_{0j} is the unique effect of tutor j on the mean level of the outcome, Θ_{10} is the average value of the treatment contrast, and u_{1j} is the unique effect of tutor j on that contrast. The random effects u_{0j} and u_{1j} are assumed multivariate normal with variances τ_{00} and τ_{11}, respectively, and covariance τ_{01}. Here the hierarchical model departs from the ANOVA model because the latter assumes that the tutor main effect and the tutor-by-practice interaction are independent. When the contrast coefficients sum to zero and the data are balanced, this covariance does not affect estimation or hypothesis testing regarding the fixed effects and variance components.

The correspondences between the hierarchical model and the ANOVA model are

$$\Theta_{00} = \mu, \qquad \Theta_{10} = \frac{\alpha_2 - \alpha_1}{2}, \qquad u_{0j} = \pi_j,$$

$$u_{1j} = \frac{(\alpha\beta)_{j2} - (\alpha\beta)_{j1}}{2}, \qquad \tau_{00} = \tau^2, \qquad \tau_{11} = \delta^2/2.$$

Combining Eqs. (13.22) and (13.23) yields the single model

$$y_{ij} = \Theta_{00} + \Theta_{10}X_{1ij} + u_{0j} + u_{1j}X_{1ij} + r_{ij}. \tag{13.24}$$

13.5.2.2 Results

The results of the analysis (Table 13.9) are again mathematically identical to the results based on the usual ANOVA. The reader will again notice that squaring the t statistic associated with the practice contrast will yield the corresponding F of the ANOVA in Table 13.8; and dividing each chi-square (for the tutor-effect variance and the tutor-by-practice variance) by its degrees of freedom will yield the corresponding F in the ANOVA table.

13.5.3 Generalization

We again turn to the analysis of the Thai primary school data, and our goal now is to examine the relationship between pre-primary school experience and total achievement. Within at least some of our 103 classrooms, there will be some children who have experienced pre-primary education and some children who have not. The proportion having pre-primary experience in the sample is .38. We may view pre-primary experience (yes, no) as crossed with classrooms, meaning that the design is a 103-by-2 crossed design. Of course, the proportion having pre-primary education will vary from class to class so that the design is unbalanced. Indeed, in quite a few classrooms there is no variation in pre-primary experience. To be meaningful, the estimate of the effect of pre-primary experience must be adjusted for other background variables. For example, past research has shown that Thai children of high social class are substantially more likely than Thai children of low social class to have the benefit of pre-primary experience (Raudenbush, Kidchanapanish, and Kang, 1991). Given

Table 13.9 Crossed Two-Factor Results Based on the Hierarchical Linear Model

Fixed effects			
Parameter	Coefficient	Standard error	t
Grand mean, Θ_{00}	106.25	3.47	—
Practice contrast, Θ_{10}	39.95	2.31	17.28

Variance components			
Parameter	Estimate	Chi-square	df
τ_{00}	111.65	127.39	9
τ_{11}	44.94	56.65	9
σ^2	33.95		

the unbalance of the design and the need to control for covariates measured both continuously and discretely, use of the mixed two-way ANOVA is inappropriate, and we have utilized a hierarchical linear model instead.

The level-one (within-classroom) model relates the outcome, Y_{ij}, to the effect of pre-primary experience, adjusting for several covariates:

$$y_{ij} = \beta_{0j} + \beta_{1j}(X_{ij} - \overline{X}_{.j}) + \beta_{2j}(SES)_{ij} + \beta_{3j}(Time)_{ij} + \beta_{4j}(Sex)_{ij}$$
$$+ \beta_{5j}(Dialect)_{ij} + \beta_{6j}(Breakfast)_{ij} + r_{ij} , \tag{13.25}$$

where the covariates SES, Time, Sex, Dialect, and Breakfast were already defined in Section 13.4, and the X_{1ij} is an indicator taking on a value of 1 if child i in classroom j has had pre-primary experience and 0 if not. Note that the key independent variable is $X_{ij} - \overline{X}_{.j}$, representing the contrast between those having and those not having pre-primary experience *within classroom*, j. By centering the indicator around its classroom mean, we guarantee that student-level predictor is orthogonal to all school-level variables and therefore represents the within-classroom contrast.

At level two (between classrooms), only the intercept and the effect of pre-primary education were allowed to be random. Other covariates could have been specified to have random coefficients, but they were not; and classroom-level predictors could have been utilized in this analysis, but they were not. Thus, the level-two model is

$$\beta_{0j} = \Theta_{00} + u_{0j} ,$$
$$\beta_{1j} = \Theta_{10} + u_{1j} , \tag{13.26}$$
$$\beta_{qj} = \Theta_{q0} \quad \text{for } q > 1,$$

where the random effects u_{0j} and u_{1j} are assumed multivariate normal as above.

The results (Table 13.10) indicate that, controlling for the covariates, children attending pre-primary school scored, on average, 10.4 points higher than did their classmates who did not attend pre-primary school, $t = 2.02$, $p < .05$. This represents a small effect of about 10% of a standard deviation overall and 16% of the within-classroom standard deviation. Results show no indication that the pre-primary education effect varies across classrooms, as indicated by a chi-square of 57.4 (to be compared to the percentiles of a central chi-square with 58 degrees of freedom) equivalent to an F of 57.4/58 < 1.00. The results do suggest that significant mean differences remain across classrooms, even after controlling for the covariates, as indicated by a chi-square of 1206.8 with 58 degrees of freedom, $p < .001$. Note that the chi-squares are based on only 59 of 103 schools because only 59 schools had sufficient variation in pre-primary experience to esti-

Table 13.10 Two-Factor Nested Results for Thailand Classroom Data Based on the Hierarchical Linear Model

	Fixed effects		
	(Unconstrained model)		
Predictors	Coeff	se	t
School/classroom level			
Intercept	−36.00	8.33	—
Student level			
Pre-primary education	10.38	5.12	2.02
SES	11.28	3.97	2.84
Time to school	−6.62	2.66	−2.48
Sex	−10.44	2.85	−3.66
Dialect	27.18	9.08	2.99
Breakfast	11.85	5.12	2.32

	Variance components			
Parameter	Estimate	Chi-square	Degrees of freedom	p
$\hat{\tau}_{00}$	3520	1206.8	58	.000
$\hat{\tau}_{11}$	236	57.4	58	.500
$\hat{\sigma}^2$	3997			

mate an ordinary least squares regression. In this situation, a better test of the significance for the interaction between classroom and pre-primary education is a likelihood ratio test.

To conduct this test, the analyst reruns the model with the variance of the pre-primary effect constrained to zero and then compares the goodness-of-fit statistics for the two models. Goodness of fit is measured by the deviance, that is, −2 times the log of likelihood function evaluated at the maximum. We did this, and the results were as follows. First, in the "unconstrained" model, the covariance component estimates were

$$\hat{\tau}_{00} = 3520,$$

$$\hat{\tau}_{11} = 236,$$

$$\hat{\tau}_{01} = 9,$$

$$\hat{\sigma}^2 = 3997,$$

yielding a deviance of 2278.2. Under the constrained model, two parameters, $\hat{\tau}_{11}$ and $\hat{\tau}_{01}$ were set to zero with the remainder of the model unchanged; and the variance components estimates were

$$\hat{\tau}_{00} = 3517,$$
$$\hat{\sigma}^2 = 4017,$$

yielding a deviance of 2782.4, meaning that the increase in deviance was only .2 when the model was simplified. This increase in deviance is approximately distributed as chi-square with degrees of freedom equal to the difference in the number of parameters estimated by the two models under the null hypothesis that the constrained model is adequate. Clearly, this difference between deviances is nonsignificant. Under the simpler model, the pre-primary contrast was unchanged at 10 points, but the t-test was a bit larger ($t = 2.33$ as opposed to 2.02), reflecting the greater precision of the simpler model.

13.6 RANDOMIZED BLOCK (AND REPEATED MEASURES) DESIGNS

Randomized block designs will typically involve mixed models having both fixed and random effects. Blocks (often subjects) will typically be viewed as having random levels, and, within blocks, there will commonly be a fixed-effects design. The fixed effects may represent experimental treatment levels, or, in longitudinal studies, they may involve polynomial trends. When the within-blocks design is identical for every block, and there are no missing data or within-block covariates, classical ANOVA procedures will often be appropriate. Under the same circumstances, multivariate ANOVA will allow more flexible assumptions regarding the variances and covariances of the within-block observations. However, when the within-blocks design varies, as (1) when the number or spacing of time series observations differs in a panel study, (2) when some blocks have missing data, or (3) when within-block covariates are present, these classical approaches are problematic. Under these circumstances, analysis by means of a hierarchical linear model offers a more flexible approach.

We first consider a simple randomized block design from the standpoint of the ANOVA and show how to formulate the hierarchical model to duplicate the results. We will then consider a generalization in which simplification of the model enables one to disentangle blocks-by-treatments variance from within-cell variance.

13.6.1 Randomized Block Design: Classical ANOVA Approach

13.6.1.1 The Model

We consider a randomized block design in which there are no between-blocks factors. This is a two-way crossed design (blocks crossed with a fixed-effects factor) identical to that studied in Section 13.5 except that there are no replications within cells. Each block is observed once and only once under each treatment or on each occasion.

The standard ANOVA model for the randomized block design may be written as

$$y_{ij} = \mu + \alpha_i + \pi_j + e_{ij} , \tag{13.27}$$

where y_{ij} is the outcome for block j under treatment i ($i = 1, \ldots, p$; $j = 1, \ldots, J$), α_i is the effect of treatment i, π_j is the effect of block j, and the error, e_{ij}, has two components: $(\alpha\pi)_{ij}$ and r_{ij}. That is, because there is only one observation per cell, the treatment-by-blocks interaction effect, $(\alpha\pi)_{ij}$, is confounded with the within-cell error, r_{ij}. Typical assumptions of the model are that the random components are mutually independent and normally distributed:

$$\pi_j \sim N(0, \tau^2), \qquad e_{ij} \sim N(0, \sigma^2),$$

although, of course, the error σ^2 confounds variance attributable to block-by-treatment interaction and variance attributable to within-cell error. We initially assume that the blocks-by-treatment interaction effects are null.

13.6.1.2 Example and Results

The data in Table 13.11 are from Kirk (p. 244) and involve eight blocks, each observed under four treatment conditions. For ease of understanding, let us assume that the treatment is the duration of instruction prior to a cognitive test and that the outcome is the number correct on the test. We assume that 32 subjects have been classified into eight blocks of four each based on a pretest of cognitive ability. Within each block, subjects are then assigned at random to have 1, 2, 3, or 4 minutes of instruction prior to the test. We view blocks as random and treatments as fixed. The table shows the results, with the variation attributable to treatment effects explained by linear, quadratic, and cubic polynomial trend components.

The results indicate a highly statistically significant effect of duration, most of which is accounted for by the linear trend. Variance estimates are given by equating the expected mean squares to their observed values as

Table 13.11   Repeated Measures Data and Source Table (Additive Effects Model)

Data

Treatment (duration)

Block	1	2	3	4
1	3	4	7	7
2	6	5	8	8
3	3	4	7	9
4	3	3	6	8
5	1	2	5	10
6	2	3	6	10
7	2	4	5	9
8	2	3	6	11

ANOVA source table

Source	df	Sum of squares	Mean square	E(mean square)	F
Duration	3	194.5	64.8	$\sigma^2 + n \, \Sigma \, \alpha_k^2/(p - 1)$	47.78
Linear	1	184.9	184.9		136.26
Quadratic	1	8.0	8.0		5.90
Cubic	1	1.6	1.6		1.18
Blocks	7	12.5	1.8	$\sigma^2 + p\tau^2$	1.32
Error	21	28.5	1.4	σ^2	
Total	31	235.5			

in the one-way ANOVA, yielding

$$\hat{\sigma}^2 = MS(\text{error}) = 1.36,$$

$$\hat{\tau}^2 = [MS(\text{blocks}) - MS(\text{error})]/p = .11.$$

A test of the null hypothesis of no block effects, i.e., H_0: $\tau^2 = 0$, is given by the ratio of mean squares (blocks to error) yielding $F(7, 21) = 1.32$, implying that little evidence of block effects exists. This inference, of course, is fragile in light of the assumption needed to justify it—that no treatment-by-blocks interaction exists.

13.6.2   Analysis by Means of a Hierarchical Linear Model

13.6.2.1   The Model

Specification of the hierarchical linear model for the randomized block design is similar to specification for the two-factor crossed design discussed in Section 13.5. The difference is that in the case of the randomized block design, there is no replication within cells. Hence, the model must be simplified.

According to the level-one (within-block) model, the outcome depends on polynomial trend components plus error:

$$y_{ij} = \beta_{0j} + \beta_{1j}(\text{LIN})_{ij} + \beta_{2j}(\text{QUAD})_{ij} \qquad (13.28)$$
$$+ \beta_{3j}(\text{CUBE})_{ij} + r_{ij} , \qquad r_{ij} \sim N(0, \sigma^2),$$

where y_{ij} is the outcome for subject i in block j; β_{0j} is the mean for block j; $(\text{LIN})_{ij}$ assigns the linear contrast values $(-1.5, -.5, .5, 1.5)$ to durations $(1, 2, 3, 4)$, respectively; $(\text{QUAD})_{ij}$ assigns the quadratic contrast values $(.5, -.5, -.5, .5)$; $(\text{CUBE})_{ij}$ assigns the cubic contrast values $(-.5, 1.5, -1.5, .5)$; β_{1j}, β_{2j}, and β_{3j} are the linear, quadratic, and cubic regression parameters, respectively; and r_{ij} is the within-cell error.

Notice that with only four observations per block and four regression coefficients (β's) in the level-one model, no degrees of freedom remain to estimate within-cell error. If we assume, however, that the contrast values do not vary across blocks (no blocks-by-duration interactions), we can treat the trend parameters as fixed, yielding the level-two (between-blocks) model

$$\beta_{0j} = \Theta_{00} + u_{0j} , \qquad u_{0j} \sim N(0, \tau^2),$$
$$\beta_{1j} = \Theta_{10} ,$$
$$\beta_{2j} = \Theta_{20} , \qquad\qquad\qquad\qquad\qquad (13.29)$$
$$\beta_{3j} = \Theta_{30} ,$$

where Θ_{00} is the grand mean and u_{0j} is the unique effect of block j assumed normally distributed with a mean of zero and a variance of τ^2. The coefficients, β_{1j}, β_{2j}, and β_{3j}, are constrained to be invariant across blocks.

The correspondences between the hierarchical model and the ANOVA model are

$$\Theta_{00} = \mu,$$
$$\Theta_{10} = (-1.5\alpha_1 - .5\alpha_2 + .5\alpha_3 + 1.5\alpha_4),$$
$$\Theta_{20} = (.5\alpha_1 - .5\alpha_2 - .5\alpha_3 + .5\alpha_4),$$
$$\Theta_{30} = (-.5\alpha_1 + 1.5\alpha_2 - 1.5\alpha_3 + .5\alpha_4), \text{ and}$$
$$u_{0j} = \pi_j .$$

Combining Eqs. (13.28) and (13.29) yields the single model

$$y_{ij} = \Theta_{00} + \Theta_{10}(\text{LIN})_{ij} + \Theta_{20}(\text{QUAD})_{ij}$$
$$+ \Theta_{30}(\text{CUBE})_{ij} + u_{0j} + r_{ij} . \qquad (13.30)$$

This model, like the ANOVA model above, assumes that the variance-covariance matrix of the repeated measures is compound symmetric: $Var(Y_{ij})$ $= \tau^2 + \sigma^2$; $Cov(Y_{ij}, Y_{i'j}) = \tau^2$.

13.6.2.2 Results

The results of the analysis via the hierarchical model (Table 13.12) are again mathematically identical to the results based on the usual ANOVA. The HLM computer program allows the investigator to specify an omnibus test for the effect of duration which is the combined null hypothesis

$$H_0: \Theta_{10} = \Theta_{20} = \Theta_{30} = 0.$$

The result is a large-sample chi-square test with 3 df, which, in this case, takes on a value of 143.33. Note that this value divided by 3 yields the omnibus $F(3, 21) = 47.78$ of the ANOVA. The reader will again notice that squaring the t statistic associated with each of the trend components yields the corresponding F in the ANOVA table (Table 13.11). Based on the hierarchical model, the $t(LIN) = 11.67$, when squared, yields the ANOVA $F(LIN) = 136.26$, and similarly, the t values for the quadratic and cubic trends can be converted into F values. The estimates of σ^2 and τ^2 are also identical. Finally, the null hypothesis that $\tau^2 = 0$ is tested in the hierarchical analysis by means of the chi-square of 9.21, which, when divided by its degrees of freedom, yields the ANOVA $F(3, 21)$ of 1.32.

Table 13.12 Randomized Block Results Based on the Hierarchical Linear Model (Additive Effects Model)

	Fixed effects		
Parameter	Coefficient	Standard error	t
Grand mean, Θ_{00}	5.38	.236	—
Linear effect, Θ_{10}	2.15	.184	11.67
Quadratic effect, Θ_{20}	1.00	.412	2.43
Cubic effect, Θ_{30}	−0.20	.184	−1.09
	Variance components		
Parameter	Estimate	Chi-square	df
τ^2	.107	9.21	7
σ^2	1.36		

13.6.3 Generalization: A More Credible Model

The hierarchical results duplicate the ANOVA results because they are based on the same statistical assumptions, the most fragile of which is that block-by-treatment interaction is null. This assumption does not endanger the statistical basis of the tests of the fixed effects (the effects of duration and the associated trends) (Kirk, p. 249). However, if block-by-treatment interactions are large and falsely assumed zero, two consequences would be apparent. First, the generality of the inferences about duration would be endangered in that, if the effect of treatment varies across blocks, the effect is less general than if the effect is invariant across blocks. Second, the estimate of the variance of the block effects and the associated test could be wrong, perhaps misleading future researchers into believing that the basis upon which the blocks were formed was irrelevant to the outcome.

One feature of the results offers an opportunity to escape this impasse. We see in Table 13.12 that there is no evidence of a cubic trend in the data. If the cubic trend is dropped from the model, both the linear and quadratic trends can be allowed to vary randomly across blocks while still leaving a degree of freedom within each block to estimate the within-cell variance. In fact, the results of such an analysis (not reported here) indicated significant main effects of both the linear and quadratic trends. Although there was quite strong evidence that the linear trend varied significantly across blocks, there was no evidence that the quadratic trend varied across blocks. These results laid the basis for estimation of a more parsimonious and informative model, the results of which are displayed in Table 13.13. In this model, there is a random linear effect and a fixed quadratic effect, while the cubic effect is dropped entirely. Using this model, one can simultaneously account for between-block variance, block-by-treatment variance, and within-cell variance.

The level-one (within-blocks) model becomes

$$y_{ij} = \beta_{0j} + \beta_{1j}(\text{LIN})_{ij} + \beta_{1j}(\text{QUAD})_{ij} + r_{ij}, \qquad (13.31)$$

$$r_{ij} \sim N(0, \sigma^2),$$

which is the same model as Eq. (13.28) except that the cubic trend has been dropped. The level-two (between-blocks) model is now

$$\beta_{0j} = \Theta_{00} + u_{0j},$$

$$\beta_{1j} = \Theta_{10} + u_{1j}, \text{ and} \qquad (13.32)$$

$$\beta_{2j} = \Theta_{20}.$$

Here the base and linear trend vary randomly around their average effects while the quadratic effect is constrained to be constant across blocks. Var-

Table 13.13 Randomized Block Results Based on the Hierarchical Linear
Model (Linear-by-Blocks Interaction Model)

Fixed effects			
Parameter	Coefficient	Standard error	t
Grand mean, Θ_{00}	5.38	.236	—
Linear effect, Θ_{10}	2.15	.263	8.19
Quadratic effect, Θ_{20}	1.00	.300	3.34

Variance-covariance components		
Parameter	Estimate	Chi-square	df
τ_{00}	.268	17.41	7
τ_{11}	.408	26.88	7
τ_{01}	− .320		
σ^2	.718		

iances to be estimated include τ_{00}, the variance of the block effects; τ_{11}, the variance of the linear trends (block-by-treatments variance); and σ^2, the within-cell variance. We also estimate τ_{01}, the covariance between u_{0j}, the block effect on the base, and u_{1j}, the block effect on the linear trend.

The results in Table 13.13 have several noteworthy features. First, we see a very strong negative covariance between the block mean and the linear effect: blocks with high means have small linear trends. The covariance $\hat{\tau}_{01} = -.32$ is equivalent to a correlation of $-.97$, which could result from a ceiling effect on the outcome. Given this relationship, it is clear that the linear trend must be modeled as varying across blocks; indeed, a likelihood ratio test, comparing a model with a fixed linear trend to a model with a random linear trend, produced clear evidence in favor of the random linear trend.

Second, with the model respecified, the between-blocks variance is significantly greater than zero (note the chi-square of 17.4 with 7 degrees of freedom). In contrast, the compound-symmetry model (Table 13.12) seemed to indicate no significant between-block variation. However, that model was based on the assumption of no blocks-by-treatment interaction, an assumption strongly undermined by our reanalysis.

Third, the level-one random error variance, σ^2, is considerably smaller in the reanalysis than it had been in Table 13.12. Recall that this variance confounds within-cell variance and block-by-treatment variance. The removal of the block-by-linear variance allowed isolation of the within-cell variance.

Other error structures may also be evaluated by means of hierarchical models. For example, Goldstein (1987) shows how to formulate and estimate a model in which the within-cell error is dropped and the linear, quadratic, and cubic effects along with the mean are assumed to vary randomly across blocks. This is really a fully multivariate model formulated via a hierarchical linear modeling approach.

13.7 A SCHEME FOR CLASSIFYING DESIGNS

This chapter has illustrated how two-level hierarchical linear models generalize analytic approaches to a number of common experimental designs. The hierarchical linear model itself is a specific case of a more general mixed linear model in that the hierarchical model applies only to random effects that are nested. Lindley and Smith (1972), Dempster, Rubin, and Tsutakawa (1981), and Goldstein (1987) have shown how a general class of crossed random effects models can be formulated and estimated. Kang (1991) has developed a general computing algorithm for such crossed random effects models. As these models become accessible, it becomes possible to develop a fairly complete taxonomy for generalizing analysis for experimental designs.

Table 13.14 Common Experimental Designs and Analytic Approaches That Generalize Their Applicability

Experimental design	Generalized analytic approach
One-way ANOVA	
Fixed-effects model	ANOVA or multiple regression
Random-effects model	Two-level hierarchical model
Two-way ANOVA (within cell replications)	
Both effects fixed	Multiple regression
One fixed, one random	Two-level hierarchical model
Both random	Crossed random effects model
Randomized block model (no within-cell replications)	
Blocks random, treatments fixed	Two-level hierarchical model
Split-plot design	
Blocks random, within- and between-block factors fixed	Two-level hierarchical model
Blocks random, within-block factor fixed, between-block factor random	Three-level hierarchical model
Blocks random, within-block factor random, between-block factor fixed	Crossed random effects model

Table 13.14 lists common experimental designs and, for each design, it identifies the statistical model that generalizes analytic possibilities. That is, the recommended approach (1) captures the relevant sources of variation and covariation, (2) allows for unbalanced data at any level, and (3) allows specification of both continuous and discrete predictors (including the possibility of both fixed and random regression coefficients for those predictors).

At present, reasonably user-friendly software is available for all the designs listed in Table 13.14 except for the crossed random effects models. Undoubtedly, that gap will soon be filled. The implication of these advances seems to be that the teaching of research designs linked to appropriate analytic models should soon be revolutionized. The conceptual strengths of current experimental and quasi-experimental design courses will be matched by the data analytic flexibility of a broad family of approaches integrated under the rubric of a general mixed linear model with crossed or nested random effects.

ACKNOWLEDGMENT

Research reported here was supported by grants from the John D. and Catherine T. MacArthur Foundation and the National Institute of Justice to the Program on Human Development and Criminal Behavior.

REFERENCES

Aitkin, M. and Longford, N. (1986). Statistical Modeling Issues in School Effectiveness Studies. *J. Roy. Statist. Soc.*, *A149*: 1–43.

Becker, B. J. (1988). Synthesizing Standardized Mean Change Measures. *Brit. J. Math. Statist. Psych.*, *41*: 257–278.

Bock, R. D. (1975). *Multivariate Statistical Methods in Behavioral Research*. McGraw-Hill, New York.

Bryk, A. S. and Raudenbush, S. W. (1987). Application of Hierarchical Linear Models to Assessing Change. *Psych. Bull.*, *101*: 147–158.

Bryk, A. S. and Raudenbush, S. W. (1988). Toward a More Appropriate Conceptualization of Research on School Effects: A Three-Level Hierarchical Linear Model. *Amer. J. Educ.*, *97*: 65–108.

Bryk, A. S. and Raudenbush, S. W. (1992). *Hierarchical Linear Models in Social and Behavioral Research: Applications and Data Analysis Methods*. Sage, Beverly Hills, California.

Bryk, A. S., Raudenbush, S. W., Seltzer, M., and Congdon, R. (1988). *An Introduction to HLM: Computer Program and Users' Manual*. University of Chicago Department of Education, Chicago.

Burstein, L. (1980). The Analysis of Multi-Level Data in Educational Research and Evaluation. *Rev. Res. Educ.*, *8*: 158–233.

Cohen, J. and Cohen, P. (1975). *Applied Multiple Regression/Correlation Analysis for the Behavioral Sciences.* Wiley, New York.

Cronbach, L. J. and Webb, N. (1975). Between- and Within-Class Effects in a Reported Aptitude-by-Treatment Interaction: Reanalysis of a Study by G. L. Anderson. *J. Educ. Psych.*, 6: 717–724.

DeLeeuw, J. and Kreft, I. (1986). Random Coefficient Models for Multilevel Analysis. *J. Educ. Statist.*, 11: 57–85.

Dempster, A. P., Laird, N. M., and Rubin, D. B. (1977). Maximum Likelihood from Incomplete Data via the EM Algorithm. *J. Roy. Statist. Soc.*, B39: 1–8.

Dempster, A. P., Rubin, D. B., and Tsutakawa, R. K. (1981). Estimation in Covariance Components Models. *J. Amer. Statist. Assoc.*, 76: 341–353.

Efron, B. and Morris, C. (1975). Data Analysis Using Stein's Estimator and Its Generalizations. *J. Amer. Statist. Assoc.*, 74: 311–319.

Goldstein, H. (1987). *Multilevel Models in Educational and Social Research.* Oxford University Press, London.

Hartley, H. O. and Rao, J. N. K. (1967). Maximum Likelihood Estimation for the Mixed Analysis of Variance Model. *Biometrika*, 54: 93–108.

Kang, S. J. (1991). A Mixed Linear Model for Unbalanced Two-Way Crossed Multilevel Data with Estimation via the EM Algorithm. Unpublished Doctoral Dissertation, Michigan State University College of Education, East Lansing.

Kirk, R. E. (1982). *Experimental Design: Procedures for the Behavioral Sciences* (2nd ed.). Brooks/Cole, Belmont, California.

Laird, N. M. and Ware, J. H. (1982). Random-Effects Models for Longitudinal Data. *Biometrics*, 38: 963–974.

Lindley, D. V. and Smith, A. F. M. (1972). Bayes Estimates for the Linear Model. *J. Roy. Statist. Soc.*, B34: 1–41.

Longford, N. T. (1987). A Fast Scoring Algorithm for Maximum Likelihood Estimation in Unbalanced Models with Nested Random Effects. *Biometrika*, 74: 817–827.

Mason, W. M., Wong, G. Y., and Entwisle, B. (1984). Contextual Analysis Through the Multilevel Linear Model. *Sociological Methodology 1983–1984*. Leinhardt, S. (ed.). Jossey Bass, San Francisco, pp. 702–103.

Morris, C. N. (1983). Parametric Empirical Bayes Inference: Theory and Applications. *J. Amer. Statist. Assoc.*, 78: 47–65.

Nelson, G. K. (1976). Concomitant Effects of Visual, Motor, and Verbal Experiences in Young Children's Concept of Development. *J. Educ. Psych.*, 68: 466–473.

Raffe, D. (1991). Assessing the Impact of a Decentralized Initiative: The British Technical and Vocational Initiative. *Pupils, Classrooms and Schools: International Studies of Schooling from a Multilevel Perspective*. Raudenbush, S. W. and Willms, J. D. (eds.). Academic Press, New York.

Raudenbush, S. W. (1984). Educational Applications of a Hierarchical Linear Model. Unpublished Doctoral Dissertation. Harvard University Graduate School of Education.

Raudenbush, S. W. (1988). Educational Applications of Hierarchical Linear Models: A Review. *J. Educ. Statist.*, 13: 85–116.

Raudenbush, S. W. and Bryk, A. S. (1985). Empirical Bayes Meta-Analysis. *J. Educ. Statist.*, *10*: 75–98.

Raudenbush, S. W. and Bryk, A. S. (1986). A Hierarchical Model for Studying School Effects. *Soc. Educ.*, *59*: 1–17.

Raudenbush, S. W., Eamsukkawat, S., Di-Ibor, I., Kamali, M., and Taoklam, W. (1991). On the Job Improvements in Teacher Competence: Policy Options and Their Effects on Teaching and Learning in Thailand. Paper presented at the Population and Human Resources Department of the World Bank, Washington, D.C.

Raudenbush, S. W., Kidchanapanish, S., and Kang, S. J. (1991). The Effects of Preprimary Access and Quality on Educational Achievement in Thailand. *Comp. Educ. Rev.*, *35*: 255–273.

Robinson, W. S. (1950). Ecological Correlations and the Behavior of Individuals. *Amer. Soc. Rev.*, *15*: 351–357.

Rosenberg, B. (1973). Linear Regression with Randomly Dispersed Parameters. *Biometrika*, *60*: 65–72.

Ware, J. H. (1985). Linear Models for the Analysis of Longitudinal Studies. *Amer. Statist.*, *39*: 95–101.

Willms, J. D. and Raudenbush, S. W. (1989). A Longitudinal Hierarchical Linear Model for Estimating School Effects and Their Stability. *J. Educ. Meas.*, *26*: 209–232.

Winer, B. J. (1971). *Statistical Principles in Experimental Design* (2nd ed.). McGraw-Hill, New York.

14
Analysis of Categorical Response Variables

KINLEY LARNTZ University of Minnesota, St. Paul, Minnesota

14.1 INTRODUCTION

Analysis of variance and multiple linear regression methods were developed for the classic response variable, a normally distributed random vector of independent observations with equal variances. For moderate departures from the ideal assumptions, the usual analysis of variance and regression methods still work well. For more extreme departures from the ideal assumptions, transformations allow us to use the standard techniques. If, however, the response variable is discrete, specially tailored techniques yield more efficient inferences. This chapter treats the special case of a categorical response variable.

14.1.1 Discrete Response Variables

Discrete response variables are those whose possible outcome values are at most a countable set. Examples include categorical responses with two or more outcomes and numerical responses that represent counts. Discrete responses may be either intrinsically discrete or simply measured as discrete. Intrinsically discrete responses are discrete by the nature of the response. For instance, if the response is a choice of candidates in an opinion survey in a Presidential race, the choice of brand in a marketing study, or the answer to a multiple choice question ("none of the above"), then the response is intrinsically discrete. (It is assumed that responses are mutually exclusive and exhaustive; that is, each subject responds with one and only one category.) Also intrinsically discrete are responses that are counts. The number of accidents at a specific intersection, the number of crimes committed by a parolee, and the number of seals trapped by a drift net are, by their nature, discrete responses.

In contrast to intrinsically discrete variables, many investigators create discrete variables from continuous responses. Blood pressure readings may be classified as "normal," "borderline," or "high." Ideally, these classifications have scientific meaning and justification, but often the categories are arbitrarily determined. In such cases, the investigator runs the risk of getting differing conclusions based on differing categorizations. Similarly, many questionnaires are constructed to elicit a discrete response for a continuous variable. Probably the most common example would be the typical question on income. Rather than ask for an exact figure, the respondent is asked to categorize income in broad intervals (less than $10,000, $10,000 to $25,000, more than $25,000). The information in the response obviously depends on choice of cutpoints and number of categories.

14.1.2 Explanatory Variables

In ordinary analysis of variance, multiple linear regression, or the analysis of covariance, the expectation of a continuous response variable is modeled as a linear function of model parameters, usually termed coefficients. The differentiation among the three techniques is typically in the nature of the explanatory variables used in each. Analysis of variance has categorical explanatory variables, multiple linear regression has continuous explanatory variables, and analysis of covariance has both categorical and continuous explanatory variables. The so-called general linear model unifies the treatment of these models by parameterizing the categorical explanatory variables as a set of dummy variables. Of course, nonlinear relationships are also used when the investigator knows (from theoretical considerations) that a nonlinear equation relates the expectation of the continuous response variable to the explanatory variables.

In modeling categorical response variables, the left-hand side of the model equation is usually taken as a function of the probabilities for the categorical responses. The right-hand side of the model equation is usually taken as a linear function of the explanatory variables in the same way as for a continuous response variable. Specifically, we want to relate functions of the probabilities for the categorical responses to continuous and/or categorical explanatory variables. Direct linear functions do not work well, but transformations of the probabilities may be modeled as linear functions of parameters in the same way as is the general linear model for continuous response variables. We will see examples of how to construct convenient, computationally tractable functional relationships.

14.1.3   Chapter Outline

This chapter gives some methods that have proved useful for analyzing categorical response variables. Section 14.2 presents models for binomial responses. Model selection issues are treated in Section 14.3. Multinomial response variables are considered in Section 14.4. Section 14.4.5 presents a special computation algorithm for estimating multinomial response models with computer programs intended for binary response models only. Section 14.5 gives some suggestions for further study.

14.2   MODELS FOR BINARY RESPONSES

Discrete responses with two outcomes are extremely common:

1. A patient may have heart disease, or may be free of heart disease;
2. An animal may develop a tumor when fed a particular diet, or may remain free from tumors;
3. A student may get the correct answer to a true-false question or not; and
4. The urine sample from a patient in a chemical dependence program may test positive or show no signs of drug use.

Suppose we have n observations. As usual, we assume for purposes of model development that the responses for the n observations are statistically independent of each other. For the ith observation, let Y_i be the response variable and, for convenience, suppose Y_i may take on the two values 1 and 2. For instance, $Y_i = 1$ may indicate a patient with heart disease, while $Y_i = 2$ indicates the patient does not have heart disease. Let p_i denote the probability that the ith observation is 1,

$$p_i = P(Y_i = 1), \qquad i = 1, \ldots, n. \tag{14.1}$$

The goal is to model p_i as a function of explanatory variables.

14.2.1   Binary Response with One Categorical Explanatory Variable

Consider first the case of one categorical explanatory variable A with C categories. For notation, let A_i be the category for the ith case. The data then consist of n independent observations, $(A_1, Y_1), (A_2, Y_2), \ldots, (A_n, Y_n)$. This case is well known to everyone, but we will set up the structure in detail and then later extend that structure to other situations. Traditionally, data for this case are arranged in a $2 \times C$ contingency table like Table 14.1. x_{rj} is the number of cases (out of n) with response $Y_i = r$ with

Table 14.1 Contingency Table for Binary Response Variable by Categorical
Explanatory Variable

Response	Explanatory category						
	1	2	$\cdots$	c	$\cdots$	C	Total
1	x_{11}	x_{12}	$\cdots$	x_{1c}	$\cdots$	x_{1C}	x_{1+}
2	x_{21}	x_{22}	$\cdots$	x_{2c}	$\cdots$	x_{2C}	x_{2+}
Total	x_{+1}	x_{+2}	$\cdots$	x_{+c}	$\cdots$	x_{+C}	n

category $A_i = j$. x_{r+} is the total number of cases with response $Y_i = r$, and x_{+j} is the number of cases with explanatory variable $A_i = j$. This is the direct analogue of the one-way analysis of variance for a continuous response variable. Letting π_j be the probability of a 1 for category j, statistical analysis for this case proceeds straightforwardly using one of the usual chi-squared tests to check homogeneity of the C proportions:

$$H_0 : \pi_1 = \pi_2 = \cdots = \pi_C . \tag{14.2}$$

Table 14.2 gives the well known estimated cell expected values $\hat{m}_{rj}$ under the null hypothesis of equal π_i's. The Pearson chi-squared statistic X^2 is calculated as

$$X^2 = \sum_{r,j} \frac{(x_{rj} - \hat{m}_{rj})^2}{\hat{m}_{rj}}. \tag{14.3}$$

An alternative statistic is the loglikelihood ratio chi-squared statistic G^2,

Table 14.2 Estimated Cell Expected Values for the Hypothesis of
Homogeneity of Proportions for Data in Table 14.1

Response	Explanatory category						
	1	2	$\cdots$	c	$\cdots$	C	Total
1	$\dfrac{x_{1+}x_{+1}}{n}$	$\dfrac{x_{1+}x_{+2}}{n}$	$\cdots$	$\dfrac{x_{1+}x_{+c}}{n}$	$\cdots$	$\dfrac{x_{1+}x_{+C}}{n}$	x_{1+}
2	$\dfrac{x_{2+}x_{+1}}{n}$	$\dfrac{x_{2+}x_{+2}}{n}$	$\cdots$	$\dfrac{x_{2+}x_{+c}}{n}$	$\cdots$	$\dfrac{x_{2+}x_{+C}}{n}$	x_{2+}
Total	x_{+1}	x_{+2}	$\cdots$	x_{+c}	$\cdots$	x_{+C}	n

which is calculated as

$$G^2 = 2 \sum_{r,j} x_{rj} \log \frac{x_{rj}}{\hat{m}_{rj}}, \tag{14.4}$$

with 0 contributed to the sum if $x_{rj} = 0$. If the null hypothesis holds, both statistics follow an asymptotic chi-squared distribution with $C - 1$ degrees of freedom. Indeed, under the null hypothesis the two statistics are asymptotically equivalent (Bishop, Fienberg, and Holland, 1975).

Following the analysis of variance analogue, the usual chi-squared tests compare two models. The first model specifies that the probability of a 1 for each case is constant, not depending on the value of the categorical variable:

$$\text{Model I}: \quad p_i = \pi, \quad i = 1, \ldots, n. \tag{14.5}$$

The second model specifies that the probability of a 1 for each case depends on the category of the explanatory variable:

$$\text{Model II}: \quad p_i = \pi_{A_i}, \quad i = 1, \ldots, n. \tag{14,6}$$

14.2.2   Example: Sherman–Berk Spouse Assault Data

As an example of a binary response with a categorical explanatory variable, consider the data from Berk and Sherman (1988) given in Table 14.3. The data are from a randomized experiment conducted in Minneapolis in 1981–82 to determine the appropriate police response for misdemeanor spousal assault. Three hundred thirty cases were randomized into three treatments labeled here as Arrest, Advise, and Separate. Seventeen cases were omitted from the analysis and are not tabulated in Table 14.3. For the Arrest treatment, police were instructed to arrest the offender and transport him

Table 14.3 Contingency Table of Evidence of Future Violence Response by Experimental Treatment Assigned for Sherman–Berk Minneapolis Spouse Assault Experiment

	Treatment			
	Separate	Advise	Arrest	Total
Future violence	26	21	10	57
No future violence	87	87	82	256
Total	113	108	92	313

or her to jail. For the Advise treatment, police were to calm the situation down and perhaps provide some form of mediation for the couple. In the Separate treatment, the police were to order the offender away from the premises for 8 hours. The random assignment to treatment is a categorical explanatory variable. The response variable was whether or not there was evidence of repeated future violence. The goodness-of-fit statistics for testing equality of proportions for the three treatments are $X^2 = 5.19$ and $G^2 = 5.53$. Comparing each to the chi-squared distribution with 2 degrees of freedom, we get p values of .075 and .063, respectively, indicating borderline statistical significance against the hypothesis of equal treatment effects. Examining Table 14.3, we find that the Arrest treatment results in the smallest proportion of cases with evidence of future violence, followed by the Advise treatment. The Separate treatment has the highest proportion of cases with evidence of future violence. We will refit these data with a logistic regression model in Section 14.2.7. Berk and Sherman (1988) and Sherman and Berk (1984) give more details on this experiment.

14.2.3   Loglikelihood Ratio Chi-Squared Statistic

A general method of comparing two models, one of which is a subset of the other, is to use a likelihood ratio test. For the normal theory general linear model, least squares estimates correspond to maximum likelihood estimates and the F test measures the relative sizes of the residual sums of squares for the two models. For discrete response models, following the likelihood ratio procedure leads to the loglikelihood ratio chi-squared test statistic, G^2. We will now develop G^2 for comparing Model I to Model II, but the development applies to any pair of models in which one model is a subset of the other.

Consider an alternative contingency table display of the response data given in Table 14.4. The row variable corresponds to cases (individual observations), and the column variable corresponds to responses. The values in the table are indicator variables that are either 1 or 0, depending on whether the response Y_i equals the category heading the column,

$$I(Y_i = r) = \begin{cases} 1 & \text{if } Y_i = r, \\ 0 & \text{otherwise.} \end{cases} \tag{14.7}$$

This $n \times 2$ table should be thought of as the basic set of observations for a binary response problem. For the Sherman–Berk data set, Table 14.4 is a 313×2 contingency table with row totals all 1.

For each model, a set of fitted values is estimated by the method of maximum likelihood. Assuming the $I(Y_i = 1)$ are independent Bernoulli

Table 14.4 Alternative Display of Response Data by
Cases for a Binary Response Variable

Case	Response 1	Response 2
1	$I(Y_1 = 1)$	$I(Y_1 = 2)$
2	$I(Y_2 = 1)$	$I(Y_2 = 2)$
.	.	.
i	$I(Y_i = 1)$	$I(Y_i = 2)$
.	.	.
n	$I(Y_n = 1)$	$I(Y_n = 2)$
Total	x_{1+}	x_{2+}

random variables with $P(Y_i = 1) = p_i$, we can construct an estimated $\hat{p}_i$ for each case under the model. For Model I,

$$\hat{p}_i = \frac{x_{1+}}{n}, \quad i = 1, \ldots, n, \tag{14.8}$$

and for Model II,

$$\hat{p}_i = \frac{x_{1A_i}}{x_{+A_i}}, \quad i = 1, \ldots, n. \tag{14.9}$$

Table 14.5 gives the set of fitted values corresponding to the set of observed values given in Table 14.4. The maximized loglikelihood, $\log \lambda$ for a model, is given by

$$\log \lambda = \sum_{i=1}^{n} \{I(Y_i = 1) \log \hat{p}_i + I(Y_i = 2) \log(1 - \hat{p}_i)\}. \tag{14.10}$$

The loglikelihood ratio chi-squared statistic is

$$G^2 = -2 \{\log \lambda(\text{Model } I) - \log \lambda(\text{Model } II)\}. \tag{14.11}$$

G^2 computed from Eq. (14.11) is identical to G^2 computed from Eq. (14.4). Again, if Model I holds, then G^2 is asymptotically distributed as a chi-squared random variable with $C - 1$ degrees of freedom. The degrees of freedom equals the difference in number of parameters for Model II (C π_j's) and Model I (1 π). Large values of G^2 favor Model II over Model I. Equation (14.11) can be applied to any pair of models in which Model I

Table 14.5 Fitted Values Corresponding to Observations in the Alternate Display of Response Data in Table 14.4

	Response	
Case	1	2
1	$\hat{p}_1$	$1 - \hat{p}_1$
2	$\hat{p}_2$	$1 - \hat{p}_2$
⋮	⋮	⋮
i	$\hat{p}_i$	$1 - \hat{p}_i$
⋮	⋮	⋮
n	$\hat{p}_n$	$1 - \hat{p}_n$
Total	x_{1+}	x_{2+}

is a special case of Model II. This is the usual likelihood ratio test for comparing two models, one of which is nested in the other.

14.2.4 Binary Response with One Continuous Explanatory Variable: Simple Logistic Regression

Consider now the case of one continuous explanatory variable for a binary response variable. For notation, let Z_i be the value of the continuous explanatory variable for the ith case. The data then consist of n independent observations, $(Z_1, Y_1), (Z_2, Y_2), \ldots, (Z_n, Y_n)$. The continuous response analogue of this situation is simple regression with one Z (usually denoted as X in simple linear regression) and one Y. In simple linear regression, the expected value of the response Y is modeled as a linear function of Z.

Since the range of a linear function of Z is the entire real line $(-\infty, +\infty)$, we consider modeling a transformation of π as a linear function. A reasonable class of transformations is the set of inverse cumulative probability functions for continuous random variables with support $(-\infty, +\infty)$. For convenience, we standardize the distributions to have median 0. Two commonly used transformations in this family are the inverse standard normal, commonly called the *probit*, and the inverse logistic, commonly called the *logit*. The model for simple probit regression is

$$\text{probit}(p_i) = \Phi^{-1}(p_i) = \beta_0 + \beta_1 Z_i , \tag{14.12}$$

where $\Phi^{-1}(p)$ is the inverse probability function for the standard normal distribution. The model for simple logistic regression is

$$\text{logit}(p_i) = \log \frac{p_i}{1 - p_i} = \beta_0 + \beta_1 Z_i . \tag{14.13}$$

The logit transformation has a direct interpretation that is quite useful in practical work. From Eq. (14.13), the logit is simply the log of the ratio of $P(Y_i = 1)$ to $P(Y_i = 2)$, or simply the log odds of a "1" response to a "2" response. Coefficients (β_j's) in a logistic regression model have a direct interpretation in terms of additive effects on log odds. The exponentiated values of coefficients have a direct interpretation in terms of multiplicative effects on odds.

Both the probit and logit transformations have easy inverses, namely the original cumulative distribution functions. For the probit, the inverse is

$$p = \Phi(\text{probit}), \tag{14.14}$$

where $\Phi(z)$ is the cumulative distribution function for a standard normal variable. For the logit, the inverse is

$$p = \frac{e^{\text{logit}}}{1 + e^{\text{logit}}}. \tag{14.15}$$

Figure 14.1 graphs the inverse transformations, scaled to agree at $p = .50$ and $p = .90$. Also on the graph is the corresponding linear transformation, which indicates the range possible for a linear model without a transformation. Note that the probit and logit transformations yield virtually identical probabilities, with the logit having a slightly longer tail. The linear transformation is totally unsatisfactory if probabilities are very small or very large. Since the probit and logit curves are so close, either generates about the same class of models. In practical problems, there is rarely any important difference in the goodness of fit of the two curves. The only difference is in interpretation. The probit scale corresponds to the z-scale associated with a standard normal variate, and many investigators find this an easy scale to understand. The logit scale corresponds to log odds, which is directly interpretable. The choice between logit and probit is more a matter of personal preference, since there is little practical statistical difference between the two. Because of the direct interpretation of the logit in terms of log odds, we will use the logit in the remainder of this chapter.

Equation (14.13) models the logit as a linear function of Z. The interpretation of the coefficient β_1 is the change in log odds of 1 versus 2 for one unit increase in Z. e^{β_1} is the corresponding odds multiplier for one unit increase in Z. β_0 is the log odds for $Z = 0$ and e^{β_0} is the corresponding odds.

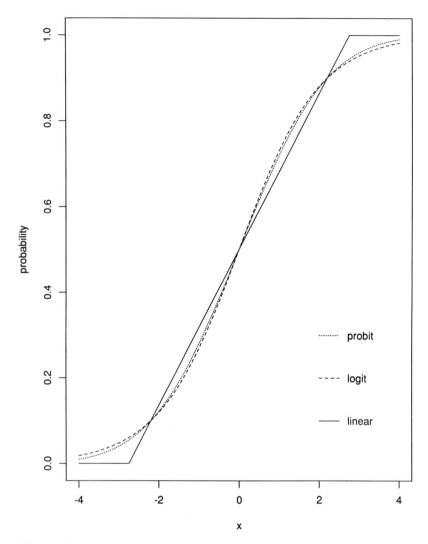

Figure 14.1 Probability as a function of probit, logit, and linear transformations.

Computer programs are typically needed to fit the logistic regression model, Eq. (14.13). Maximum likelihood estimates and approximate standard errors based on the inverse of the observed Fisher information are usually provided. For sufficiently large sample sizes, the maximum likelihood estimates are approximately normally distributed. Confidence limits for true values of the coefficients may thus be computed in the usual manner based on the approximate standard errors. Depending on the program, values of $\log \lambda$ computed from Eq. (14.10) or perhaps the "deviance," $-2 \log \lambda$, may be given.

14.2.5 Example: Illustration of the Simple Linear Logistic Regression Model

To illustrate the interpretation of model parameters for the simple linear logistic regression model, we give an example from Agresti (1990). Based on data on 1329 men from the Framingham study, Agresti gives a linear logistic regression equation relating incidence of heart disease during a 6-year follow-up period to observed blood pressure readings. Taking p_i as the probability of heart disease for the ith individual with blood pressure reading Z_i, the fitted logistic regression equation from Agresti is

$$\log \frac{\hat{p}_i}{1 - \hat{p}_i} = -6.082 + 0.0243 Z_i . \tag{14.16}$$

The fitted coefficient for blood pressure is positive, indicating that the odds of heart disease increases as blood pressure increases. The direct interpretation is that for each one unit increase in blood pressure, the odds of heart disease is multiplied by $e^{0.0243}$. For a given blood pressure reading, Eq. (14.16) can be used to get the estimated probability of heart disease as

$$\hat{p}_i = \frac{e^{-6.082 + 0.0243 Z_i}}{1 + e^{-6.082 + 0.0243 Z_i}}. \tag{14.17}$$

Table 14.6 gives the estimated logits and corresponding estimated probabilities of heart disease based on Eqs. (14.16) and (14.17) for a range of blood pressure readings within the range of the original data.

14.2.6 Logistic General Linear Model

Modeling the logit as a linear function of several continuous explanatory variables is a straightforward extension of Eq. (14.13). Suppose we have a binary response variable Y_i and associated with each case we have values of k independent variables, $Z_{i1}, Z_{i2}, \ldots, Z_{ik}$. Then the multiple linear logistic regression model is

$$\text{logit}(p_i) = \log \frac{p_i}{1 - p_i} = \beta_0 + \beta_1 Z_{i1} + \cdots + \beta_k Z_{ik} ,$$
$$i = 1, \ldots, n. \tag{14.18}$$

Table 14.6 Estimated Logits and Probabilities of
Heart Disease Based on Agresti's (1990) Fitted
Logistic Regression Model

Z_i	$\log \dfrac{\hat{p}_i}{1 - \hat{p}_i}$	$\hat{p}_i$
115	-3.288	0.036
120	-3.166	0.040
125	-3.045	0.045
130	-2.923	0.051
135	-2.802	0.057
140	-2.680	0.064
145	-2.559	0.072
150	-2.437	0.080
155	-2.316	0.090
160	-2.194	0.100
165	-2.073	0.112
170	-1.951	0.124
175	-1.830	0.138
180	-1.708	0.153
185	-1.587	0.170
190	-1.465	0.188

The independent variables may be defined in the same way as any multiple linear regression equation. By suitable choice of the independent variables, virtually any combination of categorical and continuous variables plus transformations and interactions among them may be modeled. Equation (14.18) thus represents the *logistic general linear model*.

Computer programs are necessary to get parameter estimates for Eq. (14.18). Just as for simple linear logistic regression, programs give estimates of the parameters along with associated approximate standard errors. Ideally, the program will also compute $\log \lambda$ or perhaps the deviance. Depending on the sample size n, the configuration and values of the independent variables, and the values of the Y_i's, the standard errors may or may not be useful in forming confidence limits for the true parameter values. For binary response data, the rule seems to be that more often than not there are problems with at least some of the standard errors. So extreme care must be taken when examining the output from a typical multiple linear logistic regression program.

For testing the importance of a particular independent variable, or set of independent variables, a better strategy than examining the estimated coefficients and their standard errors is as follows:

1. Fit Eq. (14.18) for the model omitting the variable, or variables, in question;
2. Fit Eq. (14.18) for the full model including all variables;
3. Form a ΔG^2 statistic as twice the difference in log λ computed by Eq. (14.10); and
4. Compare the ΔG^2 statistic to the chi-squared distribution with degrees of freedom equal to the difference in number of parameters fit in the two models. If ΔG^2 is large compared to chi-squared critical values, then the variable, or variables, should be considered as important in estimating the logit under study.

While not foolproof, this method will give a better picture of which variables are important in the model than the method of comparing the estimated coefficients to their standard errors.

14.2.6.1 Categorical Explanatory Variables

Categorical explanatory variables are included in the logistic general linear model by using dummy variables. There are many equivalent methods of choosing a set of $C - 1$ dummy variables to represent the C categories of a categorical explanatory variable. Often there is a good reason to choose one particular dummy variable representation, but no matter which representation is used, special care must be taken in interpreting the estimated coefficients. One important but often forgotten point is that the set of $C - 1$ dummy variables for a categorical explanatory variable must be considered as a group. Thus when testing the importance of the categorical explanatory variable, models must be compared with and without all $C - 1$ dummy variables as one group. Tests on coefficients for individual dummy variables may have meaning as a specific contrast between category effects, but typically the base category is arbitrary and different variables would appear important if a different base category were selected. Another alternative is to parameterize the categorical explanatory variables symmetrically, with the usual analysis of variance constraint that the sum of the category coefficients be zero.

14.2.6.2 Grouping Data

It is often convenient to group cases that have exactly the same values on all k explanatory variables. Suppose there are G distinct patterns among the explanatory variables. (For example, in the case of a single categorical explanatory variable with C categories, there are $G = C$ distinct patterns.) For each distinct pattern, $g = 1, \ldots, G$, let n_g be the number of cases with that pattern. Let x_{1g} be the number of cases with pattern g that have $Y = 1$ and let x_{2g} be the number with $Y = 2$. Table 14.7 summarizes the data grouped by distinct pattern. If π_g is the probability that $Y = 1$ for

Table 14.7 Response Data Grouped by Distinct Patterns of the Explanatory Variables

Pattern	Response 1	Response 2	Total
1	x_{11}	x_{21}	n_1
2	x_{12}	x_{22}	n_2
.	.	.	.
.	.	.	.
g	x_{1g}	x_{2g}	n_g
.	.	.	.
.	.	.	.
G	x_{1G}	x_{2G}	n_G
Total	x_{1+}	x_{2+}	n

cases with distinct pattern g, then x_{1g} follows a binomial distribution with sample size n_g and probability of success π_g. Maximum likelihood estimates for model parameters of Eq. (14.18) are identical whether or not the data are grouped. Estimated expected values corresponding to the grouped data are given in Table 14.8.

Grouping enables us to form goodness-of-fit test statistics to check for model fit. If the group sample sizes are sufficiently large, the Pearson chi-squared statistic X^2 defined by Eq. (14.3) or the loglikelihood ratio chi-squared statistic G^2 defined by Eq. (14.4) may be used to judge the appropriateness of the model, Eq. (14.18). The chi-squared statistics are

Table 14.8 Estimated Expected Values Corresponding to the Response Data Grouped by Distinct Patterns of the Explanatory Variables

Pattern	Response 1	Response 2	Total
1	$n_1\hat{\pi}_1$	$n_1(1 - \hat{\pi}_1)$	n_1
2	$n_2\hat{\pi}_2$	$n_2(1 - \hat{\pi}_2)$	n_2
.	.	.	.
.	.	.	.
g	$n_g\hat{\pi}_g$	$n_g(1 - \hat{\pi}_g)$	n_g
.	.	.	.
.	.	.	.
G	$n_G\hat{\pi}_G$	$n_G(1 - \hat{\pi}_G)$	n_G
Total	x_{1+}	x_{2+}	n

compared to the chi-squared distribution with degrees of freedom G minus the number of parameters estimated in Eq. (14.18). See Larntz (1978) and Koehler and Larntz (1980) for details on when p values based on X^2 and G^2 give valid measures of goodness of fit. In general, chi-squared based empirical rejection rates under the null hypothesis are closer to the nominal level for X^2.

14.2.7 Examples: Illustrations of Logistic General Linear Model

We give two examples illustrating the logistic general linear model. The first example returns to the Sherman–Berk Minneapolis spouse assault data and shows how different parameterizations of a categorical explanatory variable can affect interpretation of the fitted model. The second example combines a continuous and a categorical explanatory variable.

14.2.7.1 Sherman–Berk Spouse Assault Example

The simple 2×3 contingency table in Table 14.3 can be analyzed using the logistic general linear model in several ways. We first analyze it using the parameterization used by Berk and Sherman (1988) and then illustrate alternative parameterizations. As shown in Section 14.2.2, goodness-of-fit test statistics indicate borderline significance for differences among the three treatment alternatives.

Recall our notation for a categorical explanatory variable, which we denoted as A with categories $1, 2, \ldots, C$. In this example,

$$A_i = \begin{cases} 1 & \text{if case } i \text{ was assigned Separate,} \\ 2 & \text{if case } i \text{ was assigned Advise,} \\ 3 & \text{if case } i \text{ was assigned Arrest.} \end{cases} \qquad (14.19)$$

For the first parameterization, create two dummy variables as follows:

$$D_{i1} = \begin{cases} 1 & \text{if } A_i = 2 \\ 0 & \text{otherwise} \end{cases} \qquad D_{i2} = \begin{cases} 1 & \text{if } A_i = 3 \\ 0 & \text{otherwise.} \end{cases} \qquad (14.20)$$

The logistic general linear model Eq. (14.18) becomes

$$\text{logit}(p_i) = \log \frac{p_i}{1 - p_i} = \beta_0 + \beta_1 D_{i1} + \beta_2 D_{i2} . \qquad (14.21)$$

This parameterization creates particular interpretations for the coefficients as follows:

1. β_0 is the logit value for the omitted category $A_i = 1$; i.e., e^{β_0} is the odds of future violence for the Separate treatment;

2. β_1 is the difference in logits values between categories $A_i = 2$ and $A_i = 1$; i.e., e^{β_1} is the odds ratio of future violence for the Advise treatment compared to the Separate treatment; and

3. β_2 is the difference in logits values between categories $A_i = 3$ and $A_i = 1$; i.e., e^{β_2} is the odds ratio of future violence for the Arrest treatment compared to the Separate treatment.

The estimated coefficients and standard errors for this parameterization are

Term	Coef.	se(coef.)
β_0	-1.2078	0.2235
β_1	-0.2136	0.3302
β_2	-0.8963	0.4026

It is important to recognize that this parameterization does not compare the Advise and Arrest treatments directly. The coefficient of the comparison may be computed as $\hat{\beta}_2 - \hat{\beta}_1$, but the standard error must be computed from the covariance matrix of the estimated coefficients, which is not available from all programs. Alternatively, one could do another run with Advise or Arrest as the omitted category. This is highly recommended if, as is typically the case, all pairwise comparisons among treatments are of interest.

Looking at the coefficients and standard errors above, one is tempted to calculate their ratio and note the apparent statistical significance of $\hat{\beta}_2$. That may be fine if that comparison were known to be of primary interest before looking at the data. Too often, however, the omitted category is selected as the category with the highest (or lowest) failure rate in the data. In that case, it is more appropriate to look at all possible pairwise comparisons and use a multiple comparisons procedure to adjust for post hoc selection. The Bonferroni multiple comparisons procedure (Miller, 1966) can be used to get adjusted p values for the three comparisons—it simply means multiplying the nominal p value by the number of comparisons (three in this case). Table 14.9 summarizes the logit comparisons of the three treatments using the Bonferroni multiple comparisons procedure. None of the three pairwise comparisons attains an adjusted .05 level of significance.

14.2.7.2 Parole Decision Making Example

In a study of parole decision making in Minnesota, Larntz (1980) used several models to relate future felony conviction within 2 years of release

Table 14.9 Multiple Comparisons of Treatment Differences for the Sherman–Berk Spouse Assault Experiment

Comparison	Coef	se(coef)	z	Nominal p	Adjusted p
Advise vs. Separate	−0.2136	0.3302	−0.65	0.518	1.00
Arrest vs. Separate	−0.8963	0.4026	−2.23	0.026	0.08
Arrest vs. Advise	−0.6827	0.4138	−1.65	0.099	0.30

to variables available at time of release. One relatively simple model, describing the response as well or better than other models containing seven or eight variables, used two explanatory variables:

Z_1, number of parole/probation failures
Z_2, age at first adult conviction

Fitting a logistic regression to $n = 482$ cases yielded a fitted equation

$$\log \frac{\hat{p}_i}{1 - \hat{p}_i} = 0.28031 + 0.22982 Z_{i1} - 0.00708 Z_{i2} . \tag{14.22}$$

For an individual with one failure, first convicted at age 20, the estimated logit is

$$\log \frac{\hat{p}}{1 - \hat{p}} = 0.28031 + 0.22982(1) - 0.00708(20) = -1.03147$$

and the estimated rate of new felony conviction within 2 years is

$$\hat{p} = \frac{e^{-1.03147}}{1 - e^{-1.03147}} = .263.$$

Someone first convicted at age 18 with 7 prior failures would have an estimated rate of .623.

14.3 VARIABLE SELECTION CRITERIA: AKAIKE INFORMATION CRITERION

Variable selection for categorical response models has the same challenges as variable selection in multiple linear regression models. Stepwise procedures may be used, but they are often misleading or misused. A current standard practice in multiple linear regression is to compute all possible models and compare them using an objective criterion such as Mallow's C_p (Weisberg, 1985). A similar alternative for categorical response models uses the Akaike Information Criterion (*AIC*) to compare models with different sets of variables (Sakamoto and Akaike, 1978).

14.3.1 Definition of *AIC*

The use of *AIC* for contingency tables was introduced by Sakamoto and Akaike (1978). The definition used here is equivalent to their definition and has the advantage of being scaled similarly to Mallow's C_p. Several equivalent definitions of *AIC* are used in the literature [see Fowlkes, Freeny, and Landwehr (1988) for an alternate version and for insightful suggestions for fitting and interpreting logistic models in general].

The basic idea of *AIC* is to use $-2 \log \lambda$ adjusted for the number of model parameters to judge the fit of the model. Good models have a large $\log \lambda$ and a small number of parameters. We present two definitions of *AIC*, one for ungrouped data as displayed in Table 14.4 and an equivalent one for grouped data as displayed in Table 14.7. In Section 14.4, these definitions will be extended to handle comparisons of multinomial response models.

For ungrouped data with n observations, we define *AIC* as

$$AIC = -2 \log \lambda - n + 2p, \qquad (14.23)$$

where $\log \lambda$ is defined by Eq. (14.10) and p is the number of β_j's in Eq. (14.18), typically $k + 1$. For grouped data with D distinct patterns of independent variables, we define *AIC* as

$$AIC = G^2 - D + 2p, \qquad (14.24)$$

with G^2 defined by Eq. (14.4).

14.3.2 Using *AIC* to Compare Models

The goal of variable selection should not be to pick one best model, but rather to find a set of models with a relatively small number of parameters that are reasonably consistent with the data. The investigator can then examine these using substantive knowledge to determine which ones may be useful for interpreting the data in hand.

A strategy for using *AIC*, or any other objective model comparison criterion, is to fit all possible models and compute the *AIC* value for each model. Models with low *AIC* values may be compared via ΔG^2 if one model is a subset of the other, which often is the case. Since models with low numbers of parameters p are preferred, a plot of *AIC* versus p for the various models shows both *AIC* and p. (The equivalent plot in multiple regression is often referred to as a C_p plot.) Fowlkes et al. recommend plotting G^2 versus the degrees of freedom for the nominal chi-squared comparison distribution, which is essentially equivalent to the *AIC* versus p plot.

Table 14.10 Dyke–Patterson (1952) Relating Knowledge of Cancer to
[1] Lecture Attendance, [2] Newspaper Reading, [3] Doing Solid Reading,
and [4] Listening to the Radio

[1]	[2]	[3]	[4]	Good	Poor
yes	yes	yes	yes	23	8
no	yes	yes	yes	102	67
yes	no	yes	yes	1	3
no	no	yes	yes	16	16
yes	yes	no	yes	8	4
no	yes	no	yes	35	59
yes	no	no	yes	4	3
no	no	no	yes	13	50
yes	yes	yes	no	27	18
no	yes	yes	no	201	177
yes	no	yes	no	3	8
no	no	yes	no	67	83
yes	yes	no	no	7	6
no	yes	no	no	75	156
yes	no	no	no	2	10
no	no	no	no	84	393

14.3.3 Example: Illustration of *AIC* on Dyke–Patterson Cancer Knowledge Example

To illustrate use of *AIC*, we use data from Dyke and Patterson (1952).
This classic data set relates an individual's knowledge of cancer (good or
poor) to four explanatory variables:

[1]   Attends lectures or not
[2]   Reads newspapers or not
[3]   Does solid reading or not
[4]   Listens to radio or not

The data are given in Table 14.10.

To illustrate the use of *AIC* and ΔG^2 in variable selection, we fit every
possible hierarchical factorial linear model among the four explanatory
variables. Hierarchical factorial models are those in which all lower-order
combinations of any included interaction are also included in the model.
We use a shorthand notation, adapted from Fienberg (1980), to indicate
the models. For instance, the notation

[1] [2] [3] [4]

indicates a model with main effects only on the logit response. The notation

[124] [34]

indicates a model containing a three-factor interaction among factors [1], [2], and [4] and a two-factor interaction among factors [3] and [4]. Hierarchy means that all main effects are included since each of [1], [2], [3], and [4] is contained in model terms, and two-factor interactions contained in [124], namely between factors [1] and [2], [1] and [4], and [2] and [4], are also included in the model. Table 14.11 presents X^2, G^2, and AIC values for

Table 14.11 Goodness of Fit Statistics and AIC Values for All Possible Hierarchical Factorial Logistic Models for the Dyke–Patterson Cancer Knowledge Data

Model	df	X^2	G^2	p	AIC
[3]	14	68.49	65.81	2	53.81
[2]	14	113.75	110.48	2	98.48
[4]	14	185.47	192.01	2	180.01
[1]	14	191.64	199.10	2	187.10
[2][3]	13	27.69	25.91	3	15.91
[3][4]	13	53.35	51.74	3	41.74
[1][3]	13	58.60	57.69	3	47.69
[1][2]	13	104.35	100.73	3	90.73
[2][4]	13	104.75	101.55	3	91.55
[1][4]	13	173.26	178.75	3	168.75
[2][3][4]	12	19.31	18.46	4	10.46
[1][2][3]	12	20.23	20.00	4	12.00
[23]	12	24.67	22.85	4	14.85
[1][3][4]	12	46.34	45.47	4	37.47
[34]	12	52.70	51.16	4	43.16
[13]	12	53.50	53.39	4	45.39
[1][2][4]	12	97.04	93.31	4	85.31
[12]	12	103.16	99.57	4	91.57
[24]	12	104.68	101.49	4	93.49
[14]	12	172.73	178.24	4	170.24
[1][2][3][4]	11	13.61	13.59	5	7.59
[2][13]	11	15.91	16.33	5	10.33
[1][23]	11	17.24	16.92	5	10.92
[2][34]	11	19.05	18.33	5	12.33
[3][24]	11	19.33	18.46	5	12.46
[3][14]	11	45.39	43.77	5	37.77
[1][34]	11	45.88	44.98	5	38.98
[2][14]	11	95.76	92.04	5	86.04
[1][24]	11	97.02	93.27	5	87.27

Table 14.11 Continued

Model	df	X^2	G^2	p	AIC
[13][2][4]	10	9.92	10.27	6	6.27
[23][1][4]	10	10.82	10.72	6	6.72
[14][2][3]	10	11.65	11.48	6	7.48
[12][3][4]	10	12.50	11.85	6	7.85
[12][13]	10	13.45	13.20	6	9.20
[34][1][2]	10	13.46	13.48	6	9.48
[24][1][3]	10	13.61	13.58	6	9.58
[13][23]	10	13.19	13.61	6	9.61
[12][23]	10	16.76	14.95	6	10.95
[23][24]	10	16.50	15.62	6	11.62
[23][34]	10	16.41	15.62	6	11.62
[24][34]	10	19.07	18.30	6	14.30
[13][14]	10	41.14	40.48	6	36.48
[13][34]	10	41.81	41.52	6	37.52
[14][34]	10	44.84	43.22	6	39.22
[12][14]	10	94.26	90.84	6	86.84
[12][24]	10	95.45	91.91	6	87.91
[14][24]	10	95.73	91.99	6	87.99
[12][13][4]	9	7.14	7.05	7	5.05
[13][23][4]	9	7.39	7.73	7	5.73
[14][23]	9	8.81	8.55	7	6.55
[13][14][2]	9	8.41	8.57	7	6.57
[12][23][4]	9	9.62	8.64	7	6.64
[12][14][3]	9	10.15	9.91	7	7.91
[12][13][23]	9	10.65	10.23	7	8.23
[13][24]	9	9.90	10.25	7	8.25
[13][34][2]	9	9.90	10.26	7	8.26
[23][24][1]	9	10.82	10.70	7	8.70
[23][34][1]	9	10.81	10.72	7	8.72
[14][34][2]	9	11.48	11.34	7	9.34
[14][24][3]	9	11.64	11.48	7	9.48
[12][34]	9	12.28	11.71	7	9.71
[12][24][3]	9	12.49	11.85	7	9.85
[24][34][1]	9	13.45	13.46	7	11.46
[23][24][34]	9	16.44	15.60	7	13.60
[4][23]	9	16.46	15.63	7	13.63
[3][12]	9	19.71	18.38	7	16.38
[13][14][34]	9	40.87	40.18	7	38.18
[4][13]	9	42.02	41.77	7	39.77
[12][14][24]	9	94.15	90.72	7	88.72
[4][12]	9	95.54	92.01	7	90.01
[12][13][23][4]	8	4.45	4.26	8	4.26
[12][13][14]	8	5.65	5.64	8	5.64
[13][14][23]	8	5.85	5.96	8	5.96

Table 14.11 Continued

Model	df	X^2	G^2	p	AIC
[12][14][23]	8	7.05	6.68	8	6.68
[12][13][34]	8	7.10	7.02	8	7.02
[12][13][24]	8	7.13	7.05	8	7.05
[13][23][24]	8	7.35	7.69	8	7.69
[13][23][34]	8	7.38	7.72	8	7.72
[13][14][34][2]	8	8.37	8.53	8	8.53
[14][23][24]	8	8.79	8.54	8	8.54
[14][23][34]	8	8.79	8.54	8	8.54
[13][14][24]	8	8.39	8.56	8	8.56
[12][23][24]	8	9.62	8.64	8	8.64
[12][23][34]	8	9.58	8.64	8	8.64
[123]	8	9.19	9.19	8	9.19
[12][14][34]	8	9.92	9.73	8	9.73
[12][14][24][3]	8	10.13	9.90	8	9.90
[13][24][34]	8	9.86	10.22	8	10.22
[23][24][34][1]	8	10.80	10.70	8	10.70
[14][24][34]	8	11.46	11.32	8	11.32
[12][24][34]	8	12.29	11.70	8	11.70
[234]	8	15.97	15.30	8	15.30
[134]	8	40.63	39.96	8	39.96
[124]	8	93.91	90.50	8	90.50
[12][13][14][23]	7	2.84	2.80	9	4.80
[123][4]	7	3.23	3.27	9	5.27
[12][13][23][34]	7	4.46	4.25	9	6.25
[12][13][23][24]	7	4.46	4.26	9	6.26
[12][13][14][34]	7	5.59	5.59	9	7.59
[12][13][14][24]	7	5.64	5.64	9	7.64
[13][14][23][24]	7	5.80	5.93	9	7.93
[13][14][23][34]	7	5.85	5.96	9	7.96
[12][14][23][34]	7	7.00	6.66	9	8.66
[12][14][23][24]	7	7.04	6.67	9	8.67
[12][13][24][34]	7	7.10	7.02	9	9.02
[13][23][24][34]	7	7.35	7.69	9	9.69
[134][2]	7	8.06	8.20	9	10.20
[13][14][24][34]	7	8.33	8.50	9	10.50
[14][23][24][34]	7	8.76	8.53	9	10.53
[12][23][24][34]	7	9.59	8.64	9	10.64
[124][3]	7	9.63	9.49	9	11.49
[12][14][24][34]	7	9.92	9.73	9	11.73
[234][1]	7	10.42	10.41	9	12.41
[123][14]	6	1.79	1.84	10	5.84
[12][13][14][23][24]	6	2.84	2.80	10	6.80
[12][13][14][23][34]	6	2.84	2.80	10	6.80

Table 14.11 Continued

Model	df	X^2	G^2	p	AIC
[123][34]	6	3.22	3.26	10	7.26
[123][24]	6	3.23	3.27	10	7.27
[12][13][23][24][34]	6	4.46	4.25	10	8.25
[134][12]	6	5.25	5.23	10	9.23
[124][13]	6	5.28	5.31	10	9.31
[134][23]	6	5.43	5.55	10	9.55
[12][13][14][24][34]	6	5.59	5.59	10	9.59
[13][14][23][24][34]	6	5.80	5.93	10	9.93
[124][23]	6	6.38	6.11	10	10.11
[12][14][23][24][34]	6	7.00	6.65	10	10.65
[234][13]	6	7.08	7.43	10	11.43
[134][24]	6	8.02	8.16	10	12.16
[234][14]	6	8.37	8.27	10	12.27
[234][12]	6	9.04	8.30	10	12.30
[124][34]	6	9.42	9.31	10	13.31
[123][14][24]	5	1.79	1.84	11	7.84
[123][14][34]	5	1.79	1.84	11	7.84
[134][12][23]	5	2.37	2.32	11	8.32
[124][13][23]	5	2.35	2.38	11	8.38
[12][13][14][23][24][34]	5	2.84	2.80	11	8.80
[123][24][34]	5	3.22	3.26	11	9.26
[234][12][13]	5	4.09	3.94	11	9.94
[134][12][24]	5	5.25	5.23	11	11.23
[124][13][34]	5	5.23	5.26	11	11.26
[134][23][24]	5	5.38	5.51	11	11.51
[234][13][14]	5	5.54	5.71	11	11.71
[124][23][34]	5	6.35	6.09	11	12.09
[234][12][14]	5	6.56	6.32	11	12.32
[123][134]	4	1.33	1.35	12	9.35
[123][124]	4	1.50	1.53	12	9.53
[123][14][24][34]	4	1.79	1.84	12	9.84
[134][12][23][24]	4	2.37	2.32	12	10.32
[124][13][23][34]	4	2.35	2.38	12	10.38
[234][12][13][14]	4	2.54	2.52	12	10.52
[123][234]	4	3.00	3.05	12	11.05
[124][134]	4	5.03	5.04	12	13.04
[134][234]	4	5.13	5.29	12	13.29
[124][234]	4	5.89	5.75	12	13.75
[123][134][24]	3	1.33	1.34	13	11.34
[123][124][34]	3	1.50	1.53	13	11.53
[123][234][14]	3	1.61	1.66	13	11.66
[134][234][12]	3	2.06	2.04	13	12.04
[124][134][23]	3	2.07	2.09	13	12.09

Table 14.11 Continued

Model	df	X^2	G^2	p	AIC
[124][234][13]	3	2.05	2.09	13	12.09
[123][134][234]	2	1.15	1.16	14	13.16
[123][124][134]	2	1.19	1.21	14	13.21
[123][124][234]	2	1.32	1.35	14	13.35
[124][134][234]	2	1.77	1.80	14	13.80
[123][124][134][234]	1	1.01	1.02	15	15.02
[1234]	0	0.00	0.00	16	16.00

all possible hierarchical factorial logistic regression models. The models in Table 14.11 are ordered by number of parameters fit p and by AIC value to make comparison easier. Figure 14.2 plots AIC versus p for models with $AIC < 20$.

Examining Table 14.11 and Figure 14.2, we find that there are clearly many, many models with relatively small AIC. One reasonable strategy is to examine the models with low AIC values and with small numbers of parameters fit. Models with $p = 2$ and $p = 3$ are clearly inadequate, having large goodness-of-fit X^2 and G^2 values and thus large AIC values. For $p = 4$ model [2][3][4] has $AIC = 10.46$ and for $p = 5$ model [1][2][3][4] has $AIC = 7.59$. Since model [2][3][4] is nested in model [1][2][3][4], we can form $\Delta G^2 = 18.46 - 13.59 = 4.87$, which we compare to a chi-squared distribution with 1 degree of freedom. This difference is significant at level .05, but not at level .01. Table 14.12 contains the estimated logit coefficients for the model [1][2][3][4] based on dummy variable coding of 1 if the individual has the characteristic and 0 if the individual does not. The coefficients for all factors are positive, indicating that knowledge of cancer improves if an individual has the characteristic. This may be what one would expect and thus favors including [1] in the model.

Comparing model [1][2][3][4] to the best model for $p = 6$, we find that AIC decreases from 7.59 to 6.27. Again the models are nested, so we can use $\Delta G^2 = 13.59 - 10.27 = 3.32$, which is significant at level .10 but not at level .05. Again one might examine the coefficients of the model, but given that this is the most significant of six possible two-factor interactions, one might argue that we should be satisfied with the main-effects-only model. Indeed, there are many other models with low AIC values and even some that "significantly improve" on [1][2][3][4], but that is what might be expected by chance when examining so many models. Our preference here is to stay with the main-effects-only model.

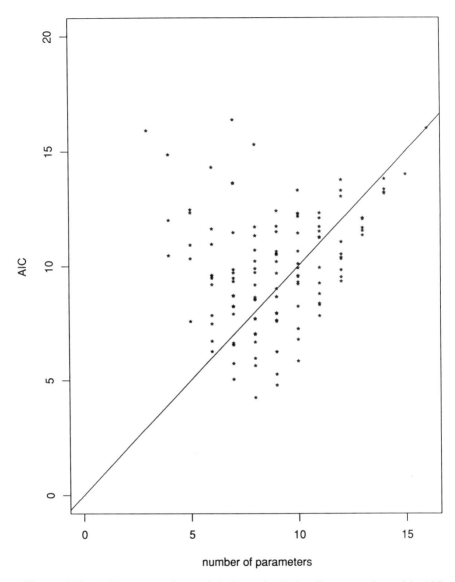

Figure 14.2 *AIC* versus *p* for models fit to the Dyke–Patterson data with *AIC* < 20.

14.4 MULTINOMIAL RESPONSES

Discrete responses with three or more possible outcomes are certainly common.

1. A survey respondent may agree, disagree, be neutral, not know, or give no response to an opinion question;
2. A firm may take several actions in response to severe financial problems such as internal reorganization, liquidation, bankruptcy, and possibly several others;
3. A doctor may classify a patient into 20 or more disease categories or classify the patient as normal;
4. A victim of sexual harassment may respond verbally, respond physically, report the incident, or ignore the incident;
5. A student may choose one of several responses to a multiple choice question; or
6. A parolee may commit a felony, commit a misdemeanor, or remain crime free.

Suppose we have n observations. As before, we assume that the responses for the n observations are statistically independent of each other. For the ith observation, let Y_i be the response variable, which will be one of R possible categories. For convenience, suppose Y_i takes on the values $1, 2, \ldots, R$. Before, when we had two possible responses, we let p_i be the probability that $Y_i = 1$ and thus $(1 - p_i)$ was the probability that $Y_i = 2$. With three or more categories, we will explicitly use separate probabilities for each of the R possible response categories. Specifically, define p_{ir} as the probability that $Y_i = r$. That is,

$$p_{ir} = P(Y_i = r), \qquad r = 1, \ldots, R, \qquad i = 1, \ldots, n, \qquad (14.25)$$

with

$$\sum_{r=1}^{R} p_{ir} = 1, \qquad i = 1, \ldots, n. \qquad (14.26)$$

For convenience, we denote the probability vector for the ith case as

$$\mathbf{p}_i = (p_{i1}, p_{i2}, \ldots, p_{iR}), \qquad i = 1, \ldots, n. \qquad (14.27)$$

The goal is to model $\mathbf{p}_i$ as a function of explanatory variables.

14.4.1 Multinomial Response with One Categorical Explanatory Variable

For the case of one categorical explanatory variable A with C categories, the analysis is identical to the binary response case. That is, create the grouped data $R \times C$ contingency table by defining x_{rj} to be the number of cases (out of n) with response $Y_i = r$ with category $A_i = j$. Letting π_j be the probability vector for category j, use a standard chi-squared test to check homogeneity of the C proportions:

$$H_0: \pi_1 = \pi_2 = \cdots = \pi_C . \tag{14.28}$$

The estimated expected values $\hat{m}_{rj}$ have the same form as presented in Table 14.2 and a test of H_0 can be constructed from either X^2, defined by Eq. (14.3), or G^2, defined by Eq. (14.4). If the null hypothesis holds, both statistics follow an asymptotic chi-squared distribution with $(R - 1)$ $(C - 1)$ degrees of freedom.

14.4.2 Multinomial Logit Transformation

If one or more of the explanatory variables is continuous, then we model transformations of $\mathbf{p}_i$ as linear functions of the explanatory variable. The problem is intrinsically multivariate in the sense that for an R category response variable, we need $R - 1$ equations to model $\mathbf{p}_i$. For one continuous explanatory variable Z, the basic data are n pairs, (Z_1, Y_1), (Z_2, Y_2), $\ldots$, (Z_n, Y_n). The direct extension of the logistic regression model transforms $\mathbf{p}_i$ to $R - 1$ logit equations as follows:

$$\log \frac{p_{i1}}{p_{i2}} = \beta_0^{(1,2)} + \beta_1^{(1,2)} Z_i ,$$

$$\log \frac{p_{i2}}{p_{i3}} = \beta_0^{(2,3)} + \beta_1^{(2,3)} Z_i ,$$

$$\vdots \tag{14.29}$$

$$\log \frac{p_{ir}}{p_{i,r+1}} = \beta_0^{(r,r+1)} + \beta_1^{(r,r+1)} Z_i ,$$

$$\vdots$$

$$\log \frac{p_{i,R-1}}{p_{iR}} = \beta_0^{(R-1,R)} + \beta_1^{(R-1,R)} Z_i .$$

This set of $R - 1$ transformations constitutes the multinomial logit transformation and Eq. (14.29) is the multinomial logit model. The interpretation of the model is that each pair of responses has a separate linear logit model. For pairs not listed in Eq. (14.29), the linear logit model is derived from the pairs listed. For instance, the logit equation for responses 1 and 3 is derived as follows:

$$\log \frac{p_{i1}}{p_{i3}} = \log \frac{p_{i1}}{p_{i2}} + \log \frac{p_{i2}}{p_{i3}}$$

$$= (\beta_0^{(1,2)} + \beta_0^{(2,3)}) + (\beta_1^{(1,2)} + \beta_1^{(2,3)})Z_i . \tag{14.30}$$

Table 14.12 Logistic Regression Main Effects Only
Coefficients for Dyke–Patterson Data

	Coef	se(coef)
Constant	− 1.4604	0.0964
Lectures	0.4204	0.1910
Newspapers	0.6498	0.1154
Solid reading	0.9806	0.1107
Radio	0.3101	0.1222

For any pair of responses r and s $(r < s)$, the logit equation for responses r and s is derived by summing the equations for r vs. $r + 1$, $r + 1$ vs. $r + 2, \ldots, s - 1$ vs. s as follows:

$$
\log \frac{p_{ir}}{p_{is}} = \log \frac{p_{ir}}{p_{i,r+1}} + \log \frac{p_{i,r+1}}{p_{i,r+2}}
$$

$$
+ \cdots + \log \frac{p_{i,s-1}}{p_{is}}
$$

$$
= \beta_0^{(r,r+1)} + \beta_0^{(r+1,r+2)} + \cdots + \beta_0^{(s-1,s)}
$$

$$
(\beta_1^{(r,r+1)} + \beta_1^{(r+1,r+2)} + \cdots + \beta_1^{(s-1,s)})Z_i . \tag{14.31}
$$

The interpretation of the multinomial logit model is best thought of in a conditional sense. Given that the response is one of r or s, then the odds of r versus s follows an ordinary linear logit model for binary response data. This is true for each pair of responses. Also, although Eq. (14.29) was defined for successive pairs of responses r versus $r + 1$, the order in which the responses appear does not matter. That is, if we apply Eq. (14.31) to get the conditional logit equation for any pair of responses r versus s, the resulting conditional logit equation will be identical for any order of responses used to construct Eq. (14.29).

14.4.3 Multinomial Logit General Linear Model

Directly analogous to the logistic general linear model, Eq. (14.18), we define the multinomial logit general linear model. Suppose we have a multinomial response variable Y_i that takes on possible categorical values $1, 2, \ldots, R$, and associated with each case we have values of k independent variables, $Z_{i1}, Z_{i2}, \ldots, Z_{ik}$. The multinomial logit general linear model is

$$\log \frac{p_{i1}}{p_{i2}} = \beta_0^{(1,2)} + \beta_1^{(1,2)} Z_{i1} + \cdots + \beta_k^{(1,2)} Z_{ik} ,$$

$$\log \frac{p_{i2}}{p_{i3}} = \beta_0^{(2,3)} + \beta_1^{(2,3)} Z_{i1} + \cdots + \beta_k^{(2,3)} Z_{ik} ,$$

$$\vdots$$

$$\log \frac{p_{ir}}{p_{i,r+1}} = \beta_0^{(r,r+1)} + \beta_1^{(r,r+1)} Z_{i1} + \cdots + \beta_k^{(r,r+1)} Z_{ik} ,$$

$$\vdots$$

$$\log \frac{p_{i,R-1}}{p_{iR}} = \beta_0^{(R-1,R)} + \beta_1^{(R-1,R)} Z_{i1} + \cdots + \beta_k^{(R-1,R)} Z_{ik} . \tag{14.32}$$

The independent variables may be defined in the same way as any multiple linear regression equation. By suitable choice of the independent variables, virtually any combination of categorical and continuous variables plus transformations and interactions among them may be modeled.

Estimation for the multinomial logistic general linear model (14.32) is, in principle, straightforward. Table 14.13 gives the basic set of observations for a multinomial response variable. Table 14.13 is analogous to Table 14.4 for the binary response case. The values in the table are indicator functions defined by Eq. (14.7). Each row of the Table 14.13 contains all zeroes, except for the value 1 indicating the response for the ith case. The column totals are x_{+r}, the number of cases with response r. For any specified set of parameters, $\beta_j^{(r,r+1)}$'s, in a given model, estimates of the prob-

Table 14.13 Display of Response Data by Cases for a Multinominal Response Variable

Case	Response					
	1	2	$\cdots$	r	$\cdots$	R
1	$I(Y_1 = 1)$	$I(Y_1 = 2)$	$\cdots$	$I(Y_1 = r)$	$\cdots$	$I(Y_1 = R)$
2	$I(Y_2 = 1)$	$I(Y_2 = 2)$	$\cdots$	$I(Y_2 = r)$	$\cdots$	$I(Y_2 = R)$
$\vdots$	$\vdots$	$\vdots$	$\vdots$	$\vdots$	$\vdots$	$\vdots$
i	$I(Y_i = 1)$	$I(Y_i = 2)$	$\cdots$	$I(Y_i = r)$	$\cdots$	$I(Y_i = R)$
$\vdots$	$\vdots$	$\vdots$	$\vdots$	$\vdots$	$\vdots$	$\vdots$
n	$I(Y_n = 1)$	$I(Y_n = 2)$	$\cdots$	$I(Y_n = r)$	$\cdots$	$I(Y_n = R)$
Total	x_{1+}	x_{2+}	$\cdots$	x_{r+}	$\cdots$	x_{R+}

Table 14.14 Fitted Values Corresponding to Observations in Display of Multinomial Data in Table 14.13

Case	Response 1	2				
1	$\hat{p}_{11}$	$\hat{p}_{12}$	$\cdots$	$\hat{p}_{1r}$	$\cdots$	$\hat{p}_{1R}$
2	$\hat{p}_{21}$	$\hat{p}_{22}$	$\cdots$	$\hat{p}_{2r}$	$\cdots$	$\hat{p}_{2R}$
i	$\hat{p}_{i1}$	$\hat{p}_{i2}$	$\cdots$	$\hat{p}_{ir}$	$\cdots$	$\hat{p}_{iR}$
n	$\hat{p}_{n1}$	$\hat{p}_{n2}$	$\cdots$	$\hat{p}_{nr}$	$\cdots$	$\hat{p}_{nR}$
Total	x_{1+}	x_{2+}	$\cdots$	x_{r+}	$\cdots$	x_{R+}

ability of each possible response for each case can be computed. These values are presented in Table 14.14. The values in Table 14.14 should be considered as estimated expected values corresponding to the observations in Table 14.13. The loglikelihood is defined in direct analogue to Eq. (14.10).

$$\log \lambda = \sum_{i=1}^{n} \sum_{r=1}^{R} I(Y_i = r) \log \hat{p}_{ir} . \tag{14.33}$$

The sum is taken over all cells in the $n \times R$ contingency table, except cells that have $\hat{p}_{ir} = 0$. The maximum likelihood estimates for the model parameters, $\beta_j^{(r,r+1)}$'s, are the set of possible model parameters that maximize $\log \lambda$. Computer programs are necessary to compute the maximum likelihood estimates for all but the simplest models. These computer programs typically provide approximate standard error estimates for the model parameters. As with binary response models, depending on the sample size n, the configurations and values of the independent variables, and the values of the Y_i's, these standard error estimates may or may not be useful for forming confidence limits or carrying out tests for true model parameters.

As for binary response models, a better method for testing the importance of a particular independent variable, or set of independent variables, is

1. Fit Eq. (14.32) for the model omitting the variable, or variables, in question;
2. Fit Eq. (14.32) for the full model including all variables;

3. Form a statistic ΔG^2 statistic as twice the difference in log λ computed by Eq. (14.33); and
4. Compare ΔG^2 to the chi-squared distribution with degrees of freedom equal to the difference in number of parameters in the two models. If ΔG^2 is large compared to chi-squared critical values, then the variable, or variables, should be considered as important in the multinomial logit model under study.

The number of degrees of freedom associated with the ΔG^2 statistic is typically $(R - 1) \times$ the degrees of freedom for the analogous test for a binary response variable. So if the difference between the two models is a single independent variable, ΔG^2 would have $R - 1$ degrees of freedom. If the difference between the two models is a categorical explanatory variable with C categories, the degrees of freedom would be $(R - 1) \times (C - 1)$.

If there are several independent observations with identical independent variable values, then data may be grouped in the same way as binary response data are grouped. This usually happens when the independent variables are categorical or when a continuous variable can be directly controlled. In such cases, grouped data tables like Table 14.7 may be constructed for multinomial response data. Note that a grouped data table is formed by summing rows of the basic set of observations Table 14.13, that have identical independent variable values. Corresponding tables of estimated fitted values may be computed under a specific model by summing the appropriate rows of Table 14.14. Equation (14.3) may be used to form the Pearson chi-squared statistic X^2 for measuring goodness of fit of the multinomial logit model. Equation (14.4) may be similarly used to form the loglikelihood ratio chi-squared statistic G^2. If sample sizes are sufficiently large, X^2 and/or G^2 may be compared to a chi-squared distribution with degrees of freedom equal to the number of cells in the grouped data contingency table minus the number of independent parameters estimated by the multinomial model (14.32). Again, more details are found in Larntz (1978) and Koehler and Larntz (1980).

14.4.4 Example: Multinomial Logit Models for Baker–Terpstra Sexual Harassment Data

As an example of the use of the multinomial logit model, consider the following example from Baker, Terpstra, and Larntz (1990). One hundred women and 143 men read a scenario describing an incident of sexual harassment, a male propositioning a female employee with the promise of job enhancement if she complied. Each person was asked to give his/her recommended action for the victim of the harassment. The responses were categorized into four response categories:

1. Leave the field,
2. Internally or externally report,
3. Physically or verbally react, and
4. Other or none.

Four variables were considered as possible explanatory variables:

1. Gender (0 = male, 1 = female),
2. Attitudes toward women scale,
3. Religiosity scale, and
4. Locus of control scale.

AIC and ΔG^2 were used to reduce the model to two variables, gender and religiosity.

For four response categories, Eq. (14.32) uses three logit equations to define the model: category 1 vs. category 2, category 2 vs. category 3, category 3 vs. category 4. Three additional logit equations can be derived from these; and for completeness in reporting the results of any multinomial logit fit, we recommend that all logit equations be reported. Table 14.15 presents the results for all six logit equations from Baker et al.

Since gender is coded 1 for females, the constant term represents the ordering of responses for males at a zero score for religiosity. Males with zero religiosity thus order the recommended responses as (2) report, (4) other or none, (1) leave the field, and (3) react. The response "report" is preferred $e^{0.56} = 1.75$ times as often as "other or none," $e^{1.48} = 4.39$ times as often as "leave the field," and $e^{1.87} = 6.49$ times as often as "react." "Other or none" is preferred $e^{0.91} = 2.48$ times as often as "leave the

Table 14.15 Multinomial Logit Coefficients for Sexual Harassment Data from Baker, Terpstra, and Larntz (1990)

Comparison	Constant		Gender		Religiosity	
	Coef.	se	Coef.	se	Coef.	se
1 vs. 2	−1.48	0.34	−0.54	0.45	0.14	0.11
1 vs. 3	0.39	0.53	−1.54	0.68	0.51	0.23
1 vs. 4	−0.91	0.38	0.23	0.52	0.21	0.14
2 vs. 3	1.87	0.46	−1.01	0.52	0.37	0.19
2 vs. 4	0.56	0.26	0.75	0.37	0.06	0.11
3 vs. 4	−1.30	0.47	1.76	0.59	−0.30	0.22

field" and $e^{1.30} = 3.67$ times as often as "react." "Leave the field" is preferred $e^{0.39} = 1.48$ times as often as "react."

The logits for females with zero religiosity is computed by adding the constant and gender coefficients. For females with zero religiosity, the order of recommended responses is (2) report, (3) react, (4) other or none, and (1) leave the field. Note that response (3) react is the second preferred recommendation by females, but the last recommended choice by males. Females preferred "report" $e^{(1.87-1.01)} = 2.36$ times as often as "react," $e^{(0.56+0.75)} = 3.71$ times as often as "other or none," and $e^{-(-1.48-0.54)} = 7.54$ times as often as "leave the field." "React" is preferred $e^{(-1.30+1.76)} = 1.58$ times as often as "other or none" and $e^{-(-0.39-1.54)} = 3.16$ times as often as "leave the field." "Other or none" is preferred $e^{-(-0.91+0.23)} = 1.97$ times as often as "leave the field."

Examining the coefficients for religiosity, we see the effect of increased religiosity is to increase the relative preference for (1) leave the field most, then for (2) report, and then for (4) other or none. Increased religiosity decreases the relative preference for (3) react.

14.4.5 Meyer's Multinomial Logit Computational Algorithm

As stated above, computations for the multinomial logit general linear model are "in principle" straightforward. If the number of observations, or number of distinct patterns for grouped data for programs that take advantage of such structure, is large, then standard algorithms such as Newton–Raphson become cumbersome. The number of parameters for a multinomial logit model may get large; we noted earlier that there are $(R - 1) \times$ as many parameters in the multinomial logit model as in the binary logit model with the same independent variable structure.

An interesting algorithm, based on the conditional binary logit structure of Eq. (14.29), finds maximum likelihood estimates for the multinomial logit general linear model with just a binary logistic regression program. The algorithm, due to Meyer (1981), is a clever application of the method of cyclic descent. It has proved useful for large problems in which memory limitations and computational costs have caused difficulties. (Multinomial logit programs in standard packages have reputations as heavy computational resource users!)

Suppose we are attempting to find maximum likelihood estimates for model (14.32) based on observations in Table 14.13 and associated independent variable values. Meyer's algorithm proceeds as follows:

1. Consider first data with positive values for responses 1 and 2 in Table 14.13. Fit a binary logit model with the same independent variable

structure as Eq. (14.32). Replace the observed data for responses 1 and 2 with the fitted values from the binary logit fit to get an updated observed table.

2. Now consider data with positive values for responses 2 and 3 in the updated observed table. (Note that data for response 2 will be the fitted values from step 2 and will not necessarily be integer valued.) Using these values as data, fit a binary logit model with the same independent variable structure as Eq. (14.32). Replace the observed data for responses 2 and 3 with the fitted values from the binary logit fit to get an updated observed table.

3. Repeat Step 2 for response 3 and 4, 4 and 5, . . . , $R - 1$ and R, and finally R and 1.

4. Steps 1, 2, and 3 constitute one cycle of Meyer's algorithm. At the end of the cycle, compute $\log \lambda$ defined by Eq. (14.33).

5. Continue cycles of Meyer's algorithm until the change in $\log \lambda$ is sufficiently small (0.001 works well).

When Meyer's algorithm has converged, the resulting updated table values are precisely the fitted values for the model as displayed in Table 14.14. In addition, the binary logit equation parameter estimates in the last cycle are the maximum likelihood estimates for the multinomial logit model (14.32), and the standard error estimates from this last cycle are identical to the standard errors estimated from the usual maximum likelihood procedure. So Meyer's algorithm gives a method of fitting the multinomial logit general linear model using only a binary logistic regression program. Binary logistic regression programs are generally more available than multinomial logit programs. Another advantage of Meyer's algorithm is that it reminds us of the conditional binary logit structure for pairs of responses under the multinomial logit model.

To illustrate Meyer's algorithm, we will use the data in Table 14.16a, which have $n = 30$ cases, $k = 2$ independent variables, and $R = 3$ responses. These data were generated from the multinomial logit model

$$\log \frac{p_{i1}}{p_{i2}} = 0.50 + 1.0Z_{i1} + 0.0Z_{i2} ,$$

$$\log \frac{p_{i2}}{p_{i3}} = 0.00 + 0.0Z_{i1} + 1.0Z_{i2} ,$$

(14.34)

where Z_{i1} and Z_{i2} were independent standard normal variables. Table 14.16b illustrates the updated data after Step 1 of Meyer's algorithm. Table 14.16c illustrates the updated data after Step 2 of Meyer's algorithm. Table 14.16d gives the updated data after one cycle of the algorithm. Note that all data values are now strictly between 0 and 1. Table 14.16e gives the

Table 14.16a Synthetic Data Used to Illustrate Meyer's
Algorithm

Case	Z_1	Z_2	Response 1	2	3
1	0.009	−0.307	0	0	1
2	−0.038	0.559	0	1	0
3	−1.017	2.693	0	1	0
4	−0.132	1.093	0	1	0
5	−0.360	0.099	0	0	1
6	−0.034	−0.918	0	0	1
7	−1.883	−1.762	0	0	1
8	0.337	0.304	0	0	1
9	0	−0.525	0	0	1
10	1.207	1.467	1	0	0
11	−0.02	0.454	1	0	0
12	−1.012	0.408	1	0	0
13	0.916	0.536	1	0	0
14	−1.383	0.076	0	1	0
15	−0.47	0.324	0	0	1
16	−0.804	−1.353	0	0	1
17	0.903	−2.423	0	0	1
18	−1.156	0.344	0	0	1
19	0.105	2.465	1	0	0
20	0.23	2.991	1	0	0
21	2.396	−1.556	1	0	0
22	0.083	1.272	0	1	0
23	−0.025	1.545	1	0	0
24	0.753	0.803	1	0	0
25	−1.108	−0.586	0	1	0
26	−2.228	0.888	0	1	0
27	1.226	−2.357	1	0	0
28	1.560	1.27	1	0	0
29	−0.524	−1.108	0	0	1
30	0.418	0.563	0	1	0

fitted values after six cycles of Meyer's algorithm. Table 14.16f gives the
true probabilities for each case, based on Eq. (14.34). The values in Table
14.16e and Table 14.16f should be comparable, although the number of
observations is only $n = 30$. Table 14.16g gives estimates and standard
errors for model parameters. These compare favorably, as they should, to
the coefficients in Eq. (14.34).

Table 14.16b Updated Data for Synthetic Example After
Step 1 of Meyer's Algorithm

Case	Response		
	1	2	3
1	0	0	1
2	0.528255	0.471745	0
3	0.224992	0.775008	0
4	0.514920	0.485080	0
5	0	0	1
6	0	0	1
7	0	0	1
8	0	0	1
9	0	0	1
10	0.935454	0.064546	0
11	0.530520	0.469480	0
12	0.146858	0.853142	0
13	0.870911	0.129089	0
14	0.073315	0.926685	0
15	0	0	1
16	0	0	1
17	0	0	1
18	0	0	1
19	0.695673	0.304327	0
20	0.765937	0.234063	0
21	0.985424	0.014576	0
22	0.624053	0.375947	0
23	0.590898	0.409102	0
24	0.840528	0.159472	0
25	0.102326	0.897674	0
26	0.019028	0.980972	0
27	0.860629	0.139371	0
28	0.964262	0.035738	0
29	0	0	1
30	0.726018	0.273982	0

Table 14.16c Updated Data for Synthetic Example After
Step 2 of Meyer's Algorithm

Case	Response		
	1	2	3
1	0	0.175952	0.824048
2	0.528255	0.26651	0.205235
3	0.224992	0.771674	0.003334
4	0.51492	0.391413	0.093668
5	0	0.400992	0.599008
6	0	0.060034	0.939966
7	0	0.052699	0.947301
8	0	0.359006	0.640994
9	0	0.121405	0.878595
10	0.935454	0.047781	0.016765
11	0.530520	0.238380	0.231100
12	0.146858	0.585901	0.267241
13	0.870911	0.045704	0.083386
14	0.073315	0.560620	0.366065
15	0	0.537612	0.462388
16	0	0.048344	0.951656
17	0	0.001336	0.998664
18	0	0.685328	0.314672
19	0.695673	0.298959	0.005368
20	0.765937	0.232472	0.001591
21	0.985424	0.000032	0.014545
22	0.624053	0.313289	0.062658
23	0.590898	0.370349	0.038753
24	0.840528	0.083018	0.076454
25	0.102326	0.214762	0.682912
26	0.019028	0.924823	0.056149
27	0.860629	0.000162	0.139209
28	0.964262	0.020900	0.014838
29	0	0.061846	0.938154
30	0.726018	0.128696	0.145286

Table 14.16d Updated Data for Synthetic Example After
One Full Cycle of Meyer's Algorithm

Case	Response 1	Response 2	Response 3
1	0.228862	0.175952	0.595186
2	0.396716	0.266510	0.336774
3	0.148518	0.771674	0.079808
4	0.403942	0.391413	0.204646
5	0.122773	0.400992	0.476235
6	0.117585	0.060034	0.822381
7	0.000292	0.052699	0.947009
8	0.441757	0.359006	0.199236
9	0.189366	0.121405	0.689229
10	0.944329	0.047781	0.007890
11	0.392184	0.238380	0.369436
12	0.027382	0.585901	0.386717
13	0.892723	0.045704	0.061574
14	0.007073	0.560620	0.432308
15	0.097240	0.537612	0.365148
16	0.009273	0.048344	0.942384
17	0.166215	0.001336	0.832449
18	0.013273	0.685328	0.301399
19	0.675653	0.298959	0.025388
20	0.757943	0.232472	0.009585
21	0.973647	0.000032	0.026322
22	0.562323	0.313289	0.124388
23	0.525157	0.370349	0.104494
24	0.854816	0.083018	0.062166
25	0.010204	0.214762	0.775033
26	0.000413	0.924823	0.074764
27	0.341911	0.000162	0.657928
28	0.974884	0.020900	0.004215
29	0.026831	0.061846	0.911323
30	0.696713	0.128696	0.174592

Table 14.16e   Estimated Expected Values for Synthetic
Example After Six Cycles of Meyer's Algorithm

Case	Response		
	1	2	3
1	0.233864	0.183138	0.582998
2	0.380074	0.323845	0.296081
3	0.065232	0.909923	0.024846
4	0.393837	0.429945	0.176218
5	0.148487	0.342402	0.509112
6	0.114314	0.105744	0.779942
7	0.000382	0.095423	0.904195
8	0.578736	0.162828	0.258436
9	0.187604	0.153062	0.659334
10	0.972375	0.018695	0.00893
11	0.372561	0.302981	0.324458
12	0.035785	0.570574	0.393641
13	0.889976	0.043405	0.066619
14	0.010440	0.517387	0.472173
15	0.135510	0.427250	0.437240
16	0.010123	0.096859	0.893018
17	0.139257	0.008569	0.852174
18	0.023287	0.574037	0.402676
19	0.656665	0.321857	0.021477
20	0.748267	0.243494	0.008239
21	0.959646	0.000635	0.039719
22	0.565656	0.320439	0.113905
23	0.514474	0.395084	0.090442
24	0.866409	0.067859	0.065732
25	0.011869	0.269066	0.719065
26	0.001418	0.840043	0.158539
27	0.285347	0.006628	0.708025
28	0.988299	0.006664	0.005037
29	0.027991	0.114111	0.857898
30	0.682107	0.148047	0.169846

Table 14.16f True Case Probabilities for Synthetic
Example

Case	Response 1	Response 2	Response 3
1	0.413380	0.248573	0.338047
2	0.502405	0.316602	0.180993
3	0.358411	0.600932	0.040657
4	0.519635	0.359809	0.120556
5	0.376285	0.327241	0.296474
6	0.312692	0.196167	0.491141
7	0.035461	0.141401	0.823138
8	0.570541	0.247008	0.182371
9	0.379898	0.230501	0.389601
10	0.817462	0.148343	0.034195
11	0.496940	0.307623	0.195438
12	0.264670	0.441606	0.293723
13	0.722271	0.175229	0.102501
14	0.176699	0.427277	0.396024
15	0.374307	0.363082	0.262611
16	0.131631	0.178322	0.690047
17	0.248814	0.061195	0.689991
18	0.232958	0.448868	0.318174
19	0.627933	0.342905	0.029162
20	0.664002	0.319920	0.016078
21	0.759130	0.041951	0.198920
22	0.583066	0.325642	0.091292
23	0.569967	0.354412	0.075621
24	0.707328	0.202134	0.090538
25	0.162984	0.299315	0.537701
26	0.111766	0.629213	0.259020
27	0.327155	0.058223	0.614622
28	0.859707	0.109530	0.030764
29	0.195022	0.199823	0.605155
30	0.614660	0.245494	0.139846

Table 14.16g Maximum Likelihood Estimates for
Parameters of Model (14.32) for Synthetic Example

Parameter	Estimate	Std. error
$\beta_0^{(1,2)}$	0.23853	0.8765
$\beta_1^{(1,2)}$	2.99826	1.5779
$\beta_2^{(1,2)}$	0.06482	0.6080
$\beta_0^{(2,3)}$	-0.7194	0.7622
$\beta_1^{(2,3)}$	-0.5088	0.8173
$\beta_2^{(2,3)}$	1.4120	0.7251

14.5 CONCLUDING COMMENTS

14.5.1 Count Responses

In this chapter, we have discussed models for categorical response variables only. Discrete responses, such as counts that may follow a Poisson distribution, are often modeled by specifying that the log of the Poisson rate follows a linear model. Such models are a subset of the much larger class of loglinear models. Indeed, the logistic regression and multinomial logit models are also included in the general class of loglinear models. Excellent sources on loglinear models and Poisson regression are Bishop et al. (1975), Haberman (1978), and Fienberg (1980). One caution must be given here, however. Standard tests concerning parameters for these models depend critically on the assumption that the response count is actually Poissonly distributed. Small departures from the Poisson assumption can result in greatly overstated statistical significance. Alternate models using the family of negative binomial distributions may be useful.

14.5.2 Choice Responses

An interesting aspect of multinomial response models occurs in taste testing and other choice situations in which not all possible choices can be presented simultaneously. These are basically incomplete multinomial response models. The Bradley–Terry model has proven extremely useful in practical modeling of such data. Fienberg and Larntz (1976) and Larntz (1975) relate loglinear models to choice models of Bradley and Terry (1952).

14.5.3 Suggested Reading

McCullagh and Nelder (1989) provide a complete source detailing generalized linear models, which include all the models discussed here plus many, many more. Cox and Snell (1989) present a wonderful statistical treatment of the analysis of binary response data, including details on multinomial response analysis. Hosmer and Lemeshow (1989) give practical details and sound advice for logistic regression models. Everitt (1977) provides a readable introduction to analyzing contingency tables and includes a lucid description of loglinear models for three-way tables. It is highly recommended as a solid starting point for nonstatisticians beginning study in this area.

REFERENCES

Agresti, A. (1990). *Categorical Data Analysis*. Wiley, New York.

Baker, D. D., Terpstra, D. E., and Larntz, K. (1990). The Influence of Individual Characteristics and Severity of Harassing Behavior on Reactions to Sexual Harassment. *Sex Roles*, *22*: 305–325.

Berk, R. A. and Sherman, L. W. (1988). Police Responses to Family Violence Incidents: An Analysis of an Experimental Design with Incomplete Randomization. *J. Amer. Statist. Assoc.*, *83*:70–76.

Bishop, Y. M. M., Fienberg, S. E., and Holland, P. W. (1975). *Discrete Multivariate Analysis: Theory and Practice*. MIT Press, Cambridge, Massachusetts.

Bradley, R. A. and Terry, M. E. (1952). The Rank Analysis of Incomplete Block Designs. I. The Method of Paired Comparisons. *Biometrika*, *39*: 324–345.

Cox, D. R., and Snell, E. J. (1989). *Analysis of Binary Data* (2nd ed.). Chapman & Hall, London.

Dyke, G. V. and Patterson, H. D. (1952). Analysis of Factorial Arrangements When the Data Are Proportions. *Biometrics*, *8*: 1–12.

Everitt, B. S. (1977). *The Analysis of Contingency Tables*. Chapman & Hall, London.

Fienberg, S. E. (1980). *The Analysis of Cross-Classified Categorical Data* (2nd ed.). MIT Press, Cambridge, Massachusetts.

Fienberg, S. E. and Larntz, K. (1976). Log Linear Representation for Paired and Multiple Comparisons Models. *Biometrika*, *63*: 245–254.

Fowlkes, E. B., Freeny, A. E., and Landwehr, J. M. (1988). Evaluating Logistic Models for Large Contingency Tables. *J. Amer. Statist. Assoc.*, *83*: 611–622.

Haberman, S. J. (1978). *Analysis of Qualitative Data*, Vols. 1 and 2. Academic Press, New York.

Hosmer, D. W. and Lemeshow, S. (1989). *Applied Logistic Regression*. Wiley, New York.

Koehler, K. and Larntz, K. (1980). An Empirical Investigation of Goodness-of-Fit Statistics for Sparse Multinomials. *J. Amer. Statist. Assoc.*, *75*: 336–344.

Larntz, K. (1975). Reanalysis of Vidmar's Data on the Effects of Decision Alternatives on Verdicts of Simulated Jurors. *J. Pers. Soc. Psych.*, *31*: 123–125.

Larntz, K. (1978). Small Sample Comparisons of Exact Levels of Chi-Squared Goodness of Fit Statistics. *J. Amer. Statist. Assoc.*, *73*: 253–263.

Larntz, K. (1980). Linear Logistic models for the Parole Decisionmaking Problem. *Indicators of Crime and Criminal Justice: Quantitative Studies*. Fienberg, S. E. and Reiss, A. J., Jr. (eds.). U.S. Government Printing Office, Washington, D.C., pp. 63–69.

McCullagh, P. and Nelder, J. A. (1989). *Generalized Linear Models*. Chapman & Hall, London.

Meyer, M. (1981). Applications and Generalizations of the Iterative Proportional Fitting Procedure. Unpublished Ph.D. dissertation, School of Statistics, University of Minnesota, Minneapolis.

Miller, R. (1966). *Simultaneous Statistical Inference*. McGraw-Hill, New York.

Sakamoto, Y. and Akaike, H. (1978). Analysis of Cross Classified Data by AIC. *Ann. Inst. Statist. Math.*, *B*, *30*: 185–197.

Sherman, L. W. and Berk, R. A. (1984). The Specific Deterrent Effects of Arrest for Domestic Assault. *Amer. Soc. Rev.*, *49*: 261–271.

Weisberg, S. (1985). *Applied Linear Regression* (2nd ed.). Wiley, New York.

15
Variance Component Estimation in Mixed Linear Models

RONALD R. HOCKING   Texas A&M University, College Station, Texas

15.1   INTRODUCTION

The classical, fixed-effects, linear model extends in a natural way to include random effects, i.e., models in which some of the parameters are assumed to be random variables. Models including both fixed and random effects are known as mixed models and include, as special cases, the fixed-effects model in which all factors are assumed fixed and the random model in which all factors are assumed random. Applications of such models are found in a wide range of fields. The application to genetics and animal breeding has had dramatic results in improving food production, and much of the theoretical development has been done by researchers in the agricultural sciences. However, applications in econometrics, the social sciences, ecology, and quality control have been equally impressive. In a broad sense, the mixed model represents a general approach to a substantial portion of statistical analysis of data from designed and undesigned experiments. Stroup (1989) and McLean, Sanders, and Stroup (1991) emphasize this point, suggesting that even broader application is possible but is limited only by the lack of understanding of the associated methodology and of adequate computer programs.

The analysis of data from a mixed linear model includes estimation of the variances of the random effects, (called variance components), estimation of the fixed-effects parameters, tests of hypotheses and specification of confidence intervals on both sets of parameters, and prediction of the random effects. Estimation of the variance components is central to the analysis of the mixed-effects model. In general, estimation of the fixed effects and the subsequent inferences about them depend on the estimates of the variance components.

In this chapter, we shall focus on this estimation problem for a broad, but not completely general class of mixed models. There is an extensive literature on the problem, as evidenced in the review papers by Searle (1971), Harville (1977), Sahai (1979), Khuri and Sahai (1985), and Sahai, Khuri, and Kapadia (1985). Classical estimation methods, except in special cases, present major computational problems with iterative techniques involving large matrices. In many situations, constraints such as nonnegativity are placed on the parameters, and satisfaction of such constraints greatly increases the computational burden. The effort required to obtain the estimates and the complexity of the procedures are such that little attention has been given to data diagnostics, i.e., to a detailed study of the observations and how they influence the estimates.

To motivate the discussion, we begin by describing models for some simple but common experimental situations and then give a general statement of the model that we shall consider.

15.1.1   Random One-Way Classification Model

Consider first a study of the response time of preschool children to a particular stimulus. Children from several different day-care centers are selected to participate. Within a center, the children are assumed to be fairly homogeneous with regard to their anticipated response, but there may be differences between centers. Since we are not interested in these particular centers, which may be viewed as a random sample from the large population of such centers, the center effect is treated as random. Suppose that there are a_1 centers in the study and that we have n_i children at the ith center. Letting y_{ij} denote the response time for the jth child at the ith center, we write the model as follows:

$$y_{ij} = \mu + \alpha_i + e_{ij}, \qquad i = 1, \ldots, a_1, \qquad j = 1, \ldots, n_i. \qquad (15.1)$$

Here μ denotes the mean response time to the stimulus for the age group of children in the study, and α_i is a random variable which represents the contribution of the ith center over and beyond the mean, μ. Note that in this model, the amount α_i is added to the response of each child in the ith center. Finally, e_{ij} is a random variable which represents the contribution of the jth child in the ith center. Assume that the α_i, $i = 1, \ldots, n$, are a random sample from a normal distribution with mean zero and variance ϕ_1, that is, $\alpha_i \sim NID(0, \phi_1)$. Also assume that the $e_{ij} \sim NID(0, \phi_0)$ and that α_i and e_{ij} are independent for all i and j. The use of ϕ_0 and ϕ_1, rather than the conventional σ_e^2 and σ_α^2, to represent these variance components is primarily for notational convenience since numerical subscripts are easier to generalize. We shall also see that these parameters need not be interpreted as variances but merely as elements of the covariance matrix. The term *variance component* will be interpreted in this general sense.

To clarify this last point, note that an equivalent statment of the random one-way classification model is obtained by specifying the mean and covariance structure. That is, the observations are assumed to be normally distributed with

$$
\begin{aligned}
E(y_{ij}) &= \mu & &\text{for all } i\,j, \\
\text{Var}(y_{ij}) &= \phi_0 + \phi_1 & &\text{for all } i, j, \\
\text{Cov}(y_{ij}, y_{ij}{}^*) &= \phi_1 & &\text{for } j \neq j^*, \\
\text{Cov}(y_{ij}, y_i{}^*{}_j{}^*) &= 0 & &\text{for } i \neq i^*.
\end{aligned}
\tag{15.2}
$$

Upon examining this covariance structure, we are struck by the fact that we might have modeled the data directly in terms of the first two moments without the use of a linear model. In particular, we assume that observations from different centers have the same variance ($\phi_0 + \phi_1$) and that the within-center observations are allowed to be correlated [$\text{Corr}(y_{ij}, y_{ij}{}^*)$ $= \phi_1/(\phi_0 + \phi_1)$]. In this formulation, there is no inherent requirement that the correlation be positive. This may often be the case, since children in the same center might be expected to respond similarly to the stimulus, but that need not be the case if there is competition within a center. We shall see that negative estimates are often a consequence of spurious data and the diagnostic methods will be useful in the detection of such problems.

15.1.2 Randomized Block Design

As a second example, consider a situation in which we wish to compare the response to several different treatments. Ideally, we would like to have a large number of similar experimental units (e.g., subjects) to which we would randomly allocate treatments. Frequently, this is not possible but we do have groups or blocks of similar units. The treatments may then be allocated at random within the blocks in a logical manner to allow treatment comparisons which are free from block differences. For example, if the block size is equal to the number of treatments, we randomly assign treatments to experimental units within a block. If the blocks are viewed as a random sample from a large population of blocks, analogous to the daycare centers in the one-way classification model, then, letting y_{ij} denote the response to the ith treatment in the jth block, we may write the model as follows:

$$
y_{ij} = \mu_i + b_j + e_{ij}, \qquad i = 1, \ldots, a_1, \qquad j = 1, \ldots, a_2. \tag{15.3}
$$

Here, μ_i is the mean response for the ith treatment, and b_j is a random variable denoting the contribution of the jth block. The b_j are assumed to be $NID(0, \phi_2)$ and independent of the e_{ij}, which are assumed to be $NID(0, \phi_0)$. We note that the covariance structure, which is implicit in this model, is similar to that in (15.2). In particular, the moment structure for this model is

$$
\begin{aligned}
E(y_{ij}) &= \mu_i &&\text{for all } i, j, \\
\mathrm{Var}(y_{ij}) &= \phi_0 + \phi_2 &&\text{for all } i, j, \\
\mathrm{Cov}(y_{ij}, y_{i^*j}) &= \phi_2 &&\text{for } i \neq i^*, \\
\mathrm{Cov}(y_{ij}, y_{i^*j^*}) &= 0 &&\text{for } j \neq j^*.
\end{aligned}
\tag{15.4}
$$

Apart from notation, this is the same covariance structure seen in the one-way classification model. In particular, we see that the covariance between observations in the same block is ϕ_2 and that the covariance between observations in different blocks is zero. This again suggests an alternative approach to modeling the data. That is, rather than saying that observations in a block are similar because they all have the same constant, b_j, added to them, we model the similarity by assuming that they are correlated. The model defined in (15.3) implicitly assumes that this correlation is positive and further assumes that this correlation is the same for all blocks. A model in which these assumptions are relaxed is known as a *repeated measures design*. For example, the blocks might be subjects and their response to the a_1 treatments is a vector of length a_1 which is assumed to be normal with some covariance structure. If we assume that this matrix has constant diagonals and positive, constant off-diagonals, then the structure is reduced to that of (15.4). We shall examine the implications of these observations in Sections 15.3 and 15.4.

For the purpose of later generalization, we write the model in a slightly modified form as

$$
y_{ij} = \mu + (\alpha_1)_i + (\alpha_2)_j + e_{ij}.
\tag{15.5}
$$

To relate to (15.3), the treatment means are denoted by $\mu_i = \mu + (\alpha_1)_i$ and the random block effects by $(\alpha_2)_j$. The use of ϕ_2 as the variance of the block effect refers to the fact that it is the second factor in (15.5). In a different application in which the first factor is also random, we would use ϕ_1 to denote the variance of the $(\alpha_1)_i$. We shall see that the numerical subscripts will be useful in generalizing our results. Finally, we note that imbalance in the randomized block design occurs when not all treatments are observed in each block. This is commonly referred to as the missing cell situation.

15.1.3 A Nested-Factorial (Hierarchical-Factorial) Design

For our third example, consider a situation in which the treatments are described in terms of two factors and all combinations of each level of each factor are of interest. For example, we may be interested in a study of the effectiveness of diet and exercise for the control of cholesterol. Suppose we have a_1 different diets and a_2 different exercise programs. Subjects from different health centers are available for the experiment, which is conducted as follows: within a center, the subjects are randomly assigned to a_2 groups, with n subjects per group, and the exercise programs are randomly assigned to those groups. For convenience, all subjects within a center receive the same diet. Each diet is used at a_3 different health centers.

In this example, the two treatment factors are in a factorial arrangement, but the third factor, health centers, is nested in diets. In this case, we have a combination of crossed and nested factors in which the two crossed factors may be assumed to be fixed effects and the nested factor would be random. A linear model for this study may be written as

$$y_{ijkt} = \mu_{ij} + (\alpha_{3(1)})_{ik} + (\alpha_{23(1)})_{ijk} + e_{ijkt} . \tag{15.6}$$

Here, μ_{ij} denotes the mean response when the ith diet is combined with the jth exercise program. The remaining factors are assumed to be independent random effects which are *NID* with zero means and variances $\phi_{3(1)}, \phi_{23(1)},$ and $\phi_0,$ respectively. The factor $\alpha_{3(1)}$ is called the center-within-diet effect and $\alpha_{23(1)}$ is the center-within-diet by exercise program interaction. A clearer understanding of the meaning of these random factors is obtained by spelling out the covariance structure which is implicit in this model. From (15.6) we have

$$\mathrm{Var}(y_{ijkt}) = \phi_{3(1)} + \phi_{23(1)} + \phi_0,$$
$$\mathrm{Cov}(y_{ijkt}, y_{ijkt^*}) = \phi_{3(1)} + \phi_{23(1)}, \qquad t \neq t^*,$$
$$\mathrm{Cov}(y_{ijkt}, y_{ij^*kt^*}) = \phi_{3(1)}, \qquad j \neq j^*,$$
$$\mathrm{Cov}(y_{ijkt}, y_{ijk^*t^*}) = 0, \qquad k \neq k^*. \tag{15.7}$$

Thus, $\phi_{3(1)}$ is the covariance between responses on individuals in the same center with a different exercise program and $\phi_{3(1)} + \phi_{23(1)}$ is the covariance between responses on different individuals in the same center receiving the same exercise program. Responses on individuals from different centers are uncorrelated. To conform to our general notation, the treatment combination means are written in classical, overparameterized form as

$$\mu_{ij} = \mu + (\alpha_1)_i + (\alpha_2)_j + (\alpha_{12})_{ij} . \tag{15.8}$$

Imbalance in this design can occur in several ways. We may have different numbers of subjects in exercise programs either within or between centers, and we may have an unequal number of centers on the various diets.

15.1.4   A Split-Plot Design

The diet-exercise study might have been conducted in a different way. Suppose that at each health center there are a_1 exercise classes, each class using a different exercise program. Within a class, we allocate the subjects to a_2 groups to receive different diets. This is known as a split-plot design recognizing that subjects are not randomly assigned to treatment combinations, but rather to diets within an exercise class. The centers are viewed as blocks, the exercise programs as whole-plot treatments, and the diets as subplot treatments. The linear model for this study is

$$y_{ijkt} = \mu_{ij} + (\alpha_3)_k + (\alpha_{13})_{ik} + (\alpha_{123})_{ijk} + e_{ijkt} . \tag{15.9}$$

Again, μ_{ij} denotes the mean of the ijth exercise-diet combination and may be written as in (15.8). The remaining factors are independent random variables which are NID with zero means and variances ϕ_3, ϕ_{13}, and ϕ_{123}, and ϕ_0, respectively. Thus, the difference between the response and its mean value is the sum of four random variables. The first is added to each observation in the kth block, the second to each observation in that block receiving the ith exercise program and the third to each observation in that block receiving the ijth treatment combination. The final random variable reflects the tth individual's contribution. There is disagreement in the literature (see Harter, 1961) as to the inclusion of the block-by-subplot interaction term, α_{23}. An examination of the covariance structure implied by (15.9) provides an explanation for this disagreement. The second moments are given as

$$\text{Var}(y_{ijkt}) = \phi_3 + \phi_{13} + \phi_{123} + \phi_0,$$
$$\text{Cov}(y_{ijkt}, y_{ijkt^*}) = \phi_3 + \phi_{13} + \phi_{123}, \qquad t \neq t^*,$$
$$\text{Cov}(y_{ijkt}, y_{ij^*kt^*}) = \phi_3 + \phi_{13}, \qquad j \neq j^*,$$
$$\text{Cov}(y_{ijkt}, y_{i^*j^*kt^*}) = \phi_3, \qquad i \neq i^*,$$
$$\text{Cov}(y_{ijkt}, y_{ijk^*t^*}) = 0, \qquad k \neq k^*. \tag{15.10}$$

Note that ϕ_3 is the covariance between observations in the same center under different exercise programs; $\phi_3 + \phi_{13}$ is the covariance between observations in the same center on the same exercise program but with different diets; and $\phi_3 + \phi_{13} + \phi_{123}$ is the covariance between observations in the same center with the same treatment combination. Observations in different centers are uncorrelated. Inclusion of the block-by-subplot interaction term, α_{23}, in (15.9) would imply that observations in the same center on the same diet but on different exercise programs would have covariance $\phi_3 + \phi_{23} \neq \phi_3$. Since diets are assigned at random within exercise programs, a nonzero ϕ_{23} does not seem likely. Again, imbalance can occur if we have different numbers of individuals per treatment combination, that is, if not all diets are used in each exercise group, or if not all exercise programs are used in each center.

15.1.5   General *k* Factor Mixed Model

These examples function as a motivation for the general model. Extending the notation of our examples, we write the general *k*-factor mixed model as follows:

$$y_{i_1 i_2 \cdots i_k t} = \mu + (\alpha_1)_{i_1} + (\alpha_2)_{i_2} + (\alpha_{12})_{i_1 i_2}$$
$$+ \cdots + (\alpha_{12 \cdots k})_{i_1 \cdots i_k} + e_{i_1 \cdots i_k t} . \tag{15.11}$$

We have written the model in the classical, overparameterized, fixed-effect form. The mixed model arises when we designate certain effects as random. In the simplest case, called the random model, μ is the only fixed effect, and the remaining effects are independent random variables which are *NID* with zero means and variances $\phi_1, \phi_2, \phi_{12}, \ldots \phi_{12 \cdots k}, \phi_0$. The subscripts on the variance components are related in the obvious way to the terms in (15.11). It is convenient to refer to the variance components, excluding ϕ_0, collectively as ϕ_τ, $\tau \in T$ where T is the collection of subsets of the integers $1, \ldots, k$. That is, $T = \{1, 2, 12, 3, 13, 23, 123, \ldots, 12 \ldots k\}$. We shall use T_0 to denote the collection of subscripts when we wish to include ϕ_0. The range on the subscripts is $i_j = 1, 2, \ldots a_j$ for $j = 1, 2, \ldots k$ and $t = 1, 2, \ldots n$ in what is called the balanced data case. In general, the range on a subscript may depend on the value of other subscripts, and it may happen that not all combinations are observed.

    In the mixed model, some of these random variables are assumed to be parameters and are called fixed effects. The mean response is the sum of these parameters, and the associated variance components are set to zero. It is convenient to assume that lower subscripted terms are fixed. Expressing the expected values in terms of these parameters results in an overparameterized model for the fixed effects. It is well known that this degeneracy can be removed (Hocking, 1985). In addition, if a term is assumed fixed, then all terms whose subscripts are subsets of the subscripts of that term are also fixed. For example, if α_{12} is fixed, then α_1 and α_2 are also fixed. As we have seen in our examples, not all terms need be included. This is equivalent to setting certain of the ϕ_τ to zero or, analogously, making certain assumptions about the covariance structure on the data. Thus, the model includes the general class of mixed factorial, nested, and nested-factorial models.

The covariance between observations which is implied by (15.11) may be described by a new set of parameters. Let

$$\theta_\tau = \sum_{s \subseteq \tau} \phi_s \qquad (15.12)$$

where the sum is over all sets of integers which are proper subsets of the integers in T. For example, $\theta_{12} = \phi_1 + \phi_2 + \phi_{12}$. We may then write

$$\text{Cov}(y_{i_1 i_2 \cdots i_k t}, y_{i_1^* i_2^* \cdots i_k^* t^*}) = \theta_\tau , \qquad (15.13)$$

where $i_j = i_{j^*}$ for $j \in \tau$ and $i_j \neq i_{j^*}$ for $j \notin \tau$. As we have noted in our examples, it may be convenient, or even more appropriate in some cases, to think about modeling the data in terms of their mean and covariance structure rather than as a linear model. In this case, the only natural constraint on the parameters is that the covariance matrix is positive definite. We shall see in Section 15.3 that this approach leads us to an alternative way of computing the estimates in the balanced data case which, in turn, provides a useful diagnostic analysis. Furthermore, this concept has recently been extended to the general case of unbalanced data, including the case of missing cells.

The mixed model in (15.11) may be written in matrix notation as

$$\mathbf{Y} = \mathbf{X}\boldsymbol{\beta} + \sum_{\tau \in T^*} \mathbf{Z}_\tau \, \boldsymbol{\alpha}_\tau + \mathbf{e}. \qquad (15.14)$$

Here $\mathbf{Y}$ is the vector of observations of length N; $\mathbf{X}$ is a known matrix of size $N \times p$, with rank p; and $\boldsymbol{\beta}$ denotes the vector of fixed-effect parameters. The matrices $\mathbf{Z}_\tau$ are also known and the random vectors, $\boldsymbol{\alpha}_\tau$ and $\mathbf{e}$, are independent, $N(\mathbf{0}, \phi_\tau \mathbf{I})$ and $N(\mathbf{0}, \phi_0 \mathbf{I})$, respectively, with T^* denoting a subset of T. Specific expressions for $\mathbf{X}$ and $\mathbf{Z}_\tau$ are implied by the particular model in question (Hocking, 1985). Although we have written (15.14) as the matrix form of (15.11), which implies that $\mathbf{X}$ is an indicator matrix, the notation extends to include covariates in the $\mathbf{X}$ matrix. In our theoretical development in the next section, we shall allow this generality unless otherwise noted. The covariance matrix of $\mathbf{Y}$ is given as

$$\mathbf{V} = \text{Var}(Y) = \sum_{\tau \in T^*} \phi_\tau \mathbf{Z}_\tau \mathbf{Z}_\tau^{\mathsf{T}} + \phi_0 \mathbf{I}_N$$

$$= \sum_{\tau \in T^*} \phi_\tau \mathbf{V}_\tau + \phi_0 \mathbf{I}_N , \qquad (15.15)$$

where we define $\mathbf{V}_\tau = \mathbf{Z}_\tau \mathbf{Z}_\tau^{\mathsf{T}}$ and superscript T denotes matrix transpose. This expression is notationally convenient, but specific expressions for the covariances, such as those given in our examples, are more informative.

Finally, this model may be generalized to allow for more complex assumptions about the covariance matrix. Extending (15.15), we may write

$$\mathbf{Y} = \mathbf{X}\boldsymbol{\beta} + \mathbf{Z}\boldsymbol{\alpha} + \mathbf{e}, \qquad (15.16)$$

where $\text{Var}(\boldsymbol{\alpha}) = \mathbf{V}_\alpha$ and $\text{Var}(\mathbf{e}) = \mathbf{V}_0$, where $\mathbf{V}_\alpha$ and $\mathbf{V}_0$ are general covariance matrices. It is usually assumed that α and $\mathbf{e}$ are not correlated. In this case,

$$\text{Var}(\mathbf{Y}) = \mathbf{Z}\mathbf{V}_\alpha\mathbf{Z}^\mathsf{T} + \mathbf{V}_0 . \qquad (15.17)$$

In the discussion in Section 15.2, some of our results will apply to (15.16) but we are primarily interested in (15.14) and the special case (15.11).

15.2 METHODS OF ESTIMATION

We begin our discussion of estimating the variance components by assuming the balanced data case as defined by (15.11). Thus, in the one-way classification example, we assume the same number of observations per group. In the randomized block design, we assume one observation on each treatment in each block. In the nested-factorial design, we assume the same number of centers for each diet and the same number of subjects for each treatment combination, and so on. The advantage of these assumptions is that, under normality, we can identify a complete set of sufficient statistics for the parameters. The impact is that we can obtain estimates of the variance components and the fixed effects, which have minimum variance in the class of unbiased estimators. These statistics are easily identified. If we write the standard AOV table for the model, assuming that all effects are fixed, the mean squares in this table are functions of the sufficient statistics. Equating the mean squares for the random effects to their expected values yields a set of linear equations whose solution yields these optimal estimates of the variance components. Simple expressions for these mean squares and their expected values are given in Hocking (1990). Without the normality assumption, these estimates can be shown to have minimum variance in the class of estimates which are quadratic functions of the data. The estimates of the fixed effects are not dependent on the variance components and are given by

$$\hat{\boldsymbol{\beta}} = (\mathbf{X}^\mathsf{T}\mathbf{X})^{-1} \mathbf{X}^\mathsf{T}\mathbf{Y}. \qquad (15.18)$$

It can be shown that the estimate of the covariance matrix $\mathbf{V}$, obtained by substituting these estimates of the variance components, is positive definite and hence is acceptable under our covariance formulation of the model. However, it is possible that the individual components may have negative estimates which violate the assumptions in the linear model formulation. Even in the covariance formulation, the negative estimates may lead to correlation patterns which are difficult to explain in the context of the application.

Many solutions have been proposed for dealing with negative estimates. Nelder (1954) discusses the interpretation of negative estimates. Thompson and Moore (1963) suggested a pooling of the AOV sums of squares until nonnegative estimates are obtained. Other suggestions include those of Rao and Chaubey (1978), Hartung (1981), and Matthew, Sinha, and Sutradhar (1989). A general solution, which is not restricted to balanced data, is to consider the maximization of the likelihood function subject to the constraints $\phi_\tau \geq 0$ for $\tau \in T_0$. The solution of such inequality-constrained optimization problems is not elementary and, even when the constraints are satisfied, the estimators may be biased. A general algorithm for the solution of this problem was described by Hartley and Rao (1967). An alternative form of the likelihood equations which ensures nonnegative estimates was given by Henderson (see Harville, 1977). Although these procedures yield nonnegative estimates of the variance components, it might be argued that an estimate of zero, which is the usual result, is not appreciably more informative than a negative estimate. What is lacking is an explanation for this unacceptable result. We shall return to this problem in Section 15.3.

The unbalanced data situation is far more difficult. Henderson (1953) described several methods for extending the AOV tables for balanced data to the unbalanced case and proposed equating mean squares to their expected values. The expected mean squares for an AOV table must be computed numerically using procedures such as those of Hartley (1967) or Goodnight and Speed (1980). Of even more concern is the fact that there is no unique AOV table and that different tables generally yield different estimators, none of which have been shown to be superior. The problem is that, even in the simplest unbalanced model, there is no complete set of sufficient statistics. The constrained maximization of the likelihood function mentioned above is an option, but the imbalance causes a substantial increase in computing. To appreciate the magnitude of the computations and, perhaps of more importance, to get a feeling for the nature of the estimates, it is informative to spell out the stationary equations for the likelihood function. We emphasize that the solution of these equations will not generally satisfy the nonnegativity constraint. Letting ϕ denote the vector of variance components, ϕ_τ for $\tau \in T_0$, the equations for ϕ may be written in pseudolinear form as

$$\Omega\phi = \delta \tag{15.19}$$

The elements of Ω and δ depend on ϕ and are defined as follows:

$$\omega_{st} = tr(\mathbf{V}^{-1}\mathbf{V}_s\mathbf{V}^{-1}\mathbf{V}_t) \quad \text{and}$$

$$\delta_s = (\mathbf{Y} - \mathbf{X}\boldsymbol{\beta})^{\mathrm{T}}(\mathbf{V}^{-1}\mathbf{V}_s\mathbf{V}^{-1})(\mathbf{Y} - \mathbf{X}\boldsymbol{\beta}). \tag{15.20}$$

Here, tr denotes the trace of the matrix and the elements of $\boldsymbol{\Omega}$ and $\boldsymbol{\delta}$ are indexed by the subsets of T_0. The matrices $\mathbf{V}_s$, for $s \in T_0$, are defined in (15.15) with $\mathbf{V}_o = \mathbf{I}_N$. (For details of this development, see Hocking, 1985). The equations for $\boldsymbol{\beta}$ are given by

$$(\mathbf{X}^{\mathrm{T}}\mathbf{V}^{-1}\mathbf{X})\hat{\boldsymbol{\beta}} = \mathbf{X}^{\mathrm{T}}\mathbf{V}^{-1}\mathbf{Y}. \tag{15.21}$$

Substituting (15.21) and (15.20), we have

$$\delta_s = \mathbf{Y}^{\mathrm{T}}\mathbf{M}^{\mathrm{T}}\mathbf{V}^{-1}\mathbf{V}_s\mathbf{V}^{-1}\mathbf{M}\mathbf{Y}, \tag{15.22}$$

where

$$\mathbf{M} = \mathbf{I} - \mathbf{X}(\mathbf{X}^{\mathrm{T}}\mathbf{V}^{-1}\mathbf{X})^{-1}\mathbf{X}^{\mathrm{T}}\mathbf{V}^{-1}. \tag{15.23}$$

Using (15.22) and (15.20), a simple procedure is suggested. That is, evaluate $\boldsymbol{\Omega}$ and $\boldsymbol{\delta}$ for an initial guess at $\boldsymbol{\phi}$, solve the system of equations to get a new value for $\boldsymbol{\phi}$, and iterate on this process. This is essentially the method of scoring proposed by R. A. Fisher (Tarone, 1988) in a more general setting, and it has reasonable convergence properties. The magnitude of the computational effort is evident. There are numerous matrix operations, including inversion, where the size of the matrices is the total number of observations. Hemmerle and Hartley (1973) showed that substantial savings can be achieved in computation but it is still a nontrivial effort. Harville (1977) provides us with a general discussion of the maximum likelihood (ML) estimation of variance components and gives alternative expressions for the estimates due to Henderson.

The issue of bias in the maximum likelihood estimates was addressed by Patterson and Thompson (1971). They proposed a factorization of the likelihood in which one factor is the distribution of a set of $N - p$ contrasts on $\mathbf{Y}$ and does not depend on the fixed effects. This likelihood function, say L_ϕ, is used to estimate $\boldsymbol{\phi}$. Given this estimate, the other factor yields the estimate of $\boldsymbol{\beta}$ given by (15.21). To illustrate the effect, consider the stationary equations for estimating $\boldsymbol{\phi}$. They have the same form as (15.19) with δ_s given by (15.22) and the elements of $\boldsymbol{\Omega}$ defined by

$$\omega_{st} = tr(\mathbf{V}^{-1}\mathbf{M}\mathbf{V}_s\mathbf{V}^{-1}\mathbf{M}\mathbf{V}_t). \tag{15.24}$$

To emphasize the effect of this modification, we note that, with balanced data, the solution of (15.19) using (15.24) may be expressed in closed form, yielding estimators which are identical to the unbiased AOV estimators. In general, an iterative solution is required and there is no guarantee that the estimators will be nonnegative. The fixed effects are estimated from

(15.21) with these modified estimates of ϕ. The constrained maximization of L_ϕ is known as modified maximum likelihood, partial likelihood, or, most commonly, restricted maximum likelihood (REML). We shall use the acronym, REML, because of its general acceptance (Corbeil and Searle, 1976). We emphasize that the adjective carries a double meaning. That is, the maximization is restricted to a factor of the original likelihood function and, furthermore, the maximization must be constrained so that the estimates are nonnegative if that is a part of the model statement.

Rao (1970, 1971a, 1971b, 1972) proposed a method called MINQUE (MIVQUE), the acronym standing for minimum norm (variance) quadratic unbiased estimator (see also Swallow and Searle, 1978). We shall use MIVQUE which is appropriate under the assumption of normality. LaMotte (1973) showed that this estimator is equivalent to the solution of the likelihood equations for L_ϕ, i.e., (15.19) using (15.24), where (15.24) is evaluated at some initial choice of ϕ. Rao suggested $\phi_\tau = 1$, for $\tau \in T_o$, or $\phi_0 = 1$ and $\phi_\tau = 0$, for $\tau \in T$ as two possibilities. The latter choice has become known as MIVQUE(0) and is appealing because of its simplicity. Note that in this case V is the identity matrix and hence the evaluation of (15.24) becomes particularly simple. The estimators may not be very good unless this initial choice is near optimum. Furthermore, there is no assurance that the estimators will be nonnegative. With balanced data, we obtain the AOV estimates for any initial choice of ϕ. One is clearly tempted to iterate on this idea, an algorithm for implementing this is described by Giesbrecht (1989).

The optimality properties of the AOV (MIVQUE) estimators with balanced data make it desirable to achieve balance in some way. One suggestion is to delete observations so that all cell frequencies are equal to the smallest one. This may be acceptable with modest imbalance but would generally discard too much information. It clearly breaks down if the minimum frequency is zero as in the randomized block design with missing cells. The other alternative is to achieve balance by imputing values for the missing observations. This concept, motivated by its applicability to a broad class of problems, ultimately leads to the development of the EM algorithm as described by Dempster, Laird, and Rubin (1977). The iterative algorithm consists formally of an E (expectation) step and an M (maximization) step, although, in practice, the two steps may not be so distinct. To see the idea, consider first the likelihood function for the complete, thus, balanced data vector Y which consists of Y_o, the actual observed data, and Y_m, the remaining, unobserved (missing) data. This likelihood can be factored as

$$L(Y_o, Y_m \mid \theta) = L(Y_o \mid \theta) \, L(Y_m \mid Y_o, \theta). \tag{15.25}$$

Here θ denotes a general parameter vector including both ϕ and β in our case. The ML method consists of maximizing the likelihood of the observed data. $L(\mathbf{Y}_o \mid \theta)$. The expected value of the logarithm of this likelihood conditional on the observed data and a current guess at θ, say $\theta^{(r)}$, can be written as

$$\log(L(\mathbf{Y}_o \mid \theta)) = Q(\theta \mid \theta^{(r)}) - H(\theta \mid \theta^{(r)}), \tag{15.26}$$

where

$$Q(\theta \mid \theta^{(r)}) = E\{\log(L(\mathbf{Y}_o, \mathbf{Y}_m \mid \theta)) \mid \mathbf{Y}_o, \theta^{(r)}\}. \tag{15.27}$$

This is the E step, and the M step consists of choosing $\theta^{(r+1)}$ to maximize $Q(\theta \mid \theta^{(r)})$ with respect to θ. Dempster et al. (1977) have shown that, under fairly general conditions, this procedure converges to a stationary value of the likelihood of the observed data. The EM algorithm has been successfully applied to many problems, including the estimation of variance components. The primary criticism of the procedure is that the convergence is generally slower than the scoring or Newton–Raphson algorithms (see e.g., Lindstrom and Bates, 1988). It should be noted that the method can be used to compute REML estimates by applying it to the REML likelihood, L_ϕ.

The application of the EM algorithm to the variance component estimation problem has been done in several different ways. It can be argued that the Hartley–Rao (1967) algorithm is essentially an EM procedure. Jennrich and Schluchter (1986) considered a direct application in either ML or REML estimation. For the latter, they used a form of the likelihood suggested by Harville (1974). Laird (1982) considered a different approach in which the random effects were treated as the unobserved data. This novel approach, which was used in both ML and REML form, led to a fairly simple set of iterative equations. These equations are equivalent to those proposed by Henderson (see Harville, 1977). It should be noted that all of these applications require the inversion of the estimated covariance matrix of $\mathbf{Y}_o$ at each iteration and they do not ensure nonnegative estimates. In the next section, we shall revisit the EM algorithm, as applied to REML estimation, and observe some interesting aspects of the problem with regard to computation and diagnostic analysis of the estimates.

15.3   RECENT DEVELOPMENTS

The problem of negative estimates has received much attention. Is this problem the result of an inappropriate model? Is it a natural phenomenon when the component is small? Or is it a consequence of the problematic data? The alternative form of the model, discussed in Section 15.1, where the parameters were θ_r, the variances and covariances of the observations, suggested an examination of the estimators. The idea is that since we can reparameterize in terms of the θ_r, which are variances and covariances, it seemed that the estimates of these parameters should be sample variances and covariances. Examination of these estimators led to the general results in Hocking (1990). In many situations, examination of this form of the estimators provides a simple explanation for the negative estimates. Hocking (1990) considered the general case of balanced data and showed that the estimates of a set of linear functions of the variance components can be computed as an average of sample variances and covariances on either the original data or marginal means of the original data. Estimates of the original components of variance are then obtained by a solution of the implied system of linear equations.

For example, recall the randomized block design in which we saw in (15.4) that the two essential parameter functions were ϕ_2 and $\phi_0 + \phi_2$. Suppose we array the data in a two-way table in which the a_1 columns correspond to treatments and the a_2 rows correspond to blocks. Now compute the sample covariance matrix for these data in which the diagonal elements are the variances for a given treatment and the off-diagonal elements are the covariances for pairs of treatments. It is easily shown that each of the off-diagonal elements is an unbiased estimate of ϕ_2 and that the AOV estimate of ϕ_2 is the average of these off-diagonal elements. Similarly, the diagonal elements of this matrix yield a_1 unbiased estimates of $\phi_0 + \phi_2$ whose average is the AOV estimate of this parameter function. It can be shown that this method of computing the estimates is computationally more efficient than fitting the linear model. However, more important, a natural diagnostic procedure is built into this method. The following steps briefly illustrate such a diagnostic procedure. It is expected that the off-diagonal elements should vary in a natural way about their average. If one or more of these elements is outlying, we can examine it further by inspecting the scatter plot of the two columns of data involved in the sample covariance. This examination may reveal unusual observations or groups of observations. The unusual covariances may also form a pattern which might suggest a modeling problem. Similarly, the diagonal elements should be examined for unusual observations or patterns. We must remember that unusual covariances can be caused by unusual correlations or variances. The estimate of ϕ_0 is then obtained in the obvious way by solving this simple system of equations.

We now see how negative estimates of ϕ_2 can arise. The sample covariances can surely be negative, and if the negatives dominate the positives, the estimate of ϕ_2 will be negative. Again, this may be caused by outlying data or a modeling problem. One point that is clear from this form of the estimate is that a large number of blocks is required to get a good estimate of the sample covariances. Negative estimates attract our attention, but they are not the only problem and the diagnostic analysis should be routinely applied. Hocking, Green, and Bremer (1989) describe an application in which the diagnostic analysis provided information about an unusual feature of the experimental setup and a problem with the raw material.

The question of what to do about negative or unusual estimates may be resolved by this diagnostic analysis. Alternatively, we might consider robust estimates of the sample covariance matrix. Rocke (1983) and Fellner (1986) proposed robust estimators which are useful but should not be routinely applied without a diagnostic examination of the data.

The extension of this concept to general balanced analysis of variance models of the form (15.11) was developed by Hocking (1990). Estimates of variance components for mixed models with any combination of crossed or nested factors can be obtained from averages of appropriate elements of certain sample covariance matrices. A more complex example, given in the next section, will help to illustrate the general concept. While a complete diagnostic analysis may be exhausting, it is a simple matter to assess the relative magnitudes of the estimates being averaged to obtain a given estimate. In addition, the computational savings can be significant.

An extension of this idea to the unbalanced data problem was proposed by Hocking et al. (1989) for the case of factorial models with no missing cells. The simple device of using the sample cell means, even though they are based on different cell frequencies, provided highly efficient estimators. Hocking (1988) showed that this approach reduced to the Yates (1934) method of unweighted means for generating an AOV table for unbalanced data. The expected values of the mean squares in this table are the analogs of those in the balanced table except that the coefficients of the variance components are based on harmonic means of cell frequencies. The idea readily extends to models with combinations of nested and crossed factors. The acronym AVE was used to describe these estimators, reflecting the fact that the estimates are based on averages of sample variances and covariances. The appeal of these estimators is simplicity because the procedure is noniterative. The estimators are quite efficient except when the ratio ϕ_0/ϕ_τ is small, and, as noted, they do not apply when there are empty cells. This includes nested designs in which there are variable ranges on the subscripts of nested factors.

More recently, Hocking and Gomez-Meza (1991) have combined the generality of the EM algorithm and the diagnostic features of the AVE estimators to yield a procedure which appears to be computationally efficient and intuitively appealing. The idea is to apply the EM procedure to the REML likelihood without the nonnegative constraint. A careful examination of the form of the estimating equations shows that the estimators (which are the iterative MIVQUE estimators) can be computed using the AVE concept applied iteratively. The idea is that a balanced data set is constructed at each iteration by imputing values for the missing observations. Sample covariance matrices are computed in the usual way except that the degrees of freedom associated with each element depend on the current estimates and hence must be computed at each iteration. The imputation requires a matrix inversion which can employ the Hemmerle–Hartley (1973) suggestions, and the degrees-of-freedom computation is simple. The estimates of the appropriate linear function of the variance components are then degrees-of-freedom weighted averages of the elements of the sample covariance matrices. The method apparently requires less computation than the usual iterative MIVQUE and has the advantage of providing the same diagnostic information as in the balanced case. The appropriate formulas are summarized in the following and illustrated in the next section.

Let $\mathbf{Y}$ denote the vector of observations if we had balanced data. It is convenient to think of the data as being reordered so that the missing values are separated from the observed data. Let $\mathbf{D}^T$ be a matrix which reorders the data as follows:

$$\mathbf{Y}^* = \mathbf{D}^T\mathbf{Y} = \begin{bmatrix} \mathbf{Y}_o \\ \mathbf{Y}_m \end{bmatrix}. \tag{15.28}$$

Here $\mathbf{Y}_o$ is the vector of observed values and $\mathbf{Y}_m$ is the vector of missing values. The design matrix, $\mathbf{X}$, and the covariance matrix, $\mathbf{V}$, of the complete data vector, $\mathbf{Y}^*$, are written in partitioned form, conformable with (15.28) as

$$\mathbf{X}^* = \begin{bmatrix} \mathbf{X}_o \\ \mathbf{X}_m \end{bmatrix}, \qquad \mathbf{V}^* = \begin{bmatrix} \mathbf{V}_{oo} & \mathbf{V}_{om} \\ \mathbf{V}_{mo} & \mathbf{V}_{mm} \end{bmatrix}. \tag{15.29}$$

With this notation, we can write the expected value of $\mathbf{Y}_m$, conditional on $\mathbf{Y}_o$ and the current guess at the parameters, as

$$E[\mathbf{Y}_m \mid \mathbf{Y}_o, \phi^{(r)}] = \mathbf{X}_m\hat{\boldsymbol{\beta}} + \mathbf{V}_{mo}\mathbf{V}_{oo}^{-1}(\mathbf{Y}_o - \mathbf{X}_o\hat{\boldsymbol{\beta}}), \tag{15.30}$$

where

$$\hat{\boldsymbol{\beta}} = (\mathbf{X}_o^T\mathbf{V}_{oo}^{-1}\mathbf{X}_o)^{-1}\mathbf{X}_o^T\mathbf{V}_{oo}^{-1}\mathbf{Y}_o. \tag{15.31}$$

The estimate of any linear function of the variance components, say γ, is then given iteratively by

$$\gamma^{(r+1)} = \begin{bmatrix} \mathbf{Y}_o \\ E[\mathbf{Y}_m \mid \mathbf{Y}_o] \end{bmatrix}^T \mathbf{D}^T \mathbf{A}_\gamma \mathbf{D} \begin{bmatrix} \mathbf{Y}_o \\ E[\mathbf{Y}_m \mid \mathbf{Y}_o] \end{bmatrix} / \mathrm{DF}_\gamma . \qquad (15.32)$$

Here, $E[\mathbf{Y}_m \mid \mathbf{Y}_o]$ denotes (15.30) evaluated at $\phi^{(r)}$. Thus, the estimate is just a quadratic function in terms of a data vector consisting of the observed data $\mathbf{Y}_o$ and imputed values for the missing data divided by the appropriate degrees of freedom. The degrees of freedom associated with this parameter function are given by

$$\mathrm{DF}_\gamma = \mathrm{tr}\{\mathbf{D}^T \mathbf{A}_\gamma \mathbf{D} \begin{bmatrix} \mathbf{V}_{oo} \\ \mathbf{V}_{mo} \end{bmatrix} \mathbf{V}_{oo}^{-1} \mathbf{M}_o \ (\mathbf{V}_{oo} \mathbf{V}_{om})\}/\gamma^{(r)}. \qquad (15.33)$$

The matrix $\mathbf{M}_o$ in (15.33) is the matrix $\mathbf{M}$ from (15.23) with $\mathbf{X}$ and $\mathbf{V}$ replaced by $\mathbf{X}_o$ and $\mathbf{V}_{oo}$ with all parameter functions evaluated at $\phi^{(r)}$. The matrix $\mathbf{A}_\gamma$ is determined by the parameter function of interest. Thus, $\mathbf{A}_\gamma$ might be the matrix of one of the quadratic forms in a balanced AOV table if γ is the expected mean square for that table. Alternatively, $\mathbf{A}_\gamma$ might be one of the matrices defined by the linear functions identified by Hocking (1990). In practice, we prefer the latter because this enables us to replace the direct evaluation of the quadratic form in (15.32) by the computation of a sample covariance matrix. This requires simply that the imputed data vector be arrayed in the appropriate two-way table for computation of the covariance matrix. There are similar savings in the computation of the degrees of freedom. Indeed, a separate degree of freedom may be computed for each element of the covariance matrix which is needed for the diagnostic analysis. For details of this technique, the reader is referred to Hocking and Gomez-Meza (1991). The basic idea is illustrated by the numerical example in the next section.

15.4 NUMERICAL EXAMPLES

To illustrate the estimation methods described in Sections 15.2 and 15.3, we borrow two examples from Kirk (1982). The original examples are modified to illustrate the various points made in the discussion of the diagnostic analysis.

Example 15.1
The first example (Kirk, 1982, p. 298) is concerned with the scores on three standard tests of 12 brain-damaged patients. Assuming that this is a random sample of such patients and that the order of application of the

Table 15.1 Scores on Standard Tests of 12 Brain-
Damaged Patients

	Tests		
Subjects	t_1	t_2	t_3
1	15	12	11
2	10	11	8
3	6	4	3*M
4	7	7	5
5	9	6	6
6	16	14	10*M
7	11	10M	7
8	13	9M	4
9	12	10M	8
10	10M	8	7M
11	11M	9	9
12	14M	11M	10

tests was randomized for each patient, we model these data as a randomized
block design with patients as blocks and tests as treatments. The data are
shown in Table 15.1.

For illustration, we have slightly modified the original data. In particular,
the observations for Subjects 3 and 6, indicated by asterisks in Table 15.1,
were replaced by 8 and 1, respectively. These represent common typographical errors which often go undetected in the analysis. Since these data
are balanced, the standard procedure is to use the AOV table for the fixed-
effect model and equate mean squares to their expected values. (For Example 15.1, ignore M's placed on some of the data points. These values
with M's will be treated as missing values in the next example.) The results
for Example 15.1 are shown in Table 15.2.

Equating the last two mean squares to their expected values, we obtain
the AOV estimates, $\phi_0 = 5.98$ and $\phi_2 = 2.08$. We emphasize that this

Table 15.2 Analysis of Variance Table for Example 15.1

Source	df	MS	E(MS)
Tests	2	52.2	$\phi_0 + Q(T)$
Patients	11	12.2	$\phi_0 + 3\theta_2$
Error	22	5.98	ϕ_0

analysis is appropriate for the model as we have formulated it, but the reader is referred to the recent paper by Samuels, Casella, and McCabe (1991) for a discussion of alternatives. As noted in Section 15.2, because the data are balanced, these estimates agree with MIVQUE and, since they are nonnegative, they also agree with REML. The ML estimates are $\phi_0 = 5.48$ and $\phi_2 = 1.90$, indicating the bias in this procedure. The two typographical errors we introduced had a considerable effect, since the original data yielded the AOV estimates of $\phi_0 = 1.38$ and $\phi_2 = 6.27$. The ML procedure gave $\phi_0 = 1.27$ and $\phi_2 = 5.75$ for the original data. Unfortunately, there is nothing in these procedures to indicate that there is any problem with these estimates and, in particular, there is no evidence of these errors in the data.

The AVE procedure described by Hocking et al. (1989) and, more generally, in Hocking (1990), is particularly simple for this example. We need only compute the 3×3 sample covariance matrix for the data in Table 15.1 treating the columns as variables. This covariance matrix is shown in Table 15.3, where we show the covariances above the diagonal and the correlations below the diagonal. This is a useful device for separating the problem of linear association from that of unusual variability.

As noted in Section 15.3, the diagonal elements are each unbiased estimates of $\phi_0 + \phi_2$ and their average is the AOV estimate of that parameter. We note that these estimates vary in a reasonable way about their average of 8.06. The above-diagonal elements are each unbiased estimates of ϕ_2 whose average is the AOV estimate. Here, the situation is quite unusual. Two of the estimates are negative while the third is positive. The average of these, 2.08, does not do a good job of representing these individual estimates. Indeed, the negative covariances suggest a problem with Test 3. Inspection of the scatter plots of either t_1 or t_2 against t_3 reveals the problem. For example, we show in Figure 15.1, a plot of the data for Test 1 (t_1) against test 3 (t_3) where the plotted symbol represents the subject number. Apart from Subjects 3, 6, and 8, there is a strong linear relation between these two tests. Subjects 3 and 6 were selected for modification in our example to give negative correlations and would be easily identified

Table 15.3 Sample Covariance (Correlation) Matrix for Example 15.1

	t_1	t_2	t_3
t_1	9.24	7.41	− .55
t_2	0.89	7.48	− .64
t_3	− .07	− .08	7.45

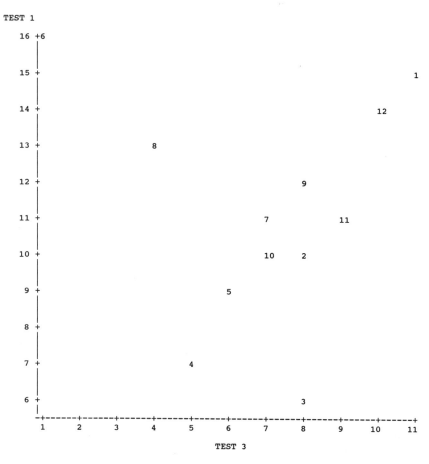

Figure 15.1 Plot of data for test 1 versus test 3 in Example 15.1.

as outliers. We shall comment below on Subject 8. In practice, we would attempt to verify these observations from original records if possible or, perhaps, repeat the analysis with these subjects deleted to assess the effect of the observations. The point we wish to make is that our approach to estimation, the AVE method, has revealed a possible problem while classical methods have failed to do so. Note that repeating the analysis with these observations deleted corresponds to the approach used in regression to assess influence of these extreme observations. In this example, the elimination of selected data is not a simple task since such deletion leads to an unbalanced problem. Beckman, Nachtsheim, and Cook (1987) have proposed likelihood-based, deletion diagnostics for detecting unusual ob-

servations, but these seem less intuitive than the simple scatter plot diagnosis given here. We shall see in the next example a solution for the
missing data situation which suggests what the correct entry should have
been. This could be used as a measure of unusualness of an observation.

We now see that negative estimates of ϕ_2 are easy to explain mathematically because sample covariances can easily be negative and a negative
estimate requires only that the negative covariances in Table 15.3 dominate
the positives. This could happen because of one or more outlying observations, as in this example, or because of small covariances that vary in a
reasonable way about zero. In the former case, we would attempt to identify
the problem with the data and take appropriate steps, while in the latter
an estimate of $\phi_2 = 0$ might be acceptable. In either case, we have a better
feel for the data than is obtained with classical methods such as AOV,
ML, or REML.

For the original data, the above-diagonal elements of the covariance
matrix are (7.4, 5.8, 5.6), which suggest that the estimate $\phi_2 = 6.37$ is
quite reasonable. Observations from Subjects 3 and 6 agreed with the bulk
of the data but, as noted before, the response of Subject 8 on Test 3 is
suspicious since it is clearly an outlier. This accounts for the slightly smaller
covariances involving Test 3. We have no way to verify this observation,
but it may well be an error. For example, a response of nine rather than
four, another common typographical error, would be more in agreement
with the rest of the data.

Example 15.2

In many problems such as the one described above, the data are unbalanced
become some cells of the table are not observed. In this case, that would
happen if some subjects did not take all of the tests. We shall assume that
the empty cells occur at random without any assignable cause. To illustrate,
we use the data in Table 15.1 with a random deletion of cells. The pattern
is shown in Table 15.1 with M denoting missing. We now see some differences in the estimates. For comparison, the results are given in Table
15.4. The differences in this example are moderate but they can be substantial. The AOV estimator presented in this table uses the Type I sums
of squares, AOV (I), (see Hocking, 1985, for a definition) but other estimates can be obtained for different choices of AOV table. The ML
method has classical appeal but REML enjoys similar properties with less
bias. The noniterative MIVQUE(0) does quite well considering that this
is only one iteration on the initial guess of $\phi_0 = 1$ and $\phi_2 = 0$. On other
occasions, the differences could be substantial. We shall comment below
on an interesting feature of this method.

The method proposed by Hocking and Gomez-Meza (1991) for computing the REML estimates is quite informative. Recall that the idea is to

Table 15.4 Estimates of Variance
Components for Example 15.2

Method	ϕ_0	ϕ_2
AOV(I)	2.026	7.152
MIVQUE(0)	1.676	7.502
ML	1.728	6.293
REML	2.002	6.867

impute data for the missing cells according to (15.30) and then estimate
parameter functions using the same quadratic forms as in the balanced
data case. The only difference is that modified degrees of freedom need
to be calculated at each iteration. For diagnostic purposes and for computational simplicity, we suggest mimicking the computations in Example
15.1, where we simply computed the sample covariance matrix for the
observed data and read off the estimates. With unbalanced data, we do
the same thing with the imputed data and, at the same time, compute
modified degrees of freedom for the elements of that matrix.

The computation of the degrees of freedom follows from (15.33), but
in this case there is a simple device for computing these values. To see
this, let

$$
\mathbf{W} = \mathbf{D}\,\mathbf{P}\begin{bmatrix}\mathbf{V}_{oo}\\\mathbf{V}_{mo}\end{bmatrix}\mathbf{V}_{oo}^{-1}\mathbf{M}_o\,(\mathbf{V}_{oo}\mathbf{V}_{om})\,\mathbf{P}\,\mathbf{D}^\mathrm{T},
\tag{15.34}
$$

where

$$
\mathbf{P} = \mathbf{I} - \mathbf{X}^*(\mathbf{X}^{*\mathrm{T}}\mathbf{X}^*)^{-1}\mathbf{X}^{*\mathrm{T}}.
\tag{15.35}
$$

Assume that $\mathbf{W}$ is written in partitioned form with submatrices $\mathbf{W}_{ij}$ for i,
$j = 1, \ldots a_1$, where each square submatrix is of size a_2. Then the degrees
of freedom associated with the elements of the sums of squares and cross-
products matrix are

$$
\mathrm{DF}_{ij} = \mathrm{tr}(\mathbf{W}_{ij})/\phi_2^{(r)} \qquad \text{for } i \neq j
$$

and

$$
\mathrm{DF}_{ii} = \mathrm{tr}(\mathbf{W}_{ii})/(\phi_0 + \phi_2)^{(r)} \qquad \text{for } i = j.
\tag{15.36}
$$

where the parameters are evaluated at the previous iteration.

For illustration, we show in Tables 15.5–15.7 the results of the final
iteration of this method. Table 15.5 gives the imputed values for this final
iteration. (It is of interest to compare these with the actual values which

Table 15.5 Imputed Data for Last Iteration of Example 15.2

Subjects	Tests		
	t_1	t_2	t_3
1	15	12	11
2	10	11	8
3	6	4	2.95
4	7	7	5
5	9	6	6
6	16	14	11.68
7	11	8.80	7
8	13	8.36	4
9	12	9.67	8
10	10.37	8	6.60
11	11.85	9	9
12	13.19	11.06	10

Table 15.6 Modified Degrees of Freedom for Example 15.2

	t_1	t_2	t_3
t_1	9.86	10.75	10.75
t_2	10.75	9.60	10.96
t_3	10.75	10.96	9.93

Table 15.7 Sample Covariance (Correlation) Matrix for Example 15.2

	t_1	t_2	t_3
t_1	9.88	7.29	6.61
t_2	0.79	8.55	6.70
t_3	0.74	0.80	8.17

were deleted from Table 15.1.) The sums of squares and cross-products matrix is computed from this data set in the usual way.

In Table 15.6, we show the degrees of freedom for the individual elements of the sums of squares and cross-products matrix. Using these degrees of freedom, we obtain the estimated sample covariance matrix as shown in Table 15.7. The estimate of ϕ_2 (6.867) is the degrees-of-freedom weighted average of the above-diagonal elements in Table 15.7. The estimate of $\phi_0 + \phi_2$ (8.869) is similarly obtained from the diagonal elements of this matrix. The diagnostic analysis proceeds as in the balanced case. As noted earlier, the data are generally consistent apart from the possible outlier on subject eight.

In this example, 11 iterations were required for convergence, starting with the initial values $\phi_\tau = 1, \tau \in T_0$. This compares favorably with 12 iterations required for the SAS$^{(R)}$ version of REML (SAS$^{(R)}$ is a registered trademark of the SAS Institute, Inc.). Limited experience with this procedure suggests that the number of iterations is about the same for either algorithm, in contrast with the usual behavior of the EM algorithm (see Lindstrom and Bates, 1988). In practice, this convergence can be improved by using initial estimates based on a degrees-of-freedom weighted average of the sample variances and covariances from the observed data.

It is of interest to examine the degrees-of-freedom matrix in Table 15.6. With balanced data, each of the entries would be 11. On the other hand, we might have computed sample variances and covariances using only the observed data. In that case, the degrees-of-freedom matrix is given in Table 15.8. The increase in degrees of freedom in Table 15.6 reflects the additional information gained by using the imputed values.

It would be nice to have a noniterative method for imputing values for the missing data. One suggestion is to replace missing observations by the mean of the elements in that column. We would expect such a method to yield inferior estimates. With only two treatments, this procedure corresponds to MINQUE(0) if we use $a_1 - 1$ degrees of freedom.

Table 15.8 Degrees of Freedom for Observed Data in Example 15.2

	t_1	t_2	t_3
t_1	8	5	6
t_2	5	7	4
t_3	6	4	8

Example 15.3

Kirk (1982, p. 564) describes an experiment for comparing three methods (a_1, a_2, a_3) for inspecting printed circuit boards. Three types of defects (b_1, b_2, b_3) may occur. For the experiment, 18 inspectors were assigned to one of three groups and the three methods were randomly assigned to the groups. Each of the inspectors was shown boards, with each type of defect, in a random order. The response is the time, in seconds, required to detect the defect. The data are shown in Table 15.9. An asterisk in the data represents the location where an error will be introduced. This experiment is in a nested-factorial design as described in Section 15.1. In this case, we assume that both methods and types of defect are fixed and that subject is a random effect, which is nested in method. The model given in (15.6) is appropriate with Factor one corresponding to methods, Factor 2 corresponding to defects, and $n = 1$, since only one observation is made on each treatment combination for each subject. Our task is to estimate the variance components $\phi_{3(1)}$ and ϕ_0. Note that in this case $\phi_{3(1)}$, the variance of the subject-within-method effect, may also be viewed as the covariance between observation on a given subject for two different types of de-

Table 15.9 Data for Example 15.3

Method	Subject	Defects		
		b_1	b_2	b_3
1	1	10.6	5.9	8.6
	2	9.7	5.7	5.4
	3	12.5	8.9*	7.9
	4	9.6	6.8	3.1
	5	7.3	2.8	5.3
	6	6.7	4.2	2.3
2	1	1.1	1.3	4.3
	2	3.2	6.3	5.4
	3	5.4	4.2	7.3
	4	0.6	1.2	1.4
	5	2.4	3.1	2.1
	6	2.4	3.3	4.4
3	1	7.8	6.9	4.7
	2	7.9	6.6	6.1
	3	10.9	8.2	7.8
	4	4.8	5.1	1.8
	5	6.2	3.8	2.6
	6	8.4	9.9	4.9

Table 15.10 Sample Covariance (Correlation) Matrix for
Method 1 of Example 15.3

	b_1	b_2	b_3
b_1	4.58	1.00	3.94
b_2	0.31	2.23	$-.26$
b_3	0.74	$-.07$	6.28

fects. With $n = 1$, the first covariance in (15.7) is not relevant, so we set
$\phi_{23(1)} = 0$.

Since the example data are balanced and the estimates are nonnegative,
the standard procedure would be to use the AOV method to estimate the
components. The resulting estimates are $\hat{\phi}_0 = 1.28$ and $\hat{\phi}_{3(1)} = 3.21$. Again,
these data are quite well behaved, so for illustration we insert an error in
the data, by changing the response for Subject 3 with $a_1 = 1$ and $b_2 = 2$
from 8.9 to 3.9. The result is that the estimates are now $\hat{\phi}_0 = 1.65$ and
$\hat{\phi}_{3(1)} = 2.59$. There is nothing in the computation of these estimates to
suggest that an error has been introduced. We then turn to the method
proposed by Hocking (1990). Hocking shows that the estimates of the
components can be obtained by simply computing sample covariance matrices and taking averages of the appropriate elements. In this example, we
compute three covariance matrices, one for each method for the data as
arrayed in Table 15.9. The results are shown in Tables 15.10–15.12.

The estimate of $\phi_{3(1)}$ is the average of the above-diagonal elements of
these three matrices, and the estimate of $\phi_{3(1)} + \phi_0$ is the average of the
diagonal elements. The results for Methods 2 and 3 show some differences,
but these are within sampling errors. With Method 1, we see substantial
difference in the sample covariances involving the second type of defect.
Examination of the scatter plots associated with those two covariances is
recommended. In Figure 15.2, we show the plot of the first type of defect
(b_1) versus the second (b_2) for Method 1 with the symbol denoting the

Table 15.11 Sample Covariance (Correlation) Matrix for
Method 2 of Example 15.3

	b_1	b_2	b_3
b_1	2.90	2.31	3.04
b_2	0.71	3.65	2.49
b_3	0.83	0.60	4.67

Table 15.12 Sample Covariance (Correlation) Matrix for
Method 3 of Example 15.3

	b_1	b_2	b_3
b_1	4.29	3.21	4.37
b_2	0.71	4.70	3.18
b_3	0.96	0.66	4.88

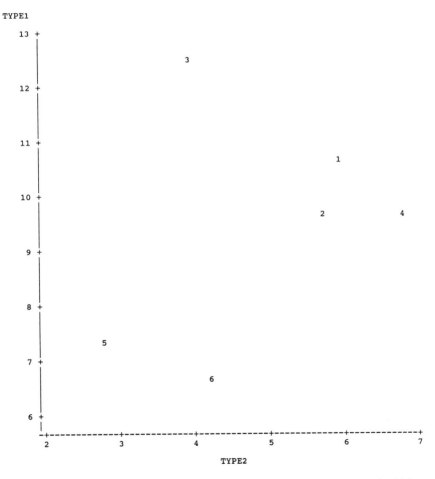

Figure 15.2 Plot of data for defects of type 1 versus type 2 for Example 15.3.

subject number. The response for Subject 3 is an obvious outlier and is responsible for these spurious results. Using the correct response, we see that the data are now consistent.

The application of Hocking's method to unbalanced data for this model is based on the same general ideas. The sample sums of squares and cross-products are computed from imputed data and then revised degrees of freedom are used to estimate the sample variances and covariances. Since no new concepts are involved, we will not provide a numerical illustration. The point to stress is that this same basic approach applies to any balanced or unbalanced AOV model. The scheme of sample covariance matrices required for a given model is given in Hocking (1990), and the procedure for imputing data and computing revised degrees of freedom is given in Hocking and Gomez-Meza (1991).

15.5 SUMMARY

The problem of estimating variance components is central to the analysis of mixed models. In applications, such as quality control, the components themselves may be of primary interest since we are interested in identifying sources of variability. In other situations, the primary interest may be in the analysis of the fixed effects. The estimation of the fixed parameters, using (15.21), and subsequent inferences on them depend on the estimates of the variance components. An important issue in such analyses is the design of the experiment. From our diagnostic analysis we obtain some information as to how we might allocate our resources. A primary consideration is that we have enough information to estimate the sample variances and covariances. We would not think of estimating a correlation with only three or four pairs of observations, but that is exactly what we are doing in a randomized block design with three or four blocks. This lack of data is the primary cause of high variability in the estimates. Since we often have limited resources, it is important that we carefully examine the existing data using diagnostic techniques such as those illustrated in Section 15.4.

With large, unbalanced experiments, the choice of method is often dictated by the amount of computation required. The AOV and MINQUE methods are noniterative, but even then the effort may not be trivial. The ML and REML methods inherit all the problems associated with constrained optimization of complicated functions. Of these methods, there seems to be a general consensus in favor of REML. The modification proposed by Hocking and Gomez-Meza (1991) shows how this method can be improved by providing diagnostic information and by reducing computation.

REFERENCES

Beckman, R. J., Nachtsheim, C. J., and Cook, R. D. (1987). Diagnostics for Mixed-Model Analysis of Variance. *Technometrics, 29*: 413–426.

Corbeil, R. R. and Searle, S. R. (1976). Restricted Maximum Likelihood (REML) Estimation of Variance Components in the Mixed Linear Model. *Technometrics, 18*: 31–38.

Dempster, A. P., Laird, N. M., and Rubin, D. B. (1977). Maximum Likelihood Estimation from Incomplete Data via the EM Algorithm. *J. Roy. Statist. Soc., B39*: 1–38.

Fellner, W. H. (1986). Robust Estimation of Variance Components. *Technometrics, 28*: 51–60.

Geisbrecht, F. G. (1989). A General Structure for the Class of Mixed Linear Models. Applications of Mixed Models in Agriculture and Related Disciplines, Southern Cooperative Series Bulletin No. 343, 183–201, Louisiana Agricultural Experiment Station, Baton Rouge.

Goodnight, J. H. and Speed, F. M. (1980). Computing Expected Mean Squares. *Biometrics, 36*: 123–125.

Harter, H. L. (1961). On the Analysis of Split-Plot Experiments. *Biometrics, 17*: 144–149.

Hartley, H. O. (1967). Expectations, Variances and Covariances of ANOVA Mean Squares by 'Synthesis'. *Biometrics, 23*: 105–114.

Hartley, H. O. and Rao, J. N. K. (1967). Maximum Likelihood Estimation for the Mixed Analysis of Variance Models. *Biometrics, 54*: 93–108.

Hartung, J. (1981). Nonnegative Minimum Biased Invariant Estimation in Variance Component Models. *Biometrics, 54*: 93–108.

Harville, D. A. (1974). Bayesian Inference for Variance Components Using only Error Contrasts. *Biometrika, 61*: 383–385.

Harville, D. A. (1977). Maximum Likelihood Approaches to Variance Component Estimation and to Related Problems. *J. Amer. Statist. Assoc., 72*: 320–340.

Hemmerle, W. J. and Hartley, H. O. (1973). Computing Maximum Likelihood Estimates for the Mixed AOV Model Using the W Transformation. *Technometrics, 15*: 819–831.

Henderson, C. R. (1953). Estimation of Variance and Covariance Components. *Biometrics, 9*: 226–252.

Hocking, R. R. (1985). *The Analysis of Linear Models*. Brooks-Cole, Monterey, California.

Hocking, R. R. (1988). A Cell Means Analysis of Mixed Linear Models. *Comm. Statist.-Theo. Meth., 17*: 983–1010.

Hocking, R. R. (1990). A New Approach to Variance Component Estimation with Diagnostic Implications. *Comm. Statist.-Theor. Meth. A19*: 4591–4618.

Hocking, R. R. and Gomez-Meza, M. V. (1991). Estimation of Variance Components and Diagnostic Analysis in Unbalanced Mixed Linear Models. (Submitted to *Biometrics*.)

Hocking, R. R., Green, J. W., and Bremer, R. H. (1989). Variance Component Estimation with Model-Based Diagnostics. *Technometrics*, *31*: 227–239.

Jennrich, R. I. and Schluchter, M. D. (1986). Unbalanced Repeated-Measures Models with Structured Covariance Matrices. *Biometrics*, *42*: 805–820.

Khuri, A. I. and Sahai, H. (1985). Variance Components Analysis: A Selective Literature Review. *Int. Statist. Rev.*, *53*: 279–300.

Kirk, R. E. (1982). *Experimental Design*. Brooks-Cole, Monterrey, California.

Laird, N. M. (1982). Computation of Variance Components using the EM Algorithm. *J. Statist. Comput. Simul.*, *14*: 295–303.

LaMotte, L. R. (1973). Quadratic Estimation of Variance Components. *Biometrics*, *29*: 311–330.

Lindstrom, M. J. and Bates, D. M. (1988). Newton–Raphson and EM Algorithms for Linear Mixed-Effects Models for Repeated-Measures Data. *J. Amer. Statist. Assoc.*, *83*: 1014–1022.

Matthew, T., Sinha, B. K., and Sutradhar, B. (1989). Nonnegative Estimation of Variance Components in Mixed Linear Models with Two Variance Components. (Submitted to *J. Amer. Statist. Assoc.*)

McLean, R. A., Sanders, W. L., and Stroup, W. W. (1991). A Unified Approach to Mixed Linear Models. *Amer. Statist.*, *45*: 54–63.

Nelder, J. A. (1954). The Interpretation of Negative Components of Variance. *Biometrika 41*: 544–548.

Patterson, H. D. and Thompson, R. (1971). Recovery of Interblock Information When Cell Sizes Are Unequal. *Biometrika*, *58*: 545–554.

Rao, C. R. (1970). Estimation of Heteroscedastic Variances in Linear Models. *J. Amer. Statist. Assoc.*, *65*: 161–172.

Rao, C. R. (1971a). Estimation of Variance and Covariance Components– MINQUE Theory. *J. Multiv. Anal. 1*: 257–275.

Rao, C. R. (1971b). Minimum Variance Quadratic Unbiased Estimation of Variance Components. *J. Multiv. Anal.*, *1*: 445–456.

Rao, C. R. (1972). Estimation of Variance and Covariance Components in Linear Models. *J. Amer. Statist. Assoc.*, *67*: 112–115.

Rao, P. S. R. S. and Chaubey, Y. P. (1978). Three Modifications of the Principles of the MINQUE. *Comm. Statist.-Simul. Comput.*, *B47*: 767–778.

Rocke, D. M. (1983). Robust Statistical Analysis of Interlaboratory Studies. *Biometrics*, *70*: 421–431.

Sahai, H. (1979). A Bibliography on Variance Components. *Int. Statist. Rev.*, *47*: 177–122.

Sahai, H., Khuri, A. I., and Kapadia, C. H. (1985). A Second Bibliography on Variance Components. *Comm. Statist.-Theor. Meth.*, *A14*: 63–115.

Samuels, M. L., Casella, G., and McCabe, G. P. (1991). Interpreting Blocks and Random Factors. *J. Amer. Statist. Assoc.*, *86*: 798–821.

Searle, S. R. (1971). Topics in Variance Component Estimation. *Biometrics*, *27*: 1–76.

Stroup, W. W. (1989). Why Mixed Models? *Applications of Mixed Models in Agriculture and Related Disciplines*, Southern Cooperative Series Bulletin No.

343, Louisiana Agricultural Experiment Station, Baton Rouge, Louisiana, pp. 183–201.

Swallow, W. H. and Searle, S. R. (1978). Minimum Variance Quadratic Unbiased Estimation (MIVQUE) of Variance Components. *Technometrics*, *20*: 265–272.

Tarone, R. E. (1988). Score Statistics. *Encyclopedia of Statistical Sciences*, Vol. 8. Kotz, S., Johnson, N. L., and Read, C. B. (eds.), Wiley, New York, pp. 304–308.

Thompson, W. A., Jr. and Moore, J. R. (1963). Non-Negative Estimates of Variance components. *Technometrics*, *5*: 441–449.

Yates, F. (1934). The Analysis of Multiple Classifications with Unequal Numbers in the Different Classes. *J. Amer. Statist. Assoc.*, *29*: 51–66.

16
Some Computer Programs for Selected ANOVA Problems

LYNNE K. EDWARDS and PATRICIA C. BLAND University of Minnesota, Minneapolis, Minnesota

16.1 INTRODUCTION

In this chapter, we will present selected numerical examples using three major statistical packages, SPSS (SPSS, 1990), SAS (SAS, 1989), and BMDP (Dixon, Brown, Engelman, Hill, and Jennrich, 1988a,b). There are other mainframe packages besides these three, in addition to special-purpose programs written for a limited number of procedures. These three packages were chosen on the basis of their availability across various mainframe manufacturers, their availability on personal computers (PCs), their wide use as referenced in various research literature, and their being characterized as general-purpose programs.

We have chosen three designs (hierarchical design, Latin square design, and confounded factorial design) and one procedure (variance component estimation) for the purpose of illustrating computer applications for selected ANOVA problems. These topics were chosen to illustrate differences among packages in analyzing a familiar design, to encourage uses of underutilized designs, and to illustrate how an irregular problem can be handled by these packages. Computer applications for designs and procedures not covered here but presented elsewhere are summarized briefly at the end of this chapter.

In the previous chapters, some of the less frequently used designs and complex statistical procedures are discussed. For some familiar designs such as hierarchical (or nested) designs, the procedures for specifying the data structure and computer commands to obtain correct results are not always straightforward. As for underutilized designs, even when readers are convinced that some of these designs are appropriate for their study, it is still difficult to use them without examples from the readily available statistical packages. We provide easy-to-follow examples in the hope that users will be less hesitant in adopting designs that fit their purposes. We also illustrate an irregular situation in variance component estimation, i.e., negative components, because it calls for careful selection of estimation procedures.

We do not intend to provide an exhaustive annotated output but, rather, to supplement what is available and to focus specifically on the issues which require additional attention. This chapter shows how the same problem can be approached by the three packages. Users may want to know which package handles the task most efficiently, in terms of parsimony in commands structure, flexibility in handling various data structures and hypotheses, and economy in processing time and/or charges. Where possible, all examples are tackled by all three packages, in order to compare these three and to provide options to readers who may have access to some but not all packages.

In this chapter, our examples were run on a VAX computer with the VMS operating system (Version 5.4). Exactly the same designs were run for all comparisons. In the case of BMDP, where two programs were used to obtain results comparable to SPSS and SAS, it is so indicated. The comparisons are given in Table 16.20. It should be noted that these comparisons are valid only for the machine on which they were run. Packages are generally developed and optimized on a specific machine. Consequently, these packages are generic in that they run on various machines, but there may be considerable variation in efficiency with respect to installation specifics and optimization. Furthermore, sometimes there is more than one way to specify the procedure within the same package. For example, specifying the model for hypothesis and error as well as specifying a certain factor to be random may result in the same desired results. However, these two approaches may not require the same amount of processing time. For these reasons, the listed figures are to be interpreted with caution.

16.2 A HIERARCHICAL DESIGN

Although a hierarchical design with only one factor nested in another is structurally simple, the default or standard procedure can provide incorrect F values in a mixed-effects model. In addition, some attention should be paid in specifying the data structure. The example from Kirk (1982, p. 485) involves an industrial psychologist interested in decreasing time needed to assemble an electronic component. Three assembly fixtures, treatment FX,

Table 16.1 ANOVA $E(MS)$ Table for the Nested Example[a]

Sources	d.f.	$E(MS)$
A	$(a - 1) = 2$	$\sigma_e^2 + n\sigma_{b(a)}^2 + nb^*\theta_a^2$
$B(A)$	$a(b^* - 1) = 3$	$\sigma_e^2 + n\sigma_{b(a)}^2$
$S(BA)$	$ab^*(n - 1) = 12$	σ_e^2

[a]A fixed; $B(A)$ random. Factors FX, $WP(FX)$, $OP(WP(FX))$ are represented by A, $B(A)$, and $S(BA)$, respectively. b^* is the number of levels of B *within* each level of A.

were evaluated. Five operators at each of six workplaces within the plant, Factor $OP(FX)$, were selected randomly. The assembly fixtures were randomly assigned to the six workplaces and the operators at those workplaces. The response variable is the number of units assembled per hour.

Let us express the statistical model for this example by mapping $A,B(A)$, and $S(BA)$ to the factors FX, $WP(FX)$, and $OP(WP\ FX)$, respectively,

$$Y_{ijk} = \mu + \alpha_i + \beta_{j(i)} + \varepsilon_{ijk},$$
$$i = 1, \ldots, a; j = 1, \ldots, b; k = 1, \ldots, n, \qquad (16.1)$$

where Y_{ijk} is the number assembled by the kth operator, working at the jth workplace nested in the ith fixture. Since workplace is randomly chosen, it is expected to be distributed normally, $\beta_j \sim N(0, \sigma_\beta^2)$; and the error is assumed random as usual, $\varepsilon_{ijk} \sim N(0, \sigma_\varepsilon^2)$. The fixtures used were considered fixed; therefore, α_i is given a usual constraint, $\Sigma_i\, \alpha_i = 0$.

Note that subscript j ranges from 1 to b, but in a crossed design expression j ranges from 1 to b^* *within* each level of A. The $E(MS)$ table for this design is Table 16.1. The test for FX is equivalent to the test on A effects, therefore $F = MSFX/MSWP(FX)$, and the test on $WP(FX)$ is equivalent to the test on $B(A)$, thus $F = MSWP(FX)/MSOP(WP(FX))$. We will test these overall tests and the nested hypotheses on $WP(FX)$.

16.2.1 Application with SPSS

To analyze this data set, first the data are read in four columns: TIME (the dependent variable), FX (fixture), WORKPLAC (workplace), and OPERATOR. The example data are presented in Table 16.2 in a format readily readable by either SPSS or SAS. To handle this problem with SPSS, we have recoded WORKPLAC in FIXTURE to make it reflect the number of levels of workplace *within* each fixture, thus emulating a crossed design. (The data were coded to match Kirk's example. Had the data been coded to reflect the number of workplace levels within each level of fixture, i.e.,

Table 16.2 A Nested ANOVA Example Data for SAS
and SPSS

TIME	FX	WORKPLAC	OPERATOR
17	1	1	1
7	1	1	2
9	1	1	3
12	1	1	4
15	1	1	5
13	1	2	6
23	1	2	7
14	1	2	8
18	1	2	9
22	1	2	10
21	2	3	11
24	2	3	12
14	2	3	13
19	2	3	14
17	2	3	15
25	2	4	16
29	2	4	17
26	2	4	18
24	2	4	19
21	2	4	20
26	3	5	21
34	3	5	22
32	3	5	23
30	3	5	24
28	3	5	25
32	3	6	26
32	3	6	27
37	3	6	28
34	3	6	29
35	3	6	30

a crossed design designation, this manipulation would have been unnecessary.) The recoding commands are:

                    RECODE   WORKPLAC
                    (1,3,5 = 1)
                    (2,4,6 = 2)
                    INTO
                    WP

Table 16.3 presents the rest of the commands followed by the overall tests and a series of nested follow-up tests of $WP(FX)$. In the MANOVA procedure, the original six-level WORKPLAC is isolated as a source of variation to facilitate the nested treatment comparisons in the second DESIGN statement. In this context, CONTRAST works in conjunction with the DESIGN statement immediately following it.

The first DESIGN statement specifies that FX is to be tested against Error1, $MSWP(FX)$, and Workplace nested in FX, $WP(FX)$, is to be tested against the within cells, $MSOP(WP(FX))$. The contrast specifications define the grand mean in the first row (which is automatically taken out of

Table 16.3 Overall Nested ANOVA and Nested Treatment Comparisons with SPSS

SPSS commands

```
MANOVA   TIME by WORKPLAC(1,6), FX(1,3), WP(1,2)/
    DESIGN = FX VS 1,
        WP WITHIN FX = 1 VS WITHIN/
    CONTRAST (WORKPLAC) = SPECIAL
            (1   1   1   1   1   1
             1   1   0   0  -1  -1
             0   0   1   1  -1  -1
             1  -1   0   0   0   0
             0   0   1  -1   0   0
             0   0   0   0   1  -1)/
    PARTITION(WORKPLAC) = (2, 1, 1, 1)/
    DESIGN =    WORKPLAC (2) VS WITHIN,
                WORKPLAC (3) VS WITHIN,
                WORKPLAC (4) VS WITHIN/
```

ANOVA summary table

Source of variation	SS	DF	MS	F	Sig of F
WITHIN CELLS	300.00	24	12.50		
WP WITHIN FX (Error 1)	220.00	3	73.33	5.87	.004
Error 1	220.00	3	73.33		
FX	1460.00	2	730.00	9.95	.047

Nested treatment comparisons					
WITHIN CELLS	300.00	24	12.50		
WORKPLAC(2)	90.00	1	90.00	7.20	.013
WORKPLAC(3)	90.00	1	90.00	7.20	.013
WORKPLAC(4)	40.00	1	40.00	3.20	.086

the *SStotal*); the second and third rows provide *FX* contrasts with a combined 2 d.f.; and a series of nested comparisons of workplaces within each level of *FX* are shown in the fourth, fifth, and sixth rows. The partition statement specifies that the *SSworkplac* is to be decomposed into four separate *SSs*: the second and third rows combined, representing *FX* with 2 d.f., and 1 d.f. each for the three comparisons of workplaces sharing the same fixture. Recall that, because it represents the grand mean, the first contrast is excluded from the *SStotal*. Subsequently, the second contrast is the first contrast of interest. In the second Design statement, we have called for a test of only the second, third, and fourth partitions, i.e., workplace 1 vs. 2, 3 vs. 4, and 5 vs. 6. The design statement specifies that the error term for the last three nested analyses is the within cells, $MSOP(WP(FX))$. The overall test and a series of nested follow-up tests of $WP(FX)$ are presented in the ANOVA summary portion (Table 16.3). In SPSS, the user cannot specify which factor is random or fixed but can specify which *MS* is to be used as the error. The nested follow-up contrasts are tested against the appropriate error term, but the reported probabilities are not controlled by any familywise error rate.

16.2.2 Application with SAS

As in SPSS, the SAS GLM procedure reads in four columns of data, TIME (the dependent variable), FX, WP, and OP for Operator (Table 16.2). First PROC GLM performs the overall ANOVA using the least squares approach, SS3 (i.e., SS type III), followed by specific hypothesis tests [i.e., TEST H = *FX* E = $WP(FX)$, and TEST H = $WP(FX)$ E = $OP(WP$ $FX)$]. The error for the second hypothesis test is $OP(WP(FX))$, the within-cells variation. However, since there is no double-parentheses format in the SAS commands, the error is presented as $OP(WP\ FX)$. For hypothesis tests, the error term may be explicitly specified, thus $OP(WP\ FX)$ in the model specification. If hypothesis tests using different error terms were not desired, $OP(WP\ FX)$ would simply be reported as ERROR by SAS, and all *F* tests and contrasts would default to that source of variation. Subsequently, if TEST (the specific hypothesis and error) is not specified, *FX* is tested against an incorrect error term, $OP(WP\ FX)$. That is, the test of *FX* in the second segment of the results (Table 16.4) with *F* value of 58.40 is an incorrect test of *FX* using $OP(WP\ FX)$ as the error. This again brings up the point of preparing the $E(MS)$ table before carrying out the statistical analysis, thereby forcing the user to know exactly which *MS* will form valid *F* tests. The SAS commands required for this analysis are:

PROC GLM;
CLASS FX WP OP;

```
MODEL TIME = FX WP(FX) OP(WP FX)/SS3;
TEST H = FX   E = WP(FX);
TEST H = WP(FX)   E = OP(WP FX);
```

A convenient feature of SAS is that we can specify a pairwise multiple comparison procedure within the same GLM run. (This is an advantage over the SPSS in the sense SPSS as of this writing has multiple comparison procedures in ONEWAY but not in the more versatile MANOVA procedure.) In SAS, if we want to conduct Tukey's HSD test on *FX*, we can simply add the following statement immediately following the above TEST:

```
MEANS FX/TUKEY LINES E = WP(FX);
```

The above specification not only allows pairwise comparisons of means but also lets the user specify the correct error term for testing *FX*. The result is well annotated and easy to interpret (Table 16.4, middle segment), where the significant difference between Fixtures 1 and 3 is indicated.

The next segment defines follow-up nested analyses by specifying contrast coefficients.

```
PROC GLM;
  CLASS FX WP;
  MODEL TIME = FX WP(FX)/SS3;
  CONTRAST 'FIXTURE 1, WP 1 VS 2'
    WP(FX) 1 –1 0 0 0 0;
  CONTRAST 'FIXTURE 2, WP 3 VS 4'
    WP(FX) 0 0 1 –1 0 0;
  CONTRAST 'FIXTURE 3, WP 5 VS 6'
    WP(FX) 0 0 0 0 1 –1;
```

The result for the second segment of PROC GLM is also in Table 16.4. Note that the correct result for *FX* is obtained in the first segment of PROC GLM. The output of interest in the second GLM segment is the contrast at each level of Fixture. (Note that the probability levels reported for these nested tests are not controlled by any familywise error rate as was the case in SPSS.)

An alternative way to program this problem is to specify that *WP(FX)* is a random factor by using the RANDOM statement, which results in a hypothesis test consistent with this mixed-effects model.

```
PROC GLM;
  CLASS FX WP OP;
  MODEL TIME = FX WP(FX)/SS3;
  RANDOM WP(FX)/TEST;
```

Table 16.4 SAS Results for the Nested Example

Source	DF	Type III SS	Mean Square	Fvalue	Pr < F
First GLM segment					
FX	2	1460.00000	730.00000000	.	.
WP(FX)	3	220.00000	73.33333333	.	.
OP(FX*WP)	24	300.00000	12.50000000	.	.
Tests of hypotheses using the Type III MS for WP(FX) as an error					
FX	2	1460.00000	730.00000000	9.95	.0474
Tests of hypotheses using the Type III MS for OP(FX*WP) as an error					
WP(FX)	3	220.00000	73.33333333	5.87	.0037

Tukey's studentized range (HSD) test for variable: TIME
      Alpha = 0.05 d.f. = 3 MSE = 73.33333
    Critical value of studentized range = 5.999
    Minimum significant difference = 16.244
Means with the same letter are not significantly different

Tukey Grouping			Mean	N	FX
	A		32.000	10	3
B	A		22.000	10	2
B			15.000	10	1

Source	DF	Type III SS	Mean Square	Fvalue	Pr < F
Second GLM segment					
FX	2	1460.00000	730.00000000	58.40	0.0001
WP(FX)	3	220.00000	73.33333333	5.87	0.0037
Contrast					
Fix1, wp1 vs 2	1	90.00000	90.00000000	7.20	0.0130
Fix2, wp3 vs 4	1	90.00000	90.00000000	7.20	0.0130
Fix3, wp5 vs 6	1	40.00000	40.00000000	3.20	0.0863

The correct hypothesis test (though not shown here), including mean squares and *F* ratios, will be presented in the Test of Hypotheses following the regular ANOVA summary table. However, this random specification does not alter the summary table, which presents results consistent with a fixed-effects model. Therefore, in SAS, a correct hypothesis test specification has to be made either by identifying the valid *F* ratio or by specifying which factor is random.

16.2.3 Application with BMDP

BMDP8V uses the statistical model and its indices to organize data for input. The way in which the data are organized is critical with respect to the order of the indices. The data for this example are to be read in six columns, each column representing five observations taken from a partic-

Table 16.5 A Nested ANOVA Using BMDP8V

BMDP8V commands

/DESIGN DEPENDENT = Fix1Wp1 to Fix3Wp6.
   LEVELS = 5, 3, 2.
   NAMES = OP, FIXTURE, WP.
   FIXED = FIXTURE.
   RANDOM = OP, WP.
   MODEL = 'F, W(F), O(FW)'.

ANOVA summary table

Source	Error term	SS	d.f.	MS	F	PROB
MEAN	W(F)	15870.000	1	15870.00	216.41	.0007
FIXTURE	W(F)	1460.000	2	730.00	9.95	.0474
W(F)	O(FW)	220.000	3	73.33	5.87	.0037
O(FW)		300.000	24	12.50		

ular fixture at a particular workplace. For example, the first column represents the observations taken at Fixture 1 at Workplace 1. To see how these data correspond to an "observation-wise" structure, compare this with the corresponding data structure for SAS and SPSS in Table 16.2.

BMDP8V allows the user to specify factors as either fixed or random, an advantageous feature shared with SAS. In addition, in P8V the summary table itself contains the appropriate F ratios for this mixed model. The BMDP8V commands and overall tests are summarized in Table 16.5. It has a clear form of presentation with each F ratio presented with its corresponding appropriate mean square.

One shortcoming is that the follow-up nested analyses cannot be handled within the same procedure but may be handled separately in BMD7D, using an observation-wise data structure. The data have to be rearranged in the format seen in Table 16.2 before we can apply BMDP7D for the follow-up tests. This is somewhat awkward for two reasons: two different procedures have to be employed in two separate runs, and the same data have to be organized differently for these runs.

Although results are not reported here, BMDP7D commands for the follow-up tests are:

/VARIABLE NAMES ARE FIXTURE, WORKPLAC, OPERATOR, TIME.
/HISTOGRAM GROUPING = WORKPLAC.
   VARIABLE = TIME.

CONTRASTS = WP_12, WP_34, WP_56.
WP_12 = 1, -1, 0, 0, 0, 0.
WP_34 = 0, 0, 1, -1, 0, 0.
WP_56 = 0, 0, 0, 0, 1, -1.

16.3 A LATIN SQUARE DESIGN

A Latin square design with repeated measures can be useful especially
when carryover effects in addition to period effects are of research interest.
Using the example in Chapter 5, in which the carryover effects were estimated by SAS and SYSTAT, we will consider applications with SPSS
and BMDP for the same data. The example involves a 4 × 4 balanced
Latin square with repeated measurements. The dependent variable is the
applause-meter reading, A factor represents Subjects, B represents Periods, C represents Treatments, and D, Carryovers. Four subjects have
provided applause ratings for four TV episodes (treatments) spread over
four periods. Here, in addition to the regular Subjects, Periods, and Treatments, the potential carryover effects are of interest. Since a SAS example
is fully documented in Chapter 5, we will not duplicate it here. Adopting
Eq. (5.25) in Chapter 5 of this book, let the statistical model for a Latin
square example be expressed as

$$Y_{ijk} = \mu + \alpha_i + \beta_j + \gamma_k + \lambda_{k'} + \varepsilon_{ijk},$$

$$i = 1, \ldots, a; j = 1, \ldots, b; k = 1, \ldots, c; k' = 1, \ldots, d, \qquad (16.2)$$

where $\alpha_i \sim N(0, \sigma_\alpha^2)$ and $\varepsilon_{ijk} \sim N(0, \sigma_\varepsilon^2)$. Y_{ijk} is the response of the ith
subject measured at the jth period viewing the kth TV episode, having
finished viewing the k'th episode during the $(j - 1)$th period. Subjects are
random. Other factors, β_j, γ_k, and $\lambda_{k'}$, are all fixed and either the sum-to-
zero constraints or parameter-set-to-zero constraints can be adopted. Al-
though for the fixed factors, Q, the quadratic functions apply [and Qs will
be printed out in the $E(MS)$ table with a proper factor specification in
SAS] and it is not necessary to express them in terms of variances, it is
helpful to understand this design vis-à-vis other examples in this chapter.
The coefficients applicable to the quadratic forms (B, C, and D) depend
on which analysis method (type I, II, or III SS decomposition) is used. In
Chapter 5, a type II solution was chosen (in which each main effect is
assessed by holding constant all other main effects). In this example, a
type II solution amounts to the same as a type III solution (i.e., the least
squares solution in which all terms excluding the one in question are held
constant) because there is no interaction in the model. In this example, a
type I solution was not selected because there is no a priori or plausible

Table 16.6 $E(MS)$ Table for a Latin Square Example with "Experimental" Decomposition[a]

Sources	d.f.				$E(MS)$	
A (Subject)	3	σ_e^2				$+\,3.6364\,\sigma_a^2$
B (Quasi-period)	2	σ_e^2			$+\,4.0\,\theta_b^2$	
C (Treatment)	3	σ_e^2		$+\,3.6364\,\theta_c^2$		
D (Carryover)	3	σ_e^2	$+\,2.5\,\theta_d^2$			
Error	3	σ_e^2				

[a]A and Error are random; B, C, D are fixed. The analysis is based on the SAS Type II SS (classical experimental) decomposition. The above components table lists coefficients obtained by specifying all factors random. If only A is specified as random and the rest are fixed, the coefficients for B, C, D will be expressed in Q (quadratic) forms. For example, for D, $\sigma_e^2 + Q(\text{Carryover})$ instead of $\sigma_e^2 + 2.5\theta_d^2$ will be reported.

ordering among the main effects. The $E(MS)$ table for this example is in Table 16.6 and the data in Table 5.1 are rearranged for input in Table 16.7.

16.3.1 Application with SPSS

Parameter estimation is requested via the PRINT PARAM(EST) specification in the second line of the MANOVA command stream. A set of special CONTRASTs with $+1$ in one classification, -1 in the last classification, and 0's in the remaining classifications are used here to adopt the last parameter-set-to-zero restriction instead of the usual sum-to-zero restriction. Note that this set of special contrasts has apparent similarity to effect coding. However, the effect coding is labeled as "deviation coding" in SPSS, and the special contrasts used here will result in actual coefficients in the contrasts being multiplied to the respective parameters. For example, the second contrast for PERIOD, $(+1, 0, 0, -1)$ will estimate $\beta_1-\beta_4$.

The data matrix in Table 16.8 is readable in SPSS. The first column represents subject identification (SUBJ), the second column PERIOD, the third TREATment, the fourth CARRYOver, and the fifth Y (the dependent variable).

First, CARRYOver and TREATment are recoded into numerical levels. In the following MANOVA procedure, contrasts are specified for the fixed factors: Period, Treatment, and Carryover. The SPSS commands and results are presented in Table 16.8. The results presented in Table 16.8 match those presented in Chapter 5 for the ANOVA summary table and the parameter estimations using the parameter-set-to-zero constraints (see Tables 5.4 and 5.5).

Table 16.7 Data Matrix for the Latin Square Example[a]

Subject	Period	Treatment	Carryover	Y
1	1	A	0	68
1	2	B	A	74
1	3	D	B	93
1	4	C	D	94
2	1	B	0	60
2	2	C	B	66
2	3	A	C	59
2	4	D	A	79
3	1	C	0	69
3	2	D	C	85
3	3	B	D	78
3	4	A	B	69
4	1	D	0	90
4	2	A	D	80
4	3	C	A	80
4	4	B	C	86

[a]Treatment levels are indicated by A, B, C, and D; and the carry-over column records treatments immediately preceding the one in question, where 0 is given to the very first treatment condition.

Table 16.8 SPSS Commands and Summary Table for the Latin Square Design Example

SPSS commands

```
RECODE   CARRYO('A'=1) ('B'=2) ('C'=3) ('D'=4) ('0'=5)
    INTO
    CARRY
RECODE   TREAT ('A'=1) ('B'=2) ('C'=3) ('D'=4)
    INTO
    TREATMENT
MANOVA   Y BY SUBJ, PERIOD, TREATMNT(1,4), CARRY(1,5)/
    PRINT=CELLINFO(MEANS) PARAM(EST)/
    METHOD=SSTYPE(UNIQUE)/
    CONTRAST(PERIOD)=SPECIAL
    (1 1  1   1
     1 0  0  -1
     0 1  0  -1
     0 0  1  -1)/
```

Table 16.8 (Continued)

SPSS commands

```
    PARTITION(PERIOD) = (1,2)/
    CONTRAST(TREATMNT) = SPECIAL
    (1  1   1    1
     1  0   0   -1
     0  1   0   -1
     0  0   1   -1)/
    CONTRAST(CARRY) = SPECIAL
    (1  1   1   1    1
     1  0   0   0   -1
     0  1   0   0   -1
     0  0   1   0   -1
     0  0   0   1   -1)/
DESIGN = SUBJ     VS RESIDUAL,
        PERIOD(1)   VS WITHIN,
        PERIOD(2)   VS RESIDUAL,
        CARRY       VS RESIDUAL,
        TREATMNT   VS RESIDUAL/
```

ANOVA summary table

Sources	SS	DF	MS	F	Sig of F
Within Cells	.00	0	.		
Period(1)	334.15	1	334.15	.	.
Residual	.35	3	.12	.	.
Subj	621.88	3	207.29	1776.79	.000
Period(2)	73.17	2	36.58	313.57	.000
Treatmnt	800.24	3	266.75	2286.40	.000
Carry	156.65	3	52.22	447.57	.000

Parameter	Coeff.	Notation from Table 5.5
		(the last parameter-set-to-zero)
Period(1)	(-16.350000)	(biased estimate; -10.25 in Table 5.5)
Period(2)	-5.75000000	$\tilde{\beta}_j - \tilde{\beta}_4$
	-4.50000000	
	0.00000000	
TREATMNT	-20.5000000	$\tilde{\gamma}_j - \tilde{\gamma}_4$
	-14.0750000	
	-11.0250000	
	0.00000000	
CARRY	-11.00000000	$\tilde{\lambda}_{k'} - \tilde{\lambda}_4$
	-7.30000000	
	-6.10000000	
	$.00000000$	

As presented in Chapter 5, SAS allows the user to specify exactly how inestimable parameters should be formed via the ESTIMATE command. The SPSS example accomplishes parameter estimation using parameter-set-to-zero constraints by means of the specification of special contrasts. These contrasts work in conjunction with partition statements, which are used to partition the total d.f. into the desired portions (e.g., PERIOD into 1 and 2 d.f.). The DESIGN specification is the last part of the MANOVA command. If no special contrasts are specified, SPSS uses deviation contrasts by default; it estimates each parameter as its deviation from the overall average for that parameter.

We have used special contrasts as shown in the previous example to use the estimations compatible with the last parameter-set-to-zero restriction. These estimates are in agreement with those reported by Cotton in Table 5.5. One possibly misleading feature of SPSS is that its output of design matrix apparently always presents the deviation coding (i.e., effect coding) regardless of the user's special contrast specifications. However, the parameter estimates do correspond to the user-specified contrasts.

16.3.2 Application with BMDP

As in SPSS default, BMDP calculates parameters on the basis of deviation contrasts (i.e., the parameter minus the overall mean for the parameter), but unlike SPSS, BMDP provides no control over the way in which the reparameterization can be specified via special contrasts. BMDP4V seems to be best suited for this Latin square example among other BMDP procedures, although it is somewhat inflexible. The BMDP4V commands and results are summarized in Table 16.9.

Table 16.9 BMDP4V Commands and Summary Table for the Latin Square Design Example

BMDP4V Commands	
/VARIABLE	NAMES = SUBJ, PERIOD, TREAT, CARRY, Y.
/TRANSFORM	IF (CARRY EQ CHAR(A)) THEN CARRY = 1.
	IF (CARRY EQ CHAR(B)) THEN CARRY = 2.
	IF (CARRY EQ CHAR(C)) THEN CARRY = 3.
	IF (CARRY EQ CHAR(D)) THEN CARRY = 4.
	IF (CARRY EQ CHAR(0)) THEN CARRY = 5.
	IF (TREAT EQ CHAR(A)) THEN TREAT = 1.
	IF (TREAT EQ CHAR(B)) THEN TREAT = 2.
	IF (TREAT EQ CHAR(C)) THEN TREAT = 3.
	IF (TREAT EQ CHAR(D)) THEN TREAT = 4.

Table 16.9 (Continued)

BMDP4V Commands

/BETWEEN	FACTORS = SUBJ, PERIOD, TREAT, CARRY.	
	CODES(SUBJ)	= 1, 2, 3, 4.
	CODES(PERIOD)	= 1, 2, 3, 4.
	CODES(TREAT)	= 1, 2, 3, 4.
	CODES(CARRY)	= 1, 2, 3, 4, 5.
/WEIGHTS	BETWEEN = EQUAL.	
/END		
DESIGN	FACTOR = SUBJ, PERIOD, TREAT, CARRY.	
	TYPE = BETWEEN.	
	REGRESSION.	
	PRINT./	
ANALYSIS PROCEDURE = STRUCTURE.		
	BFORM = 'SUBJ + PERIOD + TREAT + CARRY'.	
	ESTIMATES./	
END/		

ANOVA Summary Table

EFFECTS	VARIATE	STATISTIC F	DF	P
A|B,D,C	Subjects			
Y	SS = 621.877273			
	MS = 207.292424	1776.79	3,3	.0000
B|A,C,D	Quasi-Periods			
Y	SS = 73.166667			
	MS = 36.583333	313.57	2,3	.0003
C|A,B,D	Treatments			
Y	SS = 800.240909			
	MS = 266.746970	2286.40	3,3	.0000
D|A,B,C	Carryovers			
Y	SS = 156.650000			
	MS = 52.216667	447.57	3,3	.0002
ERROR	Residuals			
Y	SS = 0.35000000			
	MS = 0.11666667			

Parameter estimation		
Period	Inestimable	Notation from Table 5.5 (sum-to-zero)
Treatments	−9.10000	$\tilde{\gamma}_k$
	−2.67500	
	0.375000	
	11.40000000 (by subtraction)	
Carryovers	Inestimable	

Parameters are calculated using deviation contrasts in BMDP4V, but, unlike SPSS, we could find no way to control reparameterization. As Bock and Brandt (1980) pointed out, BMDP is inflexible in dealing with general linear model specifications. The summary table in Table 16.9 correctly decomposes the sums of squares. However, only Treatments (with a sum-to-zero restriction) is estimable, while Periods and Carryovers (the other two factors) are reported as inestimable.

16.4 A CONFOUNDED FACTORIAL DESIGN

Various forms of confounded factorial designs (one of the least utilized designs in applied behavioral sciences) are presented in Kirk (1982). Devised as a solution to a large and impractical block-size problem, confounded factorial designs have only a portion of the treatment combinations assigned to each block. The block size is often reduced by way of confounding group with interaction. A 2^2 randomized block confounded design with a block size of two (RBCF-2^2) can be compared to a 2×2 randomized block factorial design with a block size of four (RBF-22).

There is little advantage of a confounded factorial design over a randomized block factorial design when it is possible to assign four "matched" subjects to a block, i.e., when homogeneous sets of four subjects can be arranged. For example, with $N = 16$ in an RBCF-2^2 design, the two major factors of interest, A and B, will be tested against the pooled error term with 6 degrees of freedom, provided that the homogeneity of variance across two groups is satisfied. For the same $N = 16$, an alternative 2×2 randomized block factorial design will have 12 degrees of freedom for the error in testing these two main effects. Subsequently, a randomized block factorial design is more powerful than a confounded factorial design. In addition, the randomized block design allows a pure test of AB interaction, which is confounded with the group in the randomized block confounded design. However, when the block size is constrained to two but the number of treatment combinations exceeds two (as in a twins study, where three or more treatment combinations are examined), a confounded factorial design becomes a viable option.

An example RBCF-2^2 is taken from Kirk (1982, p. 580). Although data are presented with factor identification only in Kirk, use of a plausible research context here may motivate readers. Let us assume that we have eight sets of twins and four sets have each undergone two different treatment combinations: within the first group, half of the twins, i.e., one from each pair, were given a placebo and a quiet reading session, while the counterparts were given a pill containing oat bran and a moderate exercise session; and within the second group, half of the twins were given a placebo and a moderate exercise session, while the counterparts were given an oat bran pill and a quiet reading session. The dependent variable of interest is the change in the blood cholesterol level (the difference between the beginning and ending cholesterol reading), with large scores indicating positive and healthy results. There were A factor with two levels (dietary supplement of bran or no bran) and B factor also with two levels (moderate exercise or sedentary activity.) In this design, instead of forcing both groups go through all four crossed combinations of A and B factors, the first group undergoes half of the combinations and the second group undergoes a different half of the combinations. The statistical design for this example is

$$Y_{ijk} = \mu + \alpha_i + \beta_j + \gamma_k + \pi_{m(k)} + \alpha\beta\gamma_{ijm(k)} + \varepsilon_{ijkm},$$

$$i = 1, \ldots, a; j = 1, \ldots, b; k = 1, \ldots, c; m = 1, \ldots, n, \qquad (16.3)$$

where μ is the grand mean, α_i is the ith treatment effect A, B_j is the jth treatment effect B, γ_k is the kth group effect G, $\pi_{m(k)}$ is the mth block Z nested in the kth group G, $\alpha\beta\gamma_{ijm(k)}$ is the combined effect of AB interaction and G, and ε_{ijkm} is the error. We place the usual sum-to-zero constraints on fixed factors and assume normal distributions for the random factors. Hence,

$$\sum_i \alpha_i = 0, \qquad \sum_j \beta_j = 0, \qquad \sum_k \gamma_k = 0,$$

$$\pi_{m(k)} \sim N(0, \sigma_\pi^2), \qquad \varepsilon_{ijm(k)} \sim N(0, \sigma_\varepsilon^2).$$

To help facilitate the correct F tests, the $E(MS)$ table is presented in Table 16.10. The first level of each factor is indicated by 0, as in $g0$ for the first group, and the kth level by $k - 1$. The first group ($g0$) has undergone the treatment combination of either $a0b0$ or $a1b1$, and the

Table 16.10 ANOVA $E(MS)$ Table for the Confounded Example[a]

Sources	d.f.	$E(MS)$
G [or AB]	$(g - 1)$	$\sigma_e^2 + c\sigma_{z(g)}^2 + zc\theta_g^2$
$Z(G)$	$g(z - 1)$	$\sigma_e^2 + c\sigma_{z(g)}^2$
A	$(a - 1)$	$\sigma_e^2 + \sigma_{abz(g)}^2 + zg\theta_a^2$
B	$(b - 1)$	$\sigma_e^2 + \sigma_{abz(g)}^2 + zg\theta_b^2$
$AB \times Z(G)$	$g(z - 1)(c - 1)$	$\sigma_e^2 + \sigma_{abz(g)}^2$

[a]c is the block size and z is the number of blocks; A, B fixed; $Z(G)$ random; θ^2 is a quadratic form for a fixed factor; σ_e^2 and $\sigma_{abz(g)}^2$ are not separately estimable in this design.
Source: Modified from Kirk (1982, p. 582).

Table 16.11   A Data Matrix for a
Confounded Factorial Example

		ab 00	ab 11
g0	s0	3	16
	s1	5	14
	s2	6	17
	s3	5	15
		ab 01	ab 10
g1	s4	14	7
	s5	14	6
	s6	16	7
	s7	16	11

second group ($g1$) has undergone the combination of either $a0b1$ or $a1b0$. The first block is indicated by $s0$, and the last block is indicated by $s7$. The observed data are shown in Table 16.11 (adapted from Kirk, 1982, p. 580).

16.4.1   Application with SPSS

The data contain five columns: Group factor level, A factor level, B factor level, Block level, and Y observation (see Table 16.12). There are g number of Group levels; z number of blocks or subjects in each group, Z; A with a levels and B with b levels; and c representing the block size.

In this example, A and B factor combinations are first translated into COLUMN variable, indicating whether the responses in question appear in Column 1 (COLUMN = 1) or in Column 2 (COLUMN = 2). Although this manipulation is unnecessary, it clearly shows that, because a fraction of the treatment combinations is used, a comparison of two columns is equivalent to a comparison of A factor (or B factor) levels within each level of Group. In other words, once data are separately analyzed by Group, there are only A (or B) and Blocks in the model and there is no AB interaction. To understand this, focus on the two columns of data for the first group. If we focus on A, we are comparing $A1$ with $A2$; and if we focus on B, we are comparing $B1$ with $B2$, but the comparison is conducted on the same data. The SPSS commands and results are summarized in Table 16.13.

The first segment of the SPSS commands recode BLOCKS into Z. This operation is necessary because we have again coded blocks to reflect Kirk's

Table 16.12 A Data Matrix for a Confounded Factorial Example

G	A	B	BLOCK	Y
0	0	0	0	3
0	1	1	0	16
0	0	0	1	5
0	1	1	1	14
0	0	0	2	6
0	1	1	2	17
0	0	0	3	5
0	1	1	3	15
1	0	1	4	14
1	1	0	4	7
1	0	1	5	14
1	1	0	5	6
1	0	1	6	16
1	1	0	6	7
1	0	1	7	16
1	1	0	7	11

example; we recode BLOCKS into Z to reflect the correct number of nested levels (four) within each group.

In the next segment, $MSAB \times Z(G0)$ and $MSAB \times Z(G1)$ are examined for homogeneity. If homogeneous, they can be pooled to form the $MSAB \times Z(G)$. The temporary if commands produce these two separate error estimates followed by the summary table using the pooled error term (Table 16.13).

Because the residual variances from $G0$ and $G1$ were homogeneous (identical in this fictitious example), we are able to use the pooled residual (SPSS default). This pooled residual is used as the error for testing A and B effects. Because this is a mixed model where blocks are random, error 1 (Z within G) is used to test Group effects. Both A and B effects are shown to be significant in that both dietary supplement of bran and exercise do reduce the blood cholesterol level, while Group effect, as expected, is nonsignificant.

16.4.2 Application with SAS

SAS has a simple command structure for obtaining Group 0 and Group 1 analyses as well as the overall analysis. The data structure is the same as in SPSS. The first two GLM procedures produce the residual variance for

Table 16.13 A 2^2 Confounded Factorial Design: SPSS Commands and Results

SPSS commands

```
IF (A EQ 0 AND B EQ 0) COLUMN = 1
IF (A EQ 0 AND B EQ 1) COLUMN = 1
IF (A EQ 1 AND B EQ 1) COLUMN = 2
IF (A EQ 1 AND B EQ 0) COLUMN = 2
RECODE   BLOCK
    (0 = 0) (1 = 1) (2 = 2) (3 = 3) (4 = 0) (5 = 1) (6 = 2) (7 = 3)
    INTO
    Z
TEMPORARY
SELECT IF   (G EQ 0)
MANOVA   Y BY COLUMN(1,2), Z(0,3)/
    DESIGN = COLUMN, Z/
TEMPORARY
SELECT IF (G EQ 1)
MANOVA   Y BY COLUMN(1,2), Z(0,3)/
    DESIGN = COLUMN, Z/
MANOVA   Y BY G, A(0,1), B(0,1), Z(0,3)/
    DESIGN = G VS 1,
        Z WITHIN G = 1,
        A VS RESIDUAL,
        B VS RESIDUAL/
```

ANOVA summary table

Source of variation	SS	DF	MS	F	Sig of F
G0					
RESIDUAL	4.38	3	1.46		
COLUMN	231.12	1	231.12	158.49	.001
Z	5.37	3	1.79	1.23	.436
G1					
RESIDUAL	4.38	3	1.46		
COLUMN	105.13	1	105.13	72.09	.003
Z	14.37	3	4.79	3.29	.177
RESIDUAL	8.75	6	1.46		
A	12.25	1	12.25	8.40	.027
B	324.00	1	324.00	22.17	.000
Error 1	19.75	6	3.29		
G	6.25	1	6.25	1.90	.217

Table 16.14 A Confounded Factorial SAS Summary Table[a]

Source	DF	Type III SS	Mean square	Fvalue	Pr > F
A	1	12.25000000	12.25000000	8.40	.0274
B	1	324.00000000	324.00000000	222.17	.0001
G	1	6.25000000	6.25000000	4.29	.0839
Z(G)	6	19.75000000	3.29166667	2.26	.1725
Test of hypotheses using the Type III *MS* for Z(G) as an error					
Source	DF	Type III SS	Mean square	Fvalue	Pr > F
G	1	6.25000000	6.25000000	1.90	.2174

[a]The test for G appearing with the F value of 4.29 is incorrect. The correct test of G is performed on the bottom where G is tested against Z(G).

G0 and G1 as the error variance for the model, and the third GLM produces the summary table using the pooled residual. In the overall summary table (Table 16.14), fixed effects F ratios are reported. A and B are tested against the error from the model, $MSerror = 1.458$. In order to obtain the correct test of G, a specific hypothesis test is requested using $Z(G)$; i.e., $MSZ(G) = 3.292$ is used as the denominator term. Table 16.14 does not list the SAS result for the first two segments showing the error variances from the models with G0 and G1.

```
PROC GLM;
    WHERE G=0;
    CLASS  G A B Z;
    MODEL Y=A Z;
PROC GLM;
    WHERE G=1;
    CLASS  G A B Z;
    MODEL Y=A Z;
PROC GLM:
    CLASS  G A B Z;
    MODEL Y=A B G Z(G);
    TEST H=G  E=Z(G);
```

As in the nested design, the RANDOM statement may be employed as an alternative to a specific hypothesis test, the choice being a matter of personal preference. The command is given below but not the results.

```
PROC GLM;
    CLASS G A B Z;
    MODEL Y=A B G Z(G);
    RANDOM Z(G)/TEST;
```

16.4.3   Application with BMDP

It is rather cumbersome to analyze a mixed-effects confounded factorial design in BMDP4V. The data structure is the same as for SAS and SPSS. The BMDP4V commands for this design are shown in Table 16.15 but the results are omitted.

The first analysis procedure produces the table with all correct tests with the exception of the test of Groups. In order to produce the appropriate test for G, we resort to specifying a contrast matrix for the pooled residual $[MSAB(g0) + MSAB(g1) = 1.46]$.

The second analysis isolates the pooled Residual as a source of variation labeled $ABG(MSABG = 1.46)$ and, as a result, everything else is tested against a new error $[MSZ(G) = 3.29]$ which remains after ABG, A, B, and G have been extracted. The output is not presented here, but the test of G in the first analysis assumes a fixed model which is analogous to the SAS summary table. The complexity involved in specifying a design matrix

Table 16.15   BMDP4V Commands for the Confounded Factorial Example

```
/BETWEEN   FACTORS ARE G, A, B, Z.
/GROUP     CODES(G) = 0, 1.
     CODES(A) = 0, 1.
     CODES(B) = 0, 1.
     CODES(Z) = 0, 1, 2, 3, 4, 5, 6, 7.
/WEIGHTS   BETWEEN = EQUAL.
/END
     . . .Data are read in
/END
ANALYSIS  PROCEDURE = STRUCTURE.
     BFORMULA = 'A + B + G + Z.G'./
DESIGN   FACTORS = A, G, Z.
     TYPE = BETWEEN, CONTRAST.
     CODE = READ.
     VALUES ARE
      1,   0,   0,   0,   0,   0,
      0,   1,   0,   0,   0,   0,
      0,   0,   1,   0,   0,   0,
     -1, -1, -1,   0,   0,   0,
     . . . . . .
     . . . . . .  We omit portions of this matrix.
     . . . . . .
      0,   0,   0,   1,   1,   1./
ANALYSIS  PROCEDURE = STRUCTURE.
     BFORMULA = 'A + B + G'./
```

to isolate the pooled residual and thus test against an appropriate alternative error term for Groups precludes practical application of this strategy. Instead, the test of Groups can be obtained manually from the first summary table: $MSG/MSZ(G) = 6.25/3.29 = 1.90$. The user must know the correct error term for each source of variation in the model. Combined with the awkward and lengthy specifications of contrasts, BMDP4V is not well suited for this example.

16.5 A VARIANCE COMPONENT ESTIMATION

There are some who see variance component estimation as a dispensable component of ANOVA. The reason may be that, with most effects being fixed, teaching correct sets of MS for valid F tests is much easier and there is, therefore, no need to discuss variance component estimation at the introductory level. We see, however, in educational test and measurements, that the test items, test takers, and testing occasions are often random. Subsequently, the variance component estimation issues are relevant and are considered fundamental. An example of variance components estimation in a blood pressure study is presented in Brown and Mosteller (1991).

In this section, we use an example from a generalizability study in educational measurement to illustrate variance component estimation. The example is taken from Brennan (1983, p. 47), in which 10 persons, P, were observed under two random occasions. O, with different sets of four randomly selected Items nested in occasion, $I(O)$. That is, each person is administered a total of eight items split into two occasions of four items at each sitting. Because P is crossed with O, the effects of interests are P, O, $I(O)$, PO, and because we have only $n = 1$ in each block, the last interaction $PI(O)$ is not separately estimable from the within cells. As before, we will include this interaction term in the theoretical model.

The statistical model for this example is

$$Y_{ijk} = \mu + \alpha_i + \beta_{j(i)} + \pi_k + \alpha\pi_{ik} + \beta\pi_{j(i)k} + \varepsilon_{ijk},$$
$$i = 1, \ldots, o; j = 1, \ldots, t; k = 1, \ldots, p, \qquad (16.4)$$

where Y_{ijk} is the test score of the kth individual responding to the jth item at the ith test session, μ is the grand mean, α_i is the ith occasion effect, $\beta_{j(i)}$ is the item nested in occasion effect, π_k is the kth person or test-taker effect, and $\alpha\pi_{ik}$ and $\beta\pi_{j(i)k}$ are interactions for occasion-by-person and item nested in occasion-by-person effects, respectively. Because of the nesting structure, t^* is the total number of items *within* each level of Occasion; i.e., $t = 8$ and $t^* = 4$ in this example. All factors are assumed to be random; i.e., Items are randomly selected from a pool of equivalent items (domain referenced tests); Persons (test takers) are random as subjects are almost always considered as random; and test sessions or Occasions are random in individualized testing sessions. Subsequently, we have $\alpha_i \sim N(0, \sigma_\alpha^2)$, $\beta_{j(i)} \sim N(0, \sigma_\beta^2)$, $\pi_k \sim N(0, \sigma_\pi^2)$, and $\varepsilon_{ijk} \sim N(0, \sigma_\varepsilon^2)$. Because items are nested in occasion, there is no item-by-occasion interaction, $\alpha\beta$, in this model. The $E(MS)$ table for this model is presented in Table 16.16, where P is the person, O is the occasion, and I is the item. The error term is $PI(O)$. This example is similar to that of a repeated measures design in which each subject is measured under $2 \times 4 = 8$ treatment combinations.

The data for this measurement study are presented in Table 16.17, in which the nesting is converted to a crossed structure whereby Items 5, 6, 7, and 8 at Occasion 2 are entered as Items 1, 2, 3, and 4, respectively. (For an example of changing the data structure within a computer package, see Section 16.2.)

There are four major estimation procedures: AOV(I), which uses Type I (sequential) solution of a set of equations obtained by equating observed and expected MSs and does not involve any iteration; MIVQUE(0), minimum variance quadratic estimator with no iteration with specified values (i.e., error variance is 1 and the rest of the variances are 0's); ML, maximum likelihood estimation with iterations, in which the estimates are obtained by maximizing the full likelihood function over the parameter space; and REML, restricted maximum likelihood estimation with iterations, in which fixed-effects space is partitioned out of the random-effects space and only the latter portion of the likelihood is maximized. For more details, these procedures are fully described in Chapter 15 as well as in Searle, Caselia, and McCulloch (1992), and in Harville (1977). Empirical comparisons among these four are reported in Swallow and Monahan (1984) and in Khattree

Table 16.16 ANOVA $E(MS)$ Table for the Variance Component Estimation Problem[a]

Sources	d.f.	$E(MS)$			
P	$(p - 1)$	$\sigma_e^2 + \sigma_{pi(o)}^2 + i^*\sigma_{po}^2$			$+ i^*o\sigma_p^2$
O	$(o - 1)$	$\sigma_e^2 + \sigma_{pi(o)}^2 + i^*\sigma_{po}^2 + p\sigma_{i(o)}^2 + pi^*\sigma_o^2$			
$I(O)$	$o(i - 1)$	$\sigma_e^2 + \sigma_{pi(o)}^2 \qquad\qquad + p\sigma_{i(o)}^2$			
PO	$(p - 1)(o - 1)$	$\sigma_e^2 + \sigma_{pi(o)}^2 + i^*\sigma_{po}^2$			
$PI(O)$:error	$o(i - 1)(p - 1)$	$\sigma_e^2 + \sigma_{pi(o)}^2$			

[a]P, O, $I(O)$, Error are all random. p is the number of persons, o is the number of Occasions, i^* is the number of items *within* each occasion, $I(O)$.

Table 16.17 SAS Data for a Variance Component Estimation Example

P	O	I	Y		P	O	I	Y
1	1	1	2		1	2	1	2
2	1	1	4		2	2	1	6
3	1	1	5		3	2	1	5
4	1	1	5		4	2	1	5
5	1	1	4		5	2	1	4
6	1	1	4		6	2	1	6
7	1	1	2		7	2	1	2
8	1	1	3		8	2	1	6
9	1	1	0		9	2	1	5
10	1	1	6		10	2	1	6
1	1	2	6		1	2	2	5
2	1	2	6		2	2	2	7
3	1	2	5		3	2	2	4
4	1	2	9		4	2	2	7
5	1	2	3		5	2	2	5
6	1	2	4		6	2	2	4
7	1	2	6		7	2	2	7
8	1	2	4		8	2	2	6
9	1	2	5		9	2	2	5
10	1	2	8		10	2	2	8
1	1	3	7		1	2	3	5
2	1	3	6		2	2	3	5
3	1	3	4		3	2	3	5
4	1	3	8		4	2	3	7
5	1	3	5		5	2	3	6
6	1	3	4		6	2	3	7
7	1	3	6		7	2	3	7
8	1	3	4		8	2	3	6
9	1	3	4		9	2	3	5
10	1	3	7		10	2	3	8
1	1	4	5		1	2	4	5
2	1	4	7		2	2	4	7
3	1	4	6		3	2	4	5
4	1	4	6		4	2	4	6
5	1	4	6		5	2	4	4
6	1	4	7		6	2	4	8
7	1	4	5		7	2	4	5
8	1	4	5		8	2	4	4
9	1	4	5		9	2	4	3
10	1	4	6	(*continues to right*)	10	2	4	6

and Gill (1988). Since SPSS does not offer a variance component estimation procedure, BMDP and SAS are used for this example.

16.5.1   Application with BMDP

BMDP3V has results for either METHOD = ML or REML, but it has no option to produce AOV(I) or MIVQUE(0) within the same procedure. However, it is flexible in specifying the factors as random or fixed. By specifying PRINT (ITER), although this is not listed in the current documentation, the information at each iteration is provided. The random specification statement corresponds to each factor named in RNAMES, which takes care of the labeling.

```
/VARIABLE   NAMES = P, O, I, Y.
/PRINT      ITER.
/DESIGN     DEPENDENT = Y.
            RANDOM = P.
            RANDOM = O.
            RANDOM = I, O.
            RANDOM = P, O.
            RNAMES = 'P', 'O', 'I(O)', 'P*O'.
            METHOD = REML.
/END
. . .data are read in
/END
```

Although the results are not reported here, each component is estimated with its standard error with three decimal place accuracy in the mainframe we used. In addition, chi-square-based statistic is provided for the whole model. With an option of PRINT(ITER), the estimate at each iteration and the number of iterations needed to converge are reported. Unfortunately, because BMDP3V deals exclusively with the variance-covariance matrix, there is no way to obtain an AOV(I) and/or MIVQUE(0) solution within the P3V. Also, as noted by Bock and Brandt (1980), BMDP takes more CPU time than SAS in the ML and REML. As can be seen in Table 16.20, we found 28% more CPU time in obtaining the ML and REML solutions in BMDP3V (with three decimal place accuracy) than the comparable run in SAS (with eight decimal place accuracy).

16.5.2   Application with SAS

The SAS VARCOMP, on the other hand, produces all four major estimation procedures, TYPE1 (AOV), MIVQUE0 (MVQUE0), ML, and REML, in one run. The default assumes that the factors are all random.

By excluding $PI(O)$ from the model specification in SAS, it automatically uses the residual, i.e., what is left after taking out P, O, $I(O)$, and PO as the error term. If $PI(O)$ is specified in the model, SAS tries to produce the residual with expectation of zero and d.f. of zero. Therefore, it forces users to adopt (at least at the command level) a model without the $PI(O)$ interaction (Table 16.18). [In the example provided, leaving $PI(O)$ in the model causes little difference in component estimates in MIVQUE0, ML, and REML solutions. However, SAS does not decompose variances in the TYPE1 method. In the case of the two iterative methods, i.e., ML and REML, the overspecified model produced solutions with fewer iterations than the model excluding $PI(O)$.]

After the data are read in, PROC VARCOMP is specified. The choice of methods consists of TYPE1, which is the AOV(I) estimation, MIVQUE0, ML, and REML. The first two estimations [AOV(I) and MIVQUE0] may report negative variances, but ML and REML in VARCOMP automatically replace zeros with negative estimates. [For more information on different estimation procedures, see Hocking (1990); Hocking, Green, and Bremer (1989); and Searle et al. (1992).] In the following setup, REML solution is requested.

```
PROC VARCOMP   METHOD = REML;
       CLASS   P    O    I;
              MODEL Y  =  P
                         O
                         I(O)
                         P*O;
```

TYPE1 procedure is what Hocking calls AOV(I), the sequential SS decomposition, and it is solved on the basis of the expected mean squares listed in Table 16.18. In the terminology of Brennan (1983), TYPE1 is Algorithm 2, shown in Table 3.4.1 (p. 47). All four procedures, TYPE1, MIVQUE0, ML, and REML, are compared in Table 16.19.

Table 16.18 Expected Mean Square for TYPE1 Estimation

Source	Expected mean square
P	Var(Error) + 4 Var($P*O$) + 8 Var(P)
O	Var(Error) + 4 Var($P*O$) + 10 Var($I(O)$) + 40 Var(O)
$I(O)$	Var(Error) + 10 Var($I(O)$)
$P*O$	Var(Error) + 4 Var($P*O$)
Error	Var(Error)

Table 16.19 A Variance Components Estimation Example with Negative Estimates

Variance components	Type1[a]	ML[b]	REML[c]
Var(P)	.69652778	.62624163	.67817628
Var(O)	−.10300926	0	0
Var($I(O)$)	.59814815	.48804816	.53928705
Var(P^*O)	−.04282407	0	0
Var(Error)	1.51018519	1.48880543	1.48571359

[a]The results from Type1 and MIVQUE0 are identical in this example.
[b]The number of iterations is 7.
[c]The number of iterations is 6.

In this balanced example, the negative component estimates may be due to the sampling variability, particularly from a small sample, as Brennan suggests. Outliers or unusual observations may also result in negative estimates. For balanced and positive estimates, the simplest procedure, TYPE1, is unbiased and thus recommended. But once negative estimates are obtained, the REML may be desirable because it generally has smaller bias than the ML procedure and it converges to solutions with fewer iterations. Another possibility for negative estimates is a misspecification of the model. Hocking recommends a diagnosis of the data when faced with negative estimates (Hocking, 1990; Hocking, Green, and Bremer, 1989), e.g., a scatter plot prepared to examine possible extreme values in the data. In this example, Subject 9 at Occasion 1, Item 1, with $Y = 0$ may deviate slightly from the general cluster of data points. Because the data used in this example are artificial, it is difficult to ascertain the cause of negative estimates. Hocking contends that replacing negative estimates

Table 16.20 CPU Time and Cost Comparison Among Three Packages

	SPSS		SAS		BMDP	
Job description	CPU time	Cost	CPU time	Cost	CPU time	Cost
Hierarchical	2.75	$ .77	6.39	$1.75	2.16	$ .59(8V)
					2.41	$ .66(7D)
Confounded	2.70	$ .78	6.03	$1.69	3.29	$ .89
Variance component est. (REML and ML only)	—	—	5.78	$1.58	7.42	$1.95
Latin square	2.89	$ .83	5.54	$1.52	3.86	$1.04

with zeros is only a "fix" without understanding of the cause or nature of the data. He suggests that we use the negative estimates as a guide to diagnosing the data.

As can be seen in Table 16.19, TYPE1 (which produces the same results as MIVQUE0 in this example), ML, and REML estimates are slightly different. In noniterative methods (TYPE1 and MIVQUE0), two negative estimates, $\text{Var}(O) = -.10300926$ and $\text{VAR}(P^*O) = -.04282407$, are reported. Since some variance components are negative, TYPE1 or MIVQUE0 solutions are not recommended. The REML is recommended instead. With REML estimates, we allocate noticeably smaller variance to Items nested in Occasion, $I(O)$ than we would have with TYPE1 estimates.

The CPU and cost comparison among three packages is summarized in Table 16.2.

16.6 OTHER COMPUTER PROCEDURES

In this chapter, we have not reviewed some computer applications which can be found in other chapters of this book or in other references. For example, Keselman and Keselman in Chapter 4 present illustrations of SAS computer applications in testing specific contrasts on repeated measurements, and Games (1990) describes SAS applications for ANOVA and MANOVA approaches to repeated-measure designs. Harris in Chapter 7 shows examples of MANOVA using three major packages; Cotton in Chapter 5 gives an example of a Latin square design with carryover effects using SAS and SYSTAT; and DiGennaro and Huster (1990) and Ratkowsky, Alldredge, and Cotton (1990) show SAS applications for Latin square designs involving repeated measurements.

In Chapter 8, O'Brien and Muller give numerical examples using the power analysis algorithms developed by the authors themselves that are applicable for univariate and multivariate hypotheses. A standard power procedure along with a program is available in Cohen (1988). In Chapter 14, Larntz discusses an efficient multinominal logit computational algorithm (Meyer, 1981), which is especially useful for large memory problems. Larntz also illustrated Akaike's information criterion (AIC) for selecting variables in categorical response models, but there is no readily available statistical package that reports AIC as of this writing.

As for specialized topics, Raudenbush in Chapter 13 discusses hierarchical linear models using the HLM program (Bryk, Raudenbush, Seltzer, and Congdon, 1988), which conducts a multistage estimation procedure. In Chapter 11, Kshirsagar illustrates Tukey's nonadditivity test procedure in a numerical example using SAS. One of the multiple comparison procedures (Shaffer's modified sequentially rejective Bonferroni procedure if used fully with "logical implications") requires a recursive computer algorithm in determining the exact probability level (or the critical value) for each contrast. A Fortran program is developed by Kim-Kang and Edwards (in press) in which the maximum number of potentially true null hypotheses left in the set is computed for a given number of groups for testing each contrast.

Bock and Brandt (1980) discuss some of the program differences in handling the analysis of covariance by SAS, SPSS, and BMDP. In addition, Searle and Hudson (1982) and Searle, Hudson, and Federer (1982) present annotated computer output for this design.

Handling unbalanced data and reparameterization in ANOVA with standard packages are summarized in Searle (1987) and in Searle, Speed, and Henderson (1981). Speed, Hocking, and Hackney (1978) and Bock and Brandt (1980) discuss the issues of several different solutions (e.g., sequential, experimental, and least squares solutions) in unbalanced designs, and Searle, Speed, and Milliken (1980) focus on a method comparing marginal means and the least squares solution.

For variance component estimation, in addition to what was covered in Chapter 15 (by Hocking) and elsewhere (Hocking, 1990; Hocking, Green, and Bremer, 1989), a new algorithm for computing ML component estimates is introduced by Callanan and Harville (1989).

In helping with the first steps to appropriate data analyses, a microcomputer program is available to generate the $E(MS)$ table in Hunka (1989). Indeed, there seems to be consistent progress in the area of computer applications—be it in the illustration of using the established packages correctly, in better documentation of standard as well as nonstandard designs and analyses, or in development of computer algorithms for specialized procedures. Such progress does benefit data analysts, whose life has been made easier by these application programs. At the same time it makes us responsible for understanding the logic behind the computer analyses we have chosen.

REFERENCES

Bock, R. D. and Brandt, D. (1980). Comparison of Some Computer Programs for Univariate and Multivariate Analysis of Variance. *Handbook of Statistics*, Vol. 1. Krishnaiah, P. R. (ed.), North-Holland, New York, pp. 703–744.

Brown, C. and Mosteller, F. (1991). Components of Variance. *Fundamentals of Exploratory Analysis of Variance*, Hoaglin, D. C., Mosteller, F., and Tukey, J. W. (eds.), Wiley, New York, pp. 193–251.

Brennan, R. L. (1983). *Elements of Generalizability Theory*. ACT Publications, Iowa City, Iowa.

Bryk, A. S., Raudenbush, S. W., Seltzer, M., and Congdon, R. (1988). *An Introduction to HLM*: *Computer Program and User's Manual*. University of Chicago Department of Education, Chicago, Illinois.

Callahan, R. P. and Harville, D. A. (1989). Some New Algorithms for Computing Maximum Likelihood Estimates of Variance Components. *Proc. Interface Comput. Sci. Statist.*, *21*: 435–444.

Cohen, J. (1988). *Statistical Power Analysis for the Behavioral Sciences* (2nd ed.). Lawrence Erlbaum, Hillsdale, New Jersey.

DiGennaro, J. and Huster, W. C. (1990). The Analysis of Data from 2 × 2 Crossover Trials with Baseline Measurements. *Proceedings of the SAS Users Group*, Vol. 15, Cary, North Carolina, pp. 1102–1107.

Dixon, W. J., Brown, M. B., Engelman, L., Hill, M. A., and Jennrich, R. I. (1988a). *BMDP Statistical Software Manual*, Vol. 1. University of California Press, Berkeley.

Dixon, W. J., Brown, M. B., Engelman, L., Hill, M. A., and Jennrich, R. I. (1988b). *BMDP Statistical Software Manual*, Vol. 2. University of California Press, Berkeley.

Games, P. A. (1990). Alternative Analyses of Repeated-Measure Designs by ANOVA and MANOVA. *Statistical Method for Longitudinal Research*, Vol. 1. von Eye, A. (ed.), Academic Press, New York, pp. 81–121.

Harville, D. A. (1977). Maximum Likelihood Approaches to Variance Component Estimation and to Related Problems. *J. Amer. Statist. Assoc.*, *72*: 320–340.

Hocking, R. R. (1990). A New Approach to Variance Component Estimation with Diagnostic Implications. *Comm. Statist.–Theor. Meth.*, *A19*: 4591–4618.

Hocking, R. R., Green, J. W., and Bremer, R. H. (1989). Variance Component Estimation with Model-Based Diagnostics. *Technometrics*, *31*: 227–239.

Hunka, S. (1989). A Macintosh APL Program to Calculate Expected Mean Squares. *Amer. Statist.*, *43*: 273–273.

Kim-Kang, G. and Edwards, L. K. (in press). A Computer Algorithm for Shaffer's Modified Sequentially Rejective Bonferroni Procedure. University of Minnesota Supercomputer Research Report.

Khattree, R. and Gill, D. S. (1988). Comparison of Some Estimates of Variance Components Using Pitman Nearness Criterion. *Proceedings of the Am. Statist. Assoc.-Statist. Comput.*, pp. 133–136.

Kirk, R. E. (1982). *Experimental Design* (2nd ed.). Brooks/Cole, Belmont, California.

Meyer, M. (1981). Applications and Generalizations of the Iterative Proportional Fitting Procedure. Unpublished Ph.D. dissertation, School of Statistics, University of Minnesota, Minneapolis.

Ratkowsky, D. A., Alldredge, J. R., and Cotton, J. W. (1990). Analyzing Balanced or Unbalanced Latin Squares and Other Repeated-Measures. *Proceedings of the SAS Users Group*, Vol. 15. Cary, North Carolina, pp. 1353–1358.

SAS. (1989). *SAS User's Guide*: *Statistics* (6th ed.). SAS Institute, Cary, North Carolina.

Searle, S. R. (1987). *Linear Models for Unbalanced Data*. Wiley, New York.

Searle, S. R. and Hudson, G. F. S. (1982). Some Distinctive Features of Output from Statistical Computing Packages for Analysis of Covariance. *Biometrics*, *38*: 337–345.

Searle, S. R., Hudson, G. F. S., and Federer, W. T. (1982). Annotated Computer Output for Analysis of Covariance. *Amer. Statist.*, *37*: 172–173.

Searle, S. R., Caselia, G., and McCulloch, C. E. (1992). *Variance Components*. Wiley, New York.

Searle, S. R., Speed, F. M., and Henderson, H. V. (1981). Some Computational and Model Equivalences in Analysis of Variance of Unequal-Subclass-Numbers Data. *Amer. Statist.*, *35*: 16–33.

Searle, S. R., Speed, F. M., and Milliken, G. A. (1980). Population Marginal Means in the Linear Model: An Alternative to Least Squares Means. *Amer. Statist.*, *34*: 216–221.

Speed, F. M., Hocking, R. R., and Hackney, O. P. (1978). Methods of Analysis of Linear Models with Unbalanced Data. *J. Amer. Statist. Assoc.*, *73*: 105–112.

SPSS. (1990). *SPSS Reference Guide*. SPSS, Inc., Chicago, Illinois.

Swallow, W. H. and Monahan, J. F. (1984). Monte Carlo Comparison of ANOVA, MIVQUE, REML and ML Estimators of Variance Components. *Technometrics*, *26*: 47–58.

Appendix

Table 1 Critical Values of the F Distribution
Pr (Random Variable with F Distribution $\geq$ Tabled Value) $= \alpha$

$\alpha = .05$	Numerator d.f.									
Denominator d.f.	1	2	3	4	5	6	7	8	9	10
1	161.5	199.5	215.7	224.6	230.2	234.0	236.8	238.9	240.5	241.9
2	18.51	19.00	19.16	19.25	19.30	19.33	19.35	19.37	19.39	19.40
3	10.13	9.55	9.28	9.12	9.01	8.94	8.89	8.85	8.81	8.79
4	7.71	6.94	6.59	6.39	6.26	6.16	6.09	6.04	6.00	5.96
5	6.61	5.79	5.41	5.19	5.05	4.95	4.88	4.82	4.77	4.74
6	5.99	5.14	4.76	4.53	4.39	4.28	4.21	4.15	4.10	4.06
7	5.59	4.74	4.35	4.12	3.97	3.87	3.79	3.73	3.68	3.64
8	5.32	4.46	4.07	3.84	3.69	3.58	3.50	3.44	3.39	3.35
9	5.12	4.26	3.86	3.63	3.48	3.37	3.29	3.23	3.18	3.14
10	4.97	4.10	3.71	3.48	3.33	3.22	3.14	3.07	3.02	2.98
11	4.84	3.98	3.59	3.36	3.20	3.09	3.01	2.95	2.90	2.85
12	4.75	3.89	3.49	3.26	3.11	3.00	2.91	2.85	2.80	2.75
13	4.67	3.81	3.41	3.18	3.03	2.92	2.83	2.77	2.71	2.67
14	4.60	3.74	3.34	3.11	2.96	2.85	2.76	2.70	2.65	2.60
15	4.54	3.68	3.29	3.06	2.90	2.79	2.71	2.64	2.59	2.54
16	4.49	3.63	3.24	3.01	2.85	2.74	2.66	2.59	2.54	2.49
17	4.45	3.59	3.20	2.96	2.81	2.70	2.61	2.55	2.49	2.45
18	4.41	3.55	3.16	2.93	2.77	2.66	2.58	2.51	2.46	2.41
19	4.38	3.52	3.13	2.90	2.74	2.63	2.54	2.48	2.42	2.38
20	4.35	3.49	3.10	2.87	2.71	2.60	2.51	2.45	2.39	2.35

Table 1 (Continued)

$\alpha = .05$				Numerator d.f.					
Denominator d.f.	1	2	3	4	5	6	7	8	9	10
21	4.32	3.47	3.07	2.84	2.68	2.57	2.49	2.42	2.37	2.32
22	4.30	3.44	3.05	2.82	2.66	2.55	2.46	2.40	2.34	2.30
23	4.28	3.42	3.03	2.80	2.64	2.53	2.44	2.37	2.32	2.27
24	4.26	3.40	3.01	2.78	2.62	2.51	2.42	2.36	2.30	2.25
25	4.24	3.39	2.99	2.76	2.60	2.49	2.40	2.34	2.28	2.24
26	4.23	3.37	2.98	2.74	2.59	2.47	2.39	2.32	2.27	2.22
27	4.21	3.35	2.96	2.73	2.57	2.46	2.37	2.31	2.25	2.20
28	4.20	3.34	2.95	2.71	2.56	2.45	2.36	2.29	2.24	2.19
29	4.18	3.33	2.93	2.70	2.55	2.43	2.35	2.28	2.22	2.18
30	4.17	3.32	2.92	2.69	2.53	2.42	2.33	2.27	2.21	2.16
40	4.08	3.23	2.84	2.61	2.45	2.34	2.25	2.18	2.12	2.08
48	4.04	3.19	2.80	2.57	2.41	2.30	2.21	2.14	2.08	2.03
60	4.00	3.15	2.76	2.53	2.37	2.25	2.17	2.10	2.04	1.99
80	3.96	3.11	2.72	2.49	2.33	2.21	2.13	2.06	2.00	1.95
120	3.92	3.07	2.68	2.45	2.29	2.18	2.09	2.02	1.96	1.91
∞	3.84	3.00	2.60	2.37	2.21	2.10	2.01	1.94	1.88	1.83

$\alpha = .05$				Numerator d.f.					
Denominator d.f.	12	14	18	24	30	40	48	60	120	∞
1	243.9	245.4	247.3	249.1	250.1	251.1	251.7	252.2	253.3	254.3
2	19.41	19.42	19.44	19.45	19.46	19.47	19.48	19.48	19.49	19.50
3	8.74	8.71	8.67	8.64	8.62	8.59	8.58	8.57	8.55	8.53
4	5.91	5.87	5.82	5.77	5.75	5.72	5.70	5.69	5.66	5.63
5	4.68	4.64	4.58	4.53	4.50	4.46	4.45	4.43	4.40	4.37
6	4.00	3.96	3.90	3.84	3.81	3.77	3.76	3.74	3.70	3.67
7	3.57	3.53	3.47	3.41	3.38	3.34	3.32	3.30	3.27	3.23
8	3.28	3.24	3.17	3.12	3.08	3.04	3.02	3.01	2.97	2.93
9	3.07	3.03	2.96	2.90	2.86	2.83	2.81	2.79	2.75	2.71
10	2.91	2.86	2.80	2.74	2.70	2.66	2.64	2.62	2.58	2.54
11	2.79	2.74	2.67	2.61	2.57	2.53	2.51	2.49	2.45	2.40
12	2.69	2.64	2.57	2.51	2.47	2.43	2.41	2.38	2.34	2.30
13	2.60	2.55	2.48	2.42	2.38	2.34	2.32	2.30	2.25	2.21
14	2.53	2.48	2.41	2.35	2.31	2.27	2.24	2.22	2.18	2.13
15	2.48	2.42	2.35	2.29	2.25	2.20	2.18	2.16	2.11	2.07
16	2.42	2.37	2.30	2.24	2.19	2.15	2.13	2.11	2.06	2.01
17	2.38	2.33	2.26	2.19	2.15	2.10	2.08	2.06	2.01	1.96
18	2.34	2.29	2.22	2.15	2.11	2.06	2.04	2.02	1.97	1.92

Table 1 (Continued)

$\alpha = .05$				Numerator d.f.					
Denominator d.f.	12	14	18	24	30	40	48	60	120	∞
19	2.31	2.26	2.18	2.11	2.07	2.03	2.00	1.98	1.93	1.88
20	2.28	2.22	2.15	2.08	2.04	1.99	1.97	1.95	1.90	1.84
21	2.25	2.20	2.12	2.05	2.01	1.96	1.94	1.92	1.87	1.81
22	2.23	2.17	2.10	2.03	1.98	1.94	1.91	1.89	1.84	1.78
23	2.20	2.15	2.07	2.01	1.96	1.91	1.89	1.86	1.81	1.76
24	2.18	2.13	2.05	1.98	1.94	1.89	1.87	1.84	1.79	1.73
25	2.16	2.11	2.03	1.96	1.92	1.87	1.85	1.82	1.77	1.71
26	2.15	2.09	2.02	1.95	1.90	1.85	1.83	1.80	1.75	1.69
27	2.13	2.08	2.00	1.93	1.88	1.84	1.81	1.79	1.73	1.67
28	2.12	2.06	1.99	1.91	1.87	1.82	1.79	1.77	1.71	1.65
29	2.10	2.05	1.97	1.90	1.85	1.81	1.78	1.75	1.70	1.64
30	2.09	2.04	1.96	1.89	1.84	1.79	1.77	1.74	1.68	1.62
40	2.00	1.95	1.87	1.79	1.74	1.69	1.67	1.64	1.58	1.51
48	1.96	1.90	1.82	1.75	1.70	1.64	1.61	1.59	1.52	1.45
60	1.92	1.86	1.78	1.70	1.65	1.59	1.56	1.53	1.47	1.39
80	1.88	1.82	1.73	1.65	1.60	1.54	1.51	1.48	1.41	1.32
120	1.83	1.77	1.69	1.61	1.55	1.50	1.46	1.43	1.35	1.25
∞	1.75	1.69	1.60	1.52	1.46	1.39	1.36	1.32	1.22	1.00

$\alpha = .01$				Numerator d.f.					
Denominator d.f.	1	2	3	4	5	6	7	8	9	10
1	4052.	5000.	5403.	5625.	5764.	5859.	5928.	5981.	6023.	6056.
2	98.50	99.00	99.17	99.25	99.30	99.33	99.36	99.37	99.39	99.40
3	34.12	30.82	29.46	28.71	28.24	27.91	27.67	27.49	27.35	27.23
4	21.20	18.00	16.69	15.98	15.52	15.21	14.98	14.80	14.66	14.55
5	16.26	13.27	12.06	11.39	10.97	10.67	10.46	10.29	10.16	10.05
6	13.75	10.93	9.78	9.15	8.75	8.47	8.26	8.10	7.98	7.87
7	12.25	9.55	8.45	7.85	7.46	7.19	6.99	6.84	6.72	6.62
8	11.26	8.65	7.59	7.01	6.63	6.37	6.18	6.03	5.91	5.81
9	10.56	8.02	6.99	6.42	6.06	5.80	5.61	5.47	5.35	5.26
10	10.04	7.56	6.55	5.99	5.64	5.39	5.20	5.06	4.94	4.85
11	9.65	7.21	6.22	5.67	5.32	5.07	4.89	4.74	4.63	4.54
12	9.33	6.93	5.95	5.41	5.06	4.82	4.64	4.50	4.39	4.30
13	9.07	6.70	5.74	5.21	4.86	4.62	4.44	4.30	4.19	4.10
14	8.86	6.51	5.56	5.04	4.70	4.46	4.28	4.14	4.03	3.94
15	8.68	6.36	5.42	4.89	4.56	4.32	4.14	4.00	3.89	3.80
16	8.53	6.23	5.29	4.77	4.44	4.20	4.03	3.89	3.78	3.69

Table 1 (Continued)

$\alpha = .01$				Numerator d.f.					
Denominator d.f.	1	2	3	4	5	6	7	8	9	10
17	8.40	6.11	5.19	4.67	4.34	4.10	3.93	3.79	3.68	3.59
18	8.29	6.01	5.09	4.58	4.25	4.01	3.84	3.71	3.60	3.51
19	8.19	5.93	5.01	4.50	4.17	3.94	3.77	3.63	3.52	3.43
20	8.10	5.85	4.94	4.43	4.10	3.87	3.70	3.56	3.46	3.37
21	8.02	5.78	4.87	4.37	4.04	3.81	3.64	3.51	3.40	3.31
22	7.95	5.72	4.82	4.31	3.99	3.76	3.59	3.45	3.35	3.26
23	7.88	5.66	4.76	4.26	3.94	3.71	3.54	3.41	3.30	3.21
24	7.82	5.61	4.72	4.22	3.90	3.67	3.50	3.36	3.26	3.17
25	7.77	5.57	4.68	4.18	3.86	3.63	3.46	3.32	3.22	3.13
26	7.72	5.53	4.64	4.14	3.82	3.59	3.42	3.29	3.18	3.09
27	7.68	5.49	4.60	4.11	3.78	3.56	3.39	3.26	3.15	3.06
28	7.64	5.45	4.57	4.07	3.75	3.53	3.36	3.23	3.12	3.03
29	7.60	5.42	4.54	4.04	3.73	3.50	3.33	3.20	3.09	3.00
30	7.56	5.39	4.51	4.02	3.70	3.47	3.30	3.17	3.07	2.98
40	7.31	5.18	4.31	3.83	3.51	3.29	3.12	2.99	2.89	2.80
48	7.20	5.08	4.22	3.74	3.43	3.20	3.04	2.91	2.80	2.72
60	7.08	4.98	4.13	3.65	3.34	3.12	2.95	2.82	2.72	2.63
80	6.96	4.88	4.04	3.56	3.26	3.04	2.87	2.74	2.64	2.55
120	6.85	4.79	3.95	3.48	3.17	2.96	2.79	2.66	2.56	2.47
∞	6.63	4.61	3.78	3.32	3.02	2.80	2.64	2.51	2.41	2.32

$\alpha = .01$				Numerator d.f.					
Denominator d.f.	12	14	18	24	30	40	48	60	120	∞
1	6106.	6143.	6192.	6235.	6261.	6287.	6300.	6313.	6339.	6366.
2	99.42	99.43	99.44	99.46	99.47	99.47	99.48	99.48	99.49	99.50
3	27.05	26.92	26.75	26.60	26.51	26.41	26.36	26.32	26.22	26.13
4	14.37	14.25	14.08	13.93	13.84	13.75	13.70	13.65	13.56	13.46
5	9.89	9.77	9.61	9.47	9.38	9.29	9.25	9.20	9.11	9.02
6	7.72	7.60	7.45	7.31	7.23	7.14	7.10	7.06	6.97	6.88
7	6.47	6.36	6.21	6.07	5.99	5.91	5.87	5.82	5.74	5.65
8	5.67	5.56	5.41	5.28	5.20	5.12	5.07	5.03	4.95	4.86
9	5.11	5.00	4.86	4.73	4.65	4.57	4.52	4.48	4.40	4.31
10	4.71	4.60	4.46	4.33	4.25	4.17	4.12	4.08	4.00	3.91
11	4.40	4.29	4.15	4.02	3.94	3.86	3.82	3.78	3.69	3.60
12	4.16	4.05	3.91	3.78	3.70	3.62	3.58	3.54	3.45	3.36
13	3.96	3.86	3.71	3.59	3.51	3.43	3.38	3.34	3.25	3.17
14	3.80	3.70	3.56	3.43	3.35	3.27	3.22	3.18	3.09	3.00

Table 1 (Continued)

α = .01				Numerator d.f.						
Denominator d.f.	12	14	18	24	30	40	48	60	120	∞
15	3.67	3.56	3.42	3.29	3.21	3.13	3.09	3.05	2.96	2.87
16	3.55	3.45	3.31	3.18	3.10	3.02	2.98	2.93	2.84	2.75
17	3.46	3.35	3.21	3.08	3.00	2.92	2.88	2.83	2.75	2.65
18	3.37	3.27	3.13	3.00	2.92	2.84	2.79	2.75	2.66	2.57
19	3.30	3.19	3.05	2.92	2.84	2.76	2.72	2.67	2.58	2.49
20	3.23	3.13	2.99	2.86	2.78	2.69	2.65	2.61	2.52	2.42
21	3.17	3.07	2.93	2.80	2.72	2.64	2.59	2.55	2.46	2.36
22	3.12	3.02	2.88	2.75	2.67	2.58	2.54	2.50	2.40	2.31
23	3.07	2.97	2.83	2.70	2.62	2.54	2.49	2.45	2.35	2.26
24	3.03	2.93	2.79	2.66	2.58	2.49	2.45	2.40	2.31	2.21
25	2.99	2.89	2.75	2.62	2.54	2.45	2.41	2.36	2.27	2.17
26	2.96	2.86	2.71	2.58	2.50	2.42	2.37	2.33	2.23	2.13
27	2.93	2.82	2.68	2.55	2.47	2.38	2.34	2.29	2.20	2.10
28	2.90	2.79	2.65	2.52	2.44	2.35	2.31	2.26	2.17	2.06
29	2.87	2.77	2.62	2.49	2.41	2.33	2.28	2.23	2.14	2.03
30	2.84	2.74	2.60	2.47	2.39	2.30	2.25	2.21	2.11	2.01
40	2.66	2.56	2.42	2.29	2.20	2.11	2.07	2.02	1.92	1.80
48	2.58	2.48	2.33	2.20	2.12	2.03	1.98	1.93	1.82	1.70
60	2.50	2.39	2.25	2.12	2.03	1.94	1.89	1.84	1.73	1.60
80	2.42	2.31	2.17	2.03	1.94	1.85	1.80	1.75	1.63	1.49
120	2.34	2.23	2.09	1.95	1.86	1.76	1.71	1.66	1.53	1.38
∞	2.18	2.08	1.93	1.79	1.70	1.59	1.53	1.47	1.32	1.00

Source: Adapted from Owen, D. B. (1962). *Handbook of Statistical Tables*, Addison-Wesley, Reading, Massachusetts, with permission of the author and the publisher.

Table 2 Upper α Point of the Studentized Range Distribution with Parameter k and Degrees of Freedom d.f.

d.f.	α	2	3	4	5	k 6	7	8	9	10
1	.05	18.0	27.0	32.8	37.1	40.4	43.1	45.4	47.4	49.1
	.01	90.0	135.	164.	186.	202.	216.	227.	237.	246.
2	.05	6.09	8.33	9.80	10.9	11.7	12.4	13.0	13.5	14.0
	.01	14.0	19.0	22.3	24.7	26.6	28.2	29.5	30.7	31.7
3	.05	4.50	5.91	6.82	7.50	8.04	8.48	8.85	9.18	9.46
	.01	8.26	10.6	12.2	13.3	14.2	15.0	15.6	16.2	16.7

Table 2   (Continued)

d.f.	α	2	3	4	5	6	7	8	9	10
						k				
4	.05	3.93	5.04	5.76	6.29	6.71	7.05	7.35	7.60	7.83
	.01	6.51	8.12	9.17	9.96	10.6	11.1	11.5	11.9	12.3
5	.05	3.64	4.60	5.22	5.67	6.03	6.33	6.58	6.80	6.99
	.01	5.70	6.97	7.80	8.42	8.91	9.32	9.67	9.97	10.2
6	.05	3.46	4.34	4.90	5.31	5.63	5.89	6.12	6.32	6.49
	.01	5.24	6.33	7.03	7.56	7.97	8.32	8.61	8.87	9.10
7	.05	3.34	4.16	4.69	5.06	5.36	5.61	5.82	6.00	6.16
	.01	4.95	5.92	6.54	7.01	7.37	7.68	7.94	8.17	8.37
8	.05	3.26	4.04	4.53	4.89	5.17	5.40	5.60	5.77	5.92
	.01	4.74	5.63	6.20	6.63	6.96	7.24	7.47	7.68	7.87
9	.05	3.20	3.95	4.42	4.76	5.02	5.24	5.43	5.60	5.74
	.01	4.60	5.43	5.96	6.35	6.66	6.91	7.13	7.32	7.32
10	.05	3.15	3.88	4.33	4.65	4.91	5.12	5.30	5.46	5.60
	.01	4.48	5.27	5.77	6.14	6.43	6.67	6.87	7.05	7.21
11	.05	3.11	3.82	4.26	4.57	4.82	5.03	5.20	5.35	5.49
	.01	4.39	5.14	5.62	5.97	6.25	6.48	6.67	6.84	6.99
12	.05	3.08	3.77	4.20	4.51	4.75	4.95	5.12	5.27	5.40
	.01	4.32	5.04	5.50	5.84	6.10	6.32	6.51	6.67	6.81
13	.05	3.06	3.73	4.15	4.45	4.69	4.88	5.05	5.19	5.32
	.01	4.26	4.96	5.40	5.73	5.98	6.19	6.37	6.53	6.67
14	.05	3.03	3.70	4.11	4.41	4.64	4.83	4.99	5.13	5.25
	.01	4.21	4.89	5.32	5.63	5.88	6.08	6.26	6.41	6.54
15	.05	3.01	3.67	4.08	4.37	4.60	4.78	4.94	5.08	5.20
	.01	4.17	4.84	5.25	5.56	5.80	5.99	6.16	6.31	6.44
16	.05	3.00	3.65	4.05	4.33	4.56	4.74	4.90	5.03	5.15
	.01	4.13	4.78	5.19	5.49	5.72	5.92	6.08	6.22	6.35
17	.05	2.98	3.63	4.02	4.30	4.52	4.71	4.86	4.99	5.11
	.01	4.10	4.74	5.14	5.43	5.66	5.85	6.01	6.15	6.27
18	.05	2.97	3.61	4.00	4.28	4.49	4.67	4.82	4.96	5.07
	.01	4.07	4.70	5.09	5.38	5.60	5.79	5.94	6.08	6.20
19	.05	2.96	3.59	3.98	4.25	4.47	4.65	4.79	4.92	5.04
	.01	4.05	4.67	5.05	5.33	5.55	5.74	5.89	6.02	6.14
20	.05	2.95	3.58	3.96	4.23	4.45	4.62	4.77	4.90	5.01
	.01	4.02	4.64	5.02	5.29	5.51	5.69	5.84	5.97	6.09
24	.05	2.92	3.53	3.90	4.17	4.37	4.54	4.68	4.81	4.92
	.01	3.96	4.54	4.91	5.17	5.37	5.54	5.69	5.81	5.92
30	.05	2.89	3.49	3.84	4.10	4.30	4.46	4.60	4.72	4.83
	.01	3.89	4.45	4.80	5.05	5.24	5.40	5.54	5.56	5.76
40	.05	2.86	3.44	3.79	4.04	4.23	4.39	4.52	4.63	4.74
	.01	3.82	4.37	4.70	4.93	5.11	5.27	5.39	5.50	5.60

Appendix

611

Table 2 (Continued)

d.f.	α	__2	3	4	5	6	7	8	9	10
					k					
60	.05	2.83	3.40	3.74	3.98	4.16	4.31	4.44	4.55	4.65
	.01	3.76	4.28	4.60	4.82	4.99	5.13	5.25	5.36	5.45
120	.05	2.80	3.36	3.69	3.92	4.10	4.24	4.36	4.48	4.56
	.01	3.70	4.20	4.50	4.71	4.87	5.01	5.12	5.21	5.30
∞	.05	2.77	3.31	3.63	3.86	4.03	4.17	4.29	4.39	4.47
	.01	3.64	4.12	4.40	4.60	4.76	4.88	4.99	5.08	5.16

Source: Adapted from Harter, H. L. (1969). *Order Statistics and Their Use in Testing and Estimation*, Vol. 1, *Tests Based on Range and Studentized Range of Samples From a Normal Population*. Aerospace Research Laboratories, U.S. Air Force. Reproduced with permission of the author.

Index